Lecture Notes in Computer Sci

Edited by G. Goos, J. Hartmanis, and J. va

T0238198

Springer
Berlin
Heidelberg
New York
Hong Kong
London
Milan
Paris
Tokyo

Colin D. Walter   Çetin K. Koç
Christof Paar (Eds.)

Cryptographic Hardware and Embedded Systems – CHES 2003

5th International Workshop
Cologne, Germany, September 8-10, 2003
Proceedings

 Springer

Series Editors

Gerhard Goos, Karlsruhe University, Germany
Juris Hartmanis, Cornell University, NY, USA
Jan van Leeuwen, Utrecht University, The Netherlands

Volume Editors

Colin D. Walter
Comodo Research Lab
Bradford BD7 1DQ, UK
E-mail: colin.walter@comodogroup.com

Çetin K. Koç
Oregon State University
Corvallis, Oregon 97330, USA
E-mail: koc@ece.orst.edu

Christof Paar
Ruhr-Universität Bochum
44780 Bochum, Germany
E-mail: cpaar@crypto.rub.de

Cataloging-in-Publication Data applied for

A catalog record for this book is available from the Library of Congress.

Bibliographic information published by Die Deutsche Bibliothek
Die Deutsche Bibliothek lists this publication in the Deutsche Nationalbibliografie;
detailed bibliographic data is available in the Internet at <http://dnb.ddb.de>.

CR Subject Classification (1998): E.3, C.2, C.3, B.7.2, G.2.1, D.4.6, K.6.5, F.2.1, J.2

ISSN 0302-9743
ISBN 3-540-40833-9 Springer-Verlag Berlin Heidelberg New York

Springer-Verlag Berlin Heidelberg New York
a member of BertelsmannSpringer Science+Business Media GmbH

http://www.springer.de

© Springer-Verlag Berlin Heidelberg 2003
Printed in Germany

Typesetting: Camera-ready by author, data conversion by PTP-Berlin GmbH
Printed on acid-free paper   SPIN: 10931455   06/3142   5 4 3 2 1 0

Preface

These are the proceedings of CHES 2003, the fifth workshop on Cryptographic Hardware and Embedded Systems, held in Cologne on September 8–10, 2003. As with every previous workshop, there was a record number of submissions despite the much earlier deadline in this year's call for papers. This is a clear indication of the growing international importance of the scope of the conference and the relevance of the subject material to both industry and academia.

The increasing competition for presenting at the conference has led to many excellent papers and a higher standard overall. From the 111 submissions, time constraints meant that only 32 could be accepted. The program committee worked very hard to select the best. However, at the end of the review process there were a number of good papers – which it would like to have included but for which, sadly, there was insufficient space. In addition to the accepted papers appearing in this volume, there were three invited presentations from Hans Dobbertin (Ruhr-Universität Bochum, Germany), Adi Shamir (Weizmann Institute, Israel), and Frank Stajano (University of Cambridge, UK), and a panel discussion on the effectiveness of current hardware and software countermeasures against side channel leakage in embedded cryptosystems.

As always, the focus of the workshop is on practical aspects of cryptographic hardware and embedded system security. A number of contributions pursue ideas on the efficient use of resources (such as time, chip area, or power) within constrained devices such as smart cards. These treat a wide range of applications, including true random number generators, finite field and modular arithmetic, and symmetric ciphers. Most of the remaining papers are concerned with leakage of secret key data via side channels such as time, power, or electromagnetic radiation, or through fault induction. Some of the contributions show how to extract the secret key in particular circumstances, others are more generic methodologies. These are complemented by other papers which provide countermeasures for increased resistance against such attacks. Applications include all the standard cryptosystems, both symmetric and public key, as well as some less well known ciphers. Another point of interest is the extension to hyperelliptic cryptosystems.

The CHES workshop series is now firmly established as an international forum for intellectual exchange in creating the secure, reliable, and robust security solutions which are required nowadays. CHES will continue to deal with the pressing hardware and software implementation issues as more and more systems and applications are developed which require encryption or authentication.

We would like to thank Irmgard Kühn (Ruhr-Universität Bochum, Germany) for her help with the local organization and André Weimerskirch (also from Bochum) for his help again with the CHES website (www.chesworkshop.org) and Gökay Saldamlı and Colin van Dyke (both from Oregon State University) for their help in preparing the proceedings.

June 2003

Colin D. Walter
Çetin K. Koç
Christof Paar

Acknowledgements

The organizers express their thanks to the program committee and the external referees for their help in getting the best quality papers selected, and also the companies which provided support to the workshop.

The program committee members for CHES 2003:

- Ross Anderson, Ross.Anderson@cl.cam.ac.uk
  University of Cambridge, UK
- Beni Arazi, arazi@ece.lsu.edu
  Louisiana State University, USA
- Jean-Sébastien Coron, coron@clipper.ens.fr
  Gemplus, France
- Craig Gentry, cgentry@docomolabs-usa.com
  DoCoMo Communications Laboratories, USA
- Jim Goodman, jimg@engim.com
  Engim Canada Inc., Canada
- Louis Goubin, louis.goubin@louveciennes.sema.slb.com
  SchlumbergerSema, France
- Anwar Hasan, ahasan@ece.uwaterloo.ca
  University of Waterloo, Canada
- Kouichi Itoh, kito@flab.fujitsu.co.jp
  Fujitsu Laboratories Ltd, Japan
- Marc Joye, marc.joye@gemplus.com
  Gemplus, France
- Seungjoo Kim, skim@kisa.or.kr
  Korea Information Security Agency, Korea
- François Koeune, koeune@dice.ucl.ac.be
  Université catholique de Louvain, Belgium
- Peter Kornerup, kornerup@imada.sdu.dk
  University of Southern Denmark, Odense, Denmark
- Pil Joong Lee, pjl@postech.ac.kr
  Pohang University of Science and Technology, Korea
- Katsuyuki Okeya, ka-okeya@sdl.hitachi.co.jp
  Hitachi, Japan
- Bart Preneel, Bart.Preneel@esat.kuleuven.ac.be
  Katholieke Universiteit Leuven, Belgium
- Vincent Rijmen, Vincent.Rijmen@cryptomathic.com
  Cryptomathic, Belgium and Graz University of Technology, Austria
- Kouichi Sakurai, sakurai@csce.kyushu-u.ac.jp
  Kyushu University, Japan
- Erkay Savaş, erkays@sabanciuniv.edu
  Sabanci University, Turkey
- Werner Schindler, Werner.Schindler@bsi.bund.de
  Bundesamt für Sicherheit in der Informationstechnik, Germany

- Jean-Pierre Seifert, `Jean-Pierre.Seifert@infineon.com`
  Infineon technologies AG, Germany
- Berk Sunar, `sunar@ece.wpi.edu`
  Worcester Polytechnic Institute, USA
- Tsuyoshi Takagi, `ttakagi@cdc.informatik.tu-darmstadt.de`
  Technische Universität Darmstadt, Germany
- Elena Trichina, `etrichin@cs.uku.fi`
  University of Kuopio, Finland
- Ingrid Verbauwhede, `ingrid@ee.ucla.edu`
  University of California, Los Angeles, USA
- Sung-Ming Yen, `yensm@csie.ncu.edu.tw`
  National Central University, Taiwan

The external referees:

- Toru Akishita (Sony Corporation, Japan)
- Mehdi-Laurent Akkar (Schlumberger Smart Cards, France)
- Seigo Arita (NEC Corporation, Japan)
- Harald Baier (Technische Universität Darmstadt, Germany)
- Claude Barral (Gemplus, France)
- Régis Bevan (Oberthur Card Systems, France)
- Mike Bond (University of Cambridge, UK)
- Antoon Bosselaers (Katholieke Universiteit Leuven, Belgium)
- Eric Brier (Gemplus, France)
- Benoit Chevallier-Mames (Gemplus, France)
- Jaewook Chung (University of Waterloo, Canada)
- Charles Clancy (University of Illinois, Urbana-Champaign, USA)
- Nicolas Courtois (Schlumberger Smart Cards, France)
- Evelyne Dewitte (Katholieke Universiteit Leuven, Belgium)
- Jean-François Dhem (Gemplus, France)
- Nevine Ebeid (University of Waterloo, Canada)
- Itamar Elhanany (Ben-Gurion University, Israel)
- Wieland Fischer (Infineon Technologies AG, Germany)
- Jacques Fournier (Gemplus, France)
- Shinobu Fujita (Toshiba Corporation, Japan)
- Soichi Furuya (Hitachi Ltd., Japan)
- Berndt Gammel (Infineon Technologies AG, Germany)
- Christophe Giraud (Oberthur Card Systems, France)
- Johann Groszschaedl (Technische Universität Graz, Austria)
- Jorge Guajardo (Ruhr-Universität Bochum, Germany)
- Sang Yun Han (Pohang University of Science and Technology, Korea)
- Helena Handschuh (Gemplus, France)
- Marko Hassinen (University of Kuopio, Finland)
- Alireza Hodjat (University of California, Los Angeles, USA)
- David Hwang (University of California, Los Angeles, USA)
- Yong Ho Hwang (Pohang University of Science and Technology, Korea)

- Tetsuya Izu (Fujitsu Laboratories Ltd., Japan)
- Ji Hyun Jung (Pohang University of Science and Technology, Korea)
- Chong Hee Kim (Pohang University of Science and Technology, Korea)
- Ki Hyun Kim (Pohang University of Science and Technology, Korea)
- Masanobu Koike (Toshiba Corporation, Japan)
- Sandeep Kumar (Ruhr-Universität Bochum, Germany)
- Noboru Kunihiro (The University of Electro-Communications, Japan)
- Eonkyung Lee (Korea Information Security Agency, Korea)
- Sungjae Lee (Korea Information Security Agency, Korea)
- Philippe Loubet-Moundi (Gemplus, France)
- Jonathan Lutz (University of Waterloo, Canada)
- Raimondo Luzzi (Infineon Technologies AG, Germany)
- Bodo Möller (Technische Universität Darmstadt, Germany)
- Simon Moore (University of Cambridge, UK)
- Francis Olivier (Gemplus, France)
- Elisabeth Oswald (Technische Universität Graz, Austria)
- Dong Jin Park (Pohang University of Science and Technology, Korea)
- In Kook Park (Pohang University of Science and Technology, Korea)
- Jae Hwan Park (Pohang University of Science and Technology, Korea)
- Joon Hah Park (Pohang University of Science and Technology, Korea)
- Eric Peeters (Université catholique de Louvain, Belgium)
- Beatrice Peirani (Gemplus, France)
- Jan Pelzl (Ruhr-Universität Bochum, Germany)
- Guillaume Poupard (DCSSI Crypto Lab, France)
- Zulfikar Ramzan (IP Dynamics, Inc., USA)
- Arash Reyhani-Masoleh (University of Waterloo, Canada)
- Francisco Rodríguez-Henríquez (CINVESTAV-IPN, Mexico)
- Manfred Roth (Infineon Technologies AG, Germany)
- Gaël Rouvroy (Université catholique de Louvain, Belgium)
- Yasuyuki Sakai (Mitsubishi Electric Corporation, Japan)
- Fumihiko Sano (Toshiba Corporation, Japan)
- Akashi Satoh (IBM Japan Ltd., Japan)
- Jasper Scholten (Katholieke Universiteit Leuven, Belgium)
- Kai Schramm (Ruhr-Universität Bochum, Germany)
- Hideo Shimizu (Toshiba Corporation, Japan)
- Jong Hoon Shin (Pohang University of Science and Technology, Korea)
- Sang Gyoo Sim (Pohang University of Science and Technology, Korea)
- Toru Sorimachi (Mitsubishi Electric Corporation, Japan)
- François-Xavier Standaert (Université catholique de Louvain, Belgium)
- Daisuke Suzuki (Mitsubishi Electric Corporation, Japan)
- Masashi Takahashi (Hitachi Ltd., Japan)
- Masahiko Takenaka (Fujitsu Laboratories Ltd., Japan)
- Alexandre F. Tenca (Oregon State University, Corvallis, USA)
- Kris Tiri (University of California, Los Angeles, USA)
- Shigenori Uchiyama (NTT Laboratories, Japan)
- Frederik Vercauteren (University of Bristol, UK)

- Johannes Wolkerstorfer (Technische Universität Graz, Austria)
- Thomas Wollinger (Ruhr-Universität Bochum, Germany)
- Huapeng Wu (University of Windsor, Canada)
- YeonHyeong Yang (Pohang University of Science and Technology, Korea)
- Sung Ho Yoo (Pohang University of Science and Technology, Korea)
- Young Tae Youn (Pohang University of Science and Technology, Korea)
- Dae Hyun Yum (Pohang University of Science and Technology, Korea)

The companies which provided support to CHES 2003:

- Comodo Research Lab — http://www.comodogroup.com
- Cryptovision — http://www.cryptovision.com
- GITS AG (Gesellschaft für IT-Sicherheit) — http://www.gits-ag.de
- Ministry for Research, Landesregierung Nordrhein-Westfalen
        — http://www.bildungsportal.nrw.de
- Ph.D. school "Secure Communication"
        — www.exp-math.uni-essen.de/zahlentheorie/gkkrypto

CHES Workshop Proceedings

- Ç. K. Koç and C. Paar (Editors). *Cryptographic Hardware and Embedded Systems*, First International Workshop, Worcester, MA, USA, August 12–13, 1999, LNCS No. 1717, Springer-Verlag, Berlin, Heidelberg, New York, 1999.
- Ç. K. Koç and C. Paar (Editors). *Cryptographic Hardware and Embedded Systems – CHES 2000*, Second International Workshop, Worcester, MA, USA, August 17–18, 2000, LNCS No. 1965, Springer-Verlag, Berlin, Heidelberg, New York, 2000.
- Ç. K. Koç, D. Naccache, and C. Paar (Editors). *Cryptographic Hardware and Embedded Systems – CHES 2001*, Third International Workshop, Paris, France, May 14–16, 2001, LNCS No. 2162, Springer-Verlag, Berlin, Heidelberg, New York, 2001.
- B. Kaliski Jr., Ç. K. Koç, and C. Paar (Editors). *Cryptographic Hardware and Embedded Systems – CHES 2002*, 4th International Workshop, Redwood Shores, CA, USA, August 13–15, 2002, LNCS No. 2523, Springer-Verlag, Berlin, Heidelberg, New York, 2002.
- C. D. Walter, Ç. K. Koç, and C. Paar (Editors). *Cryptographic Hardware and Embedded Systems – CHES 2003*, 5th International Workshop, Cologne, Germany, September 8–10, 2003, LNCS No. 2779, Springer-Verlag, Berlin, Heidelberg, New York, 2003. (These proceedings).

Table of Contents

Random Number Generators

Efficient Multiplication

More on Efficient Arithmetic

Attacks on Asymmetric Cryptosystems

The Security Challenges of Ubiquitous Computing

Frank Stajano

University of Cambridge
http://www-lce.eng.cam.ac.uk/~fms27/

Ubiquitous computing, over a decade in the making, has finally graduated from whacky buzzword through fashionable research topic to something that is definitely and inevitably happening. This will mean revolutionary changes in the way computing affects our society—changes of the same magnitude and scope as those brought about by the World Wide Web.

The performance of a computer of given cost has gone up dramatically throughout the whole history of computing. Even just the last decade has brought improvements worth several orders of magnitude along such diverse dimensions as processor speed, memory capacity, disk capacity, communication bandwidth and so on. As we overtake the "a computer for everyone" milestone and march steadily towards a future in which each person owns hundreds of computing objects, we will start to explore a different region of the computer design space: keeping the performance constant and making the cost vanishingly small. Think of throw-away embedded computers inside shoes, drink cans and postage stamps.

Security engineers will face specific technical challenges such as how to provide the required cryptographic functionality within the smallest possible gate count and the smallest possible power budget: the chips to be embedded in postage stamps will be the size of a grain of sand and will be powered by the energy radiated by an external scanning device.

The more significant security challenges, however, will be the systemic ones. Ubiquitous computing is not just a wireless version of the Internet with a thousand times more computers, and it would be a naïve mistake to imagine that the traditional security solutions for distributed systems will scale to the new scenario. Authentication, authorization, and even concepts as fundamental as ownership require thorough rethinking. The security challenges of the architecture are much greater than those of the mechanisms.

At a higher level still, even goals and policies must be revised. Having hundreds of computers per person changes the situation to such an extent that even the most fundamental assumptions need reexamining. There are evident issues of privacy, but also of trust and control. One question we should keep asking is simply "Security *for whom?*" The owner of a device, for example, is no longer necessarily the party whose interests the device will attempt to safeguard.

Ubiquitous computing is happening and will affect everyone. By itself it will never be "secure" (whatever this means) if not for the dedicated efforts of people like us. We are the ones who can make the difference. So, before focusing on the implementation details, let's have a serious look at the big picture.

C.D. Walter et al. (Eds.): CHES 2003, LNCS 2779, p. 1, 2003.
© Springer-Verlag Berlin Heidelberg 2003

Multi-channel Attacks

Dakshi Agrawal, Josyula R. Rao, and Pankaj Rohatgi

IBM Watson Research Center
P.O. Box 704
Yorktown Heights, NY 10598
{agrawal,jrrao,rohatgi}@us.ibm.com

Abstract. We introduce *multi-channel attacks*, i.e., side-channel attacks which utilize multiple side-channels such as power and EM *simultaneously*. We propose an adversarial model which combines a CMOS leakage model and the maximum-likelihood principle for performing and analyzing such attacks. This model is essential for deriving the optimal and very often counter-intuitive techniques for channel selection and data analysis. We show that using multiple channels is better for template attacks by experimentally showing a three-fold reduction in the error probability. Developing sound countermeasures against multi-channel attacks requires a rigorous leakage assessment methodology. Under suitable assumptions and approximations, our model also yields a practical assessment methodology for net information leakage from the power and all available EM channels in constrained devices such as chip-cards. Classical DPA/DEMA style attacks assume an adversary weaker than that of our model. For this adversary, we apply the maximum-likelihood principle to such design new and more efficient single and multiple-channel DPA/DEMA attacks.

Keywords: Side-channel attacks, Power Analysis, EM Analysis, DPA, DEMA.

1 Introduction

1.1 The Problem

Recent research in side-channel attacks has validated and reinforced the observation that sensitive information can leak from cryptographic devices via a multitude of channels. The seminal work of [9,8] describing leakages in timing and power channels was followed by the work of [10,7,1] showing leakages via electromagnetic (EM) emanations. The work of [1] shows that even a single EM probe can yield multiple EM signals via demodulation of different carriers. Further, different EM carriers carry different information and some EM leakages exceed leakages in the power channel. All these channels provide a rich source of information for a determined adversary.

While it seems plausible that side-channel attacks can be significantly improved by capturing multiple side-channel signals such as the various EM channels and possibly the power channel, a number of questions remain. Which

C.D. Walter et al. (Eds.): CHES 2003, LNCS 2779, pp. 2–16, 2003.
© Springer-Verlag Berlin Heidelberg 2003

side-channel signals should be collected? How should information from various channels be combined? How can one quantify the advantage of using multiple channels? These issues are especially relevant to an attacker since a significant equipment cost is associated with capturing each additional side-channel signal. Furthermore, in some situations, the detection risk associated with the additional equipment/probes required to capture a particular side channel has to be weighed against the benefit provided by that channel.

1.2   Contributions

To address these issues, we present a formal adversarial model for multi-channel analysis using the power and various EM channels. [1] Our model is based on a leakage model for CMOS devices and concepts from the Signal Detection and Estimation Theory. This formal model can be used to assess how an adversary can *best exploit* the wide array of signals available to him. In theory, this model can also deal with the problem of *optimal* channel selection and data analysis. However, in practice, a straight-forward application of this model can sometimes be infeasible. We show a judicious choice of approximations that renders the model useful for most practical applications.

Formulating such an adversarial model has numerous pitfalls. Ideally, the model should capture the strongest possible multi-channel attacks on an implementation of a cryptographic algorithm involving secret data. While such a model is easy to define, using it to assess vulnerabilities and create attacks will shift the focus from multi-channel information leakage to the specifics of the algorithm and implementation.

To refocus the attention on information leakage from multiple side-channels, we will only consider *elementary leakages,* i.e., information leaked during *elementary operations* of CMOS devices. This allows us to deal with information leakage aspects of multiple channels while not losing sight of the goal of evaluating entire implementations. In fact, it can be shown that the leakage in an entire computation is just the composition of elementary leakages from all of its elementary operations[2].

We introduce an adversarial model that is based on this view of elementary leakages of CMOS devices and is phrased in terms of the maximum likelihood testing of hypotheses. The model provides a formal way of comparing efficacies of various signal selection and processing techniques that can be used by a resource limited adversary.

Applying the model to the problem of signal selection, we find that the optimal strategies for picking even two best side-channels from a set of possibilities can be complex and counter-intuitive. For instance, picking the two channels with the best signal-to-noise ratios is quite often sub-optimal. The model also shows how to best combine information from multiple channels. This can be viewed as a generalization of template attacks [4] to the case of multiple channels. We provide experimental evidence to show that multi-channel based template attacks

---

[1] Combining the timing and power channel is already known, e.g., [11].

are superior to their single channel counterparts. Specifically, for a smart-card S^2, we show that template attacks that use both an EM channel and a power signal are superior to attacks that use only a single channel.

Our model for multi-channel attacks is also valuable for the designers of cryptographic implementations since they need to know the amount of leakage from multiple sensors to select the appropriate level of countermeasures. We describe a methodology for assessing *any type of leakage* in an *information-theoretic* sense. The methodology permits the computation of bounds on the best error probability achieved by an all-powerful adversary. While such an assessment is impractical for arbitrary devices, it is feasible for the practically important case of chipcards with small word lengths.

One drawback of our model is assumption of a very powerful adversary who has full knowledge of the characteristics of the target device and is capable of performing attacks similar to template attacks on the device. In practice, such attacks are tedious to mount and often adversaries don't have knowledge about the device. Thus, DPA-style attacks continue to be important due to their simplicity and immediate applicability to unknown implementations. Using the maximum likelihood testing as a basis, we show how current single channel DPA-style attacks can be greatly improved and how multiple-channel DPA-style attacks can be designed. The key to these improvements is a relaxation of the maximum likelihood test which estimates the unknown parameters of the test on the fly. We provide empirical evidence to show that a better analysis can give a substantial reduction in the number of samples needed for a traditional DPA attack and even a better reduction factor when a multiple-channel DPA attack is carried out using a power and an EM channel with very similar leakage characteristics.

2   Adversarial Model

This section develops an adversarial model to formally address issues related to the leakage of information via multiple side-channel signals.

2.1   CMOS Side-Channel Elementary Leakages

In CMOS devices, all data processing is typically controlled by a "square-wave" shaped clock. Each clock edge triggers a short sequence of state changing events and corresponding currents in the data processing units. The events are transient and a steady state is achieved well before the next clock edge. At any clock cycle, all the events and resulting currents are determined by a comparatively small number of bits of the logic state of the device, i.e., one only needs to consider the state of *active* circuits during that clock cycle and not the entire state of the device. These bits, termed as *relevant bits* in [3], constitute the *relevant state* of the device.

---

[2] A pseudonym is used to protect vendor identity; S is a 6805-based sub-micron, double metal technology card with inbuilt noise generators

Signals found on side-channels such as power and EM result from the current flows within the device and are affected by the random thermal noise. As mentioned above, ideally, the current flows in a CMOS device are directly attributable to the relevant state of the device. However, in practice, there may be many very small leakage currents in the *inactive* parts of the circuit. These leakages can be approximated as a small Gaussian noise term having negligible correlation with any particular active part of the circuit.

Thus as a very good approximation, all side-channel emanations during a clock cycle carry information *only* about the events and the relevant state of the device that occurs during the clock cycle. This is strongly supported by the experimental results which show that algorithmic bits are significantly correlated to the power/EM signals *only* during the clock cycles when they are actively involved in a computation. Thus it is natural to model side-channel leakage from the CMOS devices in terms of the leakages of the relevant state that occur during each clock cycle. We term the operation performed by the device during each clock cycle as an *elementary operation* and define the corresponding leakage of the relevant state information from side-channels as an *elementary leakage*.

2.2   Adversarial Model for Elementary Leakages

Given the concept of elementary leakages, it is natural to formulate side-channel attacks in terms of how successful an adversary can be in obtaining information about the relevant state. For example, an adversary may be interested in the LSB of the data bus during a LOAD instruction. This has a natural formulation as a binary hypothesis testing problem for the adversary[3]. Such a formulation also makes sense as traditionally the binary hypothesis testing has been central to the notions of side-channel attack resistance and leakage immunity [3,5].

The adversarial model consists of two phases. The first phase, known as the *the profiling phase*, is a training phase for the adversary. He is given a training device identical to the target device, an elementary operation, two distinct probability distributions B_0 and B_1 on the relevant states from which the operation can be invoked and a set of sensors for monitoring side-channel signals. The adversary can invoke the elementary operation, on the training device, starting from any relevant state. It is expected that adversary uses this phase to prepare an attack.

In the second phase, known as *the hypothesis testing phase*, the adversary is given the target device and the same set of sensors. He is allowed to make a *bounded number* of invocations to the same elementary operation on the target device starting from a relevant state that is drawn *independently* for each invocation according to exactly one of the two distributions B_0 or B_1. The choice of distribution is unknown to the adversary and his task is to use the signals on the sensors to select the correct hypothesis (H_0 for B_0 and H_1 for B_1) about

---

[3] In general, the adversary faces an M-ary hypothesis testing problem on functions of relevant state, for which results are straightforward generalizations of binary hypothesis testing.

the distribution used. The utility of the side-channels can then be measured in terms of the success probability achieved by the adversary as a function of the number of invocations.

2.3   Sophisticated Attack Strategies

Assume that an adversary acquires L statistically independent sets of sensor signals $\mathbf{O}_i, i = 1, \ldots, L$. These L sets of signals may correspond to L invocations of an operation on the target device. Also assume that there are K equally likely hypotheses H_k, $k = 1, \ldots, K$, on the origin of these signals. Let $p(\mathbf{O}|H)$ be the probability distribution of the sensor signals under the hypothesis H. Under these assumptions, the *maximum likelihood hypothesis* test is optimal and it decides in favor of the hypothesis H_k if

$$k = \operatorname*{argmax}_{1 \leqslant k \leqslant K} \prod_{i=1}^{L} p(\mathbf{O}_i|H_k). \tag{1}$$

While the maximum likelihood test is optimal, it is usually impractical as an exact characterization of the probability distribution of the sensor signals $\mathbf{O}$ may be infeasible. Such a characterization has to capture the nature of each of the sensor signals and the dependencies among them. This could further be complicated by the fact that, in addition to the thermal noise, the sensor signals could also display additional structure due to the interplay between properties of the device and those of the distributions of the relevant states. For example, if the hypothesis was on the LSB of a register while the device produced widely different signals only when the MSB was different, the sensor signals will display a bimodal effect attributable to the MSB. It turns out that in practice one can obtain near optimal results by making the right assumptions about the sensor signals. Such assumptions greatly simplify the task of hypothesis testing by requiring only a partial characterization of sensor signals.

The Gaussian Assumption. One such widely applicable assumption is the *Gaussian assumption* which states that under the hypothesis H, the sensor signal $\mathbf{O}$ has a multivariate Gaussian distribution with mean μ_H and a covariance matrix Σ_H. A multivariate Gaussian distribution $p(\cdot|H)$ has the following form:

$$p(\mathbf{o}|H) = \frac{1}{\sqrt{(2\pi)^n |\Sigma_H|}} \exp(-\frac{1}{2}(\mathbf{o} - \mu_H)^T \Sigma_H^{-1}(\mathbf{o} - \mu_H)), \quad \mathbf{o} \in \mathcal{R}^n, \tag{2}$$

where $|\Sigma_H|$ denotes the determinant of Σ_H and Σ_H^{-1} denotes the inverse of Σ_H.

The Gaussian assumption holds for a large number of devices and hypotheses encountered in the practice. In fact this assumption has been used successfully in the case of chip-cards [4].

It can be shown that under the Gaussian assumption, the maximum likelihood hypothesis testing for a single observation $\mathbf{O}$ and two equally likely hypothesis H_0 and H_1[4] simplifies to the following comparison:

$$(\mathbf{O}-\mu_{H_0})^T \Sigma_{H_0}^{-1}(\mathbf{O}-\mu_{H_0}) - (\mathbf{O}-\mu_{H_1})^T \Sigma_{H_1}^{-1}(\mathbf{O}-\mu_{H_1}) \geqslant \ln(|\Sigma_{H_1}|) - \ln(|\Sigma_{H_0}|) \tag{3}$$

where a decision is made in favor of H_1 if the above comparison is true, and in favor of H_0 otherwise.

In many cases of practical interest, noise in the sensor signals does not depend on the hypothesis, that is, $\Sigma_{H_0} = \Sigma_{H_1} = \Sigma_N$. In such cases, the following well-known result from the Statistics gives the probability of error in maximum-likelihood hypothesis testing [12]:

Fact 1 *For equally likely binary hypotheses, the probability of error in the maximum likelihood testing is given by*

$$P_\epsilon = \frac{1}{2} \operatorname{erfc}\left(\frac{\Delta}{2\sqrt{2}}\right) \tag{4}$$

where $\Delta^2 = (\mu_{H_1} - \mu_{H_0})^T \Sigma_N^{-1}(\mu_{H_1} - \mu_{H_0})$ and $\operatorname{erfc}(x) = 1 - \operatorname{erf}(x)$. Note that Δ^2 has a nice interpretation as the optimal signal-to-noise ratio that an adversary can achieve under the Gaussian assumption.

In the rest of this section, we will present two applications of the theory discussed above. In the first application, a strategy for selecting multiple side channels is presented. In the second application, a template attack on multiple channels is devised.

2.4   Multiple Channel Selection

Consider a resource limited adversary who can select at most M channels for an attack. When viewed in terms of our model, this problem conceptually has a very simple solution: The adversary should choose those M channels that minimize his probability of error in the maximum likelihood testing.

This apparently simple technique can be quite subtle and tricky in practice. Clearly, in situations where a well-prepared adversary has nicely characterized/approximated signals from each of the channels under each hypothesis and the corresponding joint noise probability distribution between all the channels, the adversary can also calculate the error probability for each possible choice of M channels, at least for small M. For example, if the noise is Gaussian and independent of the hypothesis, then from Equation 4, since $\operatorname{erfc}(\cdot)$ decreases exponentially with Δ, the goal of an adversary limited to just two channels, would be to choose channels in such a manner, as to maximize the output signal-to-noise ratio Δ^2.

---

[4] Generalizations to multiple observations and more than two hypotheses are straightforward.

If instead of a rigorous approach, channels are selected by heuristic techniques, then the resulting selection could be sub-optimal for various subtle reasons. Firstly, different side-channels could leak different aspects of information relative to the hypotheses being tested and sometimes there could be value in combining channels which provide widely dissimilar information rather than combining those which provide similar but partial information. Secondly, even if many channels provide the same information, picking multiple channels from this set could still be valuable since that may be almost as good as having the ability to make multiple invocations of the device with the same data and collecting a single side-channel. Even for the case where only two side-channels can be selected, the optimal choice is quite tricky and subtle as shown by the example below where the naive approach of choosing the two signals with best signal-to-noise ratios is shown to be sub-optimal.

Example 1. Consider the case where an adversary can collect two signals $[O_1, O_2]^T$ at a single point in time, such that under the hypothesis H_0, $O_k = N_k$, for $k = 1, 2$, and under the hypothesis H_1, $O_k = S_k + N_k$. Assume that $\mathbf{N_i} = (N_1, N_2)^T$ has zero mean multivariate Gaussian distribution with

$$\Sigma_N = \begin{pmatrix} 1 & \rho \\ \rho & 1 \end{pmatrix}$$

Note that O_1 and O_2 have signal-to-noise ratios of S_1^2 and S_2^2 respectively. After some algebraic manipulations, we get

$$\Delta^2 = \frac{(S_1 + S_2)^2}{2(1 + \rho)} + \frac{(S_1 - S_2)^2}{2(1 - \rho)} \tag{5}$$

Now, consider the case of an adversary who discovers two AM modulated carrier frequencies which are close and carry compromising information, both of which have very high and equally good signal-to-noise ratios ($S_1 = S_2$) and another AM modulated carrier in a very different band with a lower signal-to-noise ratio. An intuitive approach would be to pick the two carriers with high signal-to-noise ratio. In this case $S_1 = S_2$ and we get, $\Delta^2 = 2S_1^2/(1 + \rho)$. Since both signals originate from carriers of similar frequencies, the noise that they carry will have a high correlation coefficient ρ, which reduces Δ^2 at the *output*. On the other hand, if the adversary collects one signal from a good carrier and the other from the worse quality carrier in the different band, then the noise correlation is likely to be lower or even 0. In this case:

$$\Delta^2 = \frac{(S_1 + S_2)^2}{2} + \frac{(S_1 - S_2)^2}{2} = S_1^2(1 + S_2^2/S_1^2) \tag{6}$$

It is clear that the combination of a high and a low signal-to-noise ratio signals would be *a better strategy* as long as $S_2^2/S_1^2 > (1 - \rho)/(1 + \rho)$. For example, if $\rho > 1/3$, then choosing carriers from different frequency bands with even half the signal-to-noise ratio results in better hypothesis testing. ∎

Based on above analysis, in our experiments we routinely rejected a stronger channel which is colocated with another collected channel and chose a channel further away in the spectrum even if it had a lower signal-to-noise ratio.

2.5   Multi-channel Template Attacks

In [4], the power of using the maximum likelihood principle together with the Gaussian assumption was shown to be very effective in classifying successive bytes of an RC4 key using a *single* power side-channel signal. Expanding the template approach to multiple channel is straightforward. In the template attack, at any stage, the adversary uses an identical device to build exact characterizations for the signal and noise for each of the K possibilities he has to classify. Then he uses these characterizations to classify the one signal he is given from the target device. The first step in the template approach is the identification of those time instances (or indices of sample points) where the average signals for each of the K possibilities differ significantly. The second step is to compute the joint noise distribution of the channel at those points for each of the K possibilities. The third step is to classify the given signal into the K possibilities using the maximum likelihood testing.

For multiple-channels, the template attack is identical except that the signals from the multiple channels are concatenated together to yield a larger signal, i.e., for each invocation, a combined signal is created by concatenating the signals from the individual observed channels. Notice that the process of identifying the time instances and sample points could end up selecting somewhat different time slices for each channel, depending purely on the nature of leakage in each channel. The maximum likelihood testing will pick up information from all channels (possibly at different times) for classification.

To show that multiple channels help the classification process, we invoke an operation on the smart card S with two different input bytes and look at just 3 cycles during which the input was first processed. We collected EM and power samples simultaneously and evaluated how well the template attack could classify a single EM/power trace into the two hypotheses H0 and H1 for the input byte. We did this classification first using exactly one of the power/EM channels and then performed the classification using both channels simultaneously. Figures 1 shows the mean EM and Power signals for these hypothesis during these 3 cycles side by side.[5] Fig 2 shows the error rate of our classification effort for inputs belonging to each hypothesis. One can clearly notice that using both channels simultaneously results in better classification compared to any single channel.

3   Leakage Assessment Methodology for Chipcards

The model developed in Section 2.2 can be used to derive a practical methodology for assessing information leakage from any L power and EM channels for simple CMOS devices such as 8-bit chipcards. Several key properties make such a methodology feasible. Firstly, for a fixed relevant state, the noise at any cycle is well-approximated by a Gaussian distribution. Thus, in the hypothesis testing phase, the problem becomes one of distinguishing between two distributions

---

[5] The slight offset in time is due to delay of EM signals with respect to the power signal.

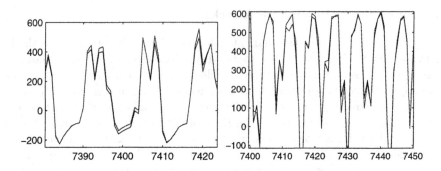

Fig. 1. Mean Power and EM signals during 3 cycles for two hypothesis

Correct Hypothesis	Error(Pwr)	Error(EM)	Error(EM+Pwr)
H0	9.5%	15.1%	2.8%
H1	20.1%	15.2%	6.6%

Fig. 2. Signal classification error using Power, EM and combination of Power and EM

B_0 and B_1 which are mixtures of Gaussians. Thus, if the number of relevant states (typically exponential in twice the word length) is small, each Gaussian in the collection can be profiled and the success probability for hypothesis testing can be computed. The problem of capturing leakages across multiple bands in multiple channels can be practically solved by splitting each channel into slightly overlapping bands upto a reasonable upper limit. Details of this assessment methodology with such practical considerations are given in the Appendix.

4   Single and Multi-channel DPA Attacks

In section 2, we assumed that the adversary had access to a test device identical to the target device and that he could carry out a profiling stage using the test device. In many circumstances, access to a test device may not be possible. In such cases, a DPA-style attack is preferred since it assumes no prior knowledge of device characteristics or implementation. In this section, we apply tools from the detection theory to optimize existing single channel DPA attacks and propose new multiple channel DPA attacks.

4.1   Improving DPA

In the traditional DPA attack, an adversary collects a set of N signals, $\mathbf{O}_i, i = 1, \ldots, N$ emanating from a given channel. Assume that the signals are normalized to have zero sample average over all N signals. For each hypothesis H under consideration, the N signals are divided into two bins, termed the 0-bin and the 1-bin with $N_{H,0}$ and $N_{H,1}$ samples respectively. Let $\mu_{H,0}[j]$ and $\mu_{H,1}[j]$ be the sample means of signals in the 0-bin and the 1-bin respectively for the hypothesis H. The next step in the DPA attack consists of computing the differences of sample means $\mu_H[j] = \mu_{H,0}[j] - \mu_{H,1}[j]$ for all hypotheses, and deciding in favor of the hypothesis H_i if $|\mu_{H_i}[j]|$ has the largest peak among all differences of means. In other words, the decision metric for the hypothesis H at time j is given by

$$M_H[j] = \left(\mu_{H,0}[j] - \mu_{H,1}[j]\right)^2,$$

(7)

and the decision is made in favor of the hypothesis H_i if for some value of j, say j_0, $M_{H_i}[j_0] >= M_H[j]$ for all H and j.

The traditional DPA attack and its variations have been successfully applied to attack several cryptographic implementations. However, by using the theory developed in the previous section, the effectiveness of traditional DPA can be increased significantly.

Before proceeding further, assume a void hypothesis H_v which corresponds to a random bifurcation of the N signals into the 0-bin and the 1-bin. Using the Gaussian assumption and Equation 3, the metric of a hypothesis H_i with respect to the null hypothesis at time j is given by

$$M_{H_i}[j] = \frac{\left(\mu_{H_i}[j] - E[\mu_{H_v}[j]]\right)^2}{V[\mu_{H_v}[j]]} - \frac{\left(\mu_{H_i}[j] - E[\mu_{H_i}[j]]\right)^2}{V[\mu_{H_i}[j]]} - \ln\left(\frac{V[\mu_{H_i}[j]]}{V[\mu_{H_v}[j]]}\right)$$

(8)

In order to compute this metric, we need the values of the following parameters: $E[\mu_{H_v}[j]]$, $V[\mu_{H_v}[j]]$, $E[\mu_H[j]]$, and $V[\mu_H[j]]$. Since in the DPA attack, the adversary skips the profiling phase of the attack, (8) is not directly applicable. In such cases, the theory suggests that unknown parameters of the test equation be estimated directly from the collected signals. If the adversary uses a maximum-likelihood estimate of these parameters, then the resulting test is referred to as the generalized maximum-likelihood testing.

For the DPA attack, calculating the maximum likelihood estimate of the test parameters involves solving a set of nonlinear coupled equations. Therefore, instead of using the maximum-likelihood estimates of these parameters, we use sample estimates as follows: Let $\sigma^2_{H,0}[j]$ and $\sigma^2_{H,1}[j]$ be the sample variances of the signals in the 0-bin and the 1-bin respectively at time j for hypothesis H. We propose the following sample estimators[6] of parameters in (8):

$$E[\mu_H[j]] = \mu_H[j]$$

$$V[\mu_H[j]] = \frac{\sigma^2_{H,0}[j]}{N_0} + \frac{\sigma^2_{H,1}[j]}{N_1}$$

(9)

---

[6] We omit the derivation of these estimators as the derivation is tedious and follows from straight-forward algebraic manipulations.

Sbox Hyp	Min Samples (Mean-diff)	Min Samples(Max-Likl)
S1,B3	640	350
S2,B3	630	210
S7,B3	110	40
S8,B3	130	90

Fig. 3. DPA results, mean-difference vs. approx. generalized maximum-likelihood

Substituting these in (8), we get the following formula for the metric:

$$M_{H_i}[j] = \frac{\left(\mu_{H_i}[j] - \mu_{H_v}[j]\right)^2}{\frac{\sigma^2_{H_v,0}[j]}{N_0} + \frac{\sigma^2_{H_v,1}[j]}{N_1}} - \ln\left(\frac{\frac{\sigma^2_{H_i,0}[j]}{N_0} + \frac{\sigma^2_{H_i,1}[j]}{N_1}}{\frac{\sigma^2_{H_v,0}[j]}{N_0} + \frac{\sigma^2_{H_v,1}[j]}{N_1}}\right) \qquad (10)$$

Intuitively, the signals in the 0-bin and 1-bin have similar distributions under the wrong hypothesis due to a random bifurcation of signals in the two bins. However, for the correct hypothesis, the distribution of signals in the 0-bin differs from the distribution of signals in the 1-bin. The traditional DPA attack only takes into account the differences in sample means. On the other hand, Equation 10 takes both the sample means and variances into account, and therefore may provide a better hypothesis test.

Figure 3 shows the results of applying this method to attacking the S-box lookup for a DES implementation. The first column shows the bit being predicted, the second shows the number of samples required for the correct key hypothesis to emerge as the winner under the traditional DPA metric while the third column shows the number of samples needed with the new metric. Clearly by using a better metric, our improvement in the DPA attack reduces the number of signals needed by a factor of 1.4–3.

4.2   Multi-channel DPA Attack

Multi-channel DPA attack is a generalization of the single-channel DPA attack. In this case, the adversary collects N signals, $\mathbf{O}_i, i = 1, \ldots, N$. In turn, each of the signals $\mathbf{O}_i$ is a collection of L signals collected from L side-channels. Thus, $\mathbf{O}_i = [\mathbf{O}_i^1, \ldots, \mathbf{O}_i^L]^T$ where $\mathbf{O}_i^l$ represents the i-th signal from the l-th channel. Note that all DPA style attacks treat each time instant independently and leakages from multiple channels can only be pooled together if they occur at the same time. Thus, in order for multi-channel DPA attacks to be effective, the selected channels must have very similar leakage characteristics.

The formulae for computing the metric for multi-channel DPA attack are generalizations of those for the single channel. The main difference is that the expected value of sample mean difference at time j under hypothesis H is a vector of length L, with the l-th entry being the sample mean difference of the l-th channel. Furthermore, the variance of the b-bin under hypothesis H at time j, is a covariance matrix of size $L \times L$ with the i, j-th entry being the correlation between signals from the i-th and j-th channels. Once again, as in the DPA attack, the adversary does not have the luxury of estimating these parameters. Therefore, we substitute sample estimates for these parameters along the same lines as in Equation 9. We skip the cumbersome formulae and directly go to the results of multi-channel DPA attacks.

Figure 4 shows sample results of an attack on the S-box lookups in a DES implementation using the power channel together with an EM channel whose leakage is similar to the power channel. The first column shows the bit being predicted, the second shows the number of signals required for the correct key hypothesis to emerge as the winner using both channels with the multi-channel metric, the last two columns show the number of signals needed for the power and EM channels separately using the new DPA/DEMA metric. From this it is clear that the number of invocations needed for two channel attacks can be significantly less compared to single-channel attacks.

Sbox Hyp	Min Samples(Pwr+EM)	Min Samples(Pwr)	Min Samples(EM)
S1,B1	150	170	640
S1,B2	60	(>1000)	340
S1,B3	110	350	160
S2,B2	30	50	230
S2,B3	120	210	340
S4,B0	60	200	340
S6,B1	180	180	190
S7,B3	30	40	520
S8,B3	60	90	140

Fig. 4. Multi-Channel DPA-style attack using Power, EM and Power&EM. and EM

4.3 Future Work on Single/Multi-channel DPA/DEMA Attacks

It is well known to DPA/DEMA practitioners that for the correct hypothesis, the correlation signal with respect to time shows multiple peaks. However, current analysis techniques, including the ones presented here, do not combine information from peaks occurring at different time instances. This problem also manifests itself when combining various Power and EM channels since peaks on different channels may not coincide. One can also view the efficacy gap between template attacks and DPA attacks as a manifestation of the same problem.

We have started work which promises to bridge this gap. The main idea is to estimate the characteristics of useful peaks on the fly given only the collected signals (without using a training set) and apply techniques based on maximum-likelihood principle to identify the correct hypothesis.

References

1. Dakshi Agrawal, Bruce Archambeault, Josyula R. Rao and Pankaj Rohatgi. The EM Side Channel(s). Proceedings of CHES 2002, Springer, LNCS 2523, pp 29–45.

2. D. Agrawal, B. Archambeault, J. R. Rao and P. Rohatgi. The EM Side Channel(s): Attacks and Assessment Methodologies. See
http://www.research.ibm.com/intsec/emf-paper.ps
3. Suresh Chari, Charanjit S. Jutla, Josyula R. Rao and Pankaj Rohatgi. Towards Sound Countermeasures to Counteract Power–Analysis Attacks. Proceedings of Crypto '99, Springer–Verlag, LNCS 1666, pp 398–412.
4. Suresh Chari, Josyula R. Rao and Pankaj Rohatgi. Template Attacks. Proceedings of CHES 2002, Springer, LNCS 2523, pp 13–28.
5. Jean–Sebastien Coron, Paul Kocher and David Naccache. Statistics and Secret Leakage. In the Proceedings of Financial Cryptography '00. Springer-Verlag, LNCS 1962, pp 157–173
6. L. Goubin and J. Patarin. DES and Differential Power Analysis. Proceedings of CHES '99, LNCS 1717, pp 158–172.
7. K. Gandolfi, C. Mourtel and F. Olivier. Electromagnetic Attacks: Concrete Results. Proceedings of CHES '01, LNCS 2162, pp 251–261.
8. P. Kocher. Timing Attacks on Implementations of Diffie-Hellman, RSA, DSS and Other Systems. Advances in Cryptology-Crypto '96, Springer-Verlag, LNCS 1109, pp 104–113.
9. P. Kocher, J. Jaffe and B. Jun. Differential Power Analysis: Leaking Secrets. Advances in Cryptology — Proceedings of Crypto '99, Springer Verlag, LNCS 1666, pp. 388–397.
10. Jean–Jacques Quisquater and David Samyde. ElectroMagnetic Analysis (EMA): Measures and Counter-Measures for Smart Cards. In Smart Card Programming and Security (E-smart 2001), LNCS 2140, pp.200-210,September 2001.
11. C. D. Walter and S. Thompson. Distinguishing Exponent Digits by Observing Modular Subtractions. In Progress in Cryptology- CT-RSA 2001, Springer, LNCS 2020, pp 192–207.
12. H. L. Van Trees. Detection, Estimation, and Modulation Theory, Part I. John Wiley & Sons. New York. 1968.

Appendix: Leakage Assessment for Chipcards

In this section, we address the question of whether one can assess and quantify the net leakage of information from multiple sensors. Can the information obtained by combining leakages from several (or even all possible) signals from available sensors be quantified regardless of the signal processing capabilities and computing power of an adversary?

Maximum likelihood testing is the optimal way to perform hypothesis testing. Thus, we use it to craft a methodology to assess information leakage from elementary operations. Our methodology takes into account signals extractable from all the given sensors across the entire EM spectrum. Results of such an assessment will enable one to bound the success probability of the optimal adversary for any given hypothesis.

Assume, that for a single invocation, the adversary captures the emanations across the entire electromagnetic spectrum from all sensors in an observation vector $\mathbf{O}$. Let Ω denote the space of all possible observation vectors $\mathbf{O}$. Since the likelihood ratio, $\Lambda(\mathbf{O})$ is a function of the random vector $\mathbf{O}$, the best achievable success probability, P_s, is given by:

$$P_s = \sum_{\mathbf{O} \in \Omega} I_{\{\Lambda(\mathbf{O})>1\}} p_{\mathbf{N1}}(\mathbf{O} - \mathbf{S}_1) + I_{\{\Lambda(\mathbf{O})<1\}} p_{\mathbf{N0}}(\mathbf{O} - \mathbf{S}_0) \tag{11}$$

where I_A denotes the indicator function of the set A, and $p_{\mathbf{N1}}(\mathbf{O} - \mathbf{S}_1)$ and $p_{\mathbf{N0}}(\mathbf{O} - \mathbf{S}_0)$ are noise distributions under the hypothesis 1 and 0.

When the adversary has access to multiple invocations, an easier way of estimating the probability of success/error involves a technique based on moment generating functions. We begin by defining the logarithm of the moment generating function of the likelihood ratio:

$$\mu(s) = \ln\left(\sum_{\mathbf{O} \in \Omega} p_{\mathbf{N1}}^s(\mathbf{O} - \mathbf{S}_1) p_{\mathbf{N0}}^{1-s}(\mathbf{O} - \mathbf{S}_0) \right) \tag{12}$$

The following is a well-known result from Information Theory:

Fact 2 *Assume we have several statistically independent observation vectors[7] $\mathbf{O}_1, \mathbf{O}_2, \ldots, \mathbf{O}_L$. For this case, the best possible exponent in the probability of error is given by the* Chernoff Information*:*

$$C \stackrel{\text{def}}{=} - \min_{0 \leqslant s \leqslant 1} \mu(s) \stackrel{\text{def}}{=} -\mu(s_m) \tag{13}$$

Note that $\mu(\cdot)$ is a smooth, infinitely differentiable, convex function and therefore it is possible to approximate s_m by interpolating in the domain of interest and finding the minima. Furthermore, under certain mild conditions on the parameters, the error probability can be approximated by:

$$P_\epsilon \approx \frac{1}{\sqrt{8\pi L \mu''(s_m)} s_m(1 - s_m)} \exp(L\mu(s_m)) \tag{14}$$

Note that in order to evaluate (11) or (14), we need to estimate $p_{\mathbf{N0}}(\cdot)$ and $p_{\mathbf{N1}}(\cdot)$. In general, this can be a difficult task. However by exploiting certain characteristics of the CMOS devices, estimation of $p_{\mathbf{N0}}(\cdot)$ and $p_{\mathbf{N1}}(\cdot)$ can be made more tractable.

Practical Considerations

We will now outline some of the practical issues associated with estimating $p_{\mathbf{N0}}(\cdot)$ and $p_{\mathbf{N1}}(\cdot)$ for any hypothesis. The key here is to estimate the noise distribution for each cycle of each elementary operation and for each relevant state R that the operation can be invoked with. This results in the signal characterization, $\mathbf{S}_R$, and the noise distribution, $p_{\mathbf{NR}}(\cdot)$ which is sufficient for evaluating $p_{\mathbf{N0}}(\cdot)$ and $p_{\mathbf{N1}}(\cdot)$.

There are two crucial assumptions that facilitate estimating $p_{\mathbf{NR}}(\cdot)$: first, on chipcards examined by us the typical clock cycle is 270 nanoseconds. For such

---

[7] For simplicity, this paper deals with *independent* elementary operation invocations. Techniques also exist for adaptive invocations.

devices, most of the compromising emanations are well below 1 GHz which can be captured by sampling the signals at a Nyquist rate of 2 GHz. This sampling rate results in a vector of 540 points per cycle per sensor. Alternatively, one can also capture all compromising emanations by sampling judiciously chosen and slightly overlapping bands of the EM spectrum. The choice of selected bands is dictated by considerations such as signal strength and limitations of the available equipment. Note that the slight overlapping of EM bands would result in a corresponding increase in the number of samples per clock cycle, however it remains in the range of 600-800 samples per sensor.

The second assumption, borne out in practice (see [4]), is that for a fixed relevant state, the noise distribution $p_{NR}(\cdot)$ can be approximated by a Gaussian distribution. This fact greatly simplifies the estimation of $p_{NR}(\cdot)$ as only about one thousand samples are needed to roughly characterize $p_{NR}(\cdot)$. Moreover, the noise density can be stored compactly in terms of the parameters of the Gaussian distribution.

These two assumptions imply that in order to estimate $p_{NR}(\cdot)$ for a fixed relevant state R, we need to repeatedly invoke (say 1000 times) an operation on the device starting in the state R, and collect samples of the emanations as described above. Subsequently, the signal characterization S_R can be obtained by averaging the collected samples. The noise characterization is obtained by first subtracting S_R from each of the samples and then using the Gaussian assumption to estimate the parameters of the noise distribution.

The assessment can now be used to bound the success of any hypothesis testing attack in our adversarial model. For any two given distributions B_0 and B_1 on the relevant states, the corresponding signal and noise characterizations, $S_0, S_1, p_{N0}(\cdot)$, and $p_{N1}(\cdot)$, are a *weighted sum* of the signal and noise assessments of the constituent relevant states S_R and $p_{NR}(\cdot)$. The error probability of maximum-likelihood testing for a single invocation or its exponent for L invocations can then be bounded using (11) and (13) respectively.

We now give a rough estimate of the effort required to obtain the leakage assessment of an elementary operation. The biggest constraint in this process is the time required to collect samples from approximately one thousand invocations for each relevant state of the elementary operation. For an r-bit machine, the relevant states of interest are approximately 2^{2r}; thus the leakage assessment requires time to perform approximately $1000 * 2^{2r}$ invocations. Assuming that the noise is Gaussian and that each sensor produces an observation vector of length 800, for n sensors the covariance matrix Σ_N has $(800 * n)^2$ entries. It follows that the computation burden of estimating the noise distribution would be proportional to $(800 * n)^2$. Such an approach is certainly feasible for an evaluation agency, from both a physical and computational viewpoint, as long as the size of the relevant state, r, is small. In our experiments, we found such assessment possible for a variety of 8-bit chipcards.

Hidden Markov Model Cryptanalysis

Chris Karlof and David Wagner

Department of Computer Science,
University of California at Berkeley, Berkeley CA 94720, USA,
{ckarlof,daw}@cs.berkeley.edu

Abstract. We present *HMM attacks*, a new type of cryptanalysis based on modeling randomized side channel countermeasures as Hidden Markov Models (HMM's). We also introduce Input Driven Hidden Markov Models (IDHMM's), a generalization of HMM's that provides a powerful and unified cryptanalytic framework for analyzing countermeasures whose operational behavior can be modeled by a probabilistic finite state machine. IDHMM's generalize previous cryptanalyses of randomized side channel countermeasures, and they also often yield better results. We present efficient algorithms for key recovery using IDHMM's. Our methods can take advantage of multiple traces of the side channel and are inherently robust to noisy measurements. Lastly, we apply IDHMM's to analyze two randomized exponentiation algorithms proposed by Oswald and Aigner. We completely recover the secret key using as few as ten traces of the side channel.

1 Introduction

Randomized countermeasures [1,2,3,4,5,6,7,8] for side channel attacks [8,9,10] are a promising, inexpensive alternative to hardware based countermeasures. In order to gain strong assurance in randomized schemes, we need some way to analyze their security properties, and ideally, we would like general-purpose techniques. To this end, we present *HMM attacks*, a new type of cryptanalysis based on modeling countermeasures as Hidden Markov Models (HMM's) [11]. We also introduce Input Driven Hidden Markov Models (IDHMM's), a generalization of HMM's. IDHMM's are particularly well suited for analyzing any randomized countermeasure whose internal operation can be modeled by a probabilistic finite state machine.

Hidden Markov Models (HMM's) [11] are a well-studied model for finite-state stochastic processes. An execution of an HMM consists of a sequence of hidden, unobserved states and a corresponding sequence of related, observable outputs. HMM's are memoryless: given the current state, the conditional probability distribution for the next state is independent of all previous states. The main problem of interest in HMM's is the *inference problem*, to infer the values of the hidden, unobserved states given only the sequence of observable outputs. The Viterbi algorithm [12] is an efficient dynamic programming algorithm for solving this problem.

At first glance, HMM's seem perfect for analyzing randomized countermeasures which can be modeled by a probabilistic finite state machine: the hidden states of the HMM represent the internal states of the countermeasures and the observable outputs represent observations of the side channel. However, HMM's have deficiencies which

C.D. Walter et al. (Eds.): CHES 2003, LNCS 2779, pp. 17–34, 2003.

Table 1. Summary of attacks on OA1 and OA2, two randomized side channel countermeasures proposed by Oswald and Aigner. Note that our new attacks are the first to work even with a noisy side channel.

Attack	Relevant countermeasure	Observation error (p_e)	Number of traces needed to recover the secret key	Workfactor
Okeya-Sakurai [13]	OA1	0	292	minimal
C.D. Walter [14]	OA1, OA2	0	2–10	minimal
HMM attacks (new)	OA1, OA2	0	10	minimal
HMM attacks (new)	OA1, OA2	0	5	2^{38}
HMM attacks (new)	OA1, OA2	0.1	10	2^{38}
HMM attacks (new)	OA1, OA2	0.25	50–500	2^{38}

prevent them from being directly applicable. Firstly, HMM's do not model inputs. HMM's model processes as a sequence of states. However, the internal operation of a randomized countermeasure is likely to be dependent on both the current internal state as well as an input: the secret key. IDHMM's extend HMM's to handle inputs so we can accurately model randomized keyed countermeasures. Secondly, standard inference techniques like Viterbi's algorithm cannot leverage multiple output traces of an HMM. However, the ability to handle multiple output traces will make our key recovery attacks more powerful. To address this, we present an efficient approximate inference algorithm for IDHMM's that handles multiple output traces.

To demonstrate how HMM attacks can be used in practice, we show how to break two randomized exponentiation algorithms proposed by Oswald and Aigner [2]. Previously known attacks [13,14] against these algorithms assume the ability to perfectly distinguish between elliptic curve point additions and doublings in the side channel. We present more powerful attacks which are robust to noise. A summary of our attacks in comparison to previous work is shown in Table 1.

2   Modeling Randomized Side Channel Countermeasures as Probabilistic Finite State Machines

Many authors have proposed randomization as a way to limit the security risks from information leaked over side channels [1,2,3,4,5,6,7,8,15]. However, the security afforded by randomization in this setting is not clear. Side channel attacks are typically successful because of the high correlation between the information leaked over the side channel and the internal state of the device, most notably a secret key used in various cryptographic operations. The hope behind randomized countermeasures is that the side channel information will become randomized as well, thus making it harder to analyze. An ideal randomized countermeasure would completely disassociate the side channel information from the internal state of the device, or more formally, for any set of measurements of the side channel, the likelihood of an adversary guessing any information about the internal state of the device would be the same as if the adversary had observed no side channel information at all. Some examples of randomized countermeasures include randomized exponentiation algorithms [1,2,3,4], random window methods [5,6],

```
Input: k, M    Output: k × M
Q = M
P = 0
for i = 1 to N
    if (kᵢ == 1) then P = P + Q
    Q = 2Q
return P
```

(a) The Binary Algorithm for ECC scalar multiplication.

```
Input: k, M    Output: k × M
Q = M
P = 0
for i = 1 to N
    R = P
    b = rand_bit()
    if (kᵢ == 0) then
        if (b == 1) then
            R = R + Q // result is discarded
    else
        P = P + Q
    Q = 2Q
return P
```

(b) A randomized variant of the Binary Algorithm for ECC multiplication.

Fig. 1. Introducing randomness into the Binary Algorithm for ECC scalar multiplication.

randomized instruction execution [7], randomized timing shifts [8], randomized blinding of the secret key [15], and randomized projective coordinates [15].

We introduce a new cryptanalytic technique based on Hidden Markov Models to analyze such randomized countermeasures. To help give the intuition behind our analysis, we first give a simple example of a fabricated countermeasure that uses randomization, show how to model its operation using a probabilistic finite state machine, and then motivate the use of Hidden Markov Models to analyze its security.

2.1 A Simple Randomized Countermeasure

Consider the randomized variant of the standard binary algorithm for doing scalar multiplications over elliptic curves shown in Figure 1(b). Assume $k = k_N k_{N-1} \ldots k_2 k_1$ is the N bit secret key and M and P are points on the elliptic curve.

The major difference between the algorithm in Figure 1(b) and the standard Binary Algorithm is as follows: in each iteration, if the next key bit is 0, then with probability $1/2$ our algorithm will execute a discarded spurious addition, but if the next key bit is 1, it behaves the same as the standard Binary Algorithm. This randomized variant of the Binary Algorithm is completely artificial and by no means secure. It was created solely to demonstrate how randomness might be used in the construction of side channel countermeasures and will serve as a running example to illustrate our techniques.

Now, assume that it is possible for an adversary observing the side channel to distinguish between elliptic curve point additions and elliptic curve point doublings in a single scalar multiplication. Then the adversary's observation of a single scalar multiplication can be represented as a sequence $(y_1, y_2, \ldots, y_N)$, $y_i \in \{D, AD\}$, where D represents an elliptic curve point doubling and A represents an elliptic curve point addition. Each y_i represents the operations observed during the processing of a single bit of the key. We refer to such a sequence as a *trace*.

Note there is no longer a one-to-one correspondence between each possible trace and each possible key. Rather, each given sequence of observable operations is consistent

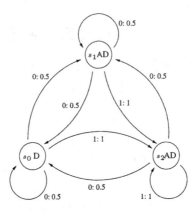

Fig. 2. A probabilistic finite state machine that models the operation of the randomized exponentiation algorithm in Figure 1(b).

with several possible keys. For example, if the trace from a scalar multiplication using the algorithm in Figure 1(b) is (AD, AD, D), then there are four possible keys consistent with this trace: namely, 011, 001, 010, or 000.

2.2 Probabilistic Finite State Machines

Although there are clearly many ad-hoc ways to break the algorithm in Figure 1(b), its primary purpose is to illustrate the development of a general technique for analyzing randomized countermeasures. Several weaknesses have been discovered in some existing randomized countermeasures [13,14,16], but the analysis techniques used are often specific to the particular countermeasure, and it is not obvious how to generalize them to a framework applicable to a larger class of algorithms. The primary benefit of a general analytical framework is that it enables the analysis of a large class of randomized countermeasures while minimizing the overhead needed to analyze any one particular algorithm. Although such a framework by itself may not in general be able to prove the security of every conceivable countermeasure, it can help quickly determine if a countermeasure is insecure, potentially saving a cryptanalyst many hours of work.

A key component for a general analytical framework is a good operational model of the countermeasures. A simple model applicable to many randomized countermeasures is a *probabilistic finite state machine*. The resulting finite state model for our running example can be easily constructed from its algorithmic description and is shown in Figure 2. Each state corresponds to a full iteration of the loop in Figure 1(b) (i.e, the processing of one key bit in its entirety) and is labeled with the operations (D or AD) that may be observed when that state is visited. Each edge is annotated with a bit from the key and a probability. In general, one of the states would be designated as initial, but in this example any state can serve as the initial state. The model of execution is simple: given the current state q_i and the next bit of the key k_{i+1}, the next state q_{i+1} is determined probabilistically according to the probabilities on those outgoing edges of q_i that are annotated with k_{i+1}.

While the edges capture the control structure of the algorithm, the states abstract away the details of the calculation. However, the label on each state indicates the observable information that leaks through the side channel when the process enters the state. In this example, the observations are what type of operations (elliptic curve point additions and/or doublings) are executed while in a particular state. Note however, since the state machine is randomized, some particular traces could arise from several different paths through the machine. For example, the trace (AD, AD, D) could arise from any of the following paths through the state machine in Figure 2: $s_2 s_2 s_0, s_2 s_1 s_0, s_1 s_2 s_0,$ or $s_1 s_1 s_0$.

More formally, we define a probabilistic finite state machine to be a sextuple

$$M = (S, I, \delta, O, s_0, \mu)$$

where

S is a finite set of internal states,

I is a finite set of input symbols,

$\delta : S \times S \times I \to [0, 1]$ is a function called the transition function,

O is a finite set of symbols that represent operations observable over the side channel,

$s_0 \in S$ is the initial state,

$\mu : S \to O$ is a function associating an observable operation with every state,

and the following condition is satisfied:

$$\forall s_i \in S, \forall b \in I, \sum_{s_j \in S} \delta(s_i, s_j, b) = 1 .$$

In our setting, the set of input symbols is $I = \{0, 1\}$, representing the bits of a secret key.

For a key $k = k_N k_{N-1} \ldots k_2 k_1$, define an *execution* q of $M = (S, \delta, O, s_0, \mu)$ on k to be a sequence $q = (q_0, q_1, \ldots, q_{N-1}, q_N)$, where $q_0 = s_0$ and $q_i \in S$, such that for $0 \le i < n$, $\delta(q_i, q_{i+1}, k_{i+1}) > 0$. Define a *trace* y of an execution q to be a sequence $y = (y_1, y_2, \ldots, y_N)$, where $y_i = \mu(q_i), \forall i \ne 0$. In the case of our example, an execution corresponds to the sequence of internal states traversed during a scalar multiplication of the secret key k, and the trace of that execution represents the sequence of observable elliptic curve point additions and doublings.

2.3   The Key Recovery Problem for Probabilistic Finite State Machines

Since one of the primary goals of side channel attacks is to recover the secret key stored within a target device, we wish to solve the following problem:

KEY RECOVERY PROBLEM FOR PROBABILISTIC FINITE STATE MACHINES

Let M be a probabilistic finite state machine. Generate a random N bit key k and an execution q of M on k. Let y be the trace of q. The Key Recovery Problem for probabilistic finite state machines is to find k given M and y.

One approach to solving the Key Recovery Problem for probabilistic finite state machines is the following: 1) Given a trace y and machine M, try to infer the execution q it resulted from, and then 2) infer k from q. Step 2 becomes easy if we restrict M to be *faithful*. A probabilistic finite state state machine $M = (S, \delta, O, s_0, \mu)$ is said to be *faithful* if it satisfies the following property:

$$\forall s_i, s_j \in S, \text{ if } \delta(s_i, s_j, 0) > 0, \text{ then } \delta(s_i, s_j, 1) = 0 .$$

For faithful machines, there is a one-to-one correspondence between an execution q and the key k used in that execution. This is because for every pair of consecutive states s_i, s_j in an execution, there is no ambiguity in what bit annotated the corresponding directed edge that was taken from s_i to s_j. Note that this condition does not limit the expressiveness of our framework by restricting M in any significant way. If for a machine M there exists s_i, s_j such that $\delta_M(s_i, s_j, 0) > 0$ and $\delta_M(s_i, s_j, 1) > 0$, then an observationally equivalent machine M' can be constructed that is identical to M except state s_j is replaced by two states s_{j_1}, s_{j_2} such that $\delta_{M'}(s_i, s_{j_1}, 0) = \delta_M(s_i, s_j, 0)$, $\delta_{M'}(s_i, s_{j_1}, 1) = 0$, $\delta_{M'}(s_i, s_{j_2}, 0) = 0$, and $\delta_{M'}(s_i, s_{j_2}, 1) = \delta_M(s_i, s_j, 1)$. Thus, without loss of generality, we will only consider probabilistic finite state machines that are faithful.

2.4   The State Inference Problem for Probabilistic Finite State Machines

We define the State Inference Problem for probabilistic finite state machines as follows:

STATE INFERENCE PROBLEM FOR PROBABILISTIC FINITE STATE MACHINES

Let M be a probabilistic finite state machine. Generate a random N bit key k and an execution q of M on k. Let y be the trace of q. The State Inference Problem for probabilistic finite state machines is to find q given M and y.

Because of the one-to-one correspondence between q and the key k used in that execution, solving the State Inference Problem for M and y is equivalent to solving the Key Recovery Problem for M and y[1].

One way an adversary might try to solve the State Inference Problem for $M = (S, I, \delta, O, s_0, \mu)$ and y of length N is to treat the unknown execution q as a random variable Q with sample space S^{N+1} and use *maximum likelihood decoding*. A simple implementation of maximum likelihood decoding involves two steps:

> **Input:** trace y, machine M
> 1. Calculate $\mathbf{Pr}[Q = s | y]$, for each $s \in S^{N+1}$.
> 2. Output $q = \underset{s \in S^{N+1}}{\operatorname{argmax}} \mathbf{Pr}[Q = s | y]$.

The adversary's output is the most likely execution q for the given trace y, which then yields the most likely key k.

---

[1] Although we have formulated both problems in a way that implies deterministic solutions, randomized algorithms with significant success probability are acceptable as well.

A naive implementation of step 1 will have a running time exponential in the length of the trace. However, we will see how to transform a probabilistic finite state machine into a Hidden Markov Model, in which there is an equivalent State Inference Problem with a polynomial running time solution. In addition to having efficient inference algorithms, we will see that Hidden Markov Models have other advantages as well.

In this section, we introduced probabilistic finite state machines, an intuitive technique for modeling the operation of randomized countermeasures, and we demonstrated the use of the model with an artificial yet instructive example. In the remainder of this paper, we will show how Hidden Markov Models not only provide a sound, well-studied framework, but also how they can be extended into even more powerful and flexible cryptanalytical tools for analyzing randomized countermeasures.

3 Assumptions

Before formally describing our analytical framework for randomized countermeasures using Hidden Markov Models, we will make our assumptions more precise. Our analysis depends on the following assumptions:

- We have collected a set of L traces from the side channel, corresponding to L executions of the countermeasure, all using the same secret key.
- Each trace of the side channel can be uniquely written as $(y_1, y_2, \ldots, y_N)$ where each y_i is an element of some finite observation set O. In the example presented in Section 2, $O = \{D, AD\}$.
- The operations in O can be probabilistically distinguished from each other.
- Each observation y_i from O can be associated with the processing of a single key bit position, and vice versa.

If the attacker is lucky, the side channel reveals exactly which action from O has been taken, and thus the observation traces $(y_1, y_2, \ldots, y_N)$ are free of errors. This is the model some previous work has used, and it does simplify analysis. However, this assumption may not always be realistic.

In the more general case, observations may only yield partial information on the actual trace, hence our measurements may contain errors. As we will see in Section 4, our techniques are still applicable in this setting. When there are only two different types of observations, a simple model of this behavior is that each observation has probability $1 - p_e$ of being correct and probability p_e of being mischaracterized. This setting may be more realistic in practice, particularly for devices that try to make all operations look alike.

4 Input Driven Hidden Markov Models as a Model for Randomized Side Channel Countermeasures

In Section 2, we outlined an approach for analyzing randomized countermeasures which infers the most likely secret key from the sequence of observable operations. However, this approach is not only intractable, but has other deficiencies as well. Four main challenges remain:

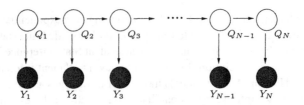

Fig. 3. An execution of a Hidden Markov Model, represented in a probabilistic graphical model. This figure depicts one execution of the HMM. Each node represents a random variable, and the directed edges indicate conditional dependencies. A shaded node indicates the corresponding variable is observed (i.e. outputs we can observe), while unshaded nodes are unobserved (i.e. what we wish to recover).

1. **Efficient inference algorithms are needed.** A naive implementation of maximum likelihood decoding for a single trace has running time exponential in the length of the trace. In order to be useful, inference algorithms must scale better.
2. **Side channel measurements may be noisy.** As we mentioned in Section 3, our measurements of the side channel may be noisy and contain errors. It is desirable to have techniques that tolerate noise.
3. **We need a model that handles inputs.** Hidden Markov Models will serve as a starting point for our techniques, but HMM's only have outputs and do not model inputs. In order to accurately model the secret key, we need a framework that models processes with both inputs and outputs.
4. **One trace is typically not enough.** For any reasonable countermeasure, the set of possible keys consistent with a single trace will be large. Hence, attacks that examine only a single trace are unlikely to be successful. However, by gathering multiple traces that result from use of the same key, we may be able to narrow the list of likely candidates. Thus, it is desirable to have techniques that can analyze an arbitrary number of traces. This will make our analysis both more general and more powerful.

First, we will show how Hidden Markov Models can be used to solve problems 1 and 2. Then, we introduce an extension to HMM's, *Input Driven Hidden Markov Models*, that address problems 3 and 4.

4.1   Hidden Markov Models

Hidden Markov Models (HMM's) [11] are a well-studied method for modeling finite-state stochastic processes. The word "hidden" indicates that the states of the process are not directly observable. Instead, related to each state is an output which is observable. One of the main problems of interest for Hidden Markov Models is the inference problem: to infer the most likely sequence of values for the hidden states given only the observations. Since there exist efficient algorithms for solving the inference problem in HMM's, this motivates trying to model randomized countermeasures as HMM's in a way so that the key recovery problem for a randomized countermeasure becomes the inference problem in an HMM.

HMM's induce two sequences of finite random variables: the hidden states, Q_1, $Q_2, \ldots, Q_N$, and the observations[2], $Y_1, Y_2, \ldots, Y_N$. Like regular Markov Models, the value of the next state is dependent only on the current state and not any previous states. That is, the distribution of Q_n is conditionally independent of $Q_1, Q_2, \ldots, Q_{n-2}$ given Q_{n-1}. In addition, it is assumed that the distribution of Y_n is conditionally independent of everything else given Q_n.

HMM's are parameterized by the local conditional distributions $\mathbf{Pr}[Q_n|Q_{n-1}]$ and $\mathbf{Pr}[Y_n|Q_n]$, both of which are assumed to be independent of n. If $S = \{s_1, s_2, \ldots, s_M\}$, the conditional distribution $\mathbf{Pr}[Q_n|Q_{n-1}]$ is parameterized by a $M \times M$ transition matrix A, where $A_{ij} = \mathbf{Pr}[Q_n = s_j|Q_{n-1} = s_i]$. Since in our setting the sample space of the observations Y_n is a finite observation set $O = \{o_1, o_2, \ldots, o_J\}$, the conditional distribution $\mathbf{Pr}[Y_n|Q_n]$ is parameterized by a $M \times J$ output matrix B, where $B_{ij} = \mathbf{Pr}[Y_n = o_j|Q_n = s_i]$.

In summary, for our setting, a Hidden Markov Model is defined by the quintuple

$$H = (S, O, A, B, s_0)$$

where

S is a finite set of internal states,

O is a finite set of symbols that represent operations observable over the side channel,

A is a $|S| \times |S|$ matrix where $A_{ij} = \mathbf{Pr}[Q_n = s_j|Q_{n-1} = s_i]$,

B is a $|S| \times |O|$ matrix where $B_{ij} = \mathbf{Pr}[Y_n = o_j|Q_n = s_i]$,

$s_0 \in S$ is the initial state.

We refer to a realization $q_1, q_2, \ldots, q_N$ of the random variables $Q_1, Q_2, \ldots, Q_N$ as an *execution* of the HMM, and to a realization $y_1, y_2, \ldots, y_N$ of the random variables $Y_1, Y_2, \ldots, Y_N$ as a *trace* of the HMM. Recall that traces are observable whereas executions are generally not.

Hidden Markov Models have a graphical representation as the probabilistic graphical model [11] shown in Figure 3. Probabilistic graphical models are a graphical representation of a set of random variables where each node represents a random variable and the directed edges indicate conditional dependencies. A shaded node indicates the corresponding variable is observed, while unshaded nodes are unobserved. In the case of HMM's, the shaded nodes correspond to the observations Y_n, and the unshaded nodes correspond to the hidden, unobserved states Q_n.

Consider a probabilistic finite state machine with no inputs, i.e., with no keys. That is, for $M = (S, \delta, O, s_0, \mu)$, the transition function is given by $\delta : S \times S \to [0, 1]$ rather than $\delta : S \times S \times I \to [0, 1]$. If we wish to infer the most likely execution for M given a trace y of length N, we can construct a Hidden Markov Model $H = (S, O, A, B, s_0)$ where the unknown states in the execution can be modeled by the random variables $Q_1, Q_2, \ldots, Q_N$ and the corresponding observations are realizations of $Y_1, Y_2, \ldots, Y_N$.

---

[2] HMM's can handle real-valued observations as well.

The transition matrix A can be easily constructed from δ, and the output matrix B is completely deterministic, i.e., each row of B has one 1 and $(|O| - 1)$ 0's.

Solving the State Inference Problem for M and y reduces to solving a similar state inference problem for H and y. The State Inference Problem for a Hidden Markov Model H given a trace $y = (y_1, y_2, \ldots, y_n)$ is as follows:

STATE INFERENCE PROBLEM FOR HIDDEN MARKOV MODELS

Let H be Hidden Markov Model. Generate an execution q of H and let y be a trace of q. The State Inference Problem for Hidden Markov Models is find q given H and y.

The standard approach for finding q in the State Inference Problem for HMM's is maximum likelihood decoding. However, as we have seen before, naive implementations for finding

$$q = \operatorname*{argmax}_{s \in S^N} \mathbf{Pr}[Q = s | Y = y]$$

will have running time exponential in N. However, the Viterbi algorithm [12] is a well-known dynamic programming solution to the State Inference Problem for HMM's with running time $O(|S|^2 \cdot N)$. This addresses the first challenge, the need for efficient inference algorithms.

HMM's can also address the second problem of noisy side channel measurements. This can be handled by proper parameterization of the distribution $\mathbf{Pr}[Y_n | Q_n]$. For example, in the toy example presented in section 2, for perfect observations, $\mathbf{Pr}[Y_n = AD | Q_n = s_2] = 1$. Noisy observations can be modeled by assuming observations are only probabilistically correct, e.g., $\mathbf{Pr}[Y_n = AD | Q_n = s_2] = 0.7$ and $\mathbf{Pr}[Y_n = D | Q_n = s_2] = 0.3$.

4.2   Input Driven Hidden Markov Models

HMM's are not completely adequate for modeling most countermeasures. In most countermeasures, the next state depends not only on the current state, but also on the next bit of the key. In the context of HMM's, we would like the key to serve as a sort of input to the HMM. However, HMM's unfortunately do not have inputs. Therefore, we extend the notion of HMM's to include the possibility of inputs by introducing *Input Driven Hidden Markov Models* (IDHMM's).

IDHMM's extend HMM's in two fundamental ways. First, the unknown input is treated as a random variable $K = (K_1, K_2, \ldots, K_N)$ such that K_n is input to the underlying HMM at step n. The local conditional distribution are updated to reflect this, i.e., we replace $\mathbf{Pr}[Q_n | Q_{n-1}]$ with $\mathbf{Pr}[Q_n | Q_{n-1}, K_n]$. Second, since one of the motivations behind developing IDHMM's was to analyze multiple traces, we need to add additional random variables to model additional execution/trace pairs. Thus, $Y^1, Y^2, \ldots, Y^L$ will represent a list of L traces, where $Y^l = (Y_1^l, Y_2^l, \ldots, Y_N^l)$. Also, $Q^1, Q^2, \ldots, Q^L$ will represent the corresponding L sequences of hidden states, where $Q^l = (Q_1^l, Q_2^l, \ldots, Q_N^l)$. IDHMM's assume the same input is used in every execution, which corresponds exactly to the assumption that the same key is used in every execution of the countermeasure.

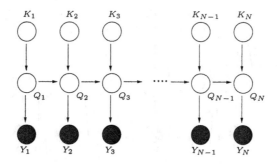

Fig. 4. Input Driven Hidden Markov Models. This figure depicts one execution of an IDHMM on input $K_1, K_2, \ldots, K_N$.

The graphical model shown in Figure 4 represents a single execution of an IDHMM. The input was applied a single time to a produce a single output trace. Figure 5 shows a graphical model representing L traces from L executions of an IDHMM in which the same input is applied in each execution, for some $L > 1$.

An Input Driven Hidden Markov Model is defined by the septuple

$$H = (S, I, O, A, B, C, s_0)$$

where

S is a finite set of internal states,

I is a finite set of input symbols,

O is a finite set of symbols that represent operations observable over the side channel,

A is a $|S| \times |I| \times |S|$ matrix where $A_{ijk} = \mathbf{Pr}[Q_n^l = s_k | Q_{n-1}^l = s_i, K_n = k_j]$,

B is a $|S| \times |O|$ matrix where $B_{ij} = \mathbf{Pr}[Y_n^l = o_j | Q_n^l = s_i]$,

C is a $N \times |I|$ matrix representing the prior distributions for $(K_1, K_2, \ldots, K_N)$,

    where $C_{jk} = \mathbf{Pr}[K_j = i_k]$,

$s_0 \in S$ is the initial state.

Since in our setting the input is a binary key chosen uniformly at random, the set of input symbols is $I = \{0, 1\}$ and the prior distributions are $\mathbf{Pr}[K_n = 0] = \mathbf{Pr}[K_n = 1] = 0.5$.

Our final goal is the inference problem for IDHMM's: we want to infer the input key K rather than the sequences of hidden states Q. We define the Key Inference Problem for Input Driven Hidden Markov Model as follows:

KEY INFERENCE PROBLEM FOR INPUT DRIVEN HIDDEN MARKOV MODELS

Let H be an Input Driven Hidden Markov Model. Generate a N bit random input key k and L executions $q = (q^1, q^2, \ldots, q^L)$ of H on k. Let $y = (y^1, y^2, \ldots, y^L)$ be the corresponding L traces. The Key Inference Problem for Input Driven Hidden Markov Models is to find k given H and y.

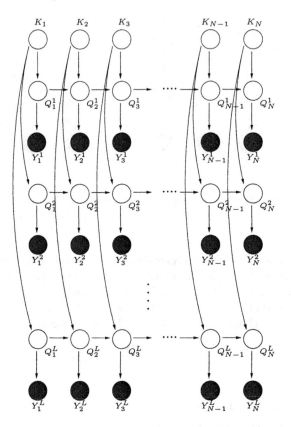

Fig. 5. Modeling Multiple Executions of an Input Driven Hidden Markov Model. This figure depicts L executions of an IDHMM on input $K_1, K_2, \ldots, K_N$.

Ideally, we would like to compute

$$k = \operatorname*{argmax}_{k \in \{0,1\}^N} \mathbf{Pr}[K = k | Y = (y^1, y^2, \ldots, y^L)] . \qquad \{*\}$$

However, we do not know how to compute this efficiently. Therefore, we introduce an approximation: we infer the posterior probabilities for each bit of the key separately, and then we use the most likely value of each bit to infer the entire key. This amounts to computing

$$k = (k_N, k_{N-1}, \ldots, k_2, k_1), \text{ where } k_n = \operatorname*{argmax}_{b \in \{0,1\}} \mathbf{Pr}[K_n = b | Y = (y^1, y^2, \ldots, y^L)] .$$

However, even this was too hard for us. Our first attempts at an algorithm to calculate $\mathbf{Pr}[K_n | y]$ using dynamic programming in a manner similar to that in the inference algorithm for HMM's encountered a significant problem: the resulting algorithm had running time exponential in L, the number of traces. Since our goal is to scale with the number of traces, this is unacceptable.

SINGLETRACEINFERENCE(H, D, y'):

    Input: An IDHMM H, a distribution D, assumed $\mathbf{Pr}[K = k] = D(k)$, and a trace
        $y' = (y'_1, y'_2, \ldots, y'_N)$

    Output: A distribution D', where $D'(k) = \mathbf{Pr}[K = k | Y' = y']$, using D as our priors on K

    1) Use a modified version of the Viterbi algorithm to compute $D'(k) := \mathbf{Pr}[K = k | Y' = y']$,
        assuming $D(k) = \mathbf{Pr}[K = k]$. Refer to the full version of this paper [17] for details.

MULTITRACEINFERENCE(H, y):

    Input: An IDHMM H and a set $y = (y^1, y^2, \ldots, y^L)$ of L traces

    Output: D_L, an approximation to the distribution $\mathbf{Pr}[K = k | Y = y]$

    1) Let $D_0 :=$ the uniform distribution on $\{0, 1\}^N$.
    2) **for** $i := 1, 2, \ldots, L$ **do**
        $D_i :=$ SINGLETRACEINFERENCE(H, D_{i-1}, y^i)
    3) Output D_L.

INFER(H, y):

    Input: An IDHMM H and a set $y = (y^1, y^2, \ldots, y^L)$ of L traces

    Output: k, a guess at the key

    1) Let $\mathbf{Pr}[K_i | Y = y]$ be as given by MULTITRACEINFERENCE(H, y).
    2) **for** $i := 1, 2, \ldots, N$ **do**
        if $\mathbf{Pr}[K_i = 1 | Y = y] > 0.5$ **then** $k_i := 1$ **else** $k_i := 0$
    3) Output $k_{\text{guess}} = k_N k_{N-1} \ldots k_2 k_1$.

Fig. 6. An approximate inference algorithm using belief propagation. Given a set of traces $y = (y^1, y^2, \ldots, y^L)$ of an Input Driven Hidden Markov Model H, we compute a guess $k_{\text{guess}} =$ INFER(H, y) at the key.

    To deal with these challenges, we introduce a new technique based on *belief propagation*. The key idea is to separate L executions of an IDHMM on the same input into L executions of an IDHMM where there are no assumptions about the input used in each execution. In terms of the graphical models, this corresponds to transforming Figure 5 into L copies of Figure 4. We can derive an efficient exact inference algorithm for a single execution of an IDHMM with running time $O(|S|^2 \cdot N)$. By applying this exact inference algorithm separately to each of the L executions, we obtain an algorithm with final running time of $O(|S|^2 \cdot N \cdot L)$.

    The problem with this approach is that we are not taking advantage of the fact that the executions all use the same key. Using L traces as input, this approach will output L separate inferences of the key, each derived independently of the others. We can link them by using belief propagation: instead of using the uniform distribution as our prior $\mathbf{Pr}[K_n]$ for each key bit in the L analyses, we use the posterior distribution $\mathbf{Pr}[K_n | y^l]$ calculated from analysis of the l-th trace as the prior distributions $\mathbf{Pr}[K_n]$ while analyzing the $l + 1$-st trace. Hence, we propagate any biases on the key bits learned from the l-th trace to the analysis of the $l + 1$-st trace. A detailed description of our belief propagation algorithm for inferring the secret key from L traces of an IDHMM is shown in Figure 6. Although the output of this algorithm is only an approximation to what we ideally want in (4.2), we have found that it works well in practice.

5   Application to Randomized Addition-Subtraction Chains

Oswald and Aigner have proposed two randomized exponentiation algorithms [2] for scalar multiplication in ECC implementations. These algorithms are based on the randomization of addition-subtraction chains. For example, instead of the usual binary decomposition $15P = 8P + 4P + 2P + 1$, $15P$ can alternatively be calculated as as

$$15P = 16P - P = 2(2(2(2(P)))) - P.$$

More generally, a series of more than two 1's in the binary representation of k can be replaced by a block of 0's and a -1, i.e., $01^a \mapsto 10^{a-1}\bar{1}$ where $\bar{1}$ represents -1. A second transformation noted by Oswald and Aigner treats isolated 0's inside a block of 1's, i.e., $01^a01^b \mapsto 10^a\bar{1}0^{b-1}\bar{1}$.

Both of these transformations can be modeled by deterministic finite state machines. Oswald and Aigner construct two randomized exponentiation algorithms by introducing randomness into these state machines while still preserving the end semantics of the two transformations. At each step where a transformation may apply, we flip a coin to decide whether or not to apply that transformation. We refer to the randomized construction based on the transformation $01^a \mapsto 10^{a-1}\bar{1}$ as OA1 and the randomized construction based on the transformation $01^a01^b \mapsto 10^a\bar{1}0^{b-1}\bar{1}$ as OA2. The randomized state machine describing the operation of OA1 (as it appears in [2]) is shown in Figure 7(a).

The randomized state machines in [2] that describe the operation of OA1 and OA2 do not conform to our definition of probabilistic state machines in Section 2, but this is easily remedied. The first hurdle is that traces cannot be parsed uniquely as words in $\{D, AD\}^*$. Although it would be convenient if our observable alphabet was $O = \{D, AD\}$, the transition from s_2 to s_1 executes a doubling first and then an addition, resulting in a DA output symbol corresponding to that key bit. This is undesirable because traces fail to be uniquely decodeable: for example, $DADD$ could be interpreted as either (DA, D, D) or (D, AD, D). We remedy this problem by interpreting the automaton in Figure 7(a) slightly differently. We relabel the DA transition from s_2 to s_1 to simply a D (i.e., $Q = 2Q$) and now associate the "owed" addition with each outgoing transition from state s_1. Our output alphabet becomes $O = \{D, AD, AAD\}$, and then each sequence of D and A operations can deterministically be decomposed into a sequence of symbols from O. The resulting state machine is shown in Figure 7(b).

A second hurdle is that Oswald and Aigner place observable operations on the edges, rather than on the states. Fortunately, edge-annotated state machines can easily be transformed into a semantically equivalent state-annotated machine (of the type defined in Section 2) by treating each edge in Figure 7(b) as a state in the probabilistic FSA. This yields a faithful probabilistic finite state machine to which our algorithms can be applied. See Figure 7(c) for the result of this process applied to OA1.

Once we have probabilistic finite state machine representations for the countermeasures, applying our techniques is straightforward. We simulated the operation of both exponentiation algorithms in software. First, we generated a random 192 bit key k. Using k, we then generated a set of traces $y = (y^1, y^2, \ldots, y^L)$. We introduced errors in the traces consistent with observation error p_e. With probability $1 - p_e$, each output symbol is observed correctly, and with probability p_e, it is changed to some other output symbol (chosenly randomly). We assumed the error probability p_e is known to the attacker,

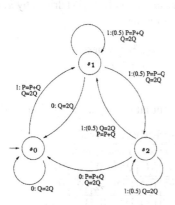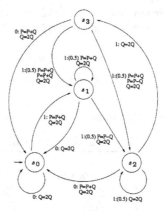

(a) A randomized state machine that represents random application of the transformation $01^a \mapsto 10^a\bar{1}$ of a key k in the scalar multiplication $k \times M$. Q is initialized to M.

(b) A reinterpretation of Figure 7(a) such that the observable operation labeling each edge is a member of $O = \{D, AD, AAD\}$.

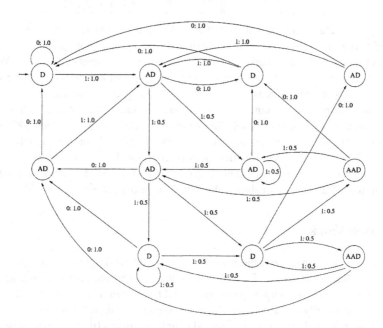

(c) A faithful probabilistic state machine (conforming to the definition in Section 2) that models the behavior of Figure 7(b).

Fig. 7. The first Oswald-Aigner construction (OA1).

and we incorporated it into the output distribution (i.e., $\mathbf{Pr}[Y_n \mid Q_n]$) of the resulting IDHMM. Treating the OA1 or OA2 countermeasure as an IDHMM driven by k, we then applied the INFER algorithm from Figure 6 to compute

$$k_{\text{guess}} = (k_N, k_{N-1}, \dots, k_2, k_1), \text{ where}$$

$$k_n = \underset{b \in \{0,1\}}{\operatorname{argmax}} \mathbf{Pr}[K_n = b \mid Y = (y^1, y^2, \dots, y^L)] .$$

The following table summarizes the results of our attacks against OA1 and OA2:

Number of key bits correctly recovered

Countermeasure	p_e	1	5	10	25	50	100	500
OA1	0	170	187	192	192	192	192	192
OA1	0.1	157	178	184	185	187	192	192
OA1	0.25	143	163	173	180	182	183	184
OA1	0.4	120	147	159	168	172	173	174
OA2	0	165	188	192	192	192	192	192
OA2	0.1	156	174	184	187	189	192	192
OA2	0.25	135	161	174	177	180	181	182
OA2	0.4	126	146	154	168	171	172	173

Each entry in the table specifies the number of key bits (out of 192) that we correctly recovered using the corresponding number of traces and given observation error p_e.

Both OA1 and OA2 are clearly insecure under our assumptions in Section 3. With a perfect side channel ($p_e = 0$), we recovered the entire secret key perfectly with as few as 10 traces; also, our techniques remain effective in the presence of noise.

One can also reduce the number of traces by combining our attack with a semi-exhaustive attack over the most likely key candidates. It suffices to recover 182 of the 192 key bits correctly, on average; then we can apply a meet-in-the-middle search over all possible 10-bit error patterns to identify the correct private key, using 2^{38} work. Hence, with a 0.1 probability of observation error, the entire key can be recovered with only 10 traces, and for a 0.25 error probability, the data complexity increases to 50-500 traces.

6   Related Work

Several authors have analyzed the security of selected randomized countermeasures against side channel attacks [13,14,16], including the Oswald-Aigner constructions. However, the analysis techniques previously used have been ad-hoc in sense that they are tailored specifically to the countermeasure being analyzed, and it is not clear how to generalize them to analyze other randomized countermeasures (if this is even possible). In contrast, our techniques are broadly applicable to randomized countermeasures whose operation can be modeled by a probabilistic finite state machine.

Based on a comprehensive case analysis, Okeya and Sakurai [13] present an attack against OA1 that with high probability recovers a 192 bit key using approximately 292 traces of the side channel. They assume the ability to perfectly distinguish between elliptic curve point additions and doublings and do not consider the case when the side channel is noisy.

C.D. Walter [14] presents an attack against OA2 based on a detailed analysis of its operation. With high probability, his attack recovers a 192 bit key using $O(10)$ traces of the side channel. This attack can be generalized to work against OA1 as well. Then, Walter also discusses how to partition traces into smaller subsections and exhaustively search (independently) for the key corresponding to each subsection. Depending on the key size, it is possible for his second technique to succeed with as few as two traces. Both his attacks assume the ability to perfectly distinguish between elliptic curve point additions and doublings.

Song et al. [18] use Hidden Markov Models to exploit weaknesses of the widely used SSH protocol. By observing the inter-keystroke timings of a user's key presses during an SSH session, the authors are able to recover significant information about the key stroke sequences. They use this technique to speed up exhaustive search for passwords by a factor of 50. Other than the work of Song et al., we are not aware of any previous work that uses HMM's for side channel cryptanalysis.

7   Conclusion

We introduced HMM attacks, a general-purpose cryptanalysis technique for evaluating the security properties of randomized countermeasures whose operation can be modeled by a probabilistic finite state machine. We also introduced Input Driven Hidden Markov Models, an extension of HMM's that model inputs, and we presented efficient approximate inference algorithms for recovering the input to an IDHMM given multiple output traces. Our work improves on existing attacks against randomized countermeasures in two fundamental ways. Firstly, previous attacks against randomized countermeasures typically consist of detailed case analyses which are not clear how to generalize to attacks on larger classes of countermeasures. We present a cryptanalytical framework applicable to a general class of randomized countermeasures. Secondly, previous attacks against randomized countermeasures assume the ability to perfectly distinguish between operations in the side channel. Our techniques are still applicable if the side channel is noisy.

We demonstrate the application of HMM attacks and IDHMM's in an analysis of Randomized Addition-Subtraction Chains proposed by Oswald and Aigner. When our observations of the side channel are perfect, we are able to completely recover the secret key using as few as 5–10 traces. Our attacks are robust to noise in the side channel as well. For instance, when the probability of each observation being incorrect is 0.25, we are still able to recover the secret key by using 50–500 traces.

References

1. Ha, J., Moon, S.: Randomized signed-scalar multiplication of ECC to resist power attacks. In: Fourth International Workshop on Cryptographic Hardware and Embedded Systems (CHES). (2002)

2. Oswald, E., Aigner, M.: Randomized addition-subtraction chains as a countermeasure against power attacks. In: Third International Workshop on Cryptographic Hardware and Embedded Systems (CHES). (2001)
3. Izu, T., Takagi, T.: A fast parallel elliptic curve multiplication resistant against side channel attacks. In: PKC2002. (2002)
4. Walter, C.: MIST: An efficient, randomized exponentiation algorithm for resisting power analysis. In: RSA 2002 - Cryptographers' Track. (2002)
5. Liardet, P.Y., Smart, N.: Preventing SPA/DPA in ECC systems using the Jacobi form. In: Third International Workshop on Cryptographic Hardware and Embedded Systems (CHES). (2001)
6. Itoh, K., Yajima, J., Takenaka, M., Torri, N.: DPA countermeasures by improving the window method. In: Fourth International Workshop on Cryptographic Hardware and Embedded Systems (CHES). (2002)
7. May, D., Muller, H., Smart, N.: Randomized register renaming to foil DPA. In: Third International Workshop on Cryptographic Hardware and Embedded Systems (CHES). (2001)
8. Goubin, L., Patarin, J.: DES and differential cryptanalysis. In: First International Workshop on Cryptographic Hardware and Embedded Systems (CHES). (1999)
9. Kocher, P., Jaffe, J., Jun, B.: Differential power analysis. Lecture Notes in Computer Science **1666** (1999) 388–397
10. Kocher, P.C.: Timing attacks on implementations of Diffie-Hellman, RSA, DSS, and other systems. Lecture Notes in Computer Science **1109** (1996) 104–113
11. Jordan, M.: An Introduction to Probabilistic Graphical Models. in preparation (2003)
12. Russell, S., Norvig, P.: Artificial Intelligence, A Modern Approach. Prentice Hall (1995)
13. Okeya, K., Sakurai, K.: On insecurity of the side channel attack countermeasure using addition-subtraction chains under distinguishability between addition and doubling. In: The 7th Australasian Conference on Information Security and Privacy. (2002)
14. Walter, C.: Security constraints on the Oswald-Aigner exponentiation algorithm. Cryptology ePrint Archive: http://eprint.iacr.org/ (2003)
15. Coron, J.S.: Resistance against differential power analysis attacks for elliptic curve cryptosystems. In: First International Workshop on Cryptographic Hardware and Embedded Systems (CHES). (1999)
16. Walter, C.: Breaking the Liardet-Smart randomized exponentiation algorithm. In: Fifth Smart Card Research and Advanced Application Conference (CARDIS). (2002)
17. Karlof, C., Wagner, D.: Hidden Markov Model Cryptanalysis. Technical Report UCB//CSD-03-1244, University of California at Berkeley (2003)
18. Song, D.X., Wagner, D., Tian, X.: Timing analysis of keystrokes and timing attacks on SSH. In: Tenth USENIX Security Symposium. (2001)

Power-Analysis Attacks on an FPGA – First Experimental Results

Sıddıka Berna Örs[1]*, Elisabeth Oswald[2,3], and Bart Preneel[1]

[1] Katholieke Universiteit Leuven, Dept. ESAT/SCD-COSIC,
Kasteelpark Arenberg 10, B–3001 Leuven-Heverlee, Belgium
[2] Institute for Applied Information Processing and Communciations (IAIK),
TU Graz, Inffeldgasse 16a, A–8010 Graz, Austria
[3] A–SIT, Technologiebeobachtung, Inffeldgasse 16a, A–8010 Graz, Austria
{siddika.bernaors, bart.preneel}@esat.kuleuven.ac.be
elisabeth.oswald@iaik.at

Abstract. Field Programmable Gate Arrays (FPGAs) are becoming increasingly popular, especially for rapid prototyping. For implementations of cryptographic algorithms, not only the speed and the size of the circuit are important, but also their security against implementation attacks such as side-channel attacks. Power-analysis attacks are typical examples of side-channel attacks, that have been demonstrated to be effective against implementations without special countermeasures. The flexibility of FPGAs is an important advantage in real applications but also in lab environments. It is therefore natural to use FPGAs to assess the vulnerability of hardware implementations to power-analysis attacks. To our knowledge, this paper is the first to describe a setup to conduct power-analysis attacks on FPGAs. We discuss the design of our hand-made FPGA-board and we provide a first characterization of the power consumption of a Virtex 800 FPGA. Finally we provide strong evidence that implementations of elliptic curve cryptosystems without specific countermeasures are indeed vulnerable to simple power-analysis attacks.

Keywords: FPGA, Power Analysis, Elliptic Curve Cryptosystems

1 Introduction

Since their publication in 1998, power-analysis attacks have attracted significant attention within the cryptographic community. So far, they have been successfully applied to different kinds of (unprotected) implementations of symmetric

---

* Sıddıka Berna Örs is funded by a research grant of the Katholieke Universiteit Leuven, Belgium. Dr. Bart Preneel is professor at the Katholieke Universiteit Leuven. Elisabeth Oswald is with the TU Graz and the A–SIT. This work was supported by Concerted Research Action GOA-MEFISTO-666 of the Flemish Government, by the FWO "Identification and Cryptography" project (G.0141.03) and by the FWF "Investigations of Simple and Differential Power Analysis" project (P16110-N04).

C.D. Walter et al. (Eds.): CHES 2003, LNCS 2779, pp. 35–50, 2003.

and public-key encryption schemes and on digital signature schemes. Most attacks which have been published in the open literature apply to software implementations of cryptographic algorithms which can be found for example in smart cards (see [KJJ99], [MDS99a] or [MDS99b]). However, modern smart cards and accelerators for cryptographic algorithms also contain hardware implementations of cryptographic algorithms.

As part of a modern design flow, FPGAs are gaining more importance. Reasons for this include their relatively low cost and the available tools. High-level descriptions (like VHDL for examples) for a circuit can easily be ported, if not directly used, for an FPGA implementation of the circuit. Naturally, it is desirable to use the resulting FPGA implementation also for an evaluation of the designed circuit against power-analysis attacks.

This article describes the first realization of power-analysis attacks on a Virtex FPGA. We can prove that this FPGA leaks a significant amount of information about its internal computations through the supply lines. We can even provide evidence that the power consumption characteristics are comparable with the power consumption characteristics of ordinary ASICs. To demonstrate how dramatic the power consumption leakage of this FPGA is, we finally perform a simple power-analysis attack on an implementation of an elliptic-curve point-multiplication.

The remainder of this article is organized as follows. We recall the principles of power-analysis attacks in Sect. 2. FPGAs are introduced in Sect. 3. For the purpose of conducting power-analysis attacks, we built a special measurement board. This measurement setup is described in Sect. 4. The results of our experiments can be found in Sect. 5. Section 6 presents the conclusion of our research.

1.1   Related Work

The characterization of the power-consumption characteristics of FPGAs has received little attention so far. Shang et al. [SKB02] is the only recent article in that field. In their article, Shang et al. analyze the dynamic power consumption of the XILINX Virtex-II family. They conclude that 60% of the dynamic power consumption is due to the interconnects, 14% is due to the clocking, 16% is due to the logic and 10% is due to the IOBs. Based on this result, it seems much more difficult to conduct power-analysis attacks on FPGAs than on ASICs. However, as we will demonstrate in this article, such attacks are feasible and can be realized in practice.

2   Power-Analysis Attacks

Power-analysis attacks are a very powerful type of side-channel attack, published first by Kocher et al. [KJJ99]. Power-analysis attacks are passive in the sense that an attacker only needs to measure the power consumption of a device without manipulating it actively, that is, an attacker uses the device in its intended mode of use. A likely scenario (in the case of attacks on smart cards) is that an attacker lets the device execute an internal authenticate command. While the device is executing this command, the attacker measures the power consumed by the device. Statistical methods allow to extract efficiently the information on the secret key that is contained in the measurements.

2.1   Power Consumption Characteristics of CMOS

Nowadays, almost all smart card processors are implemented in CMOS (complementary Metal-Oxid Silicon) technology. In CMOS technology, the values 0 and 1 are represented by V_{ss} and V_{dd}, respectively. The dominating factor for the power consumption of a CMOS gate is the dynamic power consumption [WE93]. *Transition count* leakage and *Hamming weight* leakage can typically be observed in CMOS circuits, see [MDS99b] for a detailed explanation.

The power consumption behaviour of a CMOS processor can be roughly sketched as follows. On every rising edge of the clock, the simultaneous switching of the gates causes a current flow which is visible through both V_{dd} and V_{ss}. This current flow can be observed on the outside of the device by (for example) putting a small resistor between the devices V_{ss} or V_{dd} and the true V_{dd}. The current flowing through the resistor creates a voltage which can be measured by a digital oscilloscope.

2.2   Exploiting the (Hidden) Information

Depending on how direct the information about the power consumption can be used, simple or differential power-analysis attacks (SPA or DPA) [KJJ99] have to be applied. SPA attacks are always possible when the power consumption is more or less directly related to the actions of the secret key. This is mostly the case when the instructions executed in the device give evidence about the secret key. If the instructions do not provide such information, but the processed data do so instead, then the information is typically more hidden in the overall power consumption and thus statistical methods have to be applied to bring them to light. This is the approach taken for differential power-analysis attacks.

Simple Power-Analysis Attacks. Simple power-analysis attacks exploit the relationship between the instantaneous power consumption of a device and the instructions that are executed. For simple power-analysis attacks it is assumed that every instruction has its unique power-consumption trace. An attacker simply monitors the device's power consumption while it performs a cryptographic operation. Then, the attacker carefully studies the obtained power-consumption trace to determine the sequence of instructions performed by the device. If this sequence is directly related to the secret key which was involved in the cryptographic operation, the attacker can deduce this secret key from the power-consumption trace. Such an attack typically targets implementations which use key dependent branching in the implementation.

Differential Power-Analysis Attacks. An attacker faces now the task of exploiting the hidden information about the secret key in an efficient way. For this purpose, the attacker creates a hypothetical model of the device. This hypothetical model describes, at a very abstract level, the instantaneous power consumption of the device when it executes a certain cryptographic algorithm. For this purpose, at least a small part of the unknown key has to be guessed. Fortunately (for the attacker), all algorithms deployed in practice use only small parts of the secret key at a time.

The attacker writes a simple computer program that executes the algorithm (or at least a small part of it, where a part of the key is used). The program calculates the result (of this part) for all possible key values. These values allow to predict the power consumption, which is for example related to the Hamming-weight of the internal data.

In the last stage of the attack, an attacker feeds the same input values which he used in the model to the real device and measures its power consumption. Then the attacker correlates the predictions of the model with the real power consumption values. For all the wrong key guesses, the predictions will not correlate with the real measurements, but for the correct key guess, there will be a peak visible in the correlation trace.

3   Field Programmable Logic Arrays

An FPGA consists of an array of configurable logic blocks (CLBs), surrounded by programmable I/O blocks, and connected with programmable interconnections as shown in Fig. 1 [Opt]. A typical FPGA contains from 64 to tens of thousands of logic blocks and an even greater number of flip-flops. Most FPGAs do not provide a 100% interconnect between the logic blocks. Instead, sophisticated software places and routes the logic on the device.

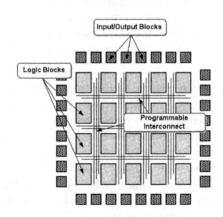

Fig. 1. The FPGA architecture

Two main classes of FPGA architectures can be distinguished. Coarse-grained architectures consist of fairly large logic blocks, often containing two or more look-up tables and two or more flip-flops. Fine-grained architectures consist of a large number of relatively simple logic blocks. Another difference in the architectures is the underlying process technology used to manufacture the device. Currently, the highest-density FPGAs are built using static memory (SRAM) technology, which is similar to microprocessors. The other common process technology is called anti-fuse, which has benefits for more plentiful programmable interconnect.

SRAM-based devices are inherently re-programmable, even in-system. After a power-up is applied to the circuit, the program data defining the logic configuration must be loaded in the SRAM [MK01].The FPGA either self-loads its configuration memory, or an external processor downloads the memory into the FPGA. The configuration time is typically less than 200 ms, depending on the device size and configuration method. In contrast, anti-fuse devices are one-time programmable (OTP). Once programmed, they cannot be modified, but they also retain their program when the power is off. Anti-fuse devices are programmed in a device programmer either by the end user or by the factory or distributor. More details on the Xilinx Virtex Architecture are provided in Appendix A.

4   The Measurement Setup

Our setup consists of essentially two boards (see Fig. 2). The main board is responsible for interfacing the PC via the parallel port. It is connected with the XILINX parallel cable in order to program the VIRTEX FPGA and it provides some LEDs, switches and buttons for testing purposes. The daughter board itself just carries the VIRTEX FPGA, it allows to access some pins for triggering and to measure the power consumption of the VIRTEX FPGA in a convenient way.

4.1   The Mother Board

The Parallel Port [Axe00] is the most commonly used port for interfacing home made projects. This port allows the input of 5 bits and the output of 12 bits. The port is composed of 4 control lines, 5 status lines and 8 data lines. The communication between the FPGA and the PC uses this parallel port. We need only 17 input/output pins to send data or commands to the FPGA and receive the result, but we designed the board in such a way that it gives us more monitoring points and thus connected 32 input/output pins of the FPGA to the board. The unused input/output pins are pulled up after the configuration.

We also designed a protocol to send and receive data to and from FPGA. When the FPGA communicates with the PC, it uses the three most significant bits of the status lines to indicate its status. The two remaining bits of status lines are used for sending the result from the FPGA to the PC. The protocol is independent from the operation executed in the FPGA. Only the length of

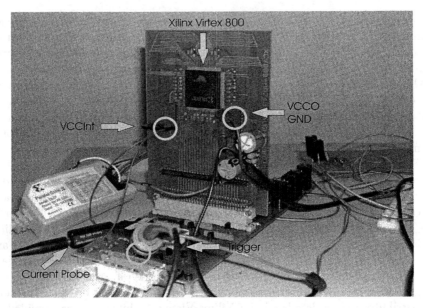

Fig. 2. The measurement setup. On the daughter board the current probe is connected to VCCINT. Alternatively it can be connected to the VCCO of the individual banks, or the GND.

the data which is communicated can be modified by the PC. This provides a flexible setup where experiments with different algorithms can be performed in a coherent manner.

4.2   The Daughter Board

We use a Xilinx XCV800 FPGA from the Virtex series in a HQ240C package. Reasons for this particular choice include:

1. The resources are sufficient to implement a 160-bit elliptic-curve point-multiplication.
2. This is the most powerful FPGA that can be used for hand-mounting on the board. This is because the pins of this FPGA are on its sides. The more powerful FPGAs have the pins underneath with a grid structure and so special machines are needed to mount them.
3. The architecture is made of combinational and memory elements. Because of this property it is a good representative of application specific integrated circuits (ASICs).

The XCV800 has 12 core voltage supply (VCCINT) pins, 16 output voltage supply (VCCO) pins and 32 ground (GND) pins. The FPGA is divided into 8 banks each with their own VCCINT and VCCO pins. After the implementation of the desired circuit and the configuration of the FPGA with the implementation data, some banks will be used more frequently than others; these banks should draw more current from their supply lines. With our setup it is possible to verify this hypothesis. In case that different parts of a design (such as an elliptic-curve addition and an elliptic-curve doubling) are mapped to different banks of the FPGA, measuring the current of the individual banks allows us to take more precise measurements for them. By measuring VCCINT and VCCO of the same bank separately, we can detect the input/output and core activity timing and power consumption separately.

Therefore we use three headers with two lines for VCCINT, VCCO and GND as shown in Fig. 2. During the normal operation of the board without measurement the two pins are connected by a jumper. When we want to measure the current flow from a specific bank, the associated jumper is replaced by a cable that is going through the hole in the current probe as shown in Fig. 2.

This setup gives the possibility of making measurements on different points at the same time and makes it easy to modify the measurement point.

Bypassing Considerations. With high-speed, high-density FPGA devices, maintaining signal integrity is the key to reliable, repeatable designs [Xil02]. Proper power bypassing and decoupling improves the overall signal integrity. Without it, power and ground voltages are affected by logic transitions and can cause operational issues.

When a logic device switches from a logic one to a logic zero, or a logic zero to a logic one, the output structure is momentarily at a low impedance across the power supply. Each transition requires that a signal line be charged or discharged, which requires energy. As a result, many electrons are suddenly needed to keep the voltage from collapsing. The function of the bypass capacitor is to provide local energy storage.

$10nF$ capacitors are placed between every VCCINT and VCCO pin of the FPGA and the nearest GND. Because we designed the setup in two different cards, the daughter card can be thought as a stand alone chip taking power from the mother board. Bypass capacitors had to be placed between the power supplies of the card and the GND line.

5   Results

We now describe the experiments conducted on the measurement setup described above. As the aim of our work was to build an alternative platform for power-analysis attacks, we decided to perform first some basic experiments to verify that the assumption which we usually make for such attacks are also valid for our FPGA setup.

5.1   Power Consumption Characteristics

As discussed in Sect. 2, we should be able to detect either transition-count leakage or Hamming-weight leakage in our setup. This is because according to Sect. 3, the CLBs consist of flip-flops (and other logic) which exhibit the power consumption characteristics of CMOS technology. The only problem can be that if the circuit, which we load into the FPGA, does not use all of the FPGAs resources, then the noise which is produced by the unused parts might be larger than the signal produced by the circuit.

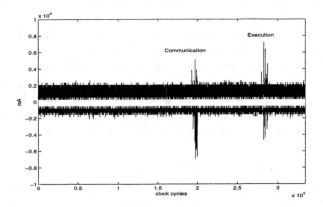

Fig. 3. Comparison between an idle bank, which corresponds to the white trace in this picture, and a bank which receives data (8 bits) and processes it, which corresponds to the dark trace in this picture.

To evaluate the behavior of the FPGA we loaded a small circuit on one of the banks of the FPGA. Then, we measured the power consumption of the whole FPGA and, at the same time, the power consumption of an empty (idle) bank (see Fig. 3 for the power consumption traces). The overall power consumption (the dark trace) shows clearly peaks when data is transmitted and when the data is processed. The light trace however does not exhibit any peaks during the whole computation. This experiment confirms that idle parts of the FPGA will not influence the overall power consumption. Moreover, even the power consumption of a very small circuit (we used only 3% of the FPGA and of the FPGAs flip-flops) can be easily detected.

With another simple set of experiments we confirmed that the amount of power consumed of the FPGA is linear in the number of switched flip-flops. We have designed registers of a specific size and loaded them on the FPGA. Then we let them repeatedly store 0 and 1 value and measured the FPGA's power consumption. Fig. 4 and 5 illustrate that the power consumed by the 6000-bit register for storing all 1s is about twice as high as the power consumed by the 3000-bit register.

A direct conclusion from such experiments is that the power consumption characteristics are essentially the same as of an ordinary CMOS circuit. Idle CLBs or even idle banks do not add too much noise to the overall power consumption.

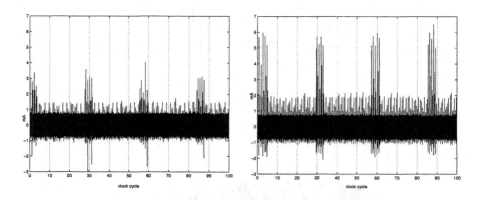

Fig. 4. Power consumption trace of a 3000-bit register.   **Fig. 5.** Power consumption trace of a 6000-bit register.

5.2 Attacking an Implementation of an Elliptic-Curve Point-Multiplication

With the experience gained from these experiments, we attacked an implementation of an EC point multiplication. We have implemented the arithmetic for a 160-bit prime field with a Montgomery modular multiplier (MMM) without final subtraction ([Mon85],[ÖBPV03], see Algorithm 1 for a description).

Algorithm 1 Montgomery modular multiplication without final subtraction

---

Require: Integers $N = (n_{l-1} \cdots n_1 n_0)_2$, $x = (x_l \cdots x_1 x_0)_2$, $y = (y_l \cdots y_1 y_0)_2$ with
    $x \in [0, 2N - 1], y \in [0, 2N - 1]$, $R = 2^{l+2}$, $gcd(N, 2) = 1$ and $N' = -N^{-1} \mod 2$
    (Notation $T = (t_l t_{l-1} ... t_0)$)
Ensure: $xyR^{-1} \mod 2N$
 1: $T \leftarrow 0$
 2: **for** i from 0 to $l + 1$ **do**
 3:   $m_i \leftarrow (t_0 + x_i y_0) N' \mod 2$
 4:    $T \leftarrow (T + x_i y + m_i N)/2$
 5: **end for**
 6: Return (T)

To obtain a linear, pipelined modular multiplier, a systolic array shown in Fig. 6 is used [Wal99]. $X(0)$ denotes the least significant bit (LSB) of the register in which the input x is stored. T denotes the intermediate value register. The carry chain is stored in the $C0$ and $C1$ registers. The Montgomery modular multiplication circuit (MMMC) consists of a controller and a data path. The data path consists of a systolic array, four internal registers, a counter and a comparator.

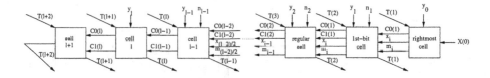

Fig. 6. Schematic view of the complete systolic array

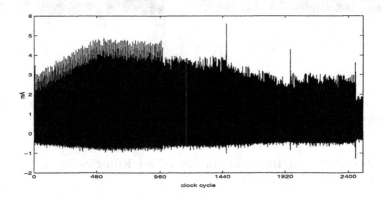

Fig. 7. Power consumption trace of 480-bit Montgomery modular multiplier from VC-CINT

The measurement of a 480-bit MMMC is depicted in Fig. 7. The three parts shown in the figure can be explained according to the algorithm and architecture used. The T register in Fig. 6 is reset in the beginning of the MMM operation and then it is being written. The number of bits in T which are updated is increasing until clock cycle l. This stage corresponds to the first part shown in power consumption trace. After l clock cycles all the bits of the T register have a value and all of them are updated before clock cycle $2l$. This stage is shown by the second part in Fig. 7. The last part in Fig. 7 corresponds to reading out the result from the pipeline. Because there is no new input on the LSB of the systolic array, starting from clock cycle $2l + 1$ the number of MSBs of the T register that are updated decreases.

5.3   Elliptic Curve Point Addition and Doubling

For the representation of the points on the elliptic curve we use modified Jacobian coordinates as proposed by Cohen *et al.* in [CMO98]. These points are represented as quadruple (X, Y, Z, aZ^4). When we convert the input point P from affine coordinates to projective coordinates we take Z as 1. Because there are both MMMC and modular addition/subtraction (MAS) circuits available, these operations can be executed in parallel. When an EC point addition is used in an EC point multiplication one of the inputs of the EC point addition circuit is always the input point P. Algorithm 2.(a). and (b)., describe the point addition and the point doubling operation, resp.

Algorithm 2 EC point addition and doubling

Require: $P_1 = (x, y, 1, a)$,
$\qquad P_2 = (X_2, Y_2, Z_2, aZ_2^4)$

Require: $P_1 = (X_1, Y_1, Z_1, aZ_1^4)$

Ensure: $P_1 + P_2 = P_3 = (X_3, Y_3, Z_3, aZ_3^4)$

Ensure: $2P_1 = P_3 = (X_3, Y_3, Z_3, aZ_3^4)$

(a)	(b)
1. $T_1 \leftarrow Z_2^2$	1. $T_1 \leftarrow Y_1^2$, $\qquad T_2 \leftarrow 2X_1$
2. $T_2 \leftarrow xT_1$	2. $T_3 \leftarrow T_1^2$, $\qquad T_2 \leftarrow 2T_2$
3. $T_1 \leftarrow T_1Z_2$, $T_3 \leftarrow X_2 - T_2$	3. $T_1 \leftarrow T_2T_1$, $\qquad T_3 \leftarrow 2T_3$
4. $T_1 \leftarrow yT_1$	4. $T_2 \leftarrow X_1^2$, $\qquad T_3 \leftarrow 2T_3$
5. $T_4 \leftarrow T_3^2$, $\quad T_5 \leftarrow Y_2 - T_1$	5. $T_4 \leftarrow Y_1Z_1$, $\qquad T_3 \leftarrow 2T_3$
6. $T_2 \leftarrow T_2T_4$,	6. $T_5 \leftarrow T_3\left(aZ_1^4\right)$, $T_6 \leftarrow 2T_2$
7. $T_4 \leftarrow T_4T_3$, $T_6 \leftarrow 2T_2$	7. $\qquad\qquad\qquad T_2 \leftarrow T_6 + T_2$
8. $Z_3 \leftarrow Z_2T_3$, $T_6 \leftarrow T_4 + T_6$	8. $\qquad\qquad\qquad T_2 \leftarrow T_2 + \left(aZ_1^4\right)$
9. $T_3 \leftarrow T_5^2$	9. $T_6 \leftarrow T_2^2$, $\qquad Z_3 \leftarrow 2T_4$
10. $T_1 \leftarrow T_1T_4$, $X_3 \leftarrow T_3 - T_6$	10. $\qquad\qquad\qquad T_4 \leftarrow 2T_1$
11. $T_6 \leftarrow Z_3^2$, $\quad T_2 \leftarrow T_2 - X_3$	11. $\qquad\qquad\qquad X_3 \leftarrow T_6 - T_4$
12. $T_3 \leftarrow T_5T_2$,	12. $\qquad\qquad\qquad T_1 \leftarrow T_1 - X_3$
13. $T_6 \leftarrow T_6^2$, $\quad Y_3 \leftarrow T_3 - T_1$	13. $T_2 \leftarrow T_2T_1$, $\qquad aZ_3^4 \leftarrow 2T_5$
14. $aZ_3^4 \leftarrow aT_6$	14. $\qquad\qquad\qquad Y_3 \leftarrow T_2 - T_3$

The multiplications and the squarings use MMMC, while the additions, doublings and subtractions employ the modular addition/subtraction circuit. The power consumption trace of one 160-bit EC point addition is shown in Fig. 9. Fourteen states can be counted easily from the trace. All the states are completed in nearly 500 clock cycles The power consumption during Step 3, 5, 7, 8, 10, 11 and 13 seems higher than for the other steps. In these steps a modular addition or subtraction is taking place as well as an MMM.

The MAS operation is performed in two steps as addition-subtraction or subtraction-addition. Depending on the result of the first operation the second operation takes place or is ignored. This behavior can be observed when we zoom in Step 3 as shown in Fig. 8. This figure shows that after 160 cycles the first subtraction ends and the next addition operation start. The addition lasts 160 clock cycles.

The power consumption trace of one 160-bit EC point doubling is shown in Fig. 10. As expected, the number of clock cycles for EC point doubling is less than the number of clock cycles for EC point addition. The main difference in power consumption between EC point addition and EC point doubling can be observed by looking at Step 7, 8, 10, 11, 12 and 14. In these steps only modular a addition/subtraction takes place. Obviously the latency and power consumption of these are smaller than the others. This means that a simple power-analysis attack is easy to perform.

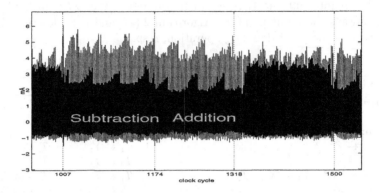

Fig. 8. Power consumption trace of Step 3 of 160-bit EC point addition

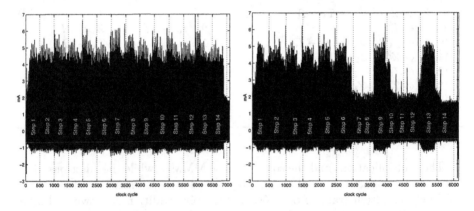

Fig. 9. Power consumption trace of 160-bit   **Fig. 10.** Power consumption trace of 160-
EC point addition from VCCINT   bit EC point doubling from VCCINT

The EC point multiplication is implemented by using a simple double-and-add algorithm. For EC point addition and EC point doubling the circuits described above are used. The power consumption trace of a 160-bit EC point multiplication is shown in Fig 11. It can be easily seen from figure 11 that the key used during this measurement is 1001100.

5.4   Applications and Future Work

Our board makes it possible to verify the effectiveness of many of the proposed countermeasures for various algorithms. In particular, we believe that countermeasures that are based on masking or blinding intermediate values (see for example [Koc96] and [Cor99] for approaches on asymmetric schemes), can be evaluated with our board. Also countermeasures for elliptic curve cryptosystems

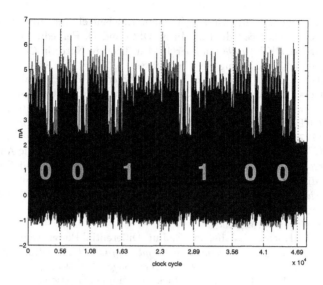

Fig. 11. Power consumption trace of a 160-bit EC point multiplication from VCCINT.

which are based on clever implementations of the elliptic curve operations [TB03] can be checked. All software based countermeasures can be evaluated with our setup. We plan to validate some of the countermeasures, such as [TB03] and to apply EM attacks [GMO01] on the FPGA.

6   Conclusion

We introduced a new platform for evaluating power analysis. Our approach consists of an FPGA, which is placed on a hand-made board which makes it very easy to conduct power-analysis attacks. We characterized the power consumption of a XILINX Virtex 800 FPGA and conclude that it is similar to the power consumption of an ordinary ASIC in CMOS technology. Therefore, it is possible to draw conclusions about the vulnerability of a certain circuit by performing power-analysis attacks on an FPGA-implementation. Since programming an FPGA is considerably cheaper than manufacturing an ASIC, assessing a devices vulnerability towards power-analysis attacks is much cheaper on our platform. Consequently, our approach describes the first cheap and efficient way to conduct power-analysis attacks on a real implementation (i.e., not on a software simulation) of a circuit in a very early stage of the design flow.

References

[Axe00]   J. Axelson. *Parallel Port Complete: Programming, Interfacing, and Using the PC's Parallel Printer Port.* Lakeview Research, Madison, WI 53704, 2000.

[CMO98]   H. Cohen, A. Miyaji, and T. Ono. Efficient elliptic curve exponentiation using mixed coordinates. In K. Ohta and D. Pei, editors, *Proceedings of ASIACRYPT 1998*, number 1514 in Lecture Notes in Computer Science, pages 51–65. Springer-Verlag, 1998.

[Cor99]   J.-S. Coron. Resistance against Differential Power Analysis for Elliptic Curve Cryptosystems. In Çetin Kaya Koç and Christof Paar, editors, *Cryptographic Hardware and Embedded Systems, First International Workshop, CHES'99, Worcester, MA, USA, August 12-13, 1999, Proceedings*, volume 1717 of *Lecture Notes in Computer Science*, pages 292–302. Springer, 1999.

[GMO01]   K. Gandolfi, Ch. Mourtel, and F. Olivier. Electromagnetic Analysis: Concrete Results. In Çetin Kaya Koç, David Naccache, and Christof Paar, editors, *Cryptographic Hardware and Embedded Systems - CHES 2001, Third International Workshop, Paris, France, May 14-16, 2001, Proceedings*, volume 2162 of *Lecture Notes in Computer Science*, pages 251–261. Springer Verlag, 2001.

[KJJ99]   P. C. Kocher, J. Jaffe, and B. Jun. Differential Power Analysis. In Michael Wiener, editor, *Advances in Cryptology-CRYPTO 1999*, volume 1666 of *Lecture Notes in Computer Science*, pages 388–397. Springer, 1999.

[Koc96]   P. C. Kocher. Timing attacks on implementations of Diffie-Hellman, RSA, DSS, and other systems. In Neal Koblitz, editor, *Proceedings of Crypto'96*, number 1109 in Lecture Notes in Computer Science, pages 104–113. Springer, 1996.

[MDS99a]   T. S. Messerges, E. A. Dabbish, and R. H. Sloan. Power Analysis Attacks of Modular Exponentiation in Smartcards. In Çetin Kaya Koç and Christof Paar, editors, *Cryptographic Hardware and Embedded Systems, First International Workshop, CHES'99, Worcester, MA, USA, August 12-13, 1999, Proceedings*, volume 1717 of *Lecture Notes in Computer Science*, pages 144–157. Springer, 1999.

[MDS99b]   T. S. Messerges, E. A. Dabbish, and R.H. Sloan. Investigations of Power Analysis Attacks on Smartcards. In *Proceedings of USENIX Workshop on Smartcard Technology*, pages 151–162, 1999.

[MK01]   M. M. Mano and C. R. Kime. *Logic and Computer Design Fundamentals*. Prentice Hall, Upper Saddle River, New Jersey 07458, second edition, 2001.

[Mon85]   P. Montgomery. Modular multiplication without trial division. *Mathematics of Computation*, Vol. 44:519–521, 1985.

[ÖBPV03]   S. B. Örs, L. Batina, B. Preneel, and J. Vandewalle. Hardware implementation of an elliptic curve processor over GF(p). In *The 10th Reconfigurable Architectures Workshop (RAW)*, Nice, France, April 2003.

[Opt]   OptiMagic, Inc. Frequently-Asked Questions (FAQ) About Programmable Logic. http://www.optimagic.com/faq.html\#FPGA.

[SKB02]   L. Shang, A. S. Kaviani, and K. Bathala. Dynamic power consumption in virtex-ii fpga family. In *Proceedings of the 2002 ACM/SIGDA 10th International Symposium on Field-Programmable Gate Arrays*, pages 157–164. ACM Press, 2002.

[TB03]   E. Trichina and A. Bellezza. Implementation of Elliptic Curve Cryptography with built-in Countermeasures against Side Channel Attacks. In Burton S. Kaliski Jr., Çetin Kaya Koç, and Christof Paar, editors, *Cryptographic Hardware and Embedded Systems - CHES 2002, 4th International Workshop, Redwood Shores, CA, USA, August 13-15, 2002, Revised Papers*, volume 2535 of *Lecture Notes in Computer Science (LNCS)*, pages 98–113. Springer, 2003.

[Wal99]   C. D. Walter. Montgomery's multiplication technique: How to make it
             smaller and faster. In Çetin Kaya Koç and Christof Paar, editors, *Crypto-*
             graphic Hardware and Embedded Systems, First International Workshop,
             CHES'99, Worcester, MA, USA, August 12-13, 1999, Proceedings, volume
             1717 of *Lecture Notes in Computer Science*, pages 80–93. Springer, 1999.

[WE93]   N. Weste and K. Eshraghian. *Principles of CMOS VLSI Design*. Addison-
             Wesley, 2nd edition, 1993.

[Xil01]   Xilinx, Inc. *Virtex 2.5 V Field Programmable Gate Arrays*, April 2 2001.
             http://direct.xilinx.com/bvdocs/publications/ds083.pdf.

[Xil02]   Xilinx, Inc. *Powering Xilinx FPGAs*, August 5 2002.
             http://support.xilinx.com/xapp/xapp158.pdf.

A   The Xilinx Virtex Architecture

Virtex devices feature a flexible, regular architecture that comprises an array
of configurable logic blocks (CLBs) surrounded by programmable input/output
blocks (IOBs), all interconnected by a rich hierarchy of fast, versatile routing
resources. Virtex FPGAs have a coarse-grained architecture, are SRAM-based,
and are customized by loading configuration data into internal memory cells.

Configurable Logic Block. The basic building block of the Virtex CLB is
the logic cell (LC) [Xil01]. A LC includes a 4-input function generator, carry
logic, and a storage element. The output from the function generator in each
LC drives both the CLB output and the D input of the flip-flop. Each Virtex
CLB contains four LCs, organized in two similar slices. Figure 12 shows a more
detailed view of a single slice. In addition to the four basic LCs, the Virtex CLB
contains logic that combines function generators to provide functions of five or
six inputs.

The Virtex function generators are implemented as 4-input look-up tables
(LUTs). In addition to operating as a function generator, each LUT can provide
a 16×1-bit synchronous RAM. The storage elements in the Virtex slice can be
configured either as edge-triggered D-type flip-flops or as level-sensitive latches.
The D inputs can be driven either by the function generators within the slice
or directly from the slice inputs, bypassing the function generators. In addition
to Clock and Clock Enable signals, each Slice has synchronous set and reset
signals (SR and BY). All the control signals can be inverted independently and
are shared by the two flip-flops within the slice.

I/O Block. The Virtex I/O Block (IOB) features SelectIO inputs and outputs
that support a wide variety of I/O signaling standards [Xil01]. The three IOB
storage elements function either as edge-triggered D-type flip-flops or as level
sensitive latches. Optional pull-up and pull-down resistors and an optional weak-
keeper circuit are attached to each pad. Prior to configuration, all pins not
involved in configuration are forced into their high-impedance state.

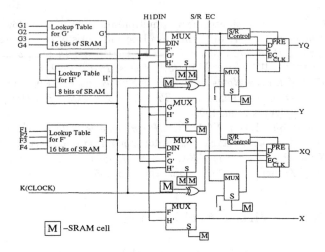

Fig. 12. Simplified diagram

I/O Banking. Some of the possible I/O standards require VCCO and/or VREF voltages. These voltages are connected to the device pins that serve groups of IOBs, called banks. Consequently, not all I/O standards can be combined within a given bank. Each bank has multiple VCCO (Output supply voltage) pins, all of which must be connected to the same voltage. This voltage is determined by the output standards in use.

Configuration of the FPGA. Virtex devices are configured by loading configuration data into the internal configuration memory. Some of the pins used for this are dedicated configuration pins, while others can be re-used as general purpose inputs and outputs once configuration is complete. Virtex supports four configuration modes which are the Slave-serial mode, the Master-serial mode, the SelectMAP mode and the Boundary-scan mode. The configuration mode pins (M2, M1, and M0) define which of these modes is used. Our board supports three of these configuration modes.

Hardware to Solve Sparse Systems of Linear Equations over GF(2)

Willi Geiselmann and Rainer Steinwandt

IAKS, Arbeitsgruppe Systemsicherheit, Prof. Dr. Th. Beth,
Fakultät für Informatik, Universität Karlsruhe, Am Fasanengarten 5,
76 131 Karlsruhe, Germany

Abstract. Bernstein [1] and Lenstra et al. [5] have proposed specialized hardware devices for speeding up the linear algebra step of the number field sieve. A key issue in the design of these devices is the question whether the required hardware fits onto a single wafer when dealing with cryprographically relevant parameters.

We describe a modification of these devices which distributes the technologically challenging single wafer design onto separate parts (chips) where the inter-chip wiring is comparatively simple. A preliminary analysis of a 'distributed variant of the proposal in [5]' suggests that the linear algebra step for 1024-bit numbers could be doable on a 23×23-network with special purpose processors in less than 19 hours at a clocking rate of 200 MHz, where each processor has about the size of a Pentium Northwood. Allowing for a 16×16 mesh of processing units with 36 mm $\times$ 36 mm, the linear algebra step might take less than 3 hours.

Keywords: Factorization, number field sieve, linear algebra, RSA

1 Introduction

Nowadays, the most common algorithm for factoring large integers is the so-called number field sieve (NFS). The NFS involves two computationally particularly expensive steps — the relation collection step and the task of solving a large sparse system of linear equations over GF(2) resp. of finding a linear dependence among binary vectors. In this contribution we deal only with the latter step. Based on the block Wiedemann algorithm [4,7], Bernstein [1] and Lenstra et al. [5] recently proposed specialized hardware devices for speeding up this part of the NFS.

In the present form, a major problem of these proposals is the size of the circuits and thereby the question of scalability: for larger parameter values, the proposed circuits do not fit onto a single wafer of diameter 300 mm any more, and high-speed communication between wafers is quite difficult to realize. But having in mind imperfections in actual manufacturing processes, already a single wafer design as proposed in [5] is rather non-trivial to realize. For circumventing this problem, in this paper we propose a technique for distributing the algorithms

C.D. Walter et al. (Eds.): CHES 2003, LNCS 2779, pp. 51–61, 2003.
© Springer-Verlag Berlin Heidelberg 2003

in [1,5] in such a way onto several wafers, that—at least for the case of 1024-bit numbers—both the performance of the algorithms does not decrease and the inter-wafer communication can be kept rather simple. It is appropriate to mention here that the idea of distributing the linear algebra step onto several 'smaller computers' is not new; e. g., in [2] ideas for implementing the linear algebra step 'in parallel on a network of relatively small machines' are described.

In Section 2 we shortly recall the essential hardware requirements of the two specialized architectures due to Bernstein and Lenstra et al., and thereafter we describe a method for overcoming the hardware limits of these approaches to a certain extent. To get a better idea of the possible use of our approach, Sections 3.2 and 3.3 analyze a 'multi-wafer' variant of the proposals in [1,5] for 512-bit and 1024-bit numbers in more detail. It turns out that even for 1024-bit numbers the linear algebra step seems to be doable within a few hours by means of a distributed hardware that can be manufactured with currently available technology.

2   Two Architectures for the Linear Algebra Step

Within the relation collection step of the NFS a (w. l. o. g. square) sparse matrix $A \in \mathrm{GF}(2)^{m \times m}$ is constructed. For 1024-bit numbers, the estimations in [5, Section 5.1] suggest values of $m \approx 4 \cdot 10^7$ or $m \approx 10^{10}$, where on average a column contains about 100 non-zero entries. For representing the matrix A throughout the computations, only the coordinates of its non-zero entries are stored.

To find the linear dependency among the columns of A needed in the NFS, the proposals in [1,5] make use of the block Wiedemann algorithm. Basically, this algorithm reduces the problem of finding a linear dependency among the columns of A to the problem of computing efficiently (long) sequences of the form

$$A \cdot v, A^2 \cdot v, \ldots, A^k \cdot v$$

where v is a—not necessarily sparse—binary vector $v \in \mathrm{GF}(2)^m$. A typical value is $k \approx 2m/K$ with a *blocking factor* $K = 1$ or $K \geq 32$ (for a blocking factor $K > 1$ several different values of the vector v are handled simultaneously).

Accordingly, for reducing the cost of the matrix step in the NFS, the devices proposed by Bernstein and Lenstra et al. aim at reducing the time required for computing such iterated (left-)multiplications with A. While the construction in [1] uses a parallel *sorting* algorithm for this purpose, the proposal in [5] relies on the use of a parallel *routing* algorithm. In the next two sections we shortly recall the respective hardware requirements of these devices; for an explanation of the algorithmic details we refer to the original papers.

2.1   Bernstein's Device for the Matrix Step

Concerning hardware requirements, the essential algorithmic tool in the proposal of [1] is Schimmler's sorting algorithm [1,6]: assume we are given a mesh of

$M \times M$ processing units $(Q_{i,j})_{1 \le i,j \le M}$ where $M := 2^n$ and each processing unit $Q_{i,j}$ stores an integer value $q_{i,j}$. Then Schimmler's sorting algorithm allows for sorting these M^2 numbers in $8M - 8$ 'steps' according to any of the following orders on the indices (i,j) of the processing units $Q_{i,j}$:

left-to-right: $(1,1) \le (1,2) \le \ldots \le (1,M) \le (2,1) \le \ldots \le (M,M)$
right-to-left: $(1,M) \le (1,M-1) \le \ldots \le (1,1) \le (2,M) \le \ldots \le (M,1)$
snakelike: $(1,1) \le (1,2) \le \ldots \le (1,M) \le (2,M) \le (2,M-1) \le \ldots \le (M,1)$

An 'elementary step' of the algorithm looks as follows: analogously as in the odd-even transposition sorting, in a single step each processing unit $Q_{i,j}$ communicates with exactly one of its horizontal or vertical neighbours. So let $\hat{Q}$, $\tilde{Q}$ be two communicating processing units, and denote by $\hat{q}$, $\tilde{q}$ the integers stored in $\hat{Q}$, $\tilde{Q}$, respectively. At the end of one 'elementary step' one of the two processing units, say $\hat{Q}$, must hold the value $\min(\hat{q}, \tilde{q})$ while the other one has to store $\max(\hat{q}, \tilde{q})$. For achieving this one can proceed as follows:

1. $\hat{Q}$ sends $\hat{q}$ to $\tilde{Q}$, and $\tilde{Q}$ sends $\tilde{q}$ to $\hat{Q}$. E.g., if the stored integers represent natural numbers $< 2^{26}$, this operation can be completed in one clock cycle via a unidirectional 26-bit bus in each direction.
2. Both $\hat{Q}$ and $\tilde{Q}$ compute the boolean value $exchange := (\tilde{q} < \hat{q})$. E.g., if $\hat{q}$ and $\tilde{q}$ are 26-bit numbers, this comparison can be done in one clock cycle.
3. If $exchange$ evaluates to true, then $\hat{Q}$ stores $\tilde{q}$ and deletes $\hat{q}$. Analogously, $\tilde{Q}$ keeps $\hat{q}$ and deletes $\tilde{q}$, in this case. If $exchange$ evaluates to false, then both $\hat{Q}$ and $\tilde{Q}$ keep their old values and delete the values received in the first step. Again, for 26-bit integers this operation does not require more than one clock cycle. In fact it is feasible to integrate this step into the previous one without requiring an additional clock cycle.

In summary, when dealing with natural numbers $< 2^{26}$, Schimmler's sorting algorithm enables us to sort M^2 numbers in less than $8M$ steps where each step takes 2 clock cycles. Assuming that each column of A contains d non-zero entries, one matrix-vector multiplication requires $m \cdot d$ processing units to store the matrix A and m processing units to store the entries of the vector v. Using Bernstein's approach, a matrix-vector multiplication can thus be realized on a mesh of size $M \times M$, provided that $M^2 \ge d \cdot m + m$. For one multiplication three sorting steps with $\approx 8 \cdot M$ exchange operations, requiring 2 clock cycles each, are necessary. Consequently, one matrix-vector multiplication can be performed in approximately $3 \cdot 2 \cdot 8 \cdot M = 48 \cdot M$ clock cycles.

For the factorization of 1024-bit numbers (using the 'small matrix' with $m \approx 4 \cdot 10^7$), in [5] the average number of transistors per processing unit is estimated to be around 2000. Assuming that a standard 0.13 μm manufacturing process is used, with [5, Table 2] we thus conclude, that one processing unit requires an area of ≈ 4760 μm^2 resp. a square of about 0.07×0.07 mm^2. Analogously, assuming a processing unit for the case of 512-bit numbers to require 1800 transistors, we obtain an estimated area of ≈ 4280 μm^2 per processing unit in this case. Here, the estimation for the number of transistors is based on a matrix of size

$6.7 \cdot 10^6 \times 6.7 \cdot 10^6$ where each column contains 63 non-zero entries (cf. [3]); for this matrix size 23 bits are sufficient to represent a column or row index.

2.2   Lenstra et al.'s Device for the Matrix Step

Similarly as Bernstein's proposal, the architecture put forward by Lenstra et al. is based on a mesh of simple processing units. But as opposed to [1], the mesh is used for *routing* rather than *sorting*. Concerning the factorization of 1024-bit numbers, Lenstra et al. discuss two possible matrix sizes ($m \approx 4 \cdot 10^7$ resp. $m \approx 10^{10}$). For describing the device that is to fit onto a single wafer of diameter 300 mm, the 'small matrix' is used, and we restrict our discussion to this case.

Depending on the precise choice of parameters, the mesh [5] uses one or two types of processing units. For the single wafer device just mentioned, only one type is used, and each node is a so-called *target node*. Basically, this means that each node stores all row coordinates of $\rho \geq 1$ non-zero entries of A as well as ρ entries of v. After having performed a complete matrix-vector multiplication $A \cdot v$, the entries storing v are replaced by the entries of the vector $A \cdot v$. Denoting again by d the number of non-zero entries per column, the non-zero entries of A can be distributed onto m/ρ processors where each processor has sufficient DRAM for storing $\rho \cdot d$ matrix entries.

The main tool utilized for the actual computation of a matrix-vector multiplication is so-called *clockwise transposition routing* which relies on the iterated application of (parallelly executed) exchange operations. Routing a single value takes about $2 \cdot \sqrt{m/\rho}$ clock cycles, and the processing of the individual matrix entries and matrix columns can overlap. As a worst-case bound we can assume a complete matrix-vector multiplication to require no more than $\rho \cdot d \cdot 2 \cdot \sqrt{m/\rho} = 2 \cdot d \cdot \sqrt{m \cdot \rho}$ clock cycles.

So far, our discussion ignored the blocking factor K: as pointed out in [5], for a given blocking factor K, Wiedemann's algorithm requires the computation of K multiplication chains

$$A \cdot v_i, A^2 \cdot v_i, \ldots, A^k \cdot v_i$$

with different vectors v_i ($1 \leq i \leq K$). Using a slightly more complicated hardware (see [5] for details), these K chains can be computed in parallel with the same routing circuit. Basically, for blocking factor K each processing unit needs $2 \cdot \rho \cdot K$ bit of memory to store the vectors v_i. In particular, the value of K is relevant when estimating the space requirement for the target units; here we assume $K = 208$ (the value chosen in [5] for the single wafer device for 1024-bit numbers).

With $m \approx 4 \cdot 10^7$ the average number of transistors necessary for one target unit—excluding DRAM—can be estimated to be around $2040 + 60 \cdot K$ (cf. Section 3.2). The area needed for a single DRAM bit is $\approx 0.7 \ \mu m^2$ resp. $0.2 \ \mu m^2$ with a specialized DRAM process. Thus, for the 'small matrix case' of 1024-bit numbers, the DRAM of a complete target cell occupies about $88700 \ \mu m^2$ resp. $25300 \ \mu m^2$ with a specialized DRAM process. The space requirement for the

$2040 + 60 \cdot 208$ transistors computes to $34600 \ \mu m^2$ and $40700 \ \mu m^2$, respectively. In total, for one target cell $123300 \ \mu m^2$ with a standard process resp. $66000 \ \mu m^2$ with a specialized DRAM process are needed in the 1024-bit case. When dealing with 512-bit numbers—and a matrix of size $6.7 \cdot 10^6 \times 6.7 \cdot 10^6$ with 63 non-zero entries per column—the DRAM per target cell occupies only $54800 \ \mu m^2$ resp. $15700 \ \mu m^2$ with a specialized DRAM process. Adding the space for $1920 + 60 \cdot 208$ transistors, we obtain a total space requirement of $89100 \ \mu m^2$ resp. $56100 \ \mu m^2$ per target cell.

2.3   Estimated Mesh Size and Performance for 512 Bit and 1024 Bit

With a standard $0.13 \ \mu m$ process, there fit $\approx 2.915 \cdot 10^{10}$ transistors on a single wafer of diameter 300 mm (cf. [5]). Using the above figures, a straightforward computation now yields the following estimation for the wafer area and time needed by Bernstein's approach, when dealing with $\log_2(n)$-bit numbers:

$\log_2(n)$	m	d	# proc	M	area in wafers	clock cycles	LA
512	$6.7 \cdot 10^6$	63	$4.3 \cdot 10^8$	2^{15}	74 (26.6)	$1.6 \cdot 10^6$	18 h
1024	$4 \cdot 10^7$	100	$4.04 \cdot 10^9$	2^{16}	295 (277.2)	$3.1 \cdot 10^6$	207 h

The area in brackets indicates the area required to store the matrix and the vector; the difference to the real area required comes from choosing M as a power of 2 with $M^2 \geq m \cdot (d + 1)$. The last column (labeled with LA) is the estimated total time of the linear algebra step; more precisely, the value given is the time for performing $3 \cdot m$ matrix-vector multiplications (see [5]) at a clocking rate of 500 MHz.

A 512-bit device with these parameters has a size of $2.14 \ m \times 2.14 \ m$; for the 1024-bit case we obtain a (wafer) area of $4.5 \ m \times 4.5 \ m$. Thus, realizing such a device seems quite hypothetical.

For the device described by Lenstra et al. [5] we assume $\rho = 42$, i. e., each target unit takes care of 42 matrix columns (cf. [5, Table 3]). For 512-bit numbers, then 402^2 target units including DRAM fit on an area of $95 \ mm \times 95 \ mm$; and for 1024-bit numbers 1026^2 target units including DRAM fit on the area of a single wafer. This results in the following estimated space and time requirements for performing the linear algebra step:

$\log_2(n)$	m	d	# targets	M	area (wafers)	clk cycles	LA
512	$6.7 \cdot 10^6$	63	$1.6 \cdot 10^5$	400	0.13	$2.1 \cdot 10^6$	17 min
1024	$4 \cdot 10^7$	100	$9.5 \cdot 10^5$	1024	1	$8.2 \cdot 10^6$	6.5 h

Here the estimatated total time of the linear algebra step is the time for performing $3 \cdot m / K = 3 \cdot m / 208$ matrix-vector multiplications at a clocking rate of 200 MHz (cf. [5]).

3   Distributing the Computation

In several of the above mentioned sizes and in several other parameter choices described in [5], the specialized hardware for the linear algebra step does no longer fit on a single wafer. But due to the practical limitations of manufacturing processes, already realizing the single wafer devices is quite challenging. In this section we want to discuss an approach for circumventing the problem of handling sophisticated, highly parallel I/O hardware for fast inter-wafer communication; at least for 1024-bit numbers this approach seems to improve the situation significantly.

3.1 Using Block Matrix Multiplication

For our discussion we adopt the assumption from [5] that the non-zero entries in the matrix $A \in \mathrm{GF}(2)^{m \times m}$ are uniformly distributed. It should be emphasized, that the original matrix A cannot be expected to have such a uniform distribution, and here we do not discuss the problem of how a preprocessing for achieving this could look like. The 'rectangular matrix blocks' we will use should allow for some leeway here, and subsequently we make the assumption that a suitable preprocessing has been done already, e. g., by having applied suitable row and column permutations to the original A.

We start by splitting the matrix A into $s \cdot s$ submatrices $A_{i,j}, 1 \leq i, j \leq s$ of approximately the same size of $m/s \times m/s$. It is not mandatory that all $A_{i,j}$ are square matrices, but we insist that for fixed $i_0 \in \{1, \ldots, s\}$ all matrices $A_{i_0,j}$ $(1 \leq j \leq s)$ have the same number of rows:

$$
A = \begin{pmatrix}
A_{1,1} & | & A_{1,2} | & \cdots & |A_{1,s} \\
\hline
A_{2,1} & |A_{2,2} & | & \cdots | & A_{2,s} \\
\vdots & & \vdots & \vdots & \vdots \\
\hline
A_{s,1} & | & A_{s,2} | & \cdots & |A_{s,s}
\end{pmatrix}
$$

The size of the hardware devices in [1,5] depends on the number of non-zero entries in the processed matrix, and the aim of the separation just mentioned is to split the matrix into s^2 submatrices with approximately the same number of non-zero elements. After splitting the vector v into appropriately sized parts

$$
v = \begin{pmatrix} v_{1,1} \\ \vdots \\ v_{1,s} \end{pmatrix} = \ldots = \begin{pmatrix} v_{s,1} \\ \vdots \\ v_{s,s} \end{pmatrix}
$$

(where the number of rows of $v_{i,j}$ is equal to the number of columns of $A_{i,j}$), the multiplication $A \cdot v$ can be realized as

$$
A \cdot v = \begin{pmatrix} \sum_{j=1}^{s} A_{1,j} \cdot v_{1,j} \\ \vdots \\ \sum_{j=1}^{s} A_{s,j} \cdot v_{s,j} \end{pmatrix}.
$$

This can be performed with s^2 multiplication circuits (preloaded with the matrices $A_{i,j}$) in the following way:

1. Load $v_{i,j}$ into the circuit corresponding to $A_{i,j}$ through a pipeline of length s and a bus of width b (if $A_{i,j}$ is not a square matrix, we can think of the missing column/row entries as being 0).
2. Perform the s^2 matrix-vector multiplications $A_{i,j} \cdot v_{i,j}$ in parallel.
3. Output the resulting vectors $A_{i,j} \cdot v_{i,j}$ through a bus of width b.
4. Perform the summations $w_i := \sum_{j=1}^{s} A_{i,j} \cdot v_{i,j}$ (for $i = 1, \ldots, s$) with s XOR-pipelines of length s. Each of these s^2 XOR-circuits is adjacent to one multiplication circuit and adds the output of this circuit to the output of the previous stage of the pipeline. Each of these XOR-circuits has two inputs and one output of width b and works during the output of the multiplication circuits in a pipeline architecture.

   These XOR-circuits are extremely simple, but have to be built up out of several chips due to the width of the bus. A different approach is to include these XORs into the adjacent multiplication circuit; then the saved hardware has to be paid for by a doubling of the I/O time. This part of the hardware should not cause a major problem and is neglected here.
5. Analogously as the vector v before, now the vector $w := (w_1, \ldots, w_s)$ is split into s^2 parts $w_{i,j}$. These $w_{i,j}$ are now ready to be loaded into the s^2 multiplication circuits to perform the multiplication $A \cdot w$ if required.

   At this stage we can also easily perform a vector-vector multiplication of the form $u \cdot Av = u \cdot w$ as needed in the block Wiedemann algorithm. The vectors u are usually chosen to be of very low Hamming weight, and we thus ignore the (marginal) computational effort of these multiplications.

   The loading of $A \cdot v$ is performed through a pipeline structure, similar to the XOR-pipeline for the outputs. If the XOR-circuits are extended with an additional register and a multiplexer (to switch between the 'horizontal' and 'vertical' bus), the same chips can be used for both pipelines.

3.2   Performance of the Distributed Device

Let us now look at the space and time requirements of the distributed architecture just described; for sake of simplicity, we assume all submatrices $A_{i,j}$ to be $(m/s) \times (m/s)$ square matrices. We also consider the choice of a blocking factor $K \geq 1$; for $K > 1$ several vectors are handled in parallel, and of course we have to take into account the additional bandwidth required here.

Step 1. Loading the K vectors $v_{i,j}$ into the multiplication circuits requires approximately $4 \cdot m \cdot K/(s \cdot b) + 4 \cdot s$ clock cycles, where the time for loading one bit is estimated to be 4 clock cycles, and $4 \cdot s$ clock cycles are needed to empty the pipeline. All but the last b input bits can be distributed to the processing units (or target cells) while the following input bits arrive. The extra time required to distribute the last b bits is neglected here.

Step 2. Each of the multiplication circuits has to perform a multiplication of a matrix with about $m \cdot d/s^2$ entries with a binary vector of size approximately m/s. With Bernstein's approach, this can be done in $48 \cdot M$ clock cycles on an $M \times M$ mesh where $M \geq \sqrt{m \cdot d/s^2 + m/s}$ is a power of 2.

With the design from [5], this matrix-vector multiplication requires M^2 target cells for ρ columns each, where $M^2 \geq m/(s \cdot \rho)$. At this, each target cell is equipped with DRAM for $\rho \cdot d/s$ matrix entries, and we can estimate the number of clock cycles required for the matrix-vector multiplication to be no larger than $2 \cdot \rho \cdot d \cdot M/s$.

Step 3. Transfering the vector w_i to the output buffer requires one sorting step in Bernstein's architecture ($\approx 2 \cdot 8 \cdot M$ clock cycles). In the architecture of Lenstra et al. the computational effort of this step is negligible (due to the known addresses of the bits of w_i, and thus the possibility to use a pipeline procedure during the output); we estimate the output to require $\approx 4 \cdot m \cdot K/(s \cdot b)$ clock cycles.

Step 4. The summation of the subvectors is performed with a pipeline structure while the outputs of the multiplication units arrive. Additional $\approx 4 \cdot s$ clock cycles are needed to empty the pipeline.

Summarizing our discussion, we have the following characterizing figures:

- With Bernstein's architecture M^2 processing units for each of the s^2 parts are required, where M is a power of 2 and $M^2 \geq m \cdot d/s^2 + m/s$. Taking into account the registers ($2 \cdot 8 \cdot \lceil \log_2(m/s) \rceil$ transistors), multiplexers ($3 \cdot 4 \cdot \lceil \log_2(m/s) \rceil$ tansistors), a subtraction unit ($5 \cdot 8 \cdot \lceil \log_2(m/s) \rceil$), and control logic (300 transistors) needed, we estimate that one processing unit consists of $\approx 68 \cdot \lceil \log_2(m/s) \rceil + 300$ transistors.[1]

  The number of clock cycles for a complete matrix-vector multiplication is approximately $8 \cdot \lceil m/(s \cdot b) \rceil + 48 \cdot M + 8 \cdot s$.

- With the architecture of Lenstra et al. $M^2 \geq m/(s \cdot \rho)$ processing units resp. target cells are used on each of the s^2 parts, if one target cell takes care of ρ matrix columns. Taking into account the required DRAM for representing the non-zero matrix entries ($\rho \cdot d \cdot \lceil \log_2(m/s) \rceil/s$ bit), the DRAM bits for storing the K processed vectors ($2 \cdot \rho \cdot K$ bit) along with an access logic ($40 \cdot K$ transistors), a register for a received 'package' from the mesh ($8 \cdot (\lceil \log_2(m/s) \rceil + K)$ bit), three multiplexers ($3 \cdot 4 \cdot (\lceil \log_2(m/s) \rceil + K)$ transistors), a subtraction unit ($(8 \cdot 5 \cdot \lceil \log_2(m/s) \rceil)/2$ transistors), and additional logic (1000 transistors) we estimate one processing unit to require $\approx 40 \cdot \lceil \log_2(m/s) \rceil + 60 \cdot K + 1000$ transistors and $\rho \cdot (d \cdot \lceil \log_2(m/s) \rceil/s + 2 \cdot K)$ bit of DRAM. The number of clock cycles for a complete matrix-vector multiplication is $\approx 8 \cdot \lceil m \cdot K/(s \cdot b) \rceil + 2 \cdot \rho \cdot d \cdot M/s + 8 \cdot s$.

In the next section we examine in more detail the performance of this distributed approach when dealing with 512-bit and 1024-bit numbers. For doing so, we consider various choices of the bus width b, the blocking factor K, the number of columns ρ handled per target cell, and the 'degree of parallelism' s.

---

[1] Note that for storing a row or column index of a submatrix $A_{i,j} \in \mathrm{GF}(2)^{m/s \times m/s}$ only $\lceil \log_2(m/s) \rceil$ bits are needed.

3.3   Application to 512-Bit and 1024-Bit Numbers

Having in mind a practical manufacturing process, it is desirable that the individual parts of the distributed circuit are significantly smaller than (the inner square of) a complete 300 mm wafer. For the distributed circuit derived from the architecture in [5], we choose the size of the individual parts to be comparable to the size of an 'ordinary' Pentium Northwood processor. To avoid problems with the available number of pins connected to the bus b, we use somewhat conservative estimations for the bus width.

For Bernstein's circuit such small processing units are not really sensible, and we choose the individual parts to be larger. For these larger parts (or in other words chips), it is sensible to allow for a larger bus width b, as more pins can be located on the chip here.

The bus width b also limits the possible choices of the blocking factor K: before performing the (next) matrix-vector multiplication by means of a (routing or sorting) mesh, we have to load the respective parts of the vector to be processed next into the processing units via the bus. However, with a simple trick we can gain some parallelism 'almost for free': assume that each part of the distributed device—in other words each chip—handles K vectors in parallel (for Bernstein's approach we have $K = 1$). Then while these K vectors are processed, we can load another K-tuple of vectors into a separate buffer on that chip. So once the result of the previous multiplication is output, we can immediately load the new vectors into the mesh. If the I/O time is about the same as the computation time, then by interleaving the processing of two 'tuples of vectors' in this way, we can in the ideal case almost halve the time needed for loading vectors onto and from the chips (of course, the cells to store these additional vectors require additional place on the chip, which has to be taken into account then).

Table 1 and 2 show the performance of the distributed device for various parameter choices; at this, a potential optimization by 'interleaving tuples of vectors' is not taken into account. As in Section 2.3, for estimating the total time of the linear algebra step, the number of multiplications is assumed to be $3 \cdot m$ in the design derived from Bernstein's proposal, and $3 \cdot m/K$ in the design derived from the proposal of Lenstra et al.

Table 1. Time for the LA step with a 'distributed Bernstein design' at a clocking rate of 500 MHz.

$\log_2(n)$	b	#proc/chip	s^2	chip size	LA time
512	(single unit)	23768^2	1	2.14 m × 2.14 m	17.6 h
512	2048	2048^2	11^2	144 mm × 144 mm	1.1 h
512	1024	1024^2	24^2	72 mm × 72 mm	0.6 h
512	1024	512^2	55^2	36 mm × 36 mm	0.3 h
1024	(single unit)	65536^2	1	4.5 m × 4.5 m	210 h
1024	2048	2048^2	37^2	144 mm × 144 mm	6.8 h
1024	1024	1024^2	84^2	72 mm × 72 mm	3.4 h

Table 2. Time for the LA step with a 'distributed Lenstra et al. design' at a clocking rate of 200 MHz.

$\log_2(n)$	b	K	ρ	#proc/chip	s^2	chip size	LA time
512	128	65	116	76^2	10^2	11.4 mm × 11.4 mm	73 min
512	128	53	51	91^2	16^2	11.4 mm × 11.4 mm	45 min
512	512	63	29	152^2	10^2	20 mm × 20 mm	19 min
512	1280	49	8	290^2	10^2	34 mm × 34 mm	8 min
512	1024	40	6	265^2	16^2	29 mm × 29 mm	6 min
1024	(single unit)	208	42	975^2	1	265 mm × 265 mm	6.1 h
1024	(single unit)	42	216	430^2	1	162 mm × 162 mm	94.7 h
1024	128	30	1086	48^2	16^2	11.4 mm × 11.4 mm	29.7 h
1024	128	70	669	51^2	23^2	11.4 mm × 11.4 mm	18.8 h
1024	512	100	250	100^2	16^2	20 mm × 20 mm	7.0 h
1024	1024	160	278	120^2	10^2	30 mm × 30 mm	5.9 h
1024	1280	135	66	195^2	16^2	36 mm × 36 mm	2.8 h

For Bernstein's approach we recognize that the distributed variant looks much more practical than the original design. Also it is worth noting, that the communication cost—i. e., the time for loading vectors onto/from the chips—is less than 5% of the overall computation time, and thus is not really relevant. For the design of Lenstra et al. the situation is quite complementary: more than 90% of the time is spent for the I/O operations. However, the obtained circuitry is much smaller and thus more realistic than the sorting based approach. In particular, with a mesh of $23^2 = 529$ Pentium Northwood sized processing units, the linear algebra step for a 1024-bit number should be doable in less than 19 hours. Note here that the overall wafer area of this distributed device is the same as for the original single wafer design of Lenstra et al. Concerning speed the distribution has to be paid for with a slow-down of more than a factor 3. However, manufacturing the small processing units is significantly simpler. Further on, already with slightly larger processing units—which allow for a broader bus—, the overall computation time can be reduced to less than 3 hours. As most of the time is spent for the I/O, the bus width should be chosen as large as possible; in our estimations we tried to be conservative here.

4   Conclusion

The above discussion suggests that for 1024-bit numbers, a 'distributed variant of the design of Lenstra et al.' could be realizable by means of current technology. Besides circumventing a technologically challenging wafer-sized circuit, also a speed-up seems to be possible, if one allows for processing units of up to, say, 36 mm × 36 mm. But already with a mesh of 23^2 processing units, where each processing unit has approximately the size of a Pentium Northwood, the linear algebra step for 1024-bit numbers seems to be doable in less than a day.

Acknowledgement. We thank Eran Tromer for valuable discussions and remarks.

References

1. Daniel J. Bernstein. Circuits for Integer Factorization: a Proposal. At the time of writing available electronically at http://cr.yp.to/papers/nfscircuit.pdf, 2001.
2. Richard P. Brent. Recent Progress and Prospects for Integer Factorisation Algorithms. In Ding-Zhu Du, Peter Eades, Vladimir Estivill-Castro, Xuemin Lin, and Arun Sharma, editors, *Computing and Combinatorics; 6th Annual International Conference, COCOON 2000*, volume 1858 of *Lecture Notes in Computer Science*, pages 3–22. Springer, 2000.
3. Stefania Cavallar, Bruce Dodson, Arjen K. Lenstra, Walter Lioen, Peter L. Montgomery, Brian Murphy, Herman te Riele, Karen Aardal, Jeff Gilchrist, Gérard Guillerm, Paul Leyland, Joël Marchand, François Morain, Alec Muffet, Chris Putnam, Craig Putnam, and Paul Zimmermann. Factorization of a 512-bit RSA Modulus. In Bart Preneel, editor, *Advances in Cryptology — EUROCRYPT 2000*, volume 1807 of *Lecture Notes in Computer Science*, pages 1–18. Springer, 2000.
4. Don Coppersmith. Solving Homogeneous Linear Equations over GF(2) via Block Wiedemann Algorithm. *Mathematics of Computation*, 62(205):333–350, 1994.
5. Arjen K. Lenstra, Adi Shamir, Jim Tomlinson, and Eran Tromer. Analysis of Bernstein's Factorization Circuit. In Yuliang Zheng, editor, *Advances in Cryptology — ASIACRYPT 2002*, volume 2501 of *Lecture Notes in Computer Science*, pages 1–26. Springer, 2002.
6. Manfred Schimmler. Fast sorting on the instruction systolic array. Technical Report 8709, Christian Albrecht Universität Kiel, Germany, 1987.
7. Douglas H. Wiedemann. Solving Sparse Linear Equations Over Finite Fields. *IEEE Transactions on Information Theory*, 32(1):54–62, 1986.

Cryptanalysis of DES Implemented on Computers with Cache

Yukiyasu Tsunoo[1], Teruo Saito[2], Tomoyasu Suzaki[2], Maki Shigeri[2], and
Hiroshi Miyauchi[1]

[1] NEC Corporation, Internet Systems Research Laboratories
4-1-1, Miyazaki, Miyamae-ku, Kawasaki, Kanagawa 216-8555, Japan
{tsunoo@bl, h-miyauchi@bc}.jp.nec.com
[2] NEC Software Hokuriku Ltd.
1, Anyoji, Tsurugi, Ishikawa 920-2141, Japan
{t-saito@qh, t-suzaki@pd, m-shigeri@pb}.jp.nec.com

Abstract. This paper presents the results of applying an attack against
the Data Encryption Standard (DES) implemented in some applications,
using side-channel information based on CPU delay as proposed in [11].
This cryptanalysis technique uses side-channel information on encryption
processing to select and collect effective plaintexts for cryptanalysis, and
infers the information on the expanded key from the collected plaintexts.
On applying this attack, we found that the cipher can be broken with
2^{23} known plaintexts and 2^{24} calculations at a success rate $> 90\%$, using
a personal computer with 600-MHz Pentium III.
We discuss the feasibility of cache attack on ciphers that need many
S-box look-ups, through reviewing the results of our experimental
attacks on the block ciphers excluding DES, such as AES.

Keywords: DES, AES, Camellia, cache, side-channel, timing attacks

1 Introduction

Recently, many proposals have been made for cryptanalysis techniques to measure physical information from a cryptographic device. These techniques are called "side-channel attacks." Typical examples are Differential Power Analysis [5], which measures the variation in power consumption caused by a cryptographic device, and Differential Fault Analysis [1], which causes some sorts of physically erroneous operation to occur in a cryptographic device and then measures resulting phenomena. Because techniques of this kind are mainly used for attacking cryptographic systems implemented on smart cards, anti-tampering measures e.g. adding noise to consumed power have been considered. "Timing attacks" [2][6] that measure the encryption time of a cryptographic application can also be treated as side-channel attacks. A countermeasure to attacks of this type is to eliminate branch processing in the implementing algorithm so that encryption times are equivalent.

Previously proposed timing attacks make use of the fact that conditional branches that occur during encryption processing cause variations in encryption

C.D. Walter et al. (Eds.): CHES 2003, LNCS 2779, pp. 62–76, 2003.

time. CPU cache misses, however, can also cause such variations. In this regard, most of the recent computers employ a "CPU cache", abbreviated simple to a "cache" from here on, between the CPU and main memory, since this type of hierarchical structure can speed program run-time on the average. If, however, the CPU accesses data that were not stored in the cache, i.e. if a cache miss occurs, a delay will be generated, as the target data must be loaded from main memory into the cache. The measurement of this delay may enable attackers to determine the occurrence and frequency of cache misses.

With the above in mind, we have focused our attention on data-access processing, i.e. the operations of the S-box commonly used by encryption algorithms, and have developed a new attack technique to infer the information on S-box input from the variations in encryption time for different plaintexts. This is classified as a side-channel attack on software-implemented ciphers, and it has already broken MISTY1 [11] successfully. It does not require specialized measuring equipment; the cipher can be broken in a relatively short time using a personal computer, if the encryption module of the cipher is available. Though Kelsey et al. described the feasibility of a cache-based attack on ciphers using a large S-box e.g. Blowfish [4], they did not refer a specific method. The first application of an attack using a cache is described in [11].

We made experimental attacks on some block ciphers including Data Encryption Standard (DES). This paper describes the cases we could break the cipher in spite of frequent S-box look-ups, or the resistance to the attack described in [11].

This paper is organized as follows. Section 2 describes the basics of the proposed attack. Section 3 then describes the method of applying this attack to DES and presents the results of our experiment. Section 4 shows the results of this attack on AES and Camellia. Lastly, section 5 concludes the paper.

2   The Basics of Attack

2.1   Cache Operation

A cache is a form of memory that allows faster reading and writing of data than those in a main memory. It is located between the CPU and main memory. When reading data from main memory, the CPU first checks the cache, and if the target data is present, it reads the data from the cache. Finding data in the cache in this way is called a "cache hit," while not finding data in the cache and reading it from main memory is called a "cache miss." In the latter case, the data read from main memory is also written to the cache [1], so that any subsequent reading of this data might speed up. In short, a delay in processing will occur even for the same instruction if target data does not exist in the cache, and this delay will appear as a variation in the program execution time.

---

[1] In reality, values near the referenced one will also be loaded into the cache.

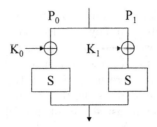

Fig. 1. Cipher With Two S-boxes

2.2   The Encryption Time and S-Box Operation

As described in Section 2.1, plaintext with the long encryption time should correspond to the frequency of cache miss. In the following, we examine the conditions for the generation of cache misses in the encryption process.

In the encryption processing, data access occurs when the S-box is referenced. What then are the conditions that would generate more cache misses when referencing the S-box? Consider that a cache miss occurs when first referencing the S-box and that the data in question is therefore loaded into the cache as described earlier. Now, when next referencing the S-box, if the S-box input value is the same as the already referenced value or its nearby one [2], data referencing can be done by accessing the cache; a cache miss does not occur. If, however, a value excluding already referenced ones and their nearby ones is referenced, the desired data will not be found in the cache and will have to be loaded from main memory; cache miss occurs. Accordingly, when making multiple S-box references during the encryption process, the number of cache misses increases proportionally with the number of different S-box input values.

Based on the above reasoning, the encryption time should be long if there are many different data referenced by the S-box during encryption. Thus, the measurement of the encryption time for a plaintext makes it possible to determine whether that plaintext is of the type that generates many cache misses in encryption (i.e., plaintext for which there are many different S-box input values).

2.3   Attack Model

The cipher with two S-boxes shown in Fig. 1 is used to explain the basics of the process of obtaining information on keys, which exploits side-channel information. The structure shown in the figure employs independent keys K_0 and K_1 in different S-boxes.

Referring to Fig. 1, we assume that the relationship between the input values of the two S-boxes under comparison is understood. The key differential value

---

[2] Values near the referenced value will be simultaneously loaded due to the characteristics of CPU.

$K_0 \oplus K_1$ (referred to below as "key difference") can therefore be inferred from the values of plaintext P_0 and P_1, using either of the following relations.

$$P_0 \oplus K_0 = P_1 \oplus K_1 \rightarrow P_0 \oplus P_1 = K_0 \oplus K_1 \tag{1}$$

$$P_0 \oplus K_0 \neq P_1 \oplus K_1 \rightarrow P_0 \oplus P_1 \neq K_0 \oplus K_1 \tag{2}$$

In other words, if the plaintexts for which S-box input values are frequently the same or frequently different are collected by measuring the encryption time, the information on the key differences can be obtained from those plaintexts. The attack comprised of 2 processes, the one for obtaining the key differences and the one for collecting cache timing data described in Section 3.2 is called a "cache attack." Obtaining key differences by a cache attack can reduce the key search space.

As the structure shown in Fig. 1 can be found in many block ciphers, it is thought that cache attacks can be widely applicable to ciphers of this type.

2.4   Non-elimination/Elimination Table Method

As described above, a correlation exists between the encryption time and the relationship between input values of separate S-boxes. We consider the following two methods of obtaining a key difference, based on such information.

The first method corresponds to the situation in which the input values of S-boxes under comparison are equivalent. In this case, Eq. (1) holds and the values for the key differences can be calculated from plaintext information. Implementing this method requires the collection of plaintexts resulting a short encryption time under the assumption that a plaintext having a small number of cache misses equals a plaintext having a short encryption time. It can therefore be guessed that most of the collected plaintexts result in equivalent input values between the S-boxes in question. Key differences can therefore be calculated for the collected plaintexts and the value counted most frequently can be regarded as the correct key difference. We call this method a "non-elimination table attack."

The second method corresponds to the situation in which the input values of S-boxes under comparison are different. In this case, Eq. (2) holds and values of improbable key differences can be excluded. Implementing this method requires the collection of plaintexts resulting a long encryption time under the assumption that a plaintext having a large number of cache misses equals a plaintext having a long encryption time. Thus, the most of the collected plaintexts are guessed to result in different input values between the S-boxes. Key differences for the collected plaintexts can therefore be calculated and the value that appears the least frequently is taken as the correct key difference. We call this method an "elimination table attack."

For DES, the number of S-box operations is 16, a rather small one, considering that each S-box has 64 entries. Therefore, it is predicted that many input values will be different between the S-boxes, making it easy to collect plaintexts. Thus, we applied an elimination table attack on DES.

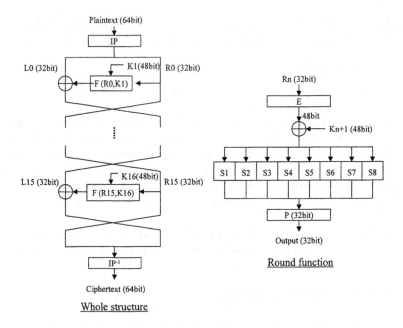

Fig. 2. Whole Structure and Round Function

3   Attack on DES

3.1   DES Structure

DES has a 16-round Feistel structure. Each round function features eight S-boxes each with a 6-bit input and a 4-bit output. An S-box operates 16 times, a small number compared to its $2^6 = 64$ entries. In the key scheduler, 48-bit of a 64-bit secret key is selected for each round, and its value is used as a expanded key for the corresponding round. Refer Fig. 2 and Fig. 3 for details.

As shown in Fig. 3, the total number of left cyclic shifts is set to 28 bits, which means that (C_0, D_0) and (C_{16}, D_{16}) have the same value. Thus, (C_1, D_1) used in the round 1 and (C_{16}, D_{16}) used in the round 16 are related by a 1-bit left cyclic shift. This relationship is used for the secret key recovery described in Section 3.2.

3.2   Attack Technique

This section describes the DES attack technique in detail. The steps making up this attack are divided into two main stages. Stage 1 is used to collect plaintexts for encryption, while Stage 2 is used to obtain key differences from the collected plaintexts. These stages are performed independently of each other.

The experiment described in this paper was done in the machine and compile environment summarized in Table 1.

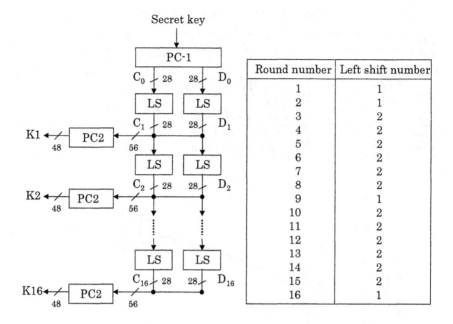

Fig. 3. Key Schedule

Collection of Plaintext. We first describe the method of collecting plaintext; Stage 1 of the attack. Here, plaintext having a long encryption time is needed to apply the elimination table attack described in Section 2.4. Our approach therefore is to encrypt a fixed amount of randomly generated plaintexts and to examine the resulting distribution of encryption time. The following method is used to measure the delay caused by cache misses as accurately as possible. The characteristics of the CPU (Pentium III) used in this experiment are also taken into account, and the DES source code that we use is the one described in [10]. In this source code, it is declared to assign 4 bytes to each entry of S-box, since S-box and bit permutation are computed simultaneously, for faster performance. The encryption time measurement method is as follows.

- Before beginning measurements, S-box data is deleted from the L1 data cache. In actuality, 16 kilobytes of random data are loaded into the 16-kilobyte data area of the L1 data cache to fill it.
- The *rdtsc* instruction, which loads the value of the processor's time stamp counter into a register, is used to measure encryption time ; the instruction is executed directly before and after encryption and the difference between the obtained values is used to compute the encryption time.

The above method enables to measure the encryption time for any plaintext and to collect plaintext/ciphertext pairs required for obtaining key differences.

Table 1. Experimental Environment

PC	NEC MateNX MA60J
CPU	Intel Pentium III(Katmai) 600MHz
L1 data cache	16-KB 4-Way Set Associative Cache 32-byte cache line
L2 cache (size / speed)	512KB / Half (300MHz)
Bus clock	100MHz
OS	Microsoft Windows2000 SP3
Compiler	Microsoft Visual C++ 6.0 SP5
Compile option	Maximize Speed (/O2)

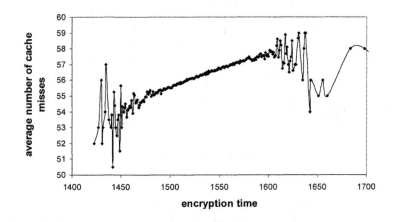

Fig. 4. Relationship Between Number of Cache Misses and Encryption Time

Relationship between Encryption Time and Cache Misses. We investigated whether the collected plaintexts actually operate as expected. Fig. 4 shows number of cache misses versus encryption time for the randomly generated plaintexts. We used a single arbitrary key for our experiment to measure the frequency of cache miss. Fig. 4 also shows that the number of cache misses increases as encryption time becomes long.

Fig. 5 shows the relationship between the number of plaintexts and the number of cache misses, for randomly generated plaintexts and plaintexts having a long encryption time. These results confirm that plaintexts having a long encryption time include significantly more plaintexts causing many cache misses than randomly generated plaintexts.

Making an Elimination Table Attack (Obtaining Key Differences). This part describes the method of obtaining key differences; Stage 2 of the attack. It is guessed that input values to the S-boxes of round 1 differ respectively from those to the corresponding S-boxes of round 16. Thus, Eq.(3) must hold.

$$K1 \oplus K16 \neq E(R0) \oplus E(R15) \qquad (3)$$

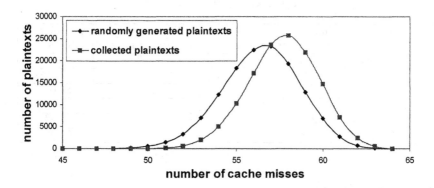

Fig. 5. Relationship Between Number of Plaintexts and Number of Cache Misses

Based on the concept presented in Section 2.4, the value appearing least frequently among those obtained by $E(R0) \oplus E(R15)$ is highly likely to be the correct key difference, when providing enough plaintexts collected in Stage 1. Thus, it can be determined that the value appearing least frequently as $E(R0) \oplus E(R15)$ is the correct key difference. This computation is performed for each pair of S-boxes S1 through S8 in rounds 1 and 16 to obtain eight key differences.

However, the 2 bits from the LSB; Least Significant Bit side of each key difference are indeterminate. This is because a cache miss does not occur if the input values of the 2 S-boxes under comparison differ to each other by the value within the range of the cache load size [3]. This means that the difference by the value less than the cache load size is ignored. Thus, the adjacent values of the value to be counted least frequently as a key difference are not counted, if the non-elimination table attack is applied. This is true to the adjacent values of the value to be counted most frequently, when elimination table attack is made. Thus, a key difference can be obtained, but the bits from the LSB side of it are still indeterminate; the 3 bits from that are be theoretically indeterminate. In our experiment, however, the 2 bits from the LSB side were found to be indeterminate because of absence of S-box addresses on a 32-byte boundary .

Considering above, we guess that the 4 bits from the MSB; Most Significant Bit side of each obtained key difference are correct, when recovering the secret key.

Recovering the Secret Key. The secret key is recovered from the 8 key differences obtained in Stage 2 in the following way.

Step 1. Prepare one plaintext/ciphertext pair by encrypting any plaintext with the actual secret key.

---

[3] The Pentium III Processor has a 32-byte cache load size i.e. 8 entries will be loaded simultaneously if it is declared to assign 4 bytes to each entry of S-box.

Table 2. Experimental Results

Number of Plaintext/Ciphertext Pairs used as 2^{n-m}	Number to be substituted for 2^{-m}	Probability of Success
2^{16}	2^{-6}	68.7%
	2^{-7}	74.7%
	2^{-8}	85.0%
2^{17}	2^{-6}	90.7%
	2^{-7}	92.3%
	2^{-8}	97.0%

Step 2. Determine 1 bit of the expanded key for round 16 by using the obtained key difference between round 1 and round 16 and then guessing any 1 bit of the expanded key for round 1. In this way, make a 32-bit (4 bit × 8 key differences) exhaustive search on the expanded key for round 1 with respect to the previously obtained key differences; this allows determining the expanded key for the corresponding round 16. 1 or more bits can be also determined by guessing 1 bit, based on the relationship between the two expanded keys for round 1 and round 16, which is described Section 3.1. Consequently, secret key is guessed by 24-bit exhaustive search on the expanded key for round 1. See the appendix for a detailed description on the secret-key recovery method.

Step 3. Encrypt the plaintext prepared in Step 1, using the secret key guessed in Step 2. If the resulting ciphertext agrees with the one obtained in Step 1, the secret key is correct. If they do not agree, return to Step 2. Note that if the secret key cannot be recovered by a 24-bit exhaustive search, the key differences guessed in Section 3.2 are mistaken.

3.3   Results of Experiment

Table 2 lists the results of DES elimination table attack described in Section 3.2. For the attack, we use 2^{n-m} out of 2^n randomly generated plaintexts which are collected in order of decreasing duration of encrypting. In reality, three numbers of 2^{-6}, 2^{-7} and 2^{-8} were taken as 2^{-m} to compare the probability of success of the attack, while two numbers of 2^{16} and 2^{17} were used as 2^{n-m}. The experiment was performed using 300 secret keys for two parameters; the number of plaintexts and the number to be substituted for 2^{-m}.

The results shown in Table 2 tell us that the secret key is recovered with a probability $> 90\%$, when collecting 2^{17} plaintext/ciphertext pairs and that setting a stricter condition for collecting plaintexts enables to collect the plaintext/ciphertext pairs having more cache misses.

3.4   Discussion

The above sections described a technique for breaking DES and the results of making attacks. Those results, however, are dependent on the experimental environment specified in Table 1. Since a cache attack is a type of side-channel attack, there is a high possibility that the results will vary significantly according to the environment of computer. It is also thought that the result and its efficiency will vary according to source code. The following discusses these factors.

A cache attack infers the frequency of cache miss from side-channel information and uses it to obtain key differences. As a consequence, S-box size can have a great effect on the attack. In the DES source code used in our experiment, it is declared to assign 4 bytes to each entry of an S-box (referred to below as *int* type). For the S-box declared as *int* type, eight entries are loaded into the cache per 1 cache load. Thus, considering that an S-box of DES has 64 entries in all, all S-box data will be loaded into the cache if eight cache misses occur. In contrast, for an S-box, for which it is declared to assign 1 byte per entry (referred to below as *char* type), 32 entries are loaded into the cache per 1 cache load; this means that only two cache misses are needed to occur, to load all S-box data. It therefore seems impossible that the duration of encryption determines the frequency of cache miss and that useful plaintexts are selected and collected. For confirmation, we applied an experimental attack on DES with the source code described in [10], after changing only the S-box declaration type from *int* to *char*, to find that the attack failed entirely. However, when a 32-bit processor such as Pentium III is used, the *int* type data is processed faster than *char* type data. Thus, the data which can be declared as *char* type will often be declared as *int* type, when implementing ciphers. The kind of implementation for faster processing can lead to the vulnerability to cache attacks.

3.5   Attack on Triple-DES

In this section, we consider whether the above cache attack on DES can be made on Triple-DES. Triple-DES performs a DES process three times in the form of

- Encryption - Decryption - Encryption, or
- Encryption - Encryption - Encryption.

In addition, there are three ways of using keys, as follows.

(a) K1 - K2 - K3
(b) K1 - K2 - K1
(c) K1 - K1 - K1

Repeating DES three times in this manner makes greater resistance to cryptanalysis techniques like differential and linear cryptanalysis that employ the correlation of round functions. At the same time, secret key variations (a) and (b) feature a longer key length than DES, making it all the more difficult to perform an exhaustive key search.

In any of the above Triple-DES variations, an S-box operates 48 times; three times as many as DES. Still, if operation delay due to cache misses can be measured, it should be possible to make a cache attack against Triple-DES in the same way as DES.

Table 3. The type of S-box of cipher and the technique of applying the cache attack. S_{size} represents the number of S-box entries while S_{num} stands for the number of the times that S-box look-up is performed. S_{miss} represents the maximum possible number of the times that cache miss is caused by S-box look-up. *Technique* shows the combination of the type of the plaintexts used for cryptanalysis and the type of the technique of applying attack

	S_{size}	S_{num}	S_{miss}	*Technique*
DES	64	16	8	plaintexts with long encryption time
MISTY1(S9)	512	48	64	and elimination table attack
Camellia	256	36	32	
AES	256	160	8	plaintexts with long encryption time
				and non-elimination table attack

It is guessed that, similarly to DES, the key difference between round 1 and 48 can be determined. For cryptanalysis on an actual computer, we can expect 2 bits from LSB side of the key difference to be indeterminate and that the actually computed key difference consists of 4-bits$\times 8 = 32$ bits. Thus, for secret key variation (a) having a key length of 168 bits, the 32 bits of K3 can first be determined by guessing the 32 bits of K1. Then, if an exhaustive key search is performed on the remaining 104 bits($= 168 - 64$), it should be possible to break the cipher in 2^{136} calculations. This concept also holds for the other key variations, that is, it should be possible to break Triple-DES in a more efficient way than applying an exhaustive key search.

4   Other Ciphers

We made experimental cache attacks on AES and Camellia. Based on the results of the attacks, this section discusses the relationship between the number of the times that S-box look-up is performed and the cache attack.

4.1   Results of the Experiment

Table 3 shows the type of S-box and the technique of applying the cache attack for each cipher. Information on DES and MISTY1 is also given in the table for comparison. The following outlines the technique of applying cache attack on Camellia and AES.

Camellia. The source code is first modified by techniques for speeding-up the cipher which are recommended by the designer and described in the specification [3]. For each of the four S-boxes declared by the speeding-up techniques, the frequency of occurrence of cache miss is directly proportional to the encryption time, as is observed for DES. The usage of this property and 2^{18} plaintexts with long encryption time provides obtaining 168-bit key differences concerning with

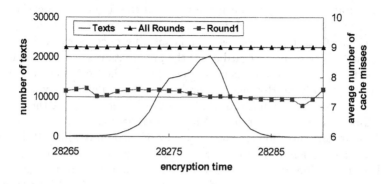

Fig. 6. Correlation Between the Encryption Time and the Average Number of Cache Misses (on AES). This graph also represents the encryption time distribution of plaintexts

a 256-bit equivalent key composed by the subkeys on round 1 through round 4, and the subkeys that are activated by the initial processing. Using the obtained key differences, approximately 2^{24} computations provides the recovery of the secret key.

AES. We employed the available source code in [7]. No correlation is found between the frequency of occurrence of cache miss and the encryption time. (See the plot labelled "All rounds" in Fig.6) However, studies on the 16 S-boxes used at the beginning of the algortithm have shown the correlation that lower frequency of cache misses implies longer encryption time. (See the plot labelled "Round 1" in Fig.6) This property provides 96-bit key differences through collecting 2^{18} plaintexts with long encryption time and regarding the value counted most often as a correct key difference. A 32-bit brute-force search using these key differences allows recovering the secret key.

4.2   Discussion

According to the paper [8] written by Ohkuma et al., the cache attack is theoretically feasible even if the number of the times that S-box look-up is performed is fairly large. The following equation represents the probability that the value of the frequency of cache miss is n, where N and M stand for the number of S-box input and the number of the times that S-box look-up is performed, respectively.

$$N^{-M} \binom{N}{n} \sum_{j=1}^{n} \binom{n}{j} (-1)^{n-j} j^M$$

This equation also indicates that it is theoretically feasible to break a cipher, if the cipher has the possibility that the value of the frequency of cache miss varies, depending on the collected plaintexts.

The accurate difference between the values of the frequency of cache miss is hard to obtain, when the attack is applied using a practical computer. When the cipher (e.g. AES) does not cause significant difference between the values of the frequency of cache miss, regardless of the plaintexts used for the attack, it is hard to perform the cryptanalysis using the values of the frequency of cache miss and the encryption time. However, we broke such kind of cipher, by utilizing the correlation between the encryption time and the probability that the values for some of S-box inputs are identical. When the ciphers cause fewer S-box look-ups and significant variations in the frequency of cache miss, like DES , we can expect that the frequency of cache miss and the encryption time correlate to each other. The correlation between the encryption time and the probability that the values for some of S-box inputs are identical, which is used for the cryptanalysis of AES, however, varies significantly, depending on the type of CPU and the method of implementing source code. For example, the durations of encrypting two sets of plaintexts with the same values of total frequency of cache miss on Intel Pentium III processor are sometimes different, depending on whether or not the cache misses occur continuously at the beginning of the encryption. In addition, which core is used for Intel Pentium III processor, Coppermine or Katmai decides which S-box to use for cryptanalysis and which attack to apply, non-elimination table attack or elimination table attack. In this case, the cipher can be broken, if we take possession of the source cord of the target cipher in advance and find the values of S-box inputs whose probability of being identical correlates to the encryption time.

5   Conclusion

We have shown that the Data Encryption Standard (DES) can be broken with 2^{23} known plaintexts and 2^{24} calculations at a success rate $> 90\%$, using a personal computer with 600-MHz Pentium III. We have also shown that a cache attack can be made against a cipher using S-boxes of different input/output widths or S-boxes of several types. Furthermore, in applying this cache attack to Triple-DES, it was found that there is a high possibility of it being broken more efficiently than an exhaustive key search.

This paper reports applying cache attack using a personal computer. In 2002, cache based cryptanalysis [9] was proposed where cache hits and/or cache misses are observed by the use of electric power or magnetic force. Since the next generation of 32-bit smartcards will use cache memories, the combination of the cache attack we proposed and Power Analysis attacks could probably be a more effective cryptanalysis technique.

We also consider countermeasures against cache attacks; a cache attack infers the number of times of occurred cache misses by observing the encryption time. Thus, if a total-data load is executed before processing, differences between the frequencies of cache misses will not be observed, making it impossible to determine the relationships between sets of S-boxes. If it is possible to clear a cache during the encryption, generating noise that has no relation with encryption at random time intervals is an effective countermeasure against cache attacks.

The cache attacks are newer technique in comparison with the timing attacks on RSA. The encryption efficiencies can be enhanced by the studies to be conducted in future.

Acknowledgement. The authors would like to thank Hiroyasu Kubo, Takeshi Kawabata, Etsuko Tsujihara and Yuya Yoshioka for their useful comments and suggestions. We are also grateful to the anonymous reviewers of CHES 2003 for their helpful comments.

References

1. E. Biham, A. Shamir, "Differential Fault Analysis of Secret Key Cryptosystems," CRYPTO'97, LNCS1294, pp. 513–525, 1997.
2. J.F. Dhem, F. Koeune, P.A. Leroux, P. Mestre, J.J. Quisquater, J.L. Willems, "A Practical Implementation of the Timing Attack", UCL Report, 1998, CG1998-1, available at http://www.dice.ucl.ac.be/crypto/techreports.html
3. Information-Technology Promotion Agency, Japan and Telecomminications AdvancementOrganization of Japan, "CRYPTREC Report 2001," 2002.
4. J. Kelsey, B. Schneier, D. Wagner, C. Hall, "Side Channel Cryptanalysis of Product Ciphers," Journal of Computer Security, vol.8, pp. 141–158, 2000.
5. P. Kocher, J. Jaffe, B. Jun, "Differential Power Analysis," CRYPTO'99, LNCS1666, pp. 388–397, Springer-Verlag, 1999.
6. F. Koeune, J.J. Quisquater, "A Timing Attack against Rijndael," UCL Report, CG1999-1, 1999, available at http://www.dice.ucl.ac.be/crypto/techreports.html
7. National Institute of Standards and Technology, "ANSI C Reference Code V2.0 (October 24, 2000)," available at http://csrc.nist.gov/CryptoToolkit/aes/rijndael/
8. K. Ohkuma, S.Kawamura, H.Shimizu, H.Muratani, "Key Inference in a Side-Channel Attack Based on Cache Miss," The 2003 Symposium on Cyptography and Information Security, 2003.(In Japanese)
9. D. Page, "Theoretical Use of Cache memory as a Cryptanalytic side-Channel," Technical Report CSTR-02-003, Department of Computer Science, University of Bristol, June 2002 available at http://www.cs.bris.ac.uk/
10. B. Schneier, "APPLIED CRYPTOGRAPHY," John Wiley & Sons, Inc. , 1996.
11. Y. Tsunoo, E. Tsujihara, K. Minematsu, H. Miyauchi, "Cryptanalysis of Block Ciphers Implemented on Computers with Cache," ISITA 2002, 2002.
12. "Data Encryption Standard (DES)," Federal Information Processing Standards Publication 46-3, 1999, available at http://csrc.nist.gov/publications/fips/

Appendix: Secret-Key Recovering Method

The following describes the process for recovering a secret key. As is described in the body of this paper, the following precondition must be satisfied to recover a secret key.

- The 4 bits from the MSB side of each key difference between S1 through S8 S-boxes of round 1 and round 16 can be obtained.

In the following, nth bit from the MSB side of the variable X is defined $X[n]$. We take the expanded key K1 of round 1 as an example:

$$K1 = K1[1]\|K1[2]\| \cdots \|K1[48]$$

Next, based on key-schedule structure, the relationship between computed key differences and secret-key information C and D can be represented in the following way.

$$
\begin{aligned}
K1 \oplus K16 &= PC2(C_1) \oplus PC2(C_{16}) \\
&= PC2(C_1) \oplus PC2(LS(C_1))
\end{aligned}
\tag{4}
$$

Using Eq. (4), C_1 and D_1 information can be computed in a step-by-step manner. The following 16 equations are Eq. (4)s expressed on a bit basis.

$$K1[7] \oplus K16[7] = C_1[3] \oplus C_1[4]$$
$$K1[16] \oplus K16[16] = C_1[4] \oplus C_1[5]$$
$$K1[10] \oplus K16[10] = C_1[6] \oplus C_1[7]$$
$$K1[20] \oplus K16[20] = C_1[7] \oplus C_1[8]$$
$$K1[3] \oplus K16[3] = C_1[11] \oplus C_1[12]$$
$$K1[15] \oplus K16[15] = C_1[12] \oplus C_1[13]$$
$$K1[1] \oplus K16[1] = C_1[14] \oplus C_1[15]$$
$$K1[9] \oplus K16[9] = C_1[15] \oplus C_1[16]$$
$$K1[19] \oplus K16[19] = C_1[16] \oplus C_1[17]$$
$$K1[2] \oplus K16[2] = C_1[17] \oplus C_1[18]$$
$$K1[14] \oplus K16[14] = C_1[19] \oplus C_1[20]$$
$$K1[22] \oplus K16[22] = C_1[20] \oplus C_1[21]$$
$$K1[13] \oplus K16[13] = C_1[23] \oplus C_1[24]$$
$$K1[4] \oplus K16[4] = C_1[24] \oplus C_1[25]$$
$$K1[21] \oplus K16[21] = C_1[27] \oplus C_1[28]$$
$$K1[8] \oplus K16[8] = C_1[28] \oplus C_1[1]$$

Using the 16 equations above to guess 7 bits of $C_1[3]$, $C_1[6]$, $C_1[11]$, $C_1[14]$, $C_1[19]$, $C_1[23]$, and $C_1[27]$ allows obtaining 16 bit of $C_1[4]$, $C_1[5]$, $C_1[7]$, $C_1[8]$, $C_1[12]$, $C_1[13]$, $C_1[15]$, $C_1[16]$, $C_1[17]$, $C_1[18]$, $C_1[20]$, $C_1[21]$, $C_1[24]$, $C_1[25]$, $C_1[28]$, and $C_1[1]$. 28 bits, i.e. all bits of C_1 are obtained by guessing 7 bits in the way described above and then guessing the remaining 5 bits of C_1; $C_1[2]$, $C_1[9]$, $C_1[10]$, $C_1[22]$, $C_1[26]$.

D_1 is treated similarly. 12-bit exhaustive search on $D_1[2]$, $D_1[5]$, $D_1[6]$, $D_1[7]$, $D_1[8]$, $D_1[11]$, $D_1[16]$ $D_1[20]$ $D_1[21]$, $D_1[26]$, $D_1[27]$, and $D_1[28]$ allows determining the 28-bit of D1 .

Overall, the above uniquely recovers 56 bits of the secret key by guessing 24 bits.

A Differential Fault Attack Technique against SPN Structures, with Application to the AES and KHAZAD

Gilles Piret and Jean-Jacques Quisquater

UCL Crypto Group,
Laboratoire de Microélectronique, Université Catholique de Louvain,
Place du Levant, 3, B-1348 Louvain-la-Neuve, Belgium
{piret, jjq}@dice.ucl.ac.be

Abstract. In this paper we describe a differential fault attack technique working against Substitution-Permutation Networks, and requiring very few faulty ciphertexts. The fault model used is realistic, as we consider random faults affecting bytes (faults affecting one only bit are much harder to induce). We implemented our attack on a PC for both the AES and KHAZAD. We are able to break the AES-128 with only 2 faulty ciphertexts, assuming the fault occurs between the antepenultimate and the penultimate MixColumn; this is better than the previous fault attacks against AES[6,10,11]. Under similar hypothesis, KHAZAD is breakable with 3 faulty ciphertexts.

Keywords: AES, Block Ciphers, Fault Attacks, Side-channel Attacks

1 Introduction

The idea of using hardware faults happening during the execution of a cryptographic algorithm for breaking it (typically, for retrieving the key) was first suggested in 1997 by D. Boneh, R.A. DeMillo, and R.J. Lipton [7,8]. They succeeded in breaking an RSA CRT with both a correct and a faulty signature of the same message. Shortly after, an adaptation of this idea on block ciphers was proposed by E. Biham and A. Shamir[5].

Application of this principle to tamper resistant devices such as smart cards is a real threat (see e.g. [1,2]): by changing the power supply voltage or the frequency of the external clock, or by applying radiations, a fault can be induced with some probability during the computation. The faults induced by most of these techniques affect one byte[1], as it is the size of a register for current smart cards; however it is often the case that the attacker cannot locate a priori at which stage of the algorithm the fault occurred.

Several authors mounted differential fault attacks against the AES[6,10,11]. In this paper we present a fault attack working against any block cipher with

---

[1] Although progresses were recently made in inducing faults affecting only one bit[13].

C.D. Walter et al. (Eds.): CHES 2003, LNCS 2779, pp. 77–88, 2003.

a Substitution-Permutation structure[2]. More precisely, its round function must have the form $\sigma[K^r] \circ \theta_r \circ \gamma_r$ (r is the round number), where:

- The γ_r layer consists in the parallel application of n 8×8 S-boxes (not necessarily identical).
- $\sigma[k]$ denotes the key addition layer:

$$\sigma[k](a) = b \Leftrightarrow b_j = a_j \oplus k_j, 1 \leq j \leq n$$

  where $\oplus$ denotes a group operation. As it is often exclusive or, in the following we will only deal with this case. But our attack could also work against other group operations.
- The diffusion layer θ_r is a mapping that is linear with respect to $\oplus$.
- K^r denotes the rth round key.

We denote the block size by $N_b = 8n$. Note that the fact that the S-boxes are 8×8 is absolutely not mandatory for our attack; we restricted to this parameter as it is common to choose such a size, well fitted with implementation considerations. 4×4 and 2×2 S-boxes can be viewed as 8×8 S-boxes as well, by considering groups of 2 (resp. 4) of them.

The last round of the cipher has the special form $\sigma[K^R] \circ \gamma_R$, as a θ layer at this stage would have no cryptographic significance. Moreover, the first round is preceded by a key addition layer. Thus the whole cipher can be described as:

$$\sigma[K^R] \circ \gamma_R \circ \left(\overset{R-1}{\underset{r=1}{\bigcirc}} \sigma[K^r] \circ \theta_r \circ \gamma_r \right) \circ \sigma[K^0]$$

Remark that the γ_r and θ_r layers need not be identical for all rounds. Only the last two rounds are important for our attack. They are depicted in Fig. 1.

2   The Attack

The Fault Model. We are dealing with *random faults*, in the sense that the faulty value is random and is assumed to be uniformly distributed. They are assumed to occur on one byte. Moreover we assume that the fault occurred somewhere between the before-last layer θ_{R-2} and the last layer θ_{R-1} (i.e. somewhere inside the frame of Fig. 1). Under this condition, the exact stage of the computation at which the fault occurred has no importance, and cannot be guessed by observing the ciphertexts either.

In the remaining of this paper, by $(C; C^*)$ we always denote a right ciphertext C and its corresponding faulty ciphertext C^*. Also, unless otherwise stated, indices will refer to byte positions (for example, C_1 denotes the left-most byte of C).

---

[2] Strictly speaking, the designation *substitution-permutation network* implies that the diffusion layer is a bit permutation. However, it becomes more and more used to refer also to ciphers with a more complex diffusion layer. So do we.

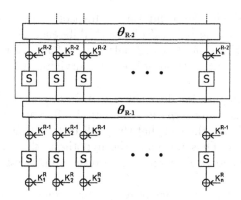

Fig. 1. Last 2,5 rounds of a Substitution-Permutation Network.

Basic Attack. Consider thus 1-byte differences at the input of the linear layer θ_{R-1}. We count $255n$ possible such differences (n different possible locations, and 255 different possible values). Because of the linearity, the number of corresponding possible differences at its output is also $255n$; but while the input difference affected one byte only, the output difference affects several ones, because of the diffusion (if the diffusion layer is optimal, all bytes are affected; with some slower diffusions only a few of them may be affected). Note that the key addition $\sigma[K^{R-1}]$ does not change the set of possible differences.

These considerations lead to a first sketch of attack. For simplicity we assume the θ_{R-1} layer achieves optimal diffusion.

1. Compute the $255n$ possible differences at the output of θ_{R-1}, i.e. the $255n$ values $\theta_{R-1}(x)$, where x has a byte hamming weight of 1. Store them in a list $\mathcal{D}$.
2. Consider a plaintext P, C its corresponding ciphertext, and C^* the faulty ciphertext.
3. Take a guess on round key K^R.
4. Compute the difference $\gamma_R^{-1} \circ \sigma[K^R](C) \oplus \gamma_R^{-1} \circ \sigma[K^R](C^*)$. Check whether it is in $\mathcal{D}$. If yes, add the round key to the list $\mathcal{L}$ of possible candidates.
5. Consider a new plaintext P (with corresponding C and C^*) and go back to step 2 (this time round key guesses only go through the list $\mathcal{L}$ of possible candidates; if the difference computed at step 4 is not in $\mathcal{D}$, remove the candidate from $\mathcal{L}$). Repeat until there remains only one candidate in $\mathcal{L}$.

If the diffusion layer is slow, only a limited number of bytes of the cipher are affected by a given fault. Thus each pair $(C; C^*)$ gives information only on a subset of the round key bytes; the guess is made only on theses bytes. The AES is a good example of this fact.

After the last round key has been found, and if it is not sufficient to retrieve the whole key, the last round is peeled off, and the attack is repeated on the reduced cipher.

Complexity Analysis. We compute the fraction of the round keys K^R that are suggested by a single pair $(C; C^*)$ with difference $\Delta = C \oplus C^*$. Suppose the number of possible differences Δ' before γ_R is 255^n. Among these, $255n$ are elements of $\mathcal{D}$. Thus the fraction of the keys surviving the test is $255n/255^n = n \cdot 255^{1-n}$.

However this computation does not take into account the fact that the number of differences Δ' before γ_R that can cause difference Δ after it is far less than 255^n; this is due to the fact that the XOR distribution table[3] of each S-box contains a lot of 0's. Thus we made the hypothesis that for the observed Δ, the fraction of elements of $\mathcal{D}$ among the corresponding possible Δ' is also about $255n/255^n$.

We conclude that the number of remaining wrong candidates for K^R after N $(C; C^*)$ pairs have been treated is about $256^n(n \cdot 255^{1-n})^N$. The conclusion (for all practical values of n) is that one pair $(C; C^*)$ is not enough to retrieve K^R, but two are (still under the hypothesis that the diffusion layer is optimal; see the AES case in section 3 for an example where it is not).

A Practical Attack. As it is presented, this attack is not really practical, as it implies a guess on the last round key, that is to say a complexity $\sim 2^{N_b}$. We show that slightly modifying the attack considerably reduces this complexity. Once again, for simplicity we assume the diffusion layer considered is optimal. A similar technique, applied only to the bytes affected by the fault, can be used when it is not.

1. Compute the $255n$ possible differences at the output of θ_{R-1}. Store them in a list $\mathcal{D}$.
2. Consider 2 right ciphertext/faulty ciphertext pairs $(C; C^*)$ and $(D; D^*)$.
3. Consider the two left-most bytes of K^R:
   - For each of the 2^{16} candidates, compute[4]:

$$\gamma_R^{-1} \circ \sigma[\langle K_1^R, K_2^R \rangle](\langle C_1, C_2 \rangle) \oplus \gamma_R^{-1} \circ \sigma[\langle K_1^R, K_2^R \rangle](\langle C_1^*, C_2^* \rangle)$$

   and

$$\gamma_R^{-1} \circ \sigma[\langle K_1^R, K_2^R \rangle](\langle D_1, D_2 \rangle) \oplus \gamma_R^{-1} \circ \sigma[\langle K_1^R, K_2^R \rangle](\langle D_1^*, D_2^* \rangle)$$

   - Compare the results with the two left-most bytes of the $255n$ differences in list $\mathcal{D}$. Make a list $\mathcal{L}$ of the $\langle K_1^R, K_2^R \rangle$ for which a match is found for both ciphertext pairs.
4. For each $K^\bullet \in \mathcal{L}$, try to extend it by one byte:
   - Remove $K^\bullet$ from $\mathcal{L}$.

---

[3] See [4] for definition of this concept.

[4] We commit a small abuse in notations by applying σ and γ_R to data of improper length. The right way to understand this is to think that e.g. $\langle C_1^*, C_2^* \rangle$ has been right-appended with 0's, and that only the 2 left-most bytes of the output are considered.

– For all 2^8 K_3^R, compute:

$$\gamma_R^{-1} \circ \sigma[\langle K_2^\bullet, K_3^R \rangle](\langle C_2, C_3 \rangle) \oplus \gamma_R^{-1} \circ \sigma[\langle K_2^\bullet, K_3^R \rangle](\langle C_2^*, C_3^* \rangle)$$

and

$$\gamma_R^{-1} \circ \sigma[\langle K_2^\bullet, K_3^R \rangle](\langle D_2, D_3 \rangle) \oplus \gamma_R^{-1} \circ \sigma[\langle K_2^\bullet, K_3^R \rangle](\langle D_2^*, D_3^* \rangle)$$

– Compare the differences obtained with bytes 2 and 3 of the $255n$ differences in $\mathcal{D}$. If a match is found (again for both ciphertext pairs), add the newly extended key $\langle K^\bullet, K_3^R \rangle$ to $\mathcal{L}$.

5. Repeat step 4 until elements of $\mathcal{L}$ have a length of n bytes.
6. Apply now the first algorithm we gave using the same pairs $(C; C^*)$ and $(D; D^*)$, but consider only the candidates K^R in $\mathcal{L}$ (their number is much smaller than 2^{N_b}).

The idea of this algorithm is that its first 5 steps compute a set of candidates of which the candidates selected by the first algorithm are a subset; otherwise stated, every candidate obtained by applying the first algorithm to pairs $(C; C^*)$ and $(D; D^*)$ will be returned by steps 1→5 of the second algorithm too, but the converse is not true. Thus, the first 5 steps of the second algorithm (that have a low complexity) perform a "first sorting" of the candidates. After that, the size of the set of candidates is quite small, and is affordable for the first algorithm.

Faults Occurring at a Wrong Location. As the attacker usually has no control on the fault location, it is important to be able to distinguish pairs $(C; C^*)$ resulting from faults occurring between θ_{R-2} and θ_{R-1} (we call such pairs *right pairs*) from other pairs (these are called *bad pairs*). It is trivial in the case of diffusion layers for which a 1-byte difference in the input implies an output difference affecting only some bytes of the output: in this case it is enough to observe whether some bytes are identical in both C and C^*.

But in the case of optimal diffusion layers, it is not possible to decide whether one only pair $(C; C^*)$ is a right or a bad one. However applying our attack to 2 pairs $(C; C^*)$ one of which is bad will very probably result in no solution for the key K^R. Thus we can indeed distinguish bad pairs $(C; C^*)$ from right ones, but only by considering *pairs* of ciphertext pairs $(C; C^*)$. Nevertheless the attack should be practical: if we consider that 1 ciphertext pair out of 100 is right, which is more than reasonable, we have 10000 pairs to examine before finding two right pairs, which is still accessible.

3 Application to the AES

3.1 Overview of the AES Structure

The AES[9] is an example of a substitution-permutation structure, as it is defined in the introduction. Both its key and block size can be 128, 192, or 256 bits. In

Table 1. The State during AES encryption

$a_{0,0}$	$a_{0,1}$	$a_{0,2}$	$a_{0,3}$
$a_{1,0}$	$a_{1,1}$	$a_{1,2}$	$a_{1,3}$
$a_{2,0}$	$a_{2,1}$	$a_{2,2}$	$a_{2,3}$
$a_{3,0}$	$a_{3,1}$	$a_{3,2}$	$a_{3,3}$

this paper we will only deal with the 128-bit block, 128-bit key variant, as it is the most widely used. Our attack can be extended trivially to other variants.

The key addition is performed using exclusive or. The γ layer (identical for all rounds) is made up of the application of 16 identical 8×8 S-boxes. The intermediate computation result, called *state* is usually represented by a 4×4 square, each cell of which is a byte (see Table 1); the θ layer (identical for all rounds) is the composition of two transformations of the state:

1. First, the ShiftRow transformation consists in shifting cyclically the rows of the state. Row 0 is not shifted, row 1 is shifted by 1 byte, row 2 is shifted by 2 bytes, and row 3 by 3 bytes. It is pictured in Table 2.
2. Then, the MixColumn transformation applies a linear transformation with optimal byte branch number(i.e. 5) to each column of the state. More precisely, application of MixColumn to the first column of the state (for example) is computed by:

$$\begin{bmatrix} b_{0,0} \\ b_{1,0} \\ b_{2,0} \\ b_{3,0} \end{bmatrix} = \begin{bmatrix} 02\ 03\ 01\ 01 \\ 01\ 02\ 03\ 01 \\ 01\ 01\ 02\ 03 \\ 03\ 01\ 01\ 02 \end{bmatrix} \cdot \begin{bmatrix} a_{0,0} \\ a_{1,0} \\ a_{2,0} \\ a_{3,0} \end{bmatrix}$$

where multiplication is performed in $GF(2^8)$ (via definition of an irreducible polynomial of degree 8 over $GF(2)$, see [9] for details).

Table 2. The ShiftRow transformation

$a_{0,0}$	$a_{0,1}$	$a_{0,2}$	$a_{0,3}$
$a_{1,0}$	$a_{1,1}$	$a_{1,2}$	$a_{1,3}$
$a_{2,0}$	$a_{2,1}$	$a_{2,2}$	$a_{2,3}$
$a_{3,0}$	$a_{3,1}$	$a_{3,2}$	$a_{3,3}$

$\xrightarrow{\text{ShiftRow}}$

$a_{0,0}$	$a_{0,1}$	$a_{0,2}$	$a_{0,3}$
$a_{1,1}$	$a_{1,2}$	$a_{1,3}$	$a_{1,0}$
$a_{2,2}$	$a_{2,3}$	$a_{2,0}$	$a_{2,1}$
$a_{3,3}$	$a_{3,0}$	$a_{3,1}$	$a_{3,2}$

We observe that a 1-byte difference in the state before the θ layer results in a 4-byte difference after it. This property is important for our attack.

Note also that the last round has no MixColumn, but well a ShiftRow. The reason behind it is purely implementation related. This last ShiftRow has no cryptographic significance, and is not relevant to our attack either.

3.2   Previous Works about Fault Analysis on the AES

Several papers have been written on the subject. We summarize here their contributions, by chronological order:

The first paper we know of is the one of Blömer and Seifert[6]. Mainly, two attacks are presented:

- The first one uses a very restrictive fault model: namely, it assumes that one can *force to 0* the value of a *chosen* bit. It is worth noting that applying this technique to the memory cells storing the key makes it trivial to retrieve, even without being able to choose precisely the location of the bit set to 0. This has been demonstrated by Biham and Shamir[5], and is true for any algorithm. [6] shows however that the fault model can be slightly relaxed. 128 faulty encryptions of plaintext **0** are required to retrieve the key using this technique.
- The second attack is implementation-dependent, and has several variants depending on the implementation. Its principle is to turn the timing attack on AES suggested by Koeune and Quisquater[12] into a fault based cryptanalysis. The fault model used also depends on the implementation. The authors claim that about 16 faulty ciphertexts (with the fault occurring at a carefully chosen location) are needed to retrieve one key byte.

In [11], Giraud presents two fault attacks on the AES. Both require the ability to obtain several faulty ciphertexts originating from the *same* plaintext (contrary to our attack):

- The first one assume it is possible to induce a fault on only one bit of an intermediate result. More precisely, it exploits faults induced on one bit before the last γ layer (while we exploit faults occurring before the last diffusion layer). Under these conditions, about 50 faulty ciphertexts are necessary to retrieve the full key (provided the location of the fault can be chosen).
- The second attack, more realistic, exploits faults on bytes. It requires the ability of inducing faults at several chosen places, including the key schedule. The author claims that 250 faulty ciphertexts are needed (it is assumed that the attacker can choose the stage of the computation where the fault takes place, but not the exact byte), and that the time needed for computation is about 5 days.

Finally, P. Dusart, G. Letourneux, and O. Vivolo[10] take advantage of byte faults occurring after the ShiftRow layer of the 9th round. Thus the fault model and the hypothesis on the fault location are exactly the same as in our attack. However the way they exploit faults is different from ours: they use the particular form of the Mixcolumn transformation and of the AES S-box to write and solve a system of equations (one by S-box) of which the unknown value is the one of the fault (i.e. of the byte difference engendered by the fault). Suggestions for 4 key bytes follow. The authors show that 5 well-located faults are necessary to retrieve 4 key bytes.

3.3 Our Results

As observed above, a 1-byte difference at the input of the θ layer of AES results in a 4-byte difference at its output. Concretely, it means that a fault on one byte before the θ_{R-1} layer will give information on only 4 bytes of the last round key (the other bytes of both ciphertexts C and C^* being identical). More precisely, with the different bytes of the state numbered as in Table 1:

- A fault on byte $a_{0,0}$, $a_{1,1}$, $a_{2,2}$, or $a_{3,3}$ will release information about round key bytes $K_{0,0}^R$, $K_{1,3}^R$, $K_{2,2}^R$, $K_{3,1}^R$.
- A fault on byte $a_{0,1}$, $a_{1,2}$, $a_{2,3}$, or $a_{3,0}$ will release information about round key bytes $K_{0,1}^R$, $K_{1,0}^R$, $K_{2,3}^R$, $K_{3,2}^R$.
- A fault on byte $a_{0,2}$, $a_{1,3}$, $a_{2,0}$, or $a_{3,1}$ will release information about round key bytes $K_{0,2}^R$, $K_{1,1}^R$, $K_{2,0}^R$, $K_{3,3}^R$.
- A fault on byte $a_{0,3}$, $a_{1,0}$, $a_{2,1}$, or $a_{3,2}$ will release information about round key bytes $K_{0,3}^R$, $K_{1,2}^R$, $K_{2,1}^R$, $K_{3,0}^R$.

Consider a fault occurring on one of the bytes $a_{0,0}$, $a_{1,1}$, $a_{2,2}$, or $a_{3,3}$. We compute that with one pair $(C; C^*)$ about 1036 candidates for $(K_{0,0}^R, K_{1,3}^R, K_{2,2}^R, K_{3,1}^R)$ remain (see complexity analysis in section 2). If two pairs are exploited, we are in principle left with the right candidate only. Thus with 8 faults at carefully chosen locations, we are able to recover the whole key.

However it is possible to do better. Suppose a fault occurs on one byte somewhere between θ_{R-3} and θ_{R-2} (rather than between θ_{R-2} and θ_{R-1}). The corresponding difference after the θ_{R-2} layer has 4 non-zero bytes. Each of them can be exploited as described previously, and releases information about a different part of the last round key. For example, a fault on $a_{0,0}$ before θ_{R-2} will result in a non-zero difference on $a_{0,0}$, $a_{1,0}$, $a_{2,0}$, and $a_{3,0}$ after it. Thus using faults occurring somewhere between θ_{R-3} and θ_{R-2} allows us to kill 4 birds with one stone. As a consequence, only 2 such faults are needed to retrieve the whole AES-128 key.

We implemented our attack on a PC. The results obtained well matched our estimates.

When one fault between θ_{R-2} and θ_{R-1} was considered, the average number of candidates for 4 bytes of K^R obtained was 1046 (instead of the expected 1036). A more surprising point (a priori) was that 2 pairs $(C; C^*)$, both giving information on the same 4 bytes of K^R, allowed to retrieve a unique value for these bytes in only 98% of the cases; otherwise two possible values for these 4 bytes remained (or even four, but it was very rare). These deviations from the expected results are due to the fact that we were making very few hypothesis on the θ layer and the S-boxes in our complexity analysis. Thus our estimations did not take into account particular features of these components. We give a more detailed explanation for the 98% figure in appendix A.

Using 2 faults between θ_{R-3} and θ_{R-2}, the number of candidates left for the whole key never exceeded 16, and we obtained one only candidate in 77% of the cases. The time needed to complete the attack is a few seconds. Also, when applying the attack to 2 ciphertext pairs one of which is bad (i.e. corresponds to a fault occurring before θ_{R-3}), the set of candidates returned by our algorithm was always empty.

4   Application to KHAZAD

4.1   Brief Description of KHAZAD

KHAZAD is a 64-bit block 128-bit key block cipher submitted to the NESSIE European project by P.S.L.M. Barreto and V. Rijmen[3]. It has 8 rounds, whose structure is the one described in section 1, with exclusive or used for key addition. Its γ layer (identical for all rounds) is made up of the application of 8 identical involutive 8×8 S-boxes. Its θ layer (also identical for all rounds) has optimal byte branch number (i.e. 9) and is also involutive.

4.2   Our Attack Applied to KHAZAD

Two faults occurring between θ_{R-1} and θ_{R-2} are enough to retrieve K^R (as each fault gives information on all bytes of K^R; remember that θ is optimal). However knowledge of K^R is not enough to retrieve the whole key. Thus once K^R is known the last round is peeled off. Then a fault occurring between θ_{R-2} and θ_{R-3} is exploited to select about $256^8 \cdot (8 \cdot 255^{-7}) \simeq 2105$ candidates for K^{R-1}. We conclude the attack by searching exhaustively among these candidates; knowledge of K^R and K^{R-1} allows to compute the main key.

Our implementation of the attack showed that using 2 right pairs $(C; C^*)$ we obtain one unique candidate for K^R in about 90% of the cases (otherwise 2 candidates remain, sometimes 4). One reason for this bad score happens to be related to the choice of the S-boxes: it seems that the worse an S-box is with respect to differential cryptanalysis, the better it resists our fault attack. As an illustration, we applied our attack to a modified version of KHAZAD using the AES S-boxes; then a unique candidate is obtained from 2 right pairs $(C; C^*)$ with probability 96%. Appendix A sketches an explanation for this.

Note that once again, the number of faulty ciphertexts needed to retrieve the key is not affected by these figures; only the time complexity of the attack (which remains small anyway) is. Also, when trying to recover K^R with 2 ciphertext pairs one of which is bad, the set of candidates returned by our algorithm was always empty.

5   Conclusion

The basic idea of our attack is to use the diffusion property of the last θ layer, in order to determine whether the difference before the last nonlinear layer γ

possibly originates in a fault or not. This provides us with a distinguishing criteria for the last round key. The fault model used is the most liberal and realistic one: we simply need random faults occurring on bytes. The ability to choose the location of the fault is not important either: of course only faults occurring at a given location (between θ_{R-2} and θ_{R-1} in the general case, between θ_{R-3} and θ_{R-1} in the case of AES) are exploitable, but those occurring elsewhere can be discarded.

We give in Table 3 a summary of existing faults attacks against the AES.

Table 3. Comparison of existing fault attacks against the AES

Ref.	Fault Model	Fault Location	# Faulty Encryptions
[6]	Force 1 bit to 0	Chosen	128
[6]	Fct of impl.	Chosen	256
[11]	Switch 1 bit	Any bit of chosen bytes	~ 50
[11]	Disturb 1 byte	Anywhere among 4 bytes	~ 250
[10]	Disturb 1 byte	Anywhere between θ_{R-2} and θ_{R-1}	~ 40
This paper	**Disturb 1 byte**	**Anywhere between θ_{R-3} and θ_{R-2}**	**2**

Amongst these attacks, the most similar to ours is the one of P. Dusart & al.[10]. The difference mainly lies in the way faults are exploited. [10] exploits the particular structure of the AES S-box and MixColumn, while we do not. The consequence is that their attack is not adaptable to other algorithms; ours can be used to attack KHAZAD(as we showed in section 4), but also ciphers like Serpent or Anubis[5]. On the other hand, note that an algorithm such as Safer++ is not directly vulnerable to our attack, due to the use of two different group operations for key mixing.

In our attack against AES, note that while 2 well-located faults are needed for easy retrieving of the key, one only well-located fault reduces the size of the key space to be explored to $1046^4 \simeq 2^{40}$.

As a final remark, it is amusing to note that it is the very simple and elegant structure of SPN structures that makes our attack so efficient... It is not clear whether ciphers with a more intricate structure could be broken with so few ciphertext pairs.

References

1. R. Anderson and M. Kuhn. Tamper resistance – a cautionary note. In *Proc. of the second USENIX workshop on electronic commerce*, pages 1–11, Oakland, California, Nov. 18-21 1996.
2. R. Anderson and M. Kuhn. Low cost attacks on tamper resistant devices. In *Proc. of 1997 Security Protocols Workshop*, volume 1361 of *Lecture Notes in Computer Science*, pages 125–136. Springer, 1997.

---

[5] But this last must be rewritten in order to comply with our description of an SPN structure.

3. P.S.L.M. Barreto and V. Rijmen. The Khazad Legacy-Level Block Cipher. Available at http://www.cryptonessie.org.
4. E. Biham and A. Shamir. Differential cryptanalysis of DES-like cryptosystems. *Journal of Cryptology*, 4(1):3–72, 1991.
5. E. Biham and A. Shamir. Differential Fault Analysis of Secret Key Cryptosystems. In B. Kaliski, editor, *Advances in Cryptology - CRYPTO '97*, volume 1294 of *Lecture Notes in Computer Science*, pages 513–525. Springer, 1997.
6. J. Blömer and J.-P. Seifert. Fault based cryptanalysis of the Advanced Encryption Standard. To appear in *Financial Cryptography '03*, LNCS. Springer, 2003. Also available at http://eprint.iacr.org/, 2002/075.
7. D. Boneh, R. A. DeMillo, and R. J. Lipton. On the Importance of Checking Cryptographic Protocols for Faults (Extended Abstract). In W. Fumy, editor, *Advances in Cryptology - EUROCRYPT '97*, volume 1233 of *Lecture Notes in Computer Science*, pages 37–51. Springer, 1997.
8. D. Boneh, R. A. DeMillo, and R. J. Lipton. On the Importance of Eliminating Errors in Cryptographic Computations. In *Journal of Cryptology 14(2)*, pages 101–120, 2001.
9. J. Daemen and V. Rijmen. AES proposal: Rijndael. In *Proc. first AES conference*, August 1998. Available on-line from the official AES page: http://csrc.nist.gov/CryptoToolkit/aes/rijndael/Rijndael.pdf.
10. P. Dusart, G. Letourneux, and O. Vivolo. Differential Fault Analysis on A.E.S. Available at http://eprint.iacr.org/, 2003/010.
11. C. Giraud. DFA on AES. Available at http://eprint.iacr.org/, 2003/008.
12. F. Koeune and J.-J. Quisquater. A timing attack against Rijndael. Technical report, available at http://www.dice.ucl.ac.be/crypto/techreports.html, 1999.
13. S. Skorobogatov and R. Anderson. Optical fault induction attacks. In Burton S. Kaliski, Çetin K. Koç, and Christof Paar, editors, *Cryptographic Hardware and Embedded Systems - CHES 2002*, volume 2523 of *Lecture Notes in Computer Science*. Springer, 2002.

A   Deeper Analysis of the AES Case

In this appendix we analyze why 2 right pairs $(C; C^*)$ and $(D; D^*)$, both releasing information on the same 4 bytes of K^R, do not allow to compute an unique value for these 4 bytes in about 2% of the cases.

Let $K_\bullet^R := \langle K_{0,0}^R, K_{1,3}^R, K_{2,2}^R, K_{3,1}^R \rangle$ denote 4 bytes of the last round key of an AES. Let $C_\bullet := \langle C_{0,0}, C_{1,3}, C_{2,2}, C_{3,1} \rangle$ and $C_\bullet^* := \langle C_{0,0}^*, C_{1,3}^*, C_{2,2}^*, C_{3,1}^* \rangle$ be a right ciphertext and its faulty counterpart, both limited to the same 4 bytes. It is easy to see that applying our attack to pair $(C; C^*)$ will return, together with $K_\bullet^R$ and other candidates, $K_\bullet^R \oplus C_\bullet \oplus C_\bullet^*$ and the 14 other candidates obtained when only some bytes of $C_\bullet \oplus C_\bullet^*$ are XORed to $K_\bullet^R$.
Consider a second pair $(D; D^*)$ with $D_\bullet := \langle D_{0,0}, D_{1,3}, D_{2,2}, D_{3,1} \rangle$ and $D_\bullet^* := \langle D_{0,0}^*, D_{1,3}^*, D_{2,2}^*, D_{3,1}^* \rangle$; $D_\bullet^*$ is the faulty counterpart of $D_\bullet$. Assume $D_\bullet \oplus D_\bullet^*$ share some bytes with $C_\bullet \oplus C_\bullet^*$; suppose for example $C_{0,0} \oplus C_{0,0}^* = D_{0,0} \oplus D_{0,0}^*$. Then, $K^R \oplus \langle C_{0,0} \oplus C_{0,0}^*, 0, 0, 0 \rangle$ will be returned by our attack (applied to both $(C; C^*)$ and $(D; D^*)$) as well as K^R.

As the probability of having the same value at a given position of $C \oplus C^*$ and $D \oplus D^*$ is $1/255$, the probability that we observe the same value at at least one position is $1 - (254/255)^4 \simeq 0,015$. So we have found the main reason why more than one key is returned in 2% of the cases. Note that this phenomenon is not specific to AES; furthermore this explanation could be generalized by referring to the XOR distribution table[4] of the S-boxes. It appears then that paradoxically good S-boxes with respect to differential cryptanalysis are also those making our fault attack the most efficient...

A New Algorithm for Switching from Arithmetic to Boolean Masking

Jean-Sébastien Coron and Alexei Tchulkine

Gemplus Card International
34 rue Guynemer, 92447 Issy-les-Moulineaux, France
{jean-sebastien.coron, alexei.tchulkine}@gemplus.com

Abstract. To protect a cryptographic algorithm against Differential Power Analysis, a general method consists in masking all intermediate data with a random value. When a cryptographic algorithm combines boolean operations with arithmetic operations, it is then necessary to perform conversions between boolean masking and arithmetic masking. A very efficient method was proposed by Louis Goubin in [6] to convert from boolean masking to arithmetic masking. However, the method in [6] for converting from arithmetic to boolean masking is less efficient. In some implementations, this conversion can be a bottleneck. In this paper, we propose an improved algorithm to convert from arithmetic masking to boolean masking. Our method can be applied to encryption schemes such as IDEA and RC6, and hashing algorithms such as SHA-1.

1 Introduction

The concept of Differential Power Analysis was introduced by Paul Kocher and al. in 1998 [7,8]. It consists in extracting information about the secret key of a cryptographic algorithm, by studying the power consumption of the electronic device during the execution of the algorithm. The attack was first described on the DES encryption scheme, then extended to other symmetrical cryptosystems such as the AES candidates [2], and also to public-key cryptosystems [5,11].

Subsequently, some countermeasures have been developed. In [3], Chari and al. proposed an approach which consists in splitting all the intermediate variables into a given number of shares, so that the power leakage of an individual share does not reveal any information to the attacker. They show that the number of power curves needed to mount an attack grows exponentially with the number of shares. A similar approach was also proposed by Goubin and al. in [5]. The drawback of this approach is that it greatly increases the computation time and the memory needed. This is a crucial issue for constrained environments such as smart-cards.

Actually, when only two shares are used, this approach consists in masking all intermediary data with a random. This technique was evaluated by Messerges in [10] for the five remaining AES candidates. For algorithms that combine boolean and arithmetic operations, two different kinds of masking must be used:

C.D. Walter et al. (Eds.): CHES 2003, LNCS 2779, pp. 89–97, 2003.

boolean masking and arithmetic masking. This is typically the case for encryption schemes such as IDEA [9] and RC6 [12], and hashing algorithms such as SHA-1 [13]. It is therefore necessary to perform conversions between boolean masking and arithmetic masking. The conversion itself must also be resistant against Differential Power Analysis. Messerges proposed in [10] an algorithm for converting between boolean masking to arithmetic masking and conversely. However, it was shown in [4] that both conversions were vulnerable to a more sophisticated Differential Power Analysis.

A new conversion algorithm was proposed by Goubin in [6]. In both directions, the conversion algorithm is such that all intermediary variable is randomly distributed; therefore, the conversion is provably resistant to first order DPA, in which no attempt is made to correlate the power consumption at different execution times. Moreover, the conversion from boolean masking to arithmetic masking is very efficient. However, the conversion from arithmetic masking to boolean masking is less efficient, as it requires a number of operations linear in the bit-size of the data to be masked. This conversion can be a bottleneck in some implementations. In this paper, we propose a secure and efficient method to convert from arithmetic masking to boolean masking.

2   Definitions

2.1   Boolean Masking and Arithmetic Masking

In this section we recall some basic definitions. We assume that the size of all intermediate variables is k bits. A typical value for k is 32 bits, as for SHA-1 and MD-5. The masking technique introduced in [3] consists in splitting each intermediate data that appears in the cryptographic algorithm. Then, an attacker must analyze multiple point distributions, which requires a number of power curves exponential in the number of shares. As in [10], we apply this technique with two shares. For algorithms that combine boolean and arithmetic functions, two different kinds of masking have to be used :

Definition 1. *We say that a data x has a boolean masking when x is written as $x = x' \oplus r$ where r is uniformly distributed.*

For example, assume that given x_1, x_2, we must compute $x_3 = x_1 \oplus x_2$ in a secure way. Then from the masked values x_1' and x_2', such that $x_1 = x_1' \oplus r_1$ and $x_2 = x_2' \oplus r_2$, we compute the two shares $x_3' = x_1' \oplus x_2'$ and $r_3 = r_1 \oplus r_2$, so that $x_3 = x_3' \oplus r_3$. Each intermediary variable is then uniformly distributed, and the procedure is resistant against first order DPA.

Definition 2. *We say that a data x has an arithmetic masking when x is written as $x = A + r \mod 2^k$ where k is the size of the register and r is uniformly distributed.*

For example, assume that given x_1, x_2, we must compute $x_3 = x_1 + x_2$ in a secure way. Then from the masked values x_1' and x_2', such that $x_1 = x_1' + r_1$ and

$x_2 = x'_2 + r_2$, we compute the two shares $x'_3 = x'_1 + x'_2$ and $r_3 = r_1 + r_2$, so that $x_3 = x'_3 + r_3$.

For algorithms that combine boolean operations and arithmetic operations, it is therefore necessary to provide a secure conversion algorithms in both directions.

2.2   From Boolean Masking to Arithmetic Masking

A very efficient method for converting from boolean masking to arithmetic masking is given in [6]. It requires a number of elementary operations which is independent from the k, the data bit-size. The method is based on the fact that for all x' the function

$$f_{x'}(r) = (x' \oplus r) - r$$

is affine in r, which means that for all x', r_1, r_2,

$$f_{x'}(r_1 \oplus r_2) = f_{x'}(r_1) \oplus f_{x'}(r_2) \oplus x'$$

Therefore, given x', r such that $x = x' \oplus r$, we generate a random k-bit integer r_1, and we can compute $A = x - r \mod 2^k$ as:

$$A = f_{x'}(r) = f_{x'}((r_1 \oplus r) \oplus r_1) = f_{x'}(r_1 \oplus r) \oplus (f_{x'}(r_1) \oplus x')$$

Since r_1 and $r_1 \oplus r$ have the uniform distribution, the conversion is resistant against DPA. We refer to [6] for the proof that f is affine and for a detailed description of the algorithm.

2.3   From Arithmetic to Boolean Masking

Louis Goubin proposed in [6] a method for converting from arithmetic to boolean masking, but the method is less efficient than from boolean to arithmetic. In particular, it requires a number of operations linear in the size of the registers; namely for a k-bit register, the number of k-bit operations is $5k + 5$.

3   Our Conversion Algorithm

We propose a new conversion algorithm from arithmetic to boolean masking which is generally more efficient than Goubin's method. Our method is based on pre-computed tables. First, we describe our method for small register size k (typically, $k = 4$).

3.1   Conversion with Small Register Size

The algorithm uses a pre-computed table G of 2^k variables of k bits.

Algorithm 1: table G generation.
Output: a table G and a random r.

1. Generate a random k-bit r.
2. For $A = 0$ to $2^k - 1$ do
   $G[A] \leftarrow (A + r) \oplus r$
3. Output G and r.

Using this table, it is easy to convert from arithmetic to boolean masking:

Algorithm 2: conversion from arithmetic to boolean masking.
Input: (A, r), such that $x = A + r$.
Output: (x', r), such that $x = x' \oplus r$.

1. Return $x' = G[A]$.

It is clear that the algorithm is resistant to first-order DPA, as all intermediary variables have the uniform distribution. In the following table, we compare our algorithm with Goubin's algorithm. The pre-computation time and conversion time is measured in number of k-bit operations.

	Algorithm 2	Goubin's method
Pre-computation time	2^{k+1}	0
Conversion time	1	$5k + 5$
Table size	2^k	0

The pre-computation time and memory required is the main limitation for algorithm 2, which is only feasible for conversion with small sizes, such as for example $k = 4$ or $k = 8$ bits. However, the table has to be computed only once for each new execution of the cryptographic algorithm; any subsequent conversion will require only one operation, instead of $5k+5$ for Goubin's method. Therefore, algorithm 2 will be more efficient when the number n of conversion during the execution of a cryptographic algorithm is greater than:

$$n > \frac{2^{k+1}}{5k + 4}$$

In this case, our method will be faster with a factor:

$$\frac{n \cdot (5k + 5)}{2^{k+1} + n}$$

For example, with $k = 8$ bits variable size, and $n = 24$ conversions, algorithm 2 is roughly two times faster than Goubin's method.

3.2   Conversion for $\ell \cdot k$-Bit Variables Using two k-Bit Tables

In this section, we show how to extend the previous algorithm in order to perform conversions for larger sizes. We consider variables of size $\ell \cdot k$ bits, and we use 2 tables with 2^k variables each. For example, for 32 bit conversions, we can take $\ell = 8$ and $k = 4$.

The idea of the algorithm is the following. We receive as input two $\ell \cdot k$-bit variables A and R, such that $x = A + R \mod 2^{\ell \cdot k}$. Our goal is to obtain x' such that $x = x' \oplus R$, in such a way that every intermediary variable has the uniform distribution. Let split R into $R_1 \| R_2$, with R_1 of size $(\ell - 1) \cdot k$ bits, and R_2 of size k bits. Then, given a random k-bit integer r, we let

$$A \leftarrow (A - r) + R_2 \mod 2^{\ell k}$$

Splitting A into $A_1 \| A_2$, where A_1 is of size $(\ell - 1) \cdot k$ bits, we now have:

$$x = (A_1 \| A_2) + (R_1 \| r) \mod 2^{\ell k}$$

Then, if $A_2 + r \geq 2^k$, we let $A_1 \leftarrow A_1 + 1 \mod 2^{(\ell-1)k}$. This is equivalent to computing the carry from the addition $A_2 + r$ and then adding this carry to A_1. Then, splitting x into $x_1 \| x_2$, where x_1 is of size $(\ell - 1) \cdot k$ bits, we have:

$$x_1 = A_1 + R_1 \mod 2^{(\ell-1)k} \quad \text{and} \quad x_2 = A_2 + r \mod 2^k$$

Then we can use the table G generated by algorithm 2 to convert x_2 from arithmetic masking to boolean masking. More precisely, we let $x_2' \leftarrow G[A_2]$, which gives:

$$x_2 = x_2' \oplus r$$

Then we let $x_2' \leftarrow (x_2' \oplus R_2) \oplus r$ so that:

$$x_2 = x_2' \oplus R_2$$

Then we apply the same method recursively to (A_1, R_1) in order to obtain x_1' such that $x_1 = x_1' \oplus R_1$, so that letting $x' = x_1' \| x_2'$, we have:

$$x = x' \oplus R$$

as required.

Actually, we can not compute the carry from $A_2 + r$ directly, because this would leak some information about x. Instead, we use a randomized carry table C, computed in the following way:

Algorithm 3: carry table C generation.
Input: a random r of k bits.
Output: a table C and a random γ of k bits.

1. Generate a random k-bit γ.
2. For $A = 0$ to $2^k - 1$ do
$$C[A] \leftarrow \begin{cases} \gamma, \text{if } A + r < 2^k \\ \gamma + 1 \mod 2^k, \text{if } A + r \geq 2^k \end{cases}$$
3. Output C and γ.

Then, instead of testing if $A_2 + r \geq 2^k$, we let:

$$A_1 \leftarrow A_1 + C[A_2] - \gamma \mod 2^{(\ell-1)k}$$

This gives the following conversion algorithm, based on the pre-computed tables G and C of algorithms 1 and 3:

Algorithm 4: Conversion with $\ell \cdot k$ bit variable:
Input: (A, R), such that $x = A + R$, and r, γ generated from algorithms 1 and 3.
Output: x', such that $x = x' \oplus R$.

1. $A \leftarrow A - r \mod 2^{\ell k}$.
2. Let denote $R = R_1 \| R_2$, where R_1 is of size $(\ell - 1)k$ bits.
3. Let $A \leftarrow A + R_2 \mod 2^{\ell k}$
4. If $\ell = 1$, then let $x' \leftarrow G[A] \oplus R_2$, then $x' \leftarrow x' \oplus r$ and return x'.
5. Otherwise, let $A = A_1 \| A_2$
6. Let $A_1 \leftarrow A_1 + C[A_2] \mod 2^{(\ell-1)k}$
7. Let $A_1 \leftarrow A_1 - \gamma \mod 2^{(\ell-1)k}$
8. Let $x'_2 \leftarrow G[A_2] \oplus R_2$.
9. Let $x'_2 \leftarrow x'_2 \oplus r$.
10. Apply algorithm 4 recursively with (A_1, R_1) to obtain x'_1.
11. Return $x' = x'_1 \| x'_2$

As previously, this conversion method is resistant to first-order DPA, because all intermediary variables have the uniform distribution. We want to compare the efficiency of our method with Goubin's method. The drawback of our method is that we need to pre-compute two tables of 2^k values. The advantage of our method is that some computation is done on small k-bits variables, whereas Goubin's method always works with full $\ell \cdot k$ bits variables. Therefore, we must take into account the register size of the micro-processor. Our method is likely to be more advantageous on a 8-bit microprocessor, which is now the most common smart-card platform, than on a 32-bit microprocessor.

To make a practical comparison, we take $k = 4$, and we distinguish two kinds of microprocessor: 8-bit and 32-bit, and two variable sizes: 8-bits and 32-bits. We take $k = 4$ because the method is easier to implement for for this value of k, but a better trade-off may be possible. We assume that an elementary operation on a 32 bit variables requires 4 elementary operations on a 8 bit microprocessor. For example, Goubin's method on 32-bit variables on a 8 bit microprocessor will require $4 \cdot (5 \cdot 32 + 5) = 660$ operations. More generally, we denote by $T_{i,j}$ (resp. $G_{i,j}$) our method (resp. Goubin's method) for i-bit variables with a j-bit microprocessor. The following table summarizes the number of steps in all possible cases:

	$T_{8,8}$	$T_{8,32}$	$T_{32,8}$	$T_{32,32}$	$G_{8,8}$	$G_{8,32}$	$G_{32,8}$	$G_{32,32}$
Pre-computation time	64	64	64	64	0	0	0	0
Conversion time	10	10	76	40	45	45	660	165
Table size	32	32	32	32	0	0	0	0

As previously, the efficiency improvement depends on how frequently we re-compute the randomized tables. If we compute the randomized tables only once at the beginning of the cryptographic algorithm, then our method will always be more efficient if there are at least two subsequent conversions. But if we choose to re-compute the tables before each conversion, then Goubin's method is more efficient for 8-bit variables, whereas our method is more efficient for 32-bit variables. Our method is particularly advantageous for 32-bit conversions on a 8-bit microprocessor: our method ($64 + 76$ operations) is then 4.7 times faster than Goubin's method (660 operations).

4   Application to SHA-1

4.1   Overview of SHA-1

SHA-1 is a hash function introduced by the American National Institute for Standards and Technology [13] in 1995. The description of SHA-1 consists of a general iteration procedure based on a compression function

$$F : \{0,1\}^{512} \times \{0,1\}^{160} \rightarrow \{0,1\}^{160}$$

In the following we give a very general overview of the algorithm (see [13] for details).

General Iteration Procedure:

1. Pad the message, so that its length is a multiple of the size of the compression function, that is 512 bits.
2. Initialize the five 32-bit chaining variables A, B, C, D, E with a given IV value.
3. For each message block M of 512 bits, let

$$(A, B, C, D, E) \leftarrow F(M, (A, B, C, D, E)) + (A, B, C, D, E)$$

   where F is the compression function.
4. Output the hash value $A\|B\|C\|D\|E$.

Compression Function F:

1. Expand the 512-bit message block M into 80 words M_i of 32 bits.
2. For $i = 0$ to 79 do:

$$(A, B, C, D, E) \leftarrow (M_i + \mathrm{rot}_5(A) + f_i(B, C, D) + E + K_i,$$
$$A, \mathrm{rot}_{30}(B), C, D)$$

where rot_j denotes left rotation by j bits, K_i are constants and:

$$\begin{aligned}
f_i(X, Y, Z) &= (X \& Y)|(\neg X \& Z), & 0 \leq i \leq 19 \\
f_i(X, Y, Z) &= X \oplus Y \oplus Z, & 20 \leq i \leq 39, 60 \leq i \leq 79 \\
f_i(X, Y, Z) &= (X \& Y)|(X \& Z)|(Y \& Z), & 40 \leq i \leq 59
\end{aligned}$$

We see that SHA-1 combines boolean operations with arithmetic operations.

4.2   Motivation

The SHA-1 hash function can be used for MAC algorithms, for example:

$$\text{MAC}_K(x) = \text{SHA-1}(K_1\|x\|K_2)$$

or for the HMAC [1] nested construction:

$$\text{HMAC}_K(x) = \text{SHA-1}(K_2\|\text{SHA-1}(x\|K_1))$$

where $K = K_1\|K_2$ is a secret-key. In this case, the implementation of SHA-1 has to be made resistant against DPA, otherwise a straightforward DPA attack would recover the secret-key K.

4.3   Implementation Result

In the following, we estimate the number of elementary operations which are required to have an implementation of SHA-1 resistant against DPA. Without DPA countermeasure, each of the 80 steps in the compression function requires roughly 15 elementary 32-bit operations. The DPA countermeasure requires to split each variable into 2 shares; this leads to 30 elementary operations. Moreover, assuming that A, B, C, D and E have initially a boolean masking, we need to convert $f_i(B, C, D)$, $\text{rot}_5(A)$ and E into arithmetic masking, then the sum $M_i + \text{rot}_5(A) + f_i(B, C, D) + E + K_i$ back to boolean masking. This gives 3 boolean to arithmetic conversions, each requiring 7 operations using [6], and one arithmetic to boolean conversion. Therefore, each step requires 51 elementary operations on 32-bit variables (or 204 operations on 8-bit variables)[1], together with one arithmetic to boolean conversion.

In the following table, we compare the efficiency of an implementation of SHA-1 resistant against DPA, using our arithmetic to boolean conversion method, and using Goubin's method, for 8-bit and 32-bit micro-processor. The time is measured in number of elementary operations for each of the 80 steps of the compression function. For our arithmetic to boolean conversion, we re-compute the randomized tables before each new conversion. This means that using our method, a 32-bit arithmetic to boolean conversion takes 140 elementary operations on a 8-bit microprocessor, and 104 operations on a 32-bit microprocessor.

	8-bit micro	32-bit micro
Our method	344	155
Goubin's method	864	216

---

[1] As previously, we assume that a 32-bit operation on a 8-bit micro-processor requires 4 elementary operations.

5 Conclusion

We have described a new conversion algorithm from arithmetic to boolean masking, which is generally more efficient than Goubin's algorithm. Our new algorithm is particularly interesting for 32-bit conversions on a 8-bit microprocessor. For example, for SHA-1 hash function, the previous table shows that an implementation secure against DPA will be roughly 2.7 times faster using our method than using Goubin's method.

References

1. M. Bellare, R. Canetti, and H. Krawczyk, *Keying hash functions for message authentication*, Advances in Cryptology - Crypto 96 Proceedings, Lecture Notes in Computer Science Vol. 1109, N. Koblitz ed, Springer-Verlag, 1996.
2. Suresh Chari, Charantjit S. Jutla, Josyula R. Rao and Pankaj Rohatgi, *A Cautionary Note Regarding Evaluation of AES Candidates on Smart-Cards*, in Proceedings of the Second Advanced Encryption Standard (AES) Candidate Conference, http://csrc.nist.gov/encryption/aes/round1/Conf2/aes2conf.htm, March 1999.
3. Suresh Chari, Charantjit S. Jutla, Josyula R. Rao and Pankaj Rohatgi, *Towards Sound Approaches to Counteract Power-Analysis Attacks*, in Proceedings of Advances in Cryptology – CRYPTO'99, Springer-Verlag, 1999, pp. 398–412.
4. Jean-Sebastien Coron and Louis Goubin, *On Boolean and Arithmetic Masking against Differential Power Analysis*, Proceedings of CHES 2000, LNCS 1965, pp. 231–237, Springer.
5. Louis Goubin and Jacques Patarin, *DES and Differential Power Analysis – The Duplication Method*, in Proceedings of CHES 99, Springer-Verlag, August 1999, pp. 158–172.
6. Louis Goubin, *A Sound Method for Switching between Boolean and Arithmetic Masking*, proceedings of CHES 2001, LNCS 2162, pp. 3–15, Springer.
7. Paul Kocher, Joshua Jaffe and Benjamin Jun, *Introduction to Differential Power Analysis and Related Attacks*, available at www.cryptography.com/dpa/technical, 1998.
8. Paul Kocher, Joshua Jaffe and Benjamin Jun, *Differential Power Analysis*, in Proceedings of Advances in Cryptology – CRYPTO'99, Springer-Verlag, 1999, pp. 388–397.
9. X. Lai and J. Massey, *A Proposal for a New Block Encryption Standard*, in Advances in Cryptology – EUROCRYPT '90 Proceedings, Springer-Verlag, 1991, pp. 389–404.
10. Thomas S. Messerges, *Securing the AES Finalists Against Power Analysis Attacks*, in Proceedings of FSE 2000, Springer-Verlag, April 2000.
11. Thomas S. Messerges, Ezzy A. Dabbish and Robert H. Sloan, "Power Analysis Attacks of Modular Exponentiation in Smartcards", in *Proceedings of Workshop on Cryptographic Hardware and Embedded Systems*, Springer-Verlag, August 1999, pp. 144–157.
12. R.L. Rivest, M.J.B. Robshaw, R. Sidney and Y.L. Yin, *The RC6 Block Cipher*, v1.1, August 20, 1998.
13. FIPS PUB 180-1, Secure Hash Standard, *U.S. department of commerce/National Institute of Standards and Technology*

DeKaRT: A New Paradigm for Key-Dependent Reversible Circuits

Jovan D. Golić

System on Chip, Telecom Italia Lab
Telecom Italia
Via Guglielmo Reiss Romoli 274, I-00148 Turin, Italy
jovan.golic@tilab.com

Abstract. A new general method for designing key-dependent reversible circuits is proposed and concrete examples are included. The method is suitable for data scrambling of internal links and memories on smart card chips in order to foil the probing attacks. It also presents a new paradigm for designing block ciphers suitable for small-size and/or high-speed hardware implementations. In particular, a concrete building block for such block ciphers with a masking countermeasure against power analysis incorporated on the logical gate level is provided.

Keywords. Keyed reversible circuits, data scrambling, block ciphers, countermeasures, probing attacks, power analysis.

1 Introduction

Probing attacks on microelectronic data-processing devices implementing cryptographic functions, such as smart cards, are invasive techniques consisting in introducing conductor microprobes into certain points of a tamper-resistant chip within the device to monitor and analyze the electrical signals at these points, in order to recover some information about the secret key used (see [1]). They can be classified as side-channel attacks if they do not change the functionality of the device. In this regard, potentially vulnerable points are those corresponding to internal links or memories that are likely to convey or contain secret information and whose hardware implementation has a regular, recognizable structure. In a microprocessor configuration, the RAM and the bus connecting it to the microprocessor are specially vulnerable, and the bus between the microprocessor and cryptoprocessor(s) may also be vulnerable. While, for a cryptographic function, it should be computationally infeasible to reconstruct the secret key from known input and output data, this need not be the case if intermediate data generated during the software execution is revealed. Therefore, there is a need to protect the sensitive data on data buses and in memories by using dedicated encryption techniques. Apart from data, one can also encrypt the memory addresses and the code instructions. This encryption may also reduce the vulnerability to other side-channel attacks such as the power analysis attacks.

C.D. Walter et al. (Eds.): CHES 2003, LNCS 2779, pp. 98–112, 2003.

The encryption/decryption of data solely on the data bus can be achieved by a fast stream cipher combining the data sequence with the keystream sequence, produced by a centralized random or pseudorandom number generator, possibly by the bitwise XOR operation. However, this solution is not satisfactory for encrypting the data to be stored in memories as the same keystream block has to be used for encrypting and decrypting any data block for a particular memory location. The keystream block can be made dependent on the location itself, but is immediately recovered from only one known pair of original and encrypted data blocks. A more satisfactory solution for encrypting the memory data, which can also be used for encrypting the bus data, is to apply a block cipher, in the electronic code book mode, and the encryption has to be performed by logical circuits typically in a single microprocessor clock cycle, as it is then transparent to the remaining components of the data-processing device. However, the usage of classical block ciphers such as DES [13] or AES [5] is not realistic due to very restrictive high-speed and small-size requirements and, also, because of small and variable block sizes involved.

Accordingly, there is a need for fast and simple techniques for key-controlled reversible transformations which, of course, cannot achieve the same security level as classical block ciphers and are therefore called data scrambling instead of encryption. Nevertheless, it has to be noted that the probing attacks are not easy to mount and consequently partial instead of full knowledge of ciphertext corresponding to known, possibly chosen, plaintext is a much more appropriate assumption for data scrambling. To this end, one can use simplified iterated designs for block ciphers, with a reduced block size and with a reduced number of rounds, but the usual method of bitwise XORing the expanded secret key with intermediate ciphertexts is not good enough for relatively small block sizes. Instead, one can incorporate a larger number of key bits by using key-controlled bit permutations to permute bits in a block. Several constructions for such permutations are proposed in [3] for block sizes being a power of 2. Nevertheless, a small number of rounds and the linearity of bit permutations do not allow one to achieve a sufficient security level especially if the block size is very small.

The main objective of this paper is to introduce a new and generic method for iterated construction of key-controlled reversible transformations which can incorporate a large number of key bits in a small number of rounds even for very small block sizes. The new, so-called DeKaRT method is based on using small elementary building blocks connected by fixed bit permutations, where all the blocks are simple and can be of the same type. The resulting hardware designs are granular, simple, and fast, and can be customized easily by choosing different building blocks or connections among them. Consequently, the method is very suitable for data scrambling to thwart the probing attacks, as well as for new designs of hardware-oriented block ciphers. Moreover, the hardware implementations of such block ciphers can easily be made resistant to differential and simple power analysis attacks on the logical gate level by applying and adapting the masking technique [10] to the building blocks used.

More details on data scrambling techniques together with some security requirements and considerations are given in Section 2. The DeKaRT method for designing key-dependent reversible transformations and logical circuits is proposed in Section 3 and some examples are presented in Section 4. The application of the DeKaRT method for the design of block ciphers is discussed in Section 5 and the application of the masking technique against power analysis is explained in Section 6. Conclusions are given in Section 7.

2 Data Scrambling

For an n-bit microprocessor, the typical block size to be dealt with by data scrambling is n bits, but can be smaller. For example, apart from the data to be stored in RAM, the address of the location where the data has to be stored can be scrambled too. In addition, the scrambling function for data can be made dependent on the original address, for example, by deriving the key for this scrambling function from the same secret key and the original address by a simple hash function. Of course, scrambling the address foils the known plaintext scenario for scrambling the data, as the address of the location where the scrambled data is stored is not known. Not only is the address size usually smaller than n, but also need not be a power of 2. Another example is scrambling the data and memory addresses within the microprocessor core (e.g., for the cache memory), where usually certain restrictions have to be respected when transmitting and storing the data. Accordingly, the block sizes as small as 8 bits or even smaller are likely to be encountered in data scrambling. For such block sizes, the key-controlled bit permutations cannot provide enough uncertainty of the key used.

For very small block sizes, data scrambling is inherently vulnerable to the dictionary attack, in the known or chosen plaintext scenario. This attack reconstructs the secret scrambling transformation used, and not the secret key itself. Secret key reconstruction attacks are more important because the same secret key bits can be used repeatedly for scrambling different data. However, it is important to have in mind that in the context of probing attacks, the known plaintext and known ciphertext scenarios are not realistic due to the fact that the ciphertext is likely to be known only partially.

If an iterated construction is used for data scrambling and if a number of secret key bits is incorporated in each of a small number of rounds (e.g., 2 to 5), then the effective key size is roughly halved due to the structural meet-in-the-middle attack in the known plaintext scenario. So, the number of key bits per round has to be relatively large, and this is not easy to achieve for very small block sizes. In particular, as proposed in [3], one can use a network composed of a small number of alternating substitution and key-controlled bit permutation layers, where the substitution layers consist of fixed and small (e.g., (4×4)-bit) reversible S-boxes. For 5 layers altogether, because the bit permutations are linear functions for a given key, the structural attack [2] in the chosen plaintext scenario is then applicable, as already noted in [3]. The attack is able of reconstructing both the S-boxes, if they are unknown to the attacker, and the bit permutations in all the layers, up to an equivalence of the total transformation, and works even for large block sizes.

The secret key used for data scrambling should better be innovated for each new execution of the cryptographic function on the chip considered, having in mind that changing the key for scrambling the data in RAM can be done only when the stored data is all used. As a consequence, the secret key is much less exposed to side-channel attacks such as the power analysis attacks and hence the data scrambling schemes need not in principle be protected by the masking technique [9], [10] which randomizes (and slows down) the key-dependent computations. The secret key should be completely independent of the secret key stored on the chip which is used for the cryptographic function itself. It can be generated by a random number generator implemented in hardware on the same chip or, alternatively, but less securely, by a relatively strong and simple pseudorandom number generator (stream cipher), implemented in hardware, from a secret seed and some innovation information, which does not have to be secret. It is important to emphasize that the secret key used for scrambling should be stored in a hardware-protected register, not in RAM.

3   DeKaRT Method

Our strategic objective stemming from data scrambling applications described in previous sections is to propose a generic method for constructing key-dependent reversible transformations $\{0,1\}^N \to \{0,1\}^N$ by logical (combinatorial) circuits that can incorporate a relatively large number of key bits with a relatively small number of logical gates arranged in a small number of levels, even if the block size N is very small, such as $N = 8$.

In iterated constructions of block ciphers, where the block size N is at least 64, the number of key bits per each round is typically at most N and the key bits are incorporated by using the bitwise XOR operation. Each round typically contains a layer of fixed nonlinear S-boxes, independent of the key, to provide confusion and may contain an extra layer of fixed affine transformations, also independent of the key, to provide diffusion. In a cipher like AES, the S-boxes have to be reversible, but act on all the input bits. If a Feistel structure is used, like in DES, then S-boxes do not have to be reversible, but act on a portion of input bits only. So, if N is small, then a large number of rounds is needed to incorporate a large number of key bits, and this is not acceptable for data scrambling as the depth of the corresponding logical circuit would be too large.

For data scrambling purposes, the number of key bits per round can be increased by using key-controlled bit permutations to incorporate the key, as suggested in [3]. However, for small N such as $N \leq 16$, this is not sufficient. Also, the key-controlled bit permutations are not cryptographically strong themselves. For example, regardless of the key, they preserve the Hamming weight of input data and, as a consequence, the XOR of all the output bits is always equal to the XOR of all the input bits. On the other hand, the computation of S-boxes does not involve any key and takes a considerable number of logical levels.

Accordingly, what we essentially need is an iterated construction composed of a number (not too large) of layers, where each layer implements a key-dependent reversible transformation, has a small logical depth, and is able to incorporate a number of key bits larger than the block size. A small logical depth in fact implies that each output bit of each layer depends on a small number of input bits to that layer. In the construction that we propose each layer consists of a number of small building blocks, each block implementing a key-dependent reversible transformation.

A generic bulding block is shown in Fig. 1. It acts on a small number of input data bits which are divided into two groups of m and n bits, respectively. The m input bits are used for control and are passed to the output intact, like in the Feistel structure. They are then used to select k out of $2^m k$ key bits by the multiplexer (MUX) circuit with m control bits, 2^m k-bit inputs, and the k-bit output $\mathbf{k}$. The MUX circuit in fact implements an $m \times k$ lookup table, i.e., k (binary) $m \times 1$ lookup tables that are specified by the key. Finally, the k bits are then used to select an $(n \times n)$-bit reversible transformation $R_{\mathbf{k}}$ acting on the remaining n input bits to produce the corresponding n output bits. The m and n bits are called the control and transformed data bits, respectively. The total number of the key bits in the building block is thus $k2^m$, which can easily be made larger than $m + n$. The set of used reversible transformations has to be chosen in a way that can easily be implemented by a logical circuit with $n + k$ input bits and n output bits. The inverse building block is the same except that the reversible transformations $R_{\mathbf{k}}$ are replaced by their inverses $R_{\mathbf{k}}^{-1}$.

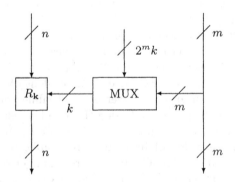

Fig. 1. The generic DeKaRT building block.

The underlying design paradigm is that a part of input data chooses a key and the key chooses a reversible transformation acting on the remaining part of input data. This justifies the name *data-chooses-key-chooses-reversible_transformation* and the notation D→K→RT for such a paradigm. For simplicity, we propose the name DeKaRT.

Consequently, in each layer, N input bits are divided into small blocks and each of them is transformed by an elementary DeKaRT block. In a uniform design all the building blocks are of the same type. The layers are connected by fixed bit permutations satisfying the following two diffusion properties. In a uniform design, the bit permutations between the layers are the same. First, the control bits in each layer should be used as the transformed bits in the next layer. Therefore, the number of control bits in each layer cannot exceed the number of the transformed bits. In a uniform design, this implies that $m \leq n$. Second, for each building block, both control bits and transformed bits should be extracted from the maximal possible number of building blocks in the preceding layer. In a uniform design, this number equals $\min(m, N/(m + n))$ for the control bits and $\min(n, N/(m+n))$ for the transformed bits. In the inverse DeKaRT network, the layers are applied in the reverse order and the inverse bit permutations are used. If $m = n$, then the used bit permutations can be made equal to their inverses.

The definition given above is quite general, but is already sufficient for proposing specific constructions. Some concrete examples for the DeKaRT building blocks are given in the next section. For cryptographic security, a number of desirable additional criteria are also proposed.

- First, regarding the choice of $m + n$, it is prudent to require that the number of building blocks per each layer is at least 2.
- Second, regarding the choice of the reversible transformations $R_\mathbf{k}$, it can be required that each output bit of $R_\mathbf{k}$ is a nonlinear function of input data bits and the key $\mathbf{k}$. Moreover, it can also be required that the algebraic normal form of this function contains at least one binary product involving both input data and key bits. In this way, the transformed and control input bits at each layer are nonlinearly combined together. This criterion implies that $n > 1$, as the only reversible functions of one binary variable are the identity and the binary complement functions, so that the single key bit has to be XORed with the input bit to obtain the output bit. Already for $n = 2$ this criterion can be satisfied, as shown in the next section. It is not satisfied if $k = n$ key bits are bitwise XORed with n input data bits, as in the usual Feistel structure.
- Third, it is desirable that the set of reversible transformations $R_\mathbf{k}$ satisfies a Shannon-type criterion that the uncertainty of n input bits provided by purely random k key bits when the output n bits are known is maximal possible, that is, n bits. For this to hold, it is necessary that $k \geq n$. This criterion can easily be satisfied by bitwise XORing a subset of n key bits with n input data bits.

Cryptanalysis of the DeKaRT networks with a small number of rounds is a problem interesting for future investigations. To this end, the method from [4] developed for Feistel networks with four rounds and randomly chosen round functions may be relevant.

4 Examples

In order to specify a concrete DeKaRT building block, one has to choose the parameters m, n, and k and to propose a logical circuit to implement the parametrized reversible transformation R_k as a function of n input data bits and k key bits. This logical circuit should be relatively simple in terms of size and depth.

A simple and sufficient method for designing such logical circuits is to use XORs with 2 input bits and (controlled) SWITCHes only, where a SWITCH has 2 input bits, 2 output bits, and 1 control bit that determines if the input bits are swapped or not. Clearly, a SWITCH can be implemented by using two MUXes in parallel, whereas only one MUX suffices for implementing each XOR. Here and throughout, unless specified differently, a MUX has 2 input bits, 1 control bit, and 1 output bit. For each XOR, one of the two input bits is a key bit, whereas for each SWITCH, the control bit is a key bit. The individual key bits are incorporated into the circuit in such a way that there are no equivalent keys, i.e., that different combinations of key bits give rise to different reversible transformations. This is not a problem for checking since the parameters n and k are small. For each fixed key, such reversible transformations are affine, and the nonlinearity is achieved by the selected key bits depending on the control input data bits. Note that for $n = 2$, all 24 reversible transformations of 2 input bits are necessarily affine. The Shannon-type criterion is not satisfied if the circuit contains the key-controlled SWITCHes only. The resulting DeKaRT building blocks thus incorporate, extend, and generalize, on an atomic level due to small block sizes, elements of known block cipher design principles such as Feistel structures [13], data-dependent bit permutations [14], [11], and key-dependent bit permutations [3].

The basic concrete example satisfying the desirable properties from Section 3 with parameters $(m, n, k) = (2, 2, 3)$ is shown in Fig. 2. Many other examples can be obtained similarly. First, by removing the 2 XORs and 1 control input we get a DeKaRT block with parameters $(1, 2, 1)$, which will be called the simplified block from Fig. 2. Second, by removing 1 control input we get a block with parameters $(1, 2, 3)$. Third, by removing the SWITCH we get a block with parameters $(2, 2, 2)$. Fourth, a very elementary block with parameters $(1, 1, 1)$ contains only 1 MUX with 2 input key bits and 1 XOR for the reversible transformation. The DeKaRT block from Fig. 2 can be implemented by using a circuit of 13 MUXes with depth 4, whereas a circuit implementing the corresponding simplified block has size 3 and depth 2, also in terms of MUXes. The two blocks incoporate the total of 12 and 2 key bits, respectively.

The blocks can readily be used for defining concrete data scrambling functions of the DeKaRT type. For example, for $N = 16$, in the uniform DeKaRT network based on the block from Fig. 2, each layer contains 4 such blocks and hence has the total of 42 MUXes, has depth 4, and incorporates 48 key bits. Accordingly, five layers like this incorporate 240 key bits and can be implemented by a circuit with 210 MUXes and depth 20. Similarly, for $N = 15$, in the uniform DeKaRT network based on the simplified block from Fig. 2, each layer contains

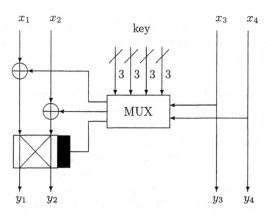

Fig. 2. An elementary DeKaRT building block.

5 such blocks and hence has the total of 15 MUXes, has depth 2, and incorporates 10 key bits. Accordingly, ten layers like this incorporate 100 key bits and can be implemented by a circuit with 150 MUXes and (the same) depth 20. It may be possible to further reduce the size and/or depth by using an optimized ASIC design, where the optimization also depends on the fixed bit permutations used between the layers. Both the networks incorporate a relatively large number of key bits and have a very small size and depth, which, for a relatively small N such as $N \leq 16$, is impossible to achieve with the networks of S-boxes and key-controlled bit permutations. Also, due to the DeKaRT paradigm, their cryptographic security is considerably improved.

In conclusion, the proposed DeKaRT method is a new and interesting tool to be used for data scrambling purposes. Depending on the size and depth constraints stemming from particular applications, one can either use the proposed concrete DeKaRT designs or easily derive new customized concrete designs by using various DeKaRT building blocks as well as various bit permutations to be used between the layers.

5   Application for Block Ciphers

The generic or concrete DeKaRT designs described in the preceding section can also be used for constructing high-speed and/or small-size block ciphers suitable for hardware implementations. For example, they may be used for the (proprietary) encryption of copyright digital data to be stored in non-volatile memories for multimedia applications. What is needed is to use sufficiently many DeKaRT building blocks per layer according to the increased block size, $N \geq 64$, and to increase the number of layers/rounds to achieve a required security level, which is higher than for data scrambling applications. Namely, it is required that it should be computationally infeasible to reconstruct the secret key faster than by exhaustive search, given any number of arbitrarily chosen plaintext-ciphertext pairs. Since the size and depth of each DeKaRT layer is considerably smaller than in a usual iterated construction of block ciphers, the number of rounds in the DeKaRT construction can be several times larger. For example, for the DeKaRT building block from Fig. 2 and $N = 128$, the number of rounds can be about 32 or larger.

For cryptographic security, one should satisfy the desirable criteria from Section 3 as well as possibly introduce two more round keys of size N to be bitwise XORed with the input and output bits. Furthermore, in view of the statistical cryptanalytic methods such as the linear cryptanalysis of block ciphers [8], instead of using only the bit permutations between the layers, it is prudent to use very simple reversible linear functions. For example, if the total numbers of transformed and control data bits per layer are equal, one can design the bit permutations as explained in Section 3 and then XOR every transformed data bit at the input to each layer with a distinct transformed data bit from the preceding layer. This usually does not increase the logical depth of the layers. For the DeKaRT building block from Fig. 2, a preliminary analysis shows that less than 32 rounds are then sufficient to achieve resistance to the linear cryptanalysis, provided that the round keys are purely random and independent.

Unlike the data scrambling functions, the encryption or decryption functions do not have to be performed in only one microprocessor clock cycle, so that they can be implemented by a combination of logical circuits and registers. For example, several layers at a time can be implemented by a logical circuit. Note that the pipelined architectures for the DeKaRT constructions are extremely fast due to the small depth of each layer.

For the DeKaRT design, the required number of key bits per round is typically larger than the block size. This is needed for data scrambling applications where the block size and the number or rounds are both relatively small. For example, the DeKaRT building block from Fig. 2 requires 3 key bits per input bit. Furthermore, as for block cipher applications the number of rounds is increased, the total number of key bits required is larger than in usual block cipher designs. These key bits can be produced from a smaller number of secret key bits, stored in RAM or in a hardware-protected register, by a key expansion algorithm.

The key expansion algorithm can produce the round keys iteratively and can itself be implemented in hardware by a combination of logical circuits and registers, so that not all the round keys have to be stored. The relations between the round keys should not facilitate the secret key reconstruction attacks in the chosen plaintext scenario and should prevent the secret key reconstruction attacks in the related key scenario. The proposed DeKaRT variant of the key expansion algorithm is as follows. Let K and K' denote the bit sizes of the secret key and the round key, respectively. The K secret key bits are first expanded by linear transformations into K' key bits by using an appropriate linear code so that any subset of K'' expanded key bits are linearly independent, where K'' is relatively large ($K'' \leq K$). In other words, the minimum distance of the dual of this linear code should be at least $K'' + 1$ (e.g., see [7]). The obtained expanded key is then used as an input to a DeKaRT network of block size K' which is parametrized by a fixed randomly generated key satisfying an additional condition that every MUX block in the network implements balanced binary lookup tables. The K' bits produced after every two layers of the DeKaRT network are successively used as round keys, together with the K' input bits. For the decryption transformation, the round keys can be produced in the reverse order starting from the round key for the last round, which can be precomputed and stored.

As the number of layers is thus doubled when compared with the DeKaRT network used for the block cipher, the DeKaRT building blocks used could be as simple as the simplified block from Fig. 2. Alternatively, the K' round key bits can be produced after each layer of the DeKaRT network, if one allows portions of successive round keys to be bit permutations of each other. In the iterated DeKaRT algorithm each round is a reversible transformation, which is important as it satisfies the criterion that each round key is purely random if the input to the first round is purely random.

The key expansion algorithm can be simplified by using only linear transformations in the following way. The K secret key bits are first expanded by linear transformations into $2K'$ key bits by using an appropriate linear code so that there are no small subsets of linearly dependent expanded key bits. The expanded $2K'$ bits are then used as the round keys for the first two rounds, whereas the subsequent pairs of successive round keys are produced by applying fixed bit permutations to the expanded key bits. Of course, other simplifications or modifications are also possible.

6   DeKaRT Construction with Masking against Power Analysis

Side-channel attacks on software or hardware implementations of various cryptosystems aim at recovering the secret key information from certain physical measurements performed on the electronic device during the computation such as the power consumption, the time, and the electromagnetic radiation. They do not change the functionality of the device and are typically not invasive. Power analysis attacks [6] are very powerful as they do not require expensive resources and as most (software) implementations without specific countermeasures incorporated are vulnerable to such attacks. Among them, the (first-order) differential power analysis (DPA) attacks are particularly interesting, because they use a relatively simple statistical technique that is almost independent of the implementation of the cryptographic algorithm. More sophisticated statistical analysis of power consumption curves may also be feasible.

The basis of power analysis attacks on cryptographic electronic devices are elementary computations within the device that depend on the secret key information and possibly on the known output and input information. If in addition the power consumption corresponding to these elementary computations depends on the input data, then it is not surprising that the power consumption curves contain information about the secret key which may be feasible to extract by statistical techniques. Software implementations, in which the operations are synchronized by the microprocessor clock, are especially vulnerable. Hardware implementations are also potentially vulnerable, but may require a higher sampling frequency for obtaining the power consumption curves. A general algorithmic strategy to counteract power analysis attacks is to randomize the computations depending on the secret key by masking the original data with random masks and by modifying the computations accordingly [9], [10]. An alternative way of dealing with power analysis attacks is making use of a special encoding of data that tends to balance the power consumption, such as the dual-rail encoding, and the corresponding self-checking logical circuits, possibly asynchronous (see [12]).

The DPA attack on a microelectronic device implementing a cryptographic algorithm can be prevented if every elementary computation involving the secret information and performed by a logical gate in the hardware implementation is randomized. More precisely, the general condition to be satisfied is that the output value of each logical gate in the protected hardware design should follow the same probability distribution for each fixed value of the secret key and input information, where the uncertainty is provided by purely random masks. In other words, it should be statistically independent of the secret key and input information. In principle, even only one logical gate violating the condition may render the hardware implementation vulnerable to DPA. It may be interesting to note that randomizing the software operations does not ensure that the underlying hardware operations are all randomized.

In principle, for a logical gate with m binary inputs only m independent masking bits are sufficient, but this number can possibly be reduced. Of course, the greater the total number of masking bits used in the whole circuit, the greater the resistance to more sophisticated power analysis attacks such as a higher-order DPA. So, the whole hardware implementation can be masked by masking individual logical gates, and masking a logical gate means finding an equivalent logical circuit that can securely compute the masked output from the masked inputs, where all the masks are binary and are XORed with the input and output bits. More precisely, for a logical gate implementing a Boolean function $f(X)$, the masked gate should implement the function $f'(X', R, r) = f(X' \oplus R) \oplus r$, where R is a binary vectorial input mask and r is a binary output mask. The computation is required to be secure if $X' = X \oplus R$ and the computed output is then $f(X) \oplus r$, as desired. The masking bits should preferably be produced by a random number generator each time the cryptographic function is executed.

A general masking technique based on using the MUX gate with 2 input bits, 1 control bit, and 1 output bit is proposed in [10]. It essentially consists in representing a Boolean function by a tree of MUXes and then in masking the MUXes. The function values as binary constants are used as inputs to the top layer of MUXes in the tree and are all masked by the same masking bit, which is also the masking bit for the output. The main observation from [10] is essentially that a MUX as an elementary logical gate can be masked by using a cascade connection of a SWITCH and a MUX, where the SWITCH is controlled by the control masking bit and the MUX is controlled by the masked control bit.

Namely, if the two input masking bits are the same, then this cascade produces the MUX output bit masked with the same mask as the two inputs. Moreover, the two outputs of the SWITCH are computed securely, that is, each of them is statistically independent of the original input bits. As a SWITCH can be implemented by a parallel connection of two MUXes, the masked MUX is thus securely implemented by using 3 MUXes and has a logical depth of 2 MUXes. Consequently, the resulting logical circuit for the masked Boolean function contains 3 times as many MUXes and has double depth in comparison with the original tree of MUXes. This is the price to pay for the protection against DPA. It is not specifically explained in [10] how to apply this technique to an arbitrary logical circuit.

An important issue which is not addressed in [10] is related to the atomicity of the MUX implementation in hardware. If the MUX output is produced in a single step by a single gate, then the masked MUX computation is secure. However, in practice the MUX is usually implemented by using the logical AND, OR, and NOT gates. Namely, if c denotes the control input and x and y the two data inputs, then the MUX implements the Boolean function $\bar{c} \wedge x \vee c \wedge y$. Now, if in the masked MUX each of the 3 MUXes is implemented in this way, it can be proven that the output of each elementary gate used is computed securely.

As mentioned in Section 2, if a new secret key for data scrambling is produced for every new execution of the cryptographic function and since the constraints regarding the speed and size are very restrictive, then the DeKaRT networks used for data scrambling need not be protected against power analysis by a masking technique. On the other hand, if the DeKaRT network is used for a block cipher, it should better be protected by masking. All what is needed is to mask the individual DeKaRT building blocks used and this can be achieved by adapting the MUX-based masking technique [10] described above. It is interesting that the masked round key bits needed can themselves be securely computed by the masked DeKaRT network used for the key expansion algorithm. Note that the key expansion algorithm is not vulnerable to DPA as it does not involve any input data, but other power analysis techniques may be applicable. A DeKaRT building block can be masked by using a representation in terms of MUXes (or SWITCHes) and by replacing each MUX by a masked MUX. The masked SWITCH is equivalently a cascade connection of two SWITCHes being controlled by the control masking bit and the masked control bit, respectively. The only condition to be respected is that the two inputs for each MUX (or SWITCH) should be masked by the same binary mask and that the control bit should be masked by an independent binary mask.

The MUX block from Fig. 1 can directly be represented in terms of MUXes. In fact, it contains k distinct trees of MUXes with a total of $k(2^m - 1)$ elementary MUXes with 2 input bits, 1 control bit, and 1 output bit. In each of the k trees there are m levels of MUXes controlled by m control data bits and 2^m key bits are used as inputs to the top level. Now, assume that m control data bits are masked by independent masking bits as well as that 2^m key bits for each of the k trees are masked by the same mask. Accordingly, if each MUX is replaced by a masked MUX, then each of produced k output bits is masked by the same bit that is used for masking the 2^m key bits at the input to the top level of the corresponding tree and all the computations are secure. The k masking bits can be independent, but other options are also possible. For example, they can all be the same, whereas all the masking bits for m control data bits can also be the same. Note that these two masking bits have to be independent in order for the computation to be secure.

In order to mask the block implementing the key-dependent reversible transformations $R_\mathbf{k}$, the corresponding logical circuit should be represented in terms of MUXes. This is already the case with any logical circuit composed of XORs and SWITCHes, as suggested in Section 4, and in particular with the circuit shown in Fig. 2. Then masking can be achieved by replacing each SWITCH by a masked SWITCH and by keeping the XORs as they are, having in mind that the output mask for an XOR is the XOR of the two input masks. The constraints to be satisfied are that the two inputs for each SWITCH should be masked by the same binary mask and that the control bit should be masked by an independent mask as well as that the two inputs for each XOR should be masked by independent masks. If needed, the mask at any point can be changed by using an extra XOR with another independent mask.

In the particular example from Fig. 2, which can be taken as an elementary building block to produce iterated block ciphers, we have $(m, n, k) = (2, 2, 3)$. The MUX block containing 3 trees each composed of 3 MUXes can be masked as explained above. The logical circuit composed of 2 XORs and 1 SWITCH can be masked by keeping the XORs and by replacing the SWITCH by a masked SWITCH, i.e., by a cascade of two SWITCHes, while the masking bits should be assigned following the general guidelines given above. Namely, the main condition to satisfy is that the two input masks for the cascade of SWITCHes should be the same. This can be achieved without introducing two extra XORs to change the masks for the two data inputs x_1 and x_2 by using the following two assignments of masking bits.

Let k_1, k_2, and k_3 denote the key bits used for the 2 XORs and the SWITCH in Fig. 2, respectively. In the first assignment, the input data bits x_i have independent masks r_i, $1 \leq i \leq 4$. If k_1 and k_2 are masked by r_2 and r_1, respectively, then the two inputs to the masked SWITCH are masked by the same mask, $r_1 \oplus r_2$, as desired. The third key bit k_3 can then be masked either by r_1 or r_2 and the output bits of the masked SWITCH are then both masked by $r_1 \oplus r_2$. Two additional XORs at the output can be used to change this mask into r_1 and r_2 for y_1 and y_2, respectively. As a result, the output four masks are the same as the input four masks, but the masking bits for the round keys should be adapted to the masking bits for the data.

In the second assignment, the input data bits x_i are all masked by the same mask r, whereas all the key bits are themselves also masked by the same mask r_0, which is independent of r. The two inputs to the masked SWITCH are then masked by the same mask, $r \oplus r_0$, and so are the two output bits y_1 and y_2. To maintain the same type of mask assignment, the mask for the inputs x_3 and x_4

can be changed into $r \oplus r_0$ by using two additional XORs. This does not increase the depth of the masked DeKaRT block which remains to be 7 MUXes. For the second assignment, it is easier to produce the masked round keys as their masks are independent of the masks used for the data. In particular, one can use only one masking bit for the whole data block and only one, independent masking bit for each round key. Then after each round, every intermediate ciphertext bit will be masked by the same masking bit and this bit changes from round to round depending on the round key masking bits. The masked round keys can be obtained analogously, by using the corresponding masked DeKaRT network.

7   Conclusions

The proposed DeKaRT method for constructing key-dependent reversible logical circuits is not only suitable for data scrambling functions, but can also be used for constructing a new general type of hardware-oriented block ciphers as well as the required key expansion algorithms. In addition, the resulting hardware designs can efficiently be protected against power analysis by a masking technique on the logical gate level.

Acknowledgment. The author is grateful to Renato Menicocci for pointing out the reference [10] as well as for fruitful discussions on the countermeasures against power analysis attacks.

References

1. R. Anderson and M. Kuhn, "Tamper resistance – a cautionary note," *Proceedings of the 2. USENIX Workshop on Electronic Commerce*, Oakland, California, pp. 1–11, Nov. 1996.
2. A. Biryukov and A. Shamir, "Structural cryptanalysis of SASAS," Advances in Cryptology - EUROCRYPT 2001, *Lecture Notes in Computer Science*, vol. 2045, pp. 394–405, 2001.
3. E. Brier, H. Handschuh, and C. Tymen, "Fast primitives for internal data scrambling in tamper resistant hardware," Cryptographic Hardware and Embedded Systems - CHES 2001, *Lecture Notes in Computer Science*, vol. 2162, pp. 16–27, 2001.
4. D. Coppersmith, "Luby-Rackoff: four rounds is not enough," Technical Report RC 20674, IBM, Dec. 1996.
5. J. Daemen and V. Rijmen, *The Design of Rijndael: AES – The Advanced Encryption Standard*. Berlin: Springer-Verlag, 2002.
6. P. Kocher, J. Jaffe, and B. Jun, "Differential power analysis," Advances in Cryptology – CRYPTO '99, *Lecture Notes in Computer Science*, vol. 1666, pp. 388–397, 1999.
7. F. J. MacWilliams and N. J. A. Sloane, *The Theory of Error-Correcting Codes*. Amsterdam: North-Holland, 1988.
8. M. Matsui, "Linear cryptanalysis method for DES cipher," Advances in Cryptology – EUROCRYPT '93, *Lecture Notes in Computer Science*, vol. 765, pp. 386–397, 1994.

9. T. Messerges, "Securing the AES finalists against power analysis attacks," Fast Software Encryption – FSE 2000, *Lecture Notes in Computer Science*, vol. 1978, pp. 150–164, 2001.

10. T. Messerges, E. Dabbish, and L. Puhl, "Method and apparatus for preventing information leakage attacks on a microelectronic assembly," US patent No. US 6,295,606 B1, Sept. 25, 2001 (filed July 26, 1999).

11. A. A. Moldovyan and N. A. Moldovyan, "A cipher based on data-dependent permutations," *Journal of Cryptology*, vol. 15(1), pp. 61–72, 2002.

12. S. Moore, R. Anderson, R. Mullins, G. Taylor, and J. Fournier, "Balanced self-checking asynchronous logic for smart card applications," *Microprocessors and Microsystems*, to appear.

13. National Bureau of Standards, "Data Encryption Standard," Federal Information Processing Standards Publication 46, Jan. 1977.

14. R. L. Rivest, "The RC5 encryption algorithm," Fast Software Encryption – FSE '94, *Lecture Notes in Computer Science*, vol. 1008, pp. 86–96, 1995.

Parity-Based Concurrent Error Detection of Substitution-Permutation Network Block Ciphers

Ramesh Karri[1], Grigori Kuznetsov[2], and Michael Goessel[2]

[1]Department of Electrical and Computer Engineering
Polytechnic University, 6 Metrotech Center
Brooklyn, NY 11201

[2]Institute of Computer Science,
Fault Tolerant Computing Group
University of Potsdam
D-14439 Potsdam, Germany
ramesh@india.poly.edu,{grigoriy,mgoessel}@cs.uni-potsdam.de

Abstract. Deliberate injection of faults into cryptographic devices is an effective cryptanalysis technique against symmetric and asymmetric encryption algorithms. In this paper we will describe parity code based concurrent error detection (CED) approach against such attacks in substitution-permutation network (SPN) symmetric block ciphers [22]. The basic idea compares a carefully modified parity of the input plain text with that of the output cipher text resulting in a simple CED circuitry. An analysis of the SPN symmetric block ciphers reveals that on one hand, permutation of the round outputs does not alter the parity from its input to its output. On the other hand, exclusive-or with the round key and the non-linear substitution function (s-box) modify the parity from their inputs to their outputs. In order to change the parity of the inputs into the parity of outputs of an SPN encryption, we exclusive-or the parity of the SPN round function output with the parity of the round key. We also add to all s-boxes an additional 1-bit binary function that implements the combined parity of the inputs and outputs to the s-box for all its (input, output) pairs. These two modifications are used only by the CED circuitry and do not impact the SPN encryption or decryption. The proposed CED approach is demonstrated on a 16-input, 16-output SPN symmetric block cipher from [1].

1   Introduction

Until recently cryptanalysts analyzed cipher systems by using rigorous mathematics based techniques such as differential cryptanalysis [2] and linear cryptanalysis [3]. Although these techniques are useful in exploring weaknesses in algorithms, they do not exploit weaknesses in their implementations. Hardware and Software implementations of (crypto) algorithms leak information via side-channels such as time consumed by the operations, power dissipated by the operators, electromagnetic radiation emitted

C.D. Walter et al. (Eds.): CHES 2003, LNCS 2779, pp. 113–124, 2003.
© Springer-Verlag Berlin Heidelberg 2003

by the device and faulty computations resulting from deliberate injection of faults into the system. Traditional cryptanalysis techniques can be combined with such side-channel attacks to uncover break the secret key and/or break the implementation details of the cipher. Even a small amount of side-channel information is sufficient to break common ciphers [4]. For example, Differential Fault Analysis (DFA) that uses deliberate injection of faults requires between 50 to 200 cipher text blocks to recover a key of symmetric block cipher Data Encryption Standard (DES); the best traditional attack requires approximately 64 terabytes of plain text and cipher text encrypted under a single key.

1.1 Fault Based Attacks: Motivation

Fault based cryptanalysis (for example, DFA) is based on the observation that faults deliberately injected into a crypto-device leak information about the implemented algorithms. These attacks are practical since elevated levels of radiation or heat, incorrect voltage, or atypical clock rate can cause a tamperproof device to malfunction. Boneh, DeMillo and Lipton [5] presented the first fault based side-channel attack against asymmetric public-key cryptography devices. More recently, Biham and Shamir [6] presented a fault-based cryptanalysis of symmetric block cipher Data Encryption Standard (DES). They presented a transient fault based Differential Fault Analysis (DFA) attack and a permanent fault based non-DFA attack to recover the round keys using a very small number of cipher texts. They then extended their fault model to show that DFA can uncover the structure of an unknown cryptosystem implemented in an EEPROM based smart card based on the observation that it is much easier to inject a $1 \rightarrow 0$ bit flip than to inject a $0 \rightarrow 1$ bit flip in an EEPROM. Using DES as the unknown cipher, they showed that (i) about 500 faulty cipher texts are sufficient to identify the bits of the right half, (ii) about 5000 faulty cipher texts are sufficient to identify the non-linear substitution operations (s-boxes) and their input and output bits, and (iii) about 10000 faulty cipher texts are sufficient to reconstruct the DES s-boxes.

Anderson and Kuhn described additional fault based side-channel attacks on software implementations of encryption algorithms [7]. In one of the attacks they assumed that the instruction memory of smart cards can be corrupted. If in a process loop, the variable controlling the number of rounds is set to 1, encryption executes just one round, thereby compromising the round key. Another attack focused on the chip writing ability of the attacker. Assuming that the attacker is familiar with the implementation, he can extract keys from the card by overwriting specific memory locations.

1.2 Fault-Based Side Channels: The Fault Models

Boneh, Demillo and Lipton [5] use a practical fault model wherein a fault is induced at a random bit location in one of the registers at some random intermediate round of a cryptographic computation. Biham and Shamir [6] use a similar realistic fault model wherein either a transient or a permanent fault is induced randomly into the device. They then adapt this basic fault model to the asymmetric property of EEPROMs: it is much easier to induce a 1→0 bit flip than to induce a 0→1 bit flip. Anderson and Kuhn used two different fault models for microcontroller based smartcards: in the first they assume that the instruction memory of smart cards can be randomly corrupted and in the second they assume that the attacker has the ability to write into specified locations in the memory. These and other fault attacks and associated fault models are summarized in [8,9]. The proposed CED approach is applicable to the practical fault models described.

1.3 CED Architectures for Symmetric Block Ciphers: Background

Concurrent error detection (CED) followed by suppression of the corresponding faulty output can thwart fault injection attacks; on detecting a faulty computation, the stored key is protected by suppressing the corresponding faulty cipher text.

Straightforward duplication and comparison of encryption and decryption hardware yields more than 100% hardware overhead. Alternatively, a spare module for each type can be used to detect faults in hardware modules of that type. Such a spares based approach has been adopted in a hardware implementation of the 128-bit symmetric block cipher IDEA [10]. Spares based approaches are suitable for block ciphers that use arithmetic operators, such as IDEA and RC6 [11]. Although hardware is not duplicated, an extra module for each operation type entails considerable hardware overhead, especially for encryption algorithms like Advanced Encryption Standard (AES) [12] and DES that use random, non-arithmetic operations such as S-Boxes.

Time redundancy based CED approach involves encrypting (decrypting) the data a second time followed by the comparison of two results. Wolter et. al. [13] developed a CED technique for symmetric block cipher IDEA wherein the test data was encrypted and then decrypted. This approach entails more than 100% time overhead. Further, it can only tolerate transient faults if the data traverses identical paths through the encryption and decryption data paths both during the normal computation and during the re-computation.

Karri et. al [14] developed a systematic CED approach for symmetric block ciphers at the register transfer level that exploits the inverse relationship between the encryption and decryption at the algorithm level, round level and individual operation level. They demonstrated this inverse-relationship principle on 128-bit symmetric block ciphers including Advanced Encryption Standard, RC6 and Serpent. The main drawback of this approach is that it assumes that the cipher device operates in a half-duplex mode (i.e. either the encryption or the decryption but not both are simultaneously active). Bertoni et. al. [15] applied this inverse-relationship principle to round key generation of the AES encryption algorithm using additional hardware for inverse round key generation and comparison.

Another CED approach involves encoding the message before encryption and checking it for errors after decryption. Wolter et. al. [13] used residue codes for fault detection in adders, multipliers, and EXCLUSIVE-ORs. Area overhead of this approach is due to the encoders at the input and decoders at the output to translate the plain and cipher texts into the internal code words. In [16] the plaintext is encoded by setting several bits of the message to a particular fixed value, 0 or 1, and then encrypted. A mismatch between these fixed bits of deciphered text and the original plain text detects an error. The simple code (all zeroes or all ones) results in significantly less area overhead when compared to other encoding schemes. This scheme has a large fault detection latency detects since faults in the encryption hardware at the transmitter end by the decryption hardware at the receiver. Further, there is an associated performance penalty since it uses some of the bits in messages for error detection.

In [17] a CED technique that predicted the inverse of the parity of the outputs was proposed for the non-linear s-box and other functions used in DES. A similar technique that predicted the parity of the outputs for the non-linear s-box and linear mixing functions used in the AES was proposed in [18,19]. In these papers one additional parity bit per byte at the outputs of the s-boxes is added. To detect errors at the inputs to the s-boxes, the inputs of the s-boxes are also parity encoded. The size of the s-boxes is doubled by proposing a 512×9-bit implementation resulting in an area overhead of over 100% for s-box CED.

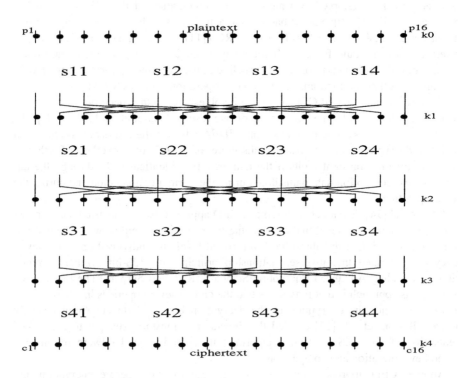

Fig. 1. Substitution Permutation Network (SPN) cryptosystem

2 Substitution-Permutation Network (SPN) Block Ciphers

The architecture of a symmetric block cipher contains a key expansion module, an encryption module and a decryption module. Key expansion module expands the user key to generate round keys and loads them into the key RAM prior to encryption or decryption. Using the round keys, the device encrypts (decrypts) the plain (cipher) text to generate the cipher (plain) text. Symmetric block ciphers have an iterative looping structure. All the rounds of encryption and decryption are identical in general, with each round using several operations and round key(s) to process the input data. Consider the well-known substitution-permutation network (SPN) cryptosystem shown in Figure 1. Such an SPN architecture consisting of a non-linear substitution layer (s-boxes) connected by output bit position permutations is an easy to understand yet realistic architecture [1]. The example SPN cryptosystem shown in Figure 1 operates on a 16-bit plaintext generating a 16-bit cipher text and four rounds. Each SPN encryption round is composed of a non-linear substitution operation (using four 4x4 s-boxes), a permutation and exclusive-or with a 16-bit round key. The sixteen 4x4 s-boxes in this example cryptosystem are different. To preserve symmetry between encryption and decryption, the first round operation is preceded by exclusive-or with the 16-bit key, key 0. Then the four 16-bit round keys (key 1, key 2, key 3, and key 4) are exclusive-ored following the permutation operation (In Figure 1 the dots on the s-box input lines represent exclusive or) in each round.

2.1 Parity-Based Concurrent Error Detection

Protection of crypto-devices entails protecting the encryption/decryption data paths as well as the key ram used to hold the round keys. Significant work has been done to protect the RAM using Parity code, Hamming code etc. In this paper we are interested in CED of the encryption data path and we do not address CED for key RAM.

The proposed CED design approach uses parity code. The specific CED implementation depends on the SPN implementation architecture. Consider the unfolded implementation architecture shown in Figure 1 (this is necessary because all 16 s-boxes are different). The parity of the inputs to the first round, $P(x)$ is determined by a parity tree of the 16 inputs. The CED structure modifies this input parity according to the successive processing steps of the SPN round function such that the modified parity is equal to the parity of the outputs of the SPN circuitry of the first round. The CED architecture shown in Figure 2 repeatedly modifies the parity in the manner discussed in each of the four rounds and compares it with the parity of the cipher text.

The operations in an SPN round are: non-linear transformations by the sixteen 4x4 substitution boxes (s-box), bit-permutation and exclusive-or with the round key. Non-linear substitution boxes used in SPN-based and other encryption algorithms have been designed to satisfy properties such as maximum non-linear order, high nonlinearity, low differential uniformity and low bias [20,21]. Satisfaction of these properties has been shown to reflect the strength of the s-box against linear and differential cryptanalysis. These s-boxes do not maintain the parity from their inputs to their outputs.

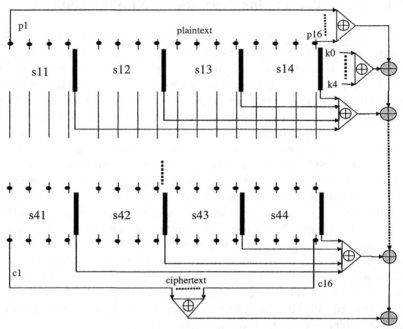

Fig. 2. CED architecture of the SPN block cipher

Table 1. 4x4 substitution box supplemented with m (i) =parity (i)⊕parity(s(i))

I	s(i)	parity(i)⊕parity(s(i))
0	E	1
1	4	0
2	D	0
3	1	1
4	2	0
5	F	0
6	B	1
7	8	0
8	3	1
9	A	0
A	6	0
B	C	1
C	5	0
D	9	1
E	0	1
F	7	1

We add to every four input, four output s-box an additional binary output for the purpose of modifying the input parity of the SPN circuitry into the output parity in the considered SPN round. Let s be an s-box with four inputs i1, i2, i3, i4 and four outputs s1(i1,i2,i3,i4), s2(i1,i2,i3,i4), s3(i1,i2,i3,i4), and s4(i1,i2,i3,i4). Then for every input i= i1,i2,i3,i4 of the s-box, the additional modifying output m(i) implements $i1 \oplus i2 \oplus i3 \oplus i4 \oplus s1 \oplus s2 \oplus s3 \oplus s4 \oplus$ =parity(i)$\oplus$parity(s(i)). Parity (i) and parity (s(i)) are the input parity and the output parity respectively of the considered s-box. The design of this modifying output m (i) for an example 4x4 s-box from [1] is shown now. The first two columns of Table 1 show the truth table of an s-box. In the first column the four-bit inputs i and in the second column the four-bit outputs s(i) of the s-boxes are given in hexadecimal representation. In the third column the binary value m(i) which imple- ments m(i) = parity(i)$\oplus$parity(s(i)) is given. Thus, for example in row 6 of Table 1 for the input i = i1,i2, i3,i4 = 0101 = 5 the functional output of the s-box is s(5) = s1(5),s2(5),s3(5), s4(5) = 1111 = F, and for the additional binary output m(5) of the s- box we have m(5) =$0 \oplus 1 \oplus 0 \oplus 1 \oplus 1 \oplus 1 \oplus 1 \oplus 1$=0.

In the complete CED architecture shown in Figure 2, a thick box appended to the right hand side of an s-box shows this modifying output. This modifying output in each s-box is used only for CED and does not impact either the encryption or the de- cryption. Since we do not change the functionality of the s-boxes, the strength of the used cryptographic algorithm, based to a large extent on the concrete form of the s- boxes, is preserved.

Next, since permutations do not change the parity no modification circuitry is nec- essary. Finally, bit-wise modulo 2 addition of the 16-bit key 0 modifies the parity of the input plain text by parity of key 0 prior to the first SPN round. Bit-wise modulo 2 addition of the 16-bit key 1 modifies the parity of the input plain text by parity of key 1 in the first SPN round. Similarly, bit-wise modulo-2 addition of the 16-bit key 2 modifies the parity of the input plain text by parity of key 2 in the second SPN round and so on. The overall modification due to all the round keys can be pre-computed during round key generation as parity of key 0 $\oplus$ parity of key 1 $\oplus$ parity of key 2 $\oplus$ parity of key 3 $\oplus$ parity of key 4. This absorbs the associated time overhead into that of round key generation.

While this architecture might apply to most of the common examples, it doesn't necessarily apply to all cases. For example, not all architectures require an explicit decryption module; some block ciphers, DES being the most noteworthy example, looks practically identical regardless of the direction as long as the round keys are reversed. Also, while many block ciphers have some internal iterative round compo- nent, often in practice, they consist of other structures (such as pre and post-whitening steps). However, this general principle can be easily adapted to these situations. The proposed CED method is also applicable to other symmetric-key primitives such as message authentication codes and stream ciphers that have an SPN structure.

3 Fault Detection Capability

In section 2 we explained how the input parity of an SPN round is modified step-by-step according to the processing steps of that round in such a way that the modified parity of the inputs is equal to the parity of the outputs of that round if no error occurs.

In this section we show how an error due to a single stuck-at-fault in a processing step of the SPN Symmetric Block Cipher is detected by the proposed CED method. This is illustrated in Figure 3 for four successive processing steps. The inputs x are processed into the outputs y in step 1 and the parity $P(x)$ of the inputs is modified into $P(x) \oplus P(x) \oplus P(y) = P(y)$. We assume now that a fault f occurs in the hardware implementing the processing step 2 with the result that the outputs of the second step are now z_f instead of the correct outputs z, with $z_f \neq z$. The parity $P(y)$ is corrected in this second step into the correct value $P(y) \oplus P(y) \oplus P(z) = P(z)$. If the error due to the fault f is detectable by parity we have $P(z_f) \neq P(z)$. In step 3 the erroneous inputs z_f (instead of the correct inputs z) are correctly processed by the fault-free hardware of this third step into the outputs u_f and now the parity $P(z)$ is modified into $P(z) \oplus P(z_f) \oplus P(u_f)$.

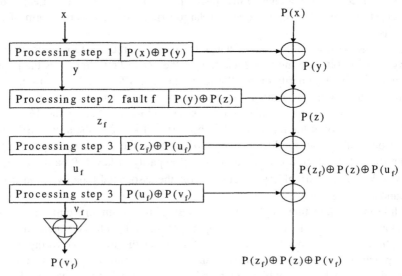

Fig. 3. Analysis of fault detection capability

Similarly in step 4 the inputs u_f are correctly processed into v_f and the parity $P(z) \oplus P(z_f) \oplus P(u_f)$ is now modified into $P(z) \oplus P(z_f) \oplus P(u_f) \oplus P(u_f) \oplus P(v_f) = P(z) \oplus P(z_f) \oplus P(v_f)$. Finally the modified parity $P(z) \oplus P(z_f) \oplus P(v_f)$ is compared with $P(v_f)$, the parity of the outputs v_f of step 4, and for $P(z) \neq P(z_f)$ the error due to the fault f will be detected. Thus, if, due to a fault f in step 2, a single bit (or an odd number of bits) is erroneous this error will be always detected by comparing the parity of the outputs of step 4 with the corresponding modified input parity.

Cryptographic algorithms are designed to satisfy the strict avalanche criterion [20,21]; even a single bit error at the inputs of an encryption step results in many different erroneous bits at the outputs of the following encryption steps. But, as explained, this property of encryption algorithms has no influence on the error detection capability of the proposed CED method.

The processing steps of the considered SPN Symmetric Block Cipher are compnent-wise exclusive-or with the round key, permutation and non-linear S-box transformation. For the operation exclusive-or with a round key every single (internal or external) stuck-at fault of an exclusive-or gate will result in a single bit error which is detected by parity. For the permutation operations a single stuck-at fault will result in a single bit error which is also detected by parity. For the 16 parity-appended S-boxes with four inputs and five outputs (four functional outputs and one parity modifying output) CED capability is described now. We designed these S-boxes using SISII logic synthesis tool from UC Berkeley. Then for all possible single stuck-at 0/1 faults the synthesized S-boxes were simulated for all possible input combinations. If all the outputs of the S-boxes are independently implemented (i.e. without sharing gates) every single stuck-at-fault results in a single bit error of the outputs of the corresponding S-box and will be obviously detected by parity. If all the five outputs of the S-boxes are jointly optimized then (in rare cases) even number of S-box output bits may be in error due to a single stuck-at fault. Then, as the experiments show, 96.3 % of the errors due to a single stuck-at fault are detected by parity. If the four functional outputs of the S-boxes are jointly optimized and if the parity-modifying bit is separately implemented 98.5% of the errors due to a single stuck-at fault are detected by parity. The area of an S-box with four inputs and four outputs without error detection in a two-level implementation is 56 units. With an additional fifth parity modifying output for CED the area, also in a two-level implementation, is 66 units. Thus, for a two-level implementation the area of the S-boxes increases by 18%. For a multi-level optimization the area of an S-box without CED is 41 units and with the additional parity modifying output it is 51 units. Thereby when the parity modifying output is separately implemented, and for a multi-level implementation the area of an S-box increases by 24.4%. Thus the overall hardware overhead is determined by an additional parity tree for computing the input parity; an 18% to 24.4% overhead for the implementation of the parity modification of the S-Boxes and some exclusive-or gates. As we have shown for a separate two-level implementation of the S-boxes with parity modification 100% error detection for all the errors due to single stuck-at faults is guaranteed by the proposed method.

3.1 Performance Penalty and Detection Latency

The parity of the input plaintext and output cipher text are computed for each plaintext and hence the associated delay should be carefully accounted for. Computing and checking of the parity can be combined with the round operations in several ways as shown in Table 2.

Table 2. Optimizing the latency of CED

Clock cycle	Approach A		Approach B		Approach C		Approach D	
	Round		Round		Round		Round	
1		P(PT)	1	P(PT)	1	P(PT)	1	P(PT)
2	1		2		2	P(1)	2	P(1)
3	2		3		3	P(2)	3	P(2)
4	3		4		4	P(3)	4	P(3)
5	4			P(CT)		P(CT)		
6		P(CT)						

In this table, we assume that encryption is implemented in four clock cycles with one round of encryption per clock cycle. In the straightforward approach A, the parity computation of the plaintext is followed by encryption (decryption) that in turn is followed by computing and checking the parity of the cipher text. This approach has a performance penalty of two clock cycles (one clock cycle for computing the parity of the input and one clock cycle for computing the parity of the cipher text + comparison with the input parity) and a fault detection latency of one complete encryption (decryption) i.e. four clock cycles.

In approach B, the parity of the plaintext is computed concurrently with the first round of encryption in clock cycle 1. This is then followed by the rest of the encryption (decryption) which in turn is followed by computation and checking of the parity of the cipher text. This approach reduces the performance penalty to one clock cycle without reducing the detection latency. In Approach C computation of the parity of the plaintext is performed concurrently with the first round of encryption in clock cycle 1. This is then followed by computation and checking of output of round one in parallel with the second round of encryption in clock cycle 2 and so on. This approach reduces the fault detection latency while maintaining the performance penalty of Approach B. Other approaches to absorbing performance penalty associated with CED are possible. Each approach has associated performance penalty, fault detection latency (the worst case duration between occurrence and detection of a fault) and fault coverage.

4 Conclusions

In this paper a new method for CED for SPN encryption Block Ciphers was proposed. More details on this general method can be found in [22]. Many of the well known symmetric block ciphers including AES [12] are SPN ciphers. According to the processing steps of the SPN network the parity of the inputs of an encryption round is modified into the parity of the outputs and compared with the actual parity of the outputs of this round. To reduce the necessary hardware overhead the parity tree for computing the parity of the inputs can be also be used to compute the parity of the outputs. If all functional outputs and the output for parity modification of the S-boxes are separately optimized a 100% error detection for all the errors due to single stuck-at faults was achieved. The additional area overhead is low. It consists of an additional parity tree for computing the parity of the inputs, a 18% to 24% increase of the area for the S-Boxes and a few exclusive-or gates only. The proposed concurrent error detection method allows detection of deliberately injected faults in addition to technical faults.

References

1. H. Heys, "A tutorial on linear and differential cryptanalysis," http://citeseer.nj.nec.com/443539.html.
2. E. Biham and A. Shamir, "Differential Cryptanalysis of DES-like Crytptosystems", *Journal of Cryptography*, Vol. 4, No. 1, pp. 3–72, 1991.
3. M. Matsui, "Linear Cryptanalysis Method for DES Cipher," *Proceedings of Advances in Cryptology-Eurocrypt*, Springer-Verlag, pp. 386–397, 1994.
4. J. Kelsey, B. Schneier, D. Wagner, and C. Hall, "Side-Channel Cryptanalysis of Product Ciphers," *Proceedings of ESORICS*, Springer, pp. 97–110, Sep 1998.
5. D. Boneh, R. DeMillo, and R. Lipton, "On the importance of checking cryptographic protocols for faults", *Proceedings of Eurocrypt*, Lecture Notes in Computer Science, Springer-Verlag, LNCS 1233, pp. 37–51, 1997.
6. E. Biham and A. Shamir, "Differential Fault Analysis of Secret Key Cryptosystems", *Proceedings of Crypto*, Aug 1997.
7. R. J. Anderson and M. Kuhn, "Low cost attack on tamper resistant devices", *Proceedings 5th International* Workshop on *Security Protocols*, Lecture Notes in Computer Sciences, Springer-Verlag, LNCS 1361, 1997.
8. C. Aumuller, P. Bier, P. Hofreiter, W. Fischer and J.-P. Seifert, "Fault attacks on RSA with CRT: concrete results and practical countermeasures," www.iacr.org/eprint/2002/072.pdf.
9. J. Bloemer and J.-P. Seifert, "Fault based cryptanalysis of the Advanced Encryption Standard," www.iacr.org/eprint/2002/075.pdf.
10. R. L. Rivest, M. J. B. Robshaw, R. Sidney, and Y. L. Yin, "The RC6 block cipher", ftp://ftp.rsasecurity.com/pub/rsalabs/aes/rc6v11.pdf
11. H. Bonnenberg, A. Curiger, N. Felber, H. Kaeslin, R. Zimmermann and W. Fichtner, "VINCI: Secure test of a VLSI high-speed encryption system", *Proceedings of IEEE International Test Conference*, pp. 782–790, Oct 1993.
12. J. Daemen and V. Rijmen, "AES proposal: Rijndael", http://www.esat.kuleuven.ac.be/~rijmen/ rijndael/ rijndaeldocV2.zip
13. S. Wolter, H. Matz, A. Schubert and R. Laur, "On the VLSI implementation of the International Data Encryption Algorithm IDEA", *IEEE International symposium on Circuits and Systems*, Vol.1, pp. 397–400, 1995.
14. R Karri, K. Wu, P. Mishra and Y. Kim, "Concurrent Error Detection of Fault Based Side-Channel Cryptanalysis of 128-Bit Symmetric Block Ciphers," *IEEE Transactions on CAD*, Dec 2002.
15. G. Bertoni, L. Breveglieri, I. Koren and V. Piuri, "On the propagation of faults and their detection in a hardware implementation of the advanced encryption standard," *Proceedings of ASAP'02*, pp. 303–312, 2002.
16. S. Fernandez-Gomez, J. J. Rodriguez-Andina and E. Mandado, "Concurrent Error Detection in Block Ciphers", *IEEE International Test Conference*, Oct 2000.
17. A. S. Butter, C. Y. Kao and J. P. Kuruts, "DES encryption and decryption unit with error checking," US patent US5432848, Jul 1995.

18. G. Bertoni, L. Breveglieri, I. Koren, P. Maistri and V. Piuri, "A parity code based fault detection for an implementation of the advanced encryption standard," *Proceedings IEEE International Symposium on Defect and Fault Tolerance in VLSI*, pp. 51–59, Nov. 2002.
19. G. Bertoni, L. Breveglieri, I. Koren, and V. Piuri, "Error Analysis and Detection Procedures for a Hardware Implementation of the Advanced Encryption Standard," *IEEE Transactions on Computers*, vol. 52, No. 4, pp. 492–505, April 2003.
20. A. F. Webster and S. E. Tavares. "On the design of S-boxes," *Proceedings of CRYPTO '85*, Springer Verlag Lecture Notes in Computer Science, LNCS 218, pp. 523–534, 1986.
21. H. Heys and S. E. Tavares, "Avalanche characteristics of substitution permutation encryption networks," *IEEE Transactions on Computers*, vol. 44, no. 9, pp. 1131–1139, Sep 1995.
22. R. Karri, M. Goessel, and G. Kousnezow, "Method for error detection in kryptographic substitution permutation networks," patent application pending.

Securing Encryption Algorithms against DPA at the Logic Level: Next Generation Smart Card Technology

Kris Tiri and Ingrid Verbauwhede

UCLA Electrical Engineering Department,
7440B Boelter Hall, P.O. Box 951594, Los Angeles, CA 90095-1594
{tiri, ingrid}@ee.ucla.edu

Abstract. This paper describes a design method to secure encryption algorithms against Differential Power Analysis at the logic level. The method employs logic gates with a power consumption, which is independent of the data signals, and therefore the technique removes the foundation for DPA. In a design experiment, a fundamental component of the DES algorithm has been implemented. Detailed transistor level simulations show a perfect security whenever the layout parasitics are not taken into account.

1 Introduction

The physical implementation of an encryption algorithm is bound to provide an attacker with important information on top of the plain- and ciphertext used in traditional cryptanalysis. Variations in, among other things, the power consumption of the encryption module and the arrival time of the encrypted data can be observed, and possibly linked to the input data and the secret key. Attacks that use this additional information and link it to the internal state, and hence to the secret key, are referred to as Side Channel Attacks (SCA's) [1].

From these SCA's, Differential Power Analysis (DPA) is the most powerful. It relies on statistical analysis and error correction to extract information from the power consumption that is correlated to the secret key [2]. Many countermeasures that conceal the supply current variations at the architectural or the algorithmic level have been put forward. Yet, they are not really effective or practicable against DPA and/or its derivatives, as the variations actually originate at the logic level.

The fact that the power consumption of a single logic gate, which is the most elementary building stone of the complete encryption module, is controlled by both the logic value and the sequence of its input signals forms the basis of DPA. Using a logic style for which a logic gate has at all times a constant power consumption that is independent of the signal transitions, removes the foundation of DPA and is therefore an effective means to halt DPA.

In this paper, we first present the basics of Differential Power Analysis. Then, we briefly discuss Sense Amplifier Based Logic, which is a logic style with signal independent power consumption. Next, we (1) introduce the design experiment, which

C.D. Walter et al. (Eds.): CHES 2003, LNCS 2779, pp. 125–136, 2003.

consists of securing a module of the DES algorithm against DPA at the logic level; (2) investigate the effectiveness of the SABL approach; and (3) discuss the effects of the layout parasitics. Finally a conclusion will be formulated.

2   Basics of Differential Power Analysis

Differential Power Analysis has been extensively described in literature. It was first introduced in [2].

A DPA is executed in two phases: data collection and data analysis. During data collection, the power consumption of the device is measured by sampling and recording the supply current for a large number of encryptions. During data analysis, a selection function, which depends on a guess of some bits of the secret key, divides the power measurements into two sets. For each set, a typical supply current is calculated and subsequently a Differential Trace (DT) is generated by computing the difference between the two typical supply currents.

The selection function D consists of predicting a state bit of the encryption module. If the correct subset of the secret key has been predicted, D is correlated with the state bit and hence with the power consumption of the logic operations that are affected by this state bit. The power consumption of the other logic operations and measurement errors however, are uncorrelated. As a result, the DT will approach the effect of the target bit on the power consumption and there are noticeable peaks in the DT. If on the other hand the guess on the secret key was incorrect, the result of the selection function is uncorrelated with the state bit: the DT will approach 0.

3   Sense Amplifier Based Logic: A CMOS Logic Style with Signal Independent Power Consumption

Every logic style can be classified into one of the two existing logic families. If the logic gate continuously draws a current from the supply and measures its state through the path the current takes, the logic style is said to be Current Mode Logic (CML). If on the other hand, the logic gate only draws a current from the supply to change state and measures its state by the amount of charge it stores on a capacitance, the logic style is said to be Voltage Mode Logic (VML).

CML has constant power consumption under the condition that the gate draws a perfectly constant current from the power supply and this independently of the in- and/or output signals. In order to build a current source capable of generating a constant current, special circuit techniques that minimize channel length modulation have to be used. The decisive drawback of CML however, is its static power consumption: even when the logic gate is not processing any data, it continuously burns the current, which makes this logic style impractical for low power applications.

A better alternative is Sense Amplifier Based Logic (SABL) [3]. SABL is a VML style that uses a fixed amount of charge for every transition, including the degenerated events in which the gate does not change state. This means that the logic gate charges in every cycle a total capacitance with a constant value, even though ultimately different capacitances are switched.

In short, SABL is based on two principles. First, it is a Dynamic and Differential Logic (DDL) and therefore has one and exactly one switching event per cycle and this independently of the input value and sequence. Second, during that switching event, it discharges and charges the sum of all the internal node capacitances together with one of the balanced output capacitances. Hence, it discharges and charges a constant capacitance value. While many DDL-styles exist [4], only SABL (1) controls exactly the contribution of the internal parasitic capacitances of the gate into the power consumption by charging and discharging each one of them in every cycle; and (2) has symmetric intrinsic in- and output capacitances at the differential signals such that it has balanced output capacitances.

In addition to the fact that every cycle the same amount of charge is switched, the charge goes through very similar charge and discharge paths during the precharge phase and during the evaluation phase respectively. As a result, the gate is subject to only minor variations in the input-output delay and in the instantaneous current. This is important since the attacker is not so much interested in the total charge per switching event, as in the instantaneous current and will sample several times per clock cycle in order to capture the instantaneous current.

4   Design Experiment

4.1 Description of Experimental Setup

Goal of the design experiment is to develop design guidelines and to identify possible hurdles for securing an encryption module against DPA at the logic level by simply removing the foundation of DPA. For this purpose, any encryption algorithm could have been chosen. The reason for choosing the DES algorithm [5] is the focus of a great part of contemporary research on how to perform and how to thwart DPA on the DES algorithm.

In order to obtain supply current traces that are as accurate as possible, simulations have been run at the transistor level using HSPICE. Simulating the complete algorithm however, is computationally unfeasible and the algorithm has been stripped-down to a minimum.

The experimental setup, which is shown in Fig. 1, is part of the last round of the DES-algorithm. The module calculates 4 bits of the ciphertext C using a subkey K of length 6 and 4 and 6 bits of the left and right plaintexts L and R respectively. The substitution box is the S1-box. The expansion of the right plaintext R, the permutation of the result of the S-box, and the inverse initial permutation, which are present in the actual DES-algorithm, have been discarded, as they do not change the power measurements: they are hardwired.

The selection function D(C,b,K) consists of calculating bit number b of the left plaintext L, using the known ciphertext C and a guess on the secret key K. The right plaintext R is also known. In the DES algorithm, R is fed into the inverse initial permutation to form part of the ciphertext C.

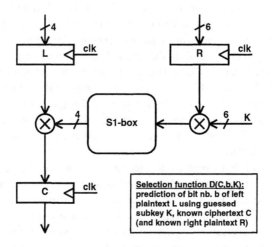

Fig. 1. Experimental setup: DPA on a submodule of the last round in the DES-algorithm

Restricting the experiment to the implementation in Fig. 1 does not simplify the task of putting a stop to DPA. On the contrary, in the implementation of the complete DES algorithm, the power consumption caused by the calculation of the other bits in the same and in the previous rounds, will act as an extra and large noise source on the power measurements. Note also that in this experiment, all measurements are 'perfect measurements'. Aside from the accuracy of HSPICE, there is no quantization error, thermal noise, jitter on the clock of the encryption module, jitter on the sampling moment or any other phenomenon that may introduce a measurement error.

To allow for a comparison, the module has been implemented both in static complementary CMOS logic (SC-CMOS), which is the default logic style in a standard cell library, and in SABL for a 0.18μm, 1.8V CMOS technology. Simulations have been done in HSPICE. In total, the supply current has been captured for 5000 clock cycles with a random input at the plaintext registers L and R, and with a fixed secret key K. The same random input, and the same secret key have been used for both implementations. In order to capture all current variation, the sampling frequency has been set to 100GHz, which corresponds to one sample every 10ps. Note that this very high level of accuracy demands massive simulations. The most time-consuming simulation required 275 hours on a HP Visualize B1000 to complete.

4.2 Effectiveness of the SABL Approach

In a first setup, the simulations are based on a netlist that does not include effects caused by the layout. The parasitic capacitances coming from the intra and inter cell routing of the data signals have been neglected. Fig. 2 shows the transient and statistical properties of the simulated supply current of the SC-CMOS and the SABL implementation of the module presented in Fig. 1.

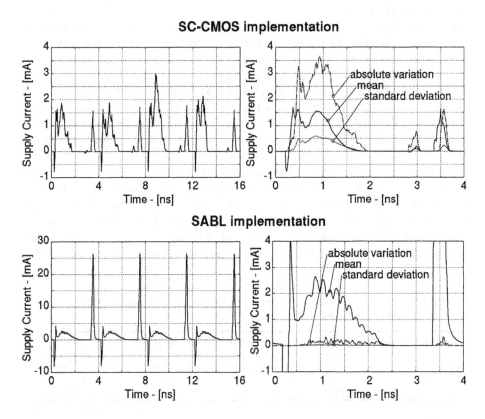

Fig. 2. Simulated supply current: supply current transient of 4 clock cycles (left) and supply current characteristics based on 5000 clock cycles (right) for SC-CMOS (top) and SABL (bottom) implementation

Fig. 2(left) depicts a snapshot of the supply current transient. In total, 4 clock cycles of each 4ns are shown. The supply current of the SABL implementation is very regular and independent of the input signals, whereas the supply current of the SC-CMOS implementation is completely different from cycle to cycle and hence highly dependent on the input signals.

Note that the supply current of the SABL module alternates between a short, high current peak and a time span with a lower current. These events correspond respectively to the precharge phase, in which all gates switch at the same moment, and the evaluation phase, in which each gate switches when its inputs arrive from preceding gates. The current in the evaluation phase is caused by the pairs of static inverters that have been inserted between the SABL gates in order to cascade these dynamic gates according to the domino design rules. We preferred the domino design rules to np design rules for ease of implementation. The pairs of static inverters however, add an extra penalty on the area and the power consumption. The mean energy consumption per clock cycle of the SABL implementation is 11.25pJ compared to 2.70pJ for the SC-CMOS implementation.

Fig. 2(right) depicts the statistical properties of the entire supply current transient. Three curves that describe the typical supply current are shown: they represent the mean supply current, the absolute variation in the mean supply current and the standard deviation on the mean supply current. The curves are generated by first folding the supply current of the 5000 clock cycles on top of each other into 1 clock cycle to generate an 'eye'-diagram and then subsequently calculating the point wise mean, absolute variation and standard deviation. The curves confirm our observations. The mean current of the SABL implementation is a representative switching event for the supply current in every clock cycle. The maximum absolute variation and the maximum standard deviation are 0.37 mA and 89.5 μA respectively. These values correspond to 2% and 4.8% of the mean current at their point of occurrence. The SC-CMOS implementation however, experiences a significant variation in the supply current from clock cycle to clock cycle. The maximum absolute variation and the maximum standard deviation are 3.66 mA and 591.2 μA respectively. These values correspond to 239% and 38.1% of the mean current at their point of occurrence. Table 1 summarizes the numbers.

Table 1. Simulated supply current: variation in the typical supply current based on 5000 clock cycles for SC-CMOS and SABL implementation

Implementation	max(abs. var.) [mA]	ratio to mean current[t]	max(std. dev.) [μA]	ratio to mean current[t]
SC-CMOS	3.66	239%	591.2	38.1%
SABL	0.37	2%	89.5	4.8%

[t] At point of occurrence.

Fig. 3 shows the Differential Traces that have been generated with 8 different key guesses in the selection function. The first bit of plaintext register L has been predicted. Note that in total 64 (=2^6) different guesses of the secret key are possible. Only the DT's of 8 of them are shown for transparency of the figure. The other 56 DT's however, are in accordance with the curves that correspond with the 7 incorrectly guessed keys. The correct secret key is 46.

For the SC-CMOS implementation, the DT of the correct secret key exhibits peaks that are significantly higher than the DT's of the incorrectly guessed keys. All peaks can be brought back to certain precise events. The first peak around 0.5ns corresponds to the rising edge of the clock. At this instant of time, the output of the first bit of the register L becomes equal to the bit that we predicted with the selection function. The second peak around 3ns corresponds to the instant that the input to the first bit of register L changes from the bit that we predicted to a new random input. The last peak around 3.5ns corresponds to the falling edge of the clock.

For the SABL implementation, one can not distinguish which DT is from the correct secret key. Moreover, the DT of the correct secret key would not even be considered as the DT of a possible correct secret key. Contrary to the SC-CMOS module, an analysis of precise events is not possible.

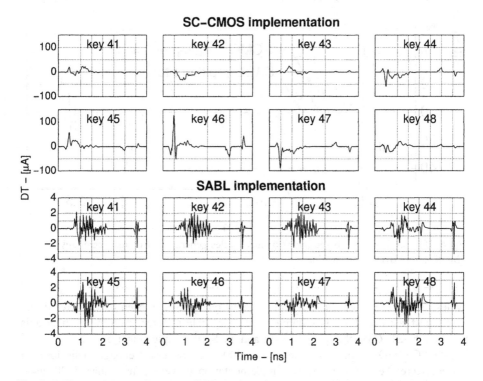

Fig. 3. Differential Traces based on 5000 clock cycles generated by 8 successive key guesses for SC-CMOS (top) and SABL (bottom) implementation. Key 46 is secret key. Please note the different scales for SC-CMOS and SABL implementations

On top of the fact that the DT of the correct secret key does not have any noticeable peaks for the SABL module, the DT's of the SABL module are much smaller than the DT's of the SC-CMOS module. As a result, to determine the DT's of the SABL implementation, the test equipment that captures the supply current in the measurement setup should have a much better accuracy than is necessary to determine the DT's of the SC-CMOS implementation, which are almost 2 orders of magnitude larger. Fig. 4 details the DT's that have been generated with the correct secret key.

Fig. 5 shows the influence of the data collection on the information content in the DT's. In each plot, the peak-to-peak value (p2p) or the root-mean-square value (rms) of the DT's generated by (1) the correct secret key; (2) an incorrect secret key; and (3) a random bit string as selection function, are shown in function of the number of clock

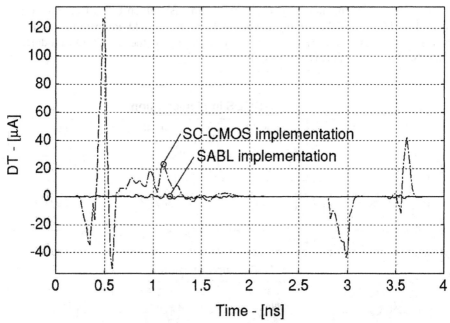

Fig. 4. Differential Traces based on 5000 clock cycles generated by correct secret key for SC-CMOS and SABL implementation

cycles used to generate the DT. The random bit string has been used to avoid any statistical biases of the S-box output. Note the scale difference on the vertical axis between the SC-CMOS implementation and the SABL implementation.

For the SC-CMOS implementation, a mere 200 clock cycles are enough to disclose the correct secret key. For the SABL implementation however, more than 5000 clock cycles have been simulated and the correct secret key does not stand out. It is very unlikely that increasing the number of clock cycles will make the correct secret key stand out. The transient response of the curves has died out and the p2p and the rms are in a steady state response: they are set by the section of the DT, which corresponds to the power consumption of the S1-box. The power consumption of the S1-box is uncorrelated with the selection function, as the bits in the left plaintext have no influence whatsoever on what happens inside the S1-box.

One could argue that occasionally the correct secret key also seems to stand out for the SABL implementation. Fig. 5 however, shows the p2p and rms of only one incorrect secret key. There are DT's of other incorrect secret keys for which the p2p or rms are comparable and/or higher than for the correct secret key.

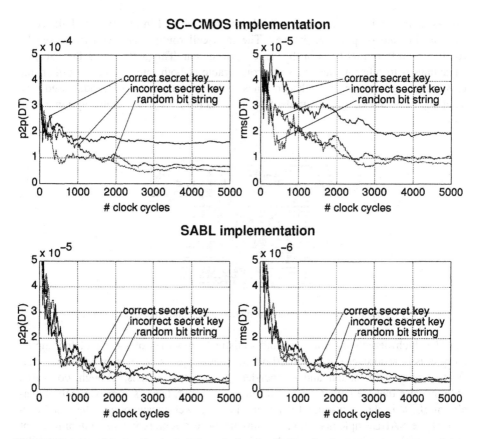

Fig. 5. Influence of data collection: peak-to-peak value (left) and root-mean-square value (right) of the Differential Traces generated by the correct secret key, an incorrect secret key and a random bit string as selection function for SC-CMOS (top) and SABL (bottom) implementation. Please note the different scales for SC-CMOS and SABL implementations

4.3 Effects of Layout Parasitics

The SABL approach has shown to be an effective remedy against DPA when the layout parasitics are not taken into account. In the next setup, the simulations are based on a netlist that does account for the effects of the layout. First, a cell library has been created that contains all cells used in the module. Then, these cells have been used to place and route the module. The complete layout in SABL is shown in Fig. 6.
The parasitic capacitances from the intra- and inter-cell routing will not only result in a performance degradation, in particular in an increase of the input-output delay and of the power consumption, but they will also result in variations in the total charge that is used per switching event if both differential output signals do not see the same parasitic capacitances.

Special attention has been given to the layout of each cell in an effort to balance its intrinsic in- and output capacitances. The inter-cell routing has been addressed by routing the differential lines in the same environment. This assures that the parasitic capacitances to other metal layers are comparable at both interconnects. Further, the cross coupling between long adjacent lines in the same layer has been addressed with shielding. The shielding has a tradeoff with an increase in power consumption and in area.

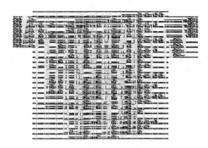

Fig. 6. Layout of SABL implementation of module presented in Fig. 1

Fig. 7, which depicts the transient and statistical properties of the simulated supply current, is in accordance with Fig. 2(bottom). The snapshot of the supply current transient remains very regular and independent of the input signals. Though, compared with the case that the layout was not taken into account, there is approximately a penalty of 100% in the input-output delay and in the power consumption. The mean current of the SABL implementation remains likewise a representative switching event for the supply current in every clock cycle. The maximum absolute variation and the maximum standard deviation are 0.26 mA and 65.8 µA respectively. These values correspond to 13% and 2% of the mean current at their point of occurrence. Note that in spite of the increase in power consumption, the absolute figures have decreased with approximately 30% compared with the case that the layout was not taken into account. The relative figure of the maximum absolute variation however, has increased by a factor of 5. The relative figure of the maximum standard deviation on the other hand, has decreased by a factor of 2.5.

Even though there does not seem to be a significant difference in the supply current characteristics between the module before layout and the one after the layout phase, the DT of the correct secret key exhibits 2 peaks that are higher than the DT's of the incorrectly guessed secret key as can be seen in Fig. 8. The first peak around 0.5ns corresponds to the rising edge of the clock. At this instant of time, the output of the first bit of the register L changes state. The peak has a value of 10.28 µA, which is a factor of 12.3 smaller than the peak at the rising edge of the SC-CMOS implementation. The latter implementation however, did not include the layout parasitics. Including the layout parasitics into the SC-CMOS implementation will increase this number. The second peak at 7.5ns corresponds to the falling edge of the clock. At this instant, the output of the XOR is read into C.

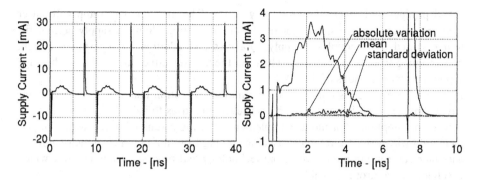

Fig. 7. Simulated supply current: supply current transient of 4 clock cycles (left) and supply current characteristics based on 5000 clock cycles (right)

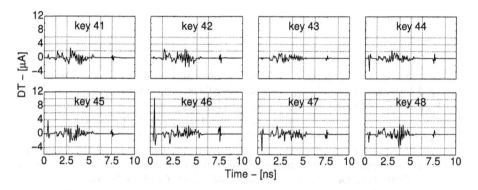

Fig. 8. Differential Traces based on 5000 clock cycles generated by 8 successive key guesses. Key 46 is secret key

5   Conclusions

We have presented a technique to thwart DPA that uses a logic style with data independent power consumption. The technique achieves perfect security whenever the layout parasitics are neglected. In our simulation setup, the secret key has not been exposed and increasing the data collection is very unlikely to help out. With parasitics, DPA is possible. Our simulations however, show that the resulting DT's are more than an order of magnitude smaller than a SC-CMOS implementation. Furthermore in our opinion, improvements are still possible. The resulting increased security will as always come in a tradeoff with some cost. Here, the cost will be an increase in power, area for a more aggressive shielding and an increase in area, initial design time for a perfect symmetric standard cell. It is still unclear however, whether a DPA on an actual product will reveal the secret key or not. The measurement setup will suffer from measurement errors, a larger resolution in the time domain, supply current filtering caused by decoupling, supply parasitics and additional large supply current noise coming from other modules, which for the non-sensitive parts will have the huge supply current variations of SC-CMOS.

In Table 1, we have also presented the minimum variation that seems achievable for any technique at the logic level or higher levels that tries to balance the instantaneous power consumption of a module implementing a logic function. Any actual implementation will suffer from larger variation coming from not only unsymmetrical intra- and inter-cell routing but as well from technology and process variations, over which absolutely no control is possible.

Acknowledgement. The authors would like to acknowledge the support of UC-MICRO (#02-079) and National Science Foundation (CCR-0098361).

References

1. Hess, E., Janssen, N., Meyer, B., Schuetze, T.:Information Leakage Attacks Against Smart Card Implementations of Cryptographic Algorithms and Countermeasures – a Survey. Proc. Of EUROSMART Security Conference (2000) 55–64

2. Kocher, P., Jaffe, J., Jun, B.: Differential Power Analysis. Proc. of Advances in Cryptology, Lecture Notes in Computer Science Vol. 1666 (1999) 388–397

3. Tiri, K., Akmal, M., Verbauwhede, I.: A Dynamic and Differential CMOS Logic with Signal Independent Power Consumption to Withstand Differential Power Analysis on Smart Cards. Proc. Of 28th European Solid-State Circuits Conference (2002) 403–406

4. Bernstein, K., Carig, M., Durham, C., Hansen, P., Hogenmiller, D., Nowak, E., Rohrer, N.: High Speed CMOS Design Styles. Kluwer Academic Publishers (1998) 111123

5. National Bureau of Standards, "Data Encryption Standard," Federal Information Processing Standards Publication 46, January 1977

Security Evaluation of Asynchronous Circuits

Jacques J.A. Fournier[1], Simon Moore[2], Huiyun Li[2], Robert Mullins[2], and
George Taylor[2]

[1] Security Technologies Department, Gemplus
Ave des Jujubiers, ZI Athelia IV,
13705 La Ciotat, CEDEX, France.
jacques.fournier@gemplus.com
[2] Computer Laboratory, University of Cambridge.
Simon.Moore@cl.cam.ac.uk

Abstract. Balanced asynchronous circuits have been touted as a superior replacement for conventional synchronous circuits. To assess these
claims, we have designed, manufactured and tested an experimental
asynchronous smart-card style device. In this paper we describe the
tests performed and show that asynchronous circuits can provide better
tamper-resistance. However, we have also discovered weaknesses with
our test chip, some of which have resulted in new designs, and others
which are more fundamental to the asynchronous design approach. This
has led us to investigate the novel approach of design-time security
analysis rather than rely on post manufacture analysis.

Keywords. Asynchronous circuits, Dual-Rail encoding, Power Analysis,
EMA, Fault Analysis, Design-time security evaluation

1 Introduction

The wide-spreading use of processors in security applications, for e.g. in smartcards or Hardware Security Modules (HSM) has increased both the financial and
social benefits that hackers would gain in tampering with such systems. During
the past seven years, there has been extensive research to enhance the security of
such systems. Most of the counter-measures developed were software-based, protecting mainly against side-channel information leakage. The performance and
cost penalties resulting from such counter-measures were affordable. However,
software protection against more recently-publicised classes of attacks, like those
involving fault injection, consume considerable memory.

There is an urgent need to put more focus on the hardware side of the
system. One attractive path is the use of *self-timed* or *asynchronous* circuits.
In this respect, we have designed, manufactured and tested an experimental
asynchronous smart-card style device. In this paper, we present the principal
results of the security analysis of a secure asynchronous processor. We highlight
the advantages brought by the *self-timed* nature of the circuit. We also analyse
some of the weaknesses that we spotted. Hence, we not only present one of the

C.D. Walter et al. (Eds.): CHES 2003, LNCS 2779, pp. 137–151, 2003.

industry's first thorough stress-testing of a clockless circuit but also propose an evaluation procedure for post manufacture analysis. Finally we introduce a concept whereby those flaws could have been identified at design level, through thorough simulation leading to what we call *design-time security analysis*.

We finish this introduction by providing motivations for using asynchronous circuits in this field, and give a brief overview of the Springbank test chip (see also [1]) and the experimental set-up used. In Section 2 we provide results for DPA and EMA before describing, in Section 3, the chip's resistance to optical probing and power glitches. Finally, in Section 4, we give an insight of our first results on Design-time analysis.

1.1 Motivation for Using Asynchronous Circuits

Speed independent (SI) asynchronous circuits are expected to offer a number of advantages over their synchronous counterparts when designing secure systems [1,2]:

Environment tolerance — SI circuits adapt to their environment which means that they should tolerate many forms of fault injection (power glitches, thermal gradients, etc). This makes fault sensing easier since only major faults need to be detected and reacted to. This is desirable since minor fluctuations in environment conditions are normal during real-world operation.

Redundant data encoding — SI circuits typically use a redundant encoding scheme (e.g. dual-rail). In the latter, each bit is encoded onto two wires **A0** & **A1** as shown in the table below. This mechanism also provides a means to encode an *alarm* signal (e.g. use $11 = alarm$ in a dual-rail scheme [1]).

A1	A0	meaning
0	0	clear
0	1	logical 0
1	0	logical 1
1	1	alarm

Balanced power consumption — Circuits comprising dual-rail (or multi-rail) codes can be balanced to reduce data dependent emissions. In the above illustration whether we have a *logical-0* or a *logical-1*, the encoding of the bit ensures that the data is transmitted and computations are performed with constant Hamming weight. This is important since side-channel analysis is based on the leakage of the Hamming weight of the sensitive data.

Fine-grained random timing variation — may be used to make correlation of repeated runs more difficult, thereby making signal averaging problematic.

Absence of a clock signal — no clock means that clock glitch attacks are removed.

Whilst dual-rail coding might be used in a clocked environment one would have to ensure that combinational circuits were balanced and glitch free. Return-to-zero (RTZ) signalling is also required to ensure data independent power emissions. Once you have gone to these lengths, it is just a small step to an SI asynchronous implementation which offers the additional benefit of better environment tolerance, i.e. tolerance to fault injection.

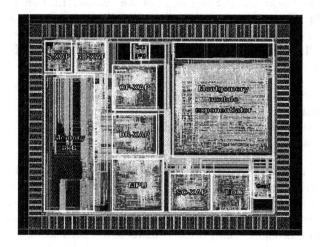

Fig. 1. Springbank Test chip

1.2 Overview of the Springbank Test Chip

The Springbank chip was fabricated in the UMC 0.18μm six metal CMOS process. It contains five 16-bit microcontroller processors, various I/O interfaces and other units as part of other projects. All five processors are based on the same 16-bit XAP architecture but with different implementations. The processors are: one synchronous XAP (S-XAP), a bundled data XAP (BD-XAP), 1-of-4 XAP (OF-XAP), 1-of-2 (dual-rail) XAP (DR-XAP) and a secure variant of the dual-rail XAP (SC-XAP). Given that all processors lie on the same chip and that we used the same standard cell library, comparisons do not need to take into account technology or foundry variations. A 1-of-4 distributed interconnect interfaces these processors to a standard single-rail SRAM holding program and data. In addition, communication between the SC-XAP and SRAM is done via a memory protection unit (MPU) with bus encryption; these were disabled in the experiments described in this paper.

Figure 1 shows a picture of the test chip. The SC-XAP is approximately twice the area of the synchronous XAP. However, the commercial standard cell library used was optimised for synchronous design and not asynchronous design. An optimised library might reduce this area penalty to 1.5 times large. Furthermore, one must remember that the clocked system requires a clock generator. Clock multipliers (PLLs), for example, can take up considerable space.

1.3 Experimental Set-Up

The aim behind these tests is to tally the gain in security while moving from a conventional clocked design (as implemented on the S-XAP) to an asynchronous dual-rail environment (an example of which is the SC-XAP which bears all the features described in section 1.1). For this reason, tests were mainly carried on the S-XAP and the SC-XAP. Since we were in a 'characterisation' phase, our aim was not to break any cryptographic algorithm. We targeted simple instructions which gave a good indication of how the hardware reacts to the several tests performed. The latter were made on the execution of a simple XOR execution whereby we:

- load the memory address at which data are found,
- load a first operand (Op1) into a register,
- perform an XOR between Op1 and the second operand (Op2),
- store result (Res) back to memory.

To monitor the above execution, after each execution of the above sequence, the three data, i.e. Op1, Op2 and Res, were retrieved via the UART port and displayed on a monitor (or stored into a file). The tests were carried out in a *white box* configuration, without any encryption mechanism activated. This allowed us to thoroughly analyse the benefits and weaknesses of asynchronism.

In the next sections, we describe the results obtained for each family of tests. The interpretation and explanation for those observations are then detailed accordingly.

2   Side Channel Analysis

In this section, we look at the information leakage through two forms of side-channels: one is by studying the power consumed by the processor (Differential Power Analysis - DPA [7]) and the other is by observing the Electro-Magnetic (EM) waves emitted by the processor (Electro-Magnetic Analysis - EMA [8,9]).

2.1   Differential Power Analysis

Power dissipation in static CMOS circuits is dominated by switching activity. As a result, the power dissipated is highly dependent on the switching activity produced by a change of input data. In the simple case of a bus, activity is observed as the Hamming weight of the state changes. Data-dependent power leakage may be exploited to reveal useful information, either by analysing single power traces (Simple Power Analysis) or by collecting many power traces and performing a statistical analysis of the power variation with respect to changes in data values (Differential Power Analysis [7]).

DPA Attacks on the Springbank Chip. Power analysis of the secure dual-rail processor revealed that small imbalances in the design of the dual-rail gates allowed some data-dependent power leakage to be observed. The XOR operation provides one example of where data-dependent power consumption may be observed. Power traces were collected for two different XOR operations, through experimentation. The operands for the first XOR instruction were 0x11 and 0x22, these were changed to 0x33 and 0x55 for the second.

Figure 2 shows the results of collecting power traces for each operation, averaging the traces over 4000 runs, and then subtracting one averaged trace from the other. The centre curve represents this difference. The small disturbance, left of centre, is the result of data-dependent differences in the power requirements for the two XOR operations.

The same kind of analysis was carried out on the S-XAP and similar data-dependant information leakages were observed. However, the extent of the leakage was more significant in the case of the clocked XAP compared with the asynchronous one. More detailed measurements showed that the data dependant information leakage of the SC-XAP was lower than that of the S-XAP by about 22 dB. This reduction is not sufficient to completely protect against DPA. However, in other cases, we have seen that a reduction in information leakage by 20-24 dB could neutralize leakage with respect to SPA.

Further reducing the data dependant power leakage. This example is a good illustration of the difficulty of designing secure processors. On paper, the SC-XAP seemed breachless thanks to its dual-rail with RTZ implementation. However, when it came down to implementing this scheme, conventional place & route tools were used. Those tools tend to optimise space which means that if a bit is encoded onto two 'wires', one wire might end up being longer than the other creating an imbalance which could produce power leakage. So

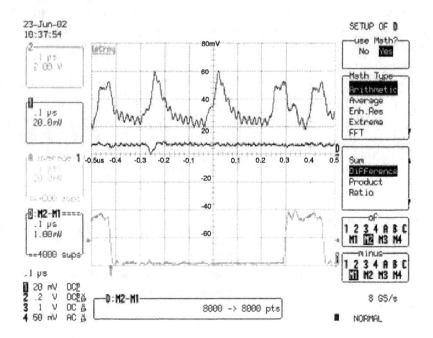

Fig. 2. Differential Power Analysis on Secure XAP (experimental graph)

a first improvement would be to either have a full-custom design or develop a place and route tool which understood how to balance signal paths. Further improvements could be made at a transistor level: current standard cell libraries typically optimise the transistor sizing of gates to minimise the delay through the gate rather than ensuring that the capacitance across all inputs is identical.

2.2   Electro-Magnetic Analysis

In this case, the tests performed were similar to the ones for the DPA, but this time, for each XOR execution, we measured the Electro-Magnetic (EM) waves emitted by the active processor (asynchronous or clocked one) [8]. For the SC-XAP, the EM signals collected were of exploitable magnitudes, which allowed successful DPA-like treatments to be carried out on the collected reference curves. Both for the SC-XAP and S-XAP, data dependant 'signatures' were obtained at three places: at the load of Op1, at the XOR execution and at the write-back of the XOR result into memory. This is illustrated in Figure 3.

The EMA results were taken *without* signal averaging since a signal was clearly visible above the noise. In Figure 3, the uppermost curve is an example of the EM signals measured and the lower three ones correspond to DEMA curves obtained by performing, on the EM curves, differential analysis [8] with respect to *Op1*, *Op2* and *Res*. The 'peaks' shown must be interpreted as leakage of the data's Hamming. We also clearly see at what instances the three different data are 'manipulated'. For the results presented here, a coil covering the processor was used, which means that we were capturing the 'global' EM waves of the entire SC-XAP. However, one could envisage using smaller coils (e.g. 40μm) to measure emissions from a smaller area [8] and target the exact region where the data is being manipulated (the data bus or the ALU for example) .

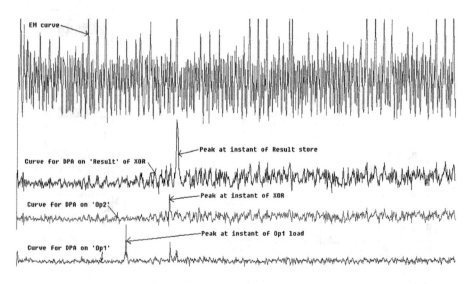

Fig. 3. DEMA Results

EMA leakage on the SC-XAP. As for DPA, balanced logic was be used as a countermeasure for EMA. And like in Section 2.1, the imbalance introduced by the design tools used has been lethal. Moreover, with EMA, we did not observe the same 22 dB reduction in the amount of information leaked because EMA is able to isolate much finer circuit areas and hence the placement and routing of components becomes far more critical in achieving a balanced design. In addition to this, the absence of the clock in asynchronous circuits eases EMA. In conventional clocked circuits, the clock usually adds noisy components to the EMA signals captured whereas in asynchronous circuits, we no longer face that inconvenience.

To make the EMA measurements more difficult, a top level metal defence grid may be used. These are seen on modern smart cards and if suitable signals are injected into them they can help mask the underlying activity.

Where operations may be performed in more than one place (e.g. if using a dual execution pipeline), non-determinism may be used to make data collection more difficult. Security evaluation of this approach is most tractable when the attacker is known to have limited resources, for example one EMA trace taken from one sensor. However, when multiple runs and multiple sensors are used the evaluation is far more complex and is dependent on the algorithm being executed. We are also investigating geometrically regular structures (e.g. PLAs) to determine if this approach to design is more secure than a conventional ASIC design flow.

3   Fault Injection Analysis

A second class of stress-testing techniques consists of injecting faults into the device in order to obtain exploitable 'abnormal' behaviours. Injecting faults into working processors can change the nature of data being treated or corrupt cryptographic computations in such a way as to unveil secret information [6].

Early forms of these so called *active* attacks were focused on the device's external interface and often involved introducing glitches on power or clock input pins [10]. Changes in temperature, either by cooling or heating the whole device or the introduction of a temperature gradient, may also be used to induce faulty behaviours. Defences against such attacks are simplified by the restricted nature of the channel by which faults are injected and can easily be detected by incorporating a suitable tamper sensor. Far greater control over the nature of the faults injected has been demonstrated recently.

Two approaches were taken to inject faults into the Springbank: the first one was by optical probing and the second one was by injecting power glitches. As for the previous side-channel analysis, we targeted an XOR operation. This time we worked only on the SC-XAP.

3.1   Optical Probing Techniques

Laser radiation with a sufficiently short wavelength (photon energy) and intensity may be used to ionise semiconductor materials. When ionisation occurs in a depletion region the production of additional carriers and the presence of an electric field (built-in field and any reverse bias) causes a current to flow. This photocurrent is capable of switching other transistors whose gates are connected to the illuminated junction. This process is a transient one where normal circuit activity resumes once the light source is removed.

In addition to what may be considered a useful attack mechanism, negative effects are also possible. These include the possibility that latch-up may be induced by the generation of photocurrents in the bulk. Of less concern, when using readily available infra-red and visible laser light sources, is the ionisation of gate- and field-oxides due to the large band gap energy of silicon dioxide (which would require a laser with a wavelength in the UV-C range). Ionisation of this type is common when higher energy forms of radiation are absorbed. The subsequent accumulation of positive charges results in a long term shift in transistor characteristics. The following sections explore the weaknesses of the dual-rail technology employed in the Springbank test chip. We then introduce a number of improvements that could secure the design against such attacks.

Optical Probing Attacks on the Springbank Chip. If the dual-rail implementation had provided a completely fault-secure design all attempts at inducing faults would have resulted in deadlock. In many cases the processor did propagate an error signal resulting in deadlock. Unfortunately, two weaknesses in the current design were revealed by the experiments.

The first involved the injection of faults into the ALU design. By targeting two different regions within the ALU two different fault behaviours were possible. The first was to disrupt the ALU operation to produce an incorrect result, the second forced the ALU to always return the result 0x0001. These results are possible as some of the dual-rail gates within the ALU do not guarantee that the presence of the error state on their inputs (in this case a *logic-1* on both dual-rail wires) is propagated. This was a known and unfortunate concession made at the design stage.

Perhaps more interesting is the second failure behaviour. In this case it was possible to set the contents of the processor's registers. The exposure of a single register cell to laser light reliably resulted in setting its value to a *logic-1*. The dual-rail register design that was used in the Springbank chip is illustrated in Figure 4. Setting the cell to '1' was made possible by its inability to store an error state (both states of the single flip-flop are valid). The precise mechanism by which the register was set first involved both the outputs of the NOR gates being pulled-low. This happened as a result of the laser producing photocurrents in the junctions of the N-type transistors in both gates. When the laser was removed, the flip-flop resolved into the *logic-1* state due to differences in the threshold values for each gate. N-type transistors in general produce much larger photocurrents due in part to the superior mobility of electrons when compared to holes (*The electrons and holes in this case are the minority carriers on the larger side of the depletion region. The depletion region extends mostly into the region of least doping.*). It is important to note that the attack was successful even with a large spot size exposing many transistors (we estimate around 100).

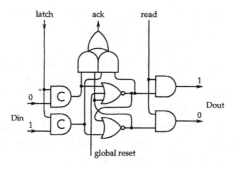

Fig. 4. Single Flip-Flop Dual-Rail Register

Optical Probing Countermeasures. The vulnerability of the ALU design may be countered by ensuring that an error state on any gate input is always propagated to its output. An approach to providing a secure dual-rail register design is shown in Figure 5. Here the number of flip-flops has been doubled. The four possible states are now split into two valid states (representing *zero-symbol* and *one-symbol*) and two error states (null encoded as 00 and error as 11). When the register is reset it is forced into the error state, this prevents the possibility that the reset signal may be used as a simple way to reset the contents of a register to a valid state (perhaps by targeting a reset signal buffer). The error state will only be propagated on the register's output if the register is read. For correct operation, the register must be written with a valid data value prior to reading. The register is also designed to produce an error signal if a 'null' state is ever stored. The ability to store a null value may assist an attacker by allowing them to inject an actual data value from another source.

We will now consider a number of different attack scenarios and how the design is able to detect the injected faults.

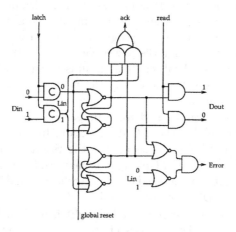

Fig. 5. Dual Flip-Flop Dual-Rail Register

Initially we consider an attack similar to the one described above. Here a large number of transistors are exposed and force all the gate outputs to a *logic-0*. By guaranteeing that both flip-flops resolve in the same direction, the resulting state of the register will always be one that represents a fault (null or error). Attempts to target and modify a single flip-flop will again always result in a fault state. A successful attack now requires greater control over the fault injection process. For example, if both flip-flops could be exposed and then the laser spot moved up away from the lowermost NOR gate, a valid data value could be written. Independent and simultaneous control over individual transistors also offers the possibility of setting registers to particular values.

Security may be further improved by including small optical tamper sensors within each standard cell. These sensors, constructed from one or two transistors, would normally play no part in normal circuit behaviour (only adding a small amount of capacitance). Their only function is to force the dual-rail outputs of the gate into an error state when illuminated. A similar approach is already taken in many standard cell libraries to protect against plasma-induced oxide damage during manufacture, in this case an antenna diode is added on every gate input. These ideas together with security-driven place-and-route would again increase the level of controllability required to perform a successful modification of register values.

3.2   Power Glitch Attacks

Power glitches may be used to inject faults at a coarse level. Tests on the SC-XAP revealed that it was resistant to short Vcc glitches which went down to the ground rail and back. For longer duration glitches we observed faulty processor behaviours which could constitute favourable conditions for the cryptanalysis of cryptographic algorithms like the DES or RSA [5,6]. By injecting the power glitch at different times, we succeeded in causing specific parts of our small program to malfunction. The interesting thing to note is that if we want to target, say, the load of Op1 instruction, we synchronize our program so as to 'cut' the power just at that instant and resume it several tens of nanoseconds later in such a way that the normal program execution resumes. The effects of the glitch are monitored through the power consumed by the processor, just like for the Differential Power Analysis. This is illustrated in Figure 6 which is a superposition of the power traces: one in the normal mode and one where we introduced the power glitch.

If we synchronize the curves and zoom in as shown in Figure 7, we see that the real impact is on the LOAD of Op1 execution. In this case, the value read as Op1 was always 0xFFFF. Consequently, the result of the XOR operation was always the logical inverse of Op2. In this case, we have targeted one precise instruction and corrupted an entire data. This scenario could be lethal if we were to attack the load of a DES key for example.

In another experiment, we generated the glitch while uploading the data's address into the address register. This led to the 'writing' of an erroneous address.

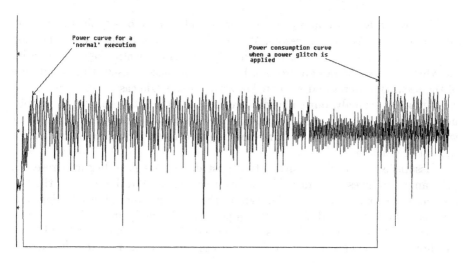

Fig. 6. Vcc Glitch on Secure XAP

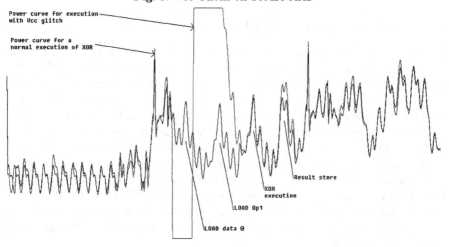

Fig. 7. Vcc glitch during LOAD Op1

Those are a few examples of how 'long' glitches can corrupt the functioning of a self-timed system. We are currently looking into this aspect.

4   Design-Time Security Analysis

We have seen that in many ways, the SC-XAP exhibits several interesting security properties like lower data-dependant power leakage and resistance to optical probing for most of the processor. We did not predict that the asynchronous nature of the circuit could facilitate attacks like EMA and voltage glitches. Moreover, other security flaws, like the low resistance of the ALU and the register bank to optical probing, are linked to unfortunate design trade-offs in the SC-XAP.

The design and evaluation of the Springbank test chip is typical of the design process for secure processors. We began with a requirements specification which included security properties. This allowed us to identify key design criteria which steered the design process. However, we lacked design time validation of the security criteria and we now know that some side cases were overlooked. Even more worryingly, our colleagues working on attack technologies developed new attacks which we had not even considered during the design process. What we seem to have recreated in our research project was a microcosm of current industrial practice.

Dissatisfied with ad hoc evaluation post design, we have begun a research programme to investigate design time security validation techniques. In the last section of this paper, we give an insight of the on-going work about *Design-time Analysis* which is bound to become important for the future design of secure processors. Design-time analysis is performed during the design process whereby we should try to simulate the behavior of the processor along with the current consumed and the energy radiated.

4.1   Simulating Side-Channel Information Leakage

To confirm the source of the imbalance observed during the side-channel analysis, we simulated the operation of the ALU executing the same XOR operations described in Section 2.1. The power simulation results were collected using Primepower^{TM} [4], a gate level power estimation tool. The power estimation includes capacitance and resistance values extracted from layout. The results of the simulation are shown in Figure 8. Even in the absence of a power model for the memory system we observe a similar data-dependent power difference during the execution of the XOR operations. Using a simple second order low pass filter model for the power distribution network provides more comparable data (Figure 9). The power dissipation curves for the XOR operations differ in shape to those measured as they include no power for memory accesses. This produces a significant drop in power at the point where one XOR operand is fetched from memory.

We hence see how data dependant leakage may be detected at design time via systematic simulation. Such simulations allow design comparisons to be made, though it is harder to predict the exact values of emissions. The simulations we have undertaken for power are based upon switching activity. In this case, capacitance masks some of the information. Similarly, for electromagnetic radiation, one has to consider wave interference. None the less, switching activity simulation gives a good approximation to the energy being consumed over time which is a good approximation for DPA and DEMA.

4.2   Design-Time Analysis of Fault Tolerance

A range of physical phenomena that can trigger faults may be modelled. We can then model a wide range of attack scenarios from single to multiple transistor failures. Given bounds on the control the attacker has, we can determine whether

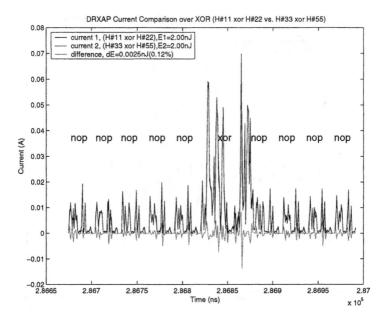

Fig. 8. Power Simulation. Secure XAP executing XOR

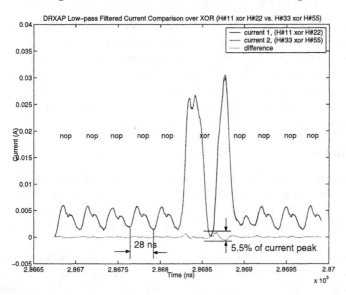

Fig. 9. Power Simulation. Secure XAP executing XOR, low pass filter applied

a fault can be injected without being detected. No matter what the source of the fault is, the end result is to somehow modify the data being manipulated. The aim of the game is to detect any attempt to cause bit flips and this is being investigated right now: had we identified, at simulation level, that no alarm signal was propagated when the ALU or the registers were tampered with, we would

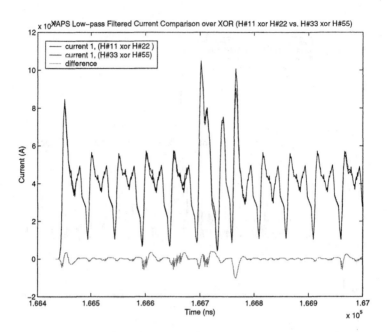

Fig. 10. Power Simulation. Synchronous XAP executing XOR, low pass filtered

have redesigned those weak parts of the circuit. Systematic testing of faults can then be undertaken at a small module level through exhaustive simulation. Such simulations can take into account a wide range of conditions (e.g. single and multiple transistor failure induced by optical probing) in much the same way that traditional fault simulation is undertaken. Where alarms are generated at the small module's level, it is then possible to reason about the propagation of alarm signals at a more abstract level for larger systems.

5   Conclusion

This paper has presented the first ever security evaluation of an asynchronous smart-card system. The secure asynchronous processor (SC-XAP) has shown interesting tamper-resistance properties. None the less we have identified weaknesses in our first attempt together with possible refinements to overcome these issues.

Asynchronous circuits could become a trustworthy platform for secure computing. Circuit area is inevitably going to be larger than a simple synchronous design, but this has to be balanced against large memory (and thus chip area) savings that are possible if fewer software countermeasures are required. The lack of ECAD tool support for asynchronous circuit design is another issue, though we were able to make use of commercial place & route tools, standard cell libraries, etc, and we were able to complete the design with just a small research team.

Finally, we mention the concept of design-time security analysis. These techniques are centered around the simulation of a wide range of measurements and fault possibilities for a wide range (preferably exhaustive) set of input data. We have demonstrated that power attacks and optical probing can be simulated. However, our longer term goal is to be able to make more general statements about the level of security attained which go far beyond current known attacks. With such an approach, we believe that security by design may become a far more powerful technique for processor designers.

Acknowledgements. The authors would like to thank Sergei Skorobogatov and Ross Anderson (University of Cambridge) provided useful insights into attack techniques. Special thanks are also due to the Gemplus Card Security Group for their testing and analysis of the Springbank test chip, in particular to Jean-François Dhem and Christophe Mourtel for useful comments on this paper. We thank the European Commission for funding some of this research as part of the G3Card project (IST-1999-13515).

References

[1] S. Moore, R. Anderson, P. Cunningham, R. Mullins and G. Taylor, "Improving Smart Card Security using Self-timed Circuits," in *Proc. 8th IEEE International Symposium on Asynchronous Circuits and Systems – ASYNC '02*, pp. 23–58, IEEE 2002.

[2] L.A. Plana, P.A. Riocreux, W.J. Bainbridge, A. Bardsley, J.D. Garside and S. Temple, "SPA - a synthesisable amulet core for smartcard applications," in *Proc. 8th IEEE International Symposium on Asynchronous Circuits and Systems – ASYNC '02*, pp. 201–210, IEEE 2002.

[3] R. Mayer-Sommer, "Smartly Analyzing the Simplicity and the Power of Simple Power Analysis on Smartcards," in *Proc. 2nd International Workshop Cryptographic Hardware and Embedded Systems – CHES '2000*, LNCS 1965, pp. 78, 2000.

[4] http://www.synopsys.com/products/power/primepower_ds.html

[5] D. Boneh, R.A. DeMillo and R.J. Lipton "On the importance of checking cryptographic protocols for faults," in *Proc. Advances in Cryptology – EUROCRYPT '97*, LNCS, pp. 274-285, 1994.

[6] E. Biham, and A. Shamir, "Differential fault analysis of secret key cryptosystems," in *Proc. 17th International Advances in Cryptology Conference – CRYPTO '97*, LNCS 1294, pp. 513–525, 1997.

[7] P. Kocher, J. Jaffe, and B. Jun, "Differential power analysis," in *Proc. 19th International Advances in Cryptology Conference – CRYPTO '99*, pp. 388–397, 1999.

[8] K. Gandolfi, C. Mourtel, and F. Olivier, "Electromagnetic analyis: Concrete results," in *Cryptographic Hardware and Embedded Systmes (CHES 2001)*, LNCS 2162, 2001.

[9] J.-J. Quisquater and D. Samyde, "ElectroMagnetic analysis EMA: Measures and coutermeasures for smart cards," in *Smart Card Programming and Security*, pp. 200–210, LNCS 2140, 2001.

[10] R. Anderson and M. Kuhn, "Tamper resistance – a cautionary note," in *Second USENIX Workshop on Electronic Commerce*, (Oakland, California), pp. 1–11, USENIX, Nov. 1996.

Design and Implementation of a True Random Number Generator Based on Digital Circuit Artifacts

Michael Epstein[1], Laszlo Hars[2], Raymond Krasinski[1], Martin Rosner[3], and Hao Zheng[4]

[1] Philips Electronics, Philips Intellectual Property and Standards, 345 Scarborough Road, Briarcliff Manor, NY 10510
{Michael.Epstein, Raymond.Krasinski}@philips.com
[2] Seagate Technology, 1251 Waterfront Place, Pittsburgh, PA 15222
Laszlo.Hars@Seagate.com
[3] Philips Electronics, Philips Research, 345 Scarborough Road, Briarcliff Manor, NY 10510
Martin.Rosner@philips.com
[4] ActiveEye, 286 Broadway, Pleasantville, NY 10570
Hao.Zheng@activeye.com

Abstract. There are many applications for true, unpredictable random numbers. For example the strength of numerous cryptographic operations is often dependent on a source of truly random numbers. Sources of random information are available in nature but are often hard to access in integrated circuits. In some specialized applications, analog noise sources are used in digital circuits at great cost in silicon area and power consumption. These analog circuits are often influenced by periodic signal sources that are in close proximity to the random number generator. We present a random number generator comprised entirely of digital circuits, which utilizes electronic noise. Unlike earlier work [11], only standard digital gates without regard to precise layout were used.

1 Introduction

True random-number generators are often desirable in many applications ranging from statistical system analysis to information security protocols and algorithms. Currently available true random number generators utilize circuitry that often consumes significant resources on integrated circuits and/or require incompatible analog and digital elements. Other, more primitive generators do not provide a convenient interface to electronic devices. In this paper we describe the design of a new type of true random number generator that is based solely on digital components (i.e. it is inexpensive to build), consumes little power, provides high throughput, and passes the DIEHARD [15] suite of tests for randomness. It is envisioned that this design of a random number

C.D. Walter et al. (Eds.): CHES 2003, LNCS 2779, pp. 152–165, 2003.

generator will provide an inexpensive alternative to generators that are currently embedded in many systems such as microprocessors and smart cards.

This paper introduces the concepts behind a simple true random number generator. The focus of this paper is on digital circuits that exhibit metastability and those that function as unstable oscillators. Others have used metastability as a source of randomness and have found some success, but only after trimming devices with a laser to achieve a precisely balanced circuit. (See, e.g. [10], [11], [12]).

We designed and implemented a wide array of this type of true random number generators. The prototype chip consists of nine distinct designs. The prototype chip was mounted on an acquisition breadboard for testing. The extracted results were analyzed to determine which designs yield the best results. The results show that even without de-biasing the resulting sequences, some of the designs provide random datasets that pass the DIEHARD suite of tests. This paper details the methodology for design, implementation and testing of a true random number generator based on digital artifacts. We conclude this paper by providing results, and outlining the necessary steps to create practical versions of these promising designs.

2 Description of Digital Artifacts

The design described in this paper yields random results by means of digital circuit artifacts [13], [14]. The design utilizes a pair of oscillators that are permitted to free-run. At some point, the free-running oscillators are coerced to matching states via a bistable device. While we believe that the circuit utilizes the metastability artifact, this conjecture has not been proven. However, we have experimental results from a similar circuit composed of discrete components, which does show metastability. A discussion of the two possible causes of randomness, metastability and oscillator drift and jitter, is presented below.

2.1 Metastability

Digital circuits, by their nature, are designed to be predictable. If the same logic levels in the same order are presented to digital circuit, the same result should always occur. However if the rules governing the inputs of digital circuits are violated then the results can become unpredictable. In particular, the metastability phenomenon may occur in a digital flip-flop or in a latch. A flip-flop, which is a memory element that can store one bit, is one type of bi-stable device. At the heart of every flip-flop is a pair of logic gates that are fed back to each other. This electrical feedback is what preserves the logical bit stored in the circuit. If the setup and hold conditions of the flip-flop are violated [1] then the pair of gates will behave unpredictably or even oscillate about some intermediate voltage [2]. During the oscillatory or metastable state, the output is neither a logical zero nor a logical one. After some time the oscillations will die out and the flip-flop will settle into a logical state of a zero or a one as shown in figure 1.

There are three uncertainties at work here. The first uncertainty is if the circuit will behave normally i.e. attain a logical state after the usual delay for the flip-flop, or will the circuit enter the metastable state [3]. The second uncertainty is the state that the flip-flop will settle into. The third uncertainty is the length of time that the circuit will remain metastable [1]. Some of these uncertainties have been measured [4] and modeled [5] for various kinds of circuitry and environmental conditions. It has also been shown that this phenomenon cannot be avoided in any flip-flop [4], [5]. Thus, the metastability phenomenon provides a source of randomness that can be used to construct a true random number generator without the need for specialized analog circuits.

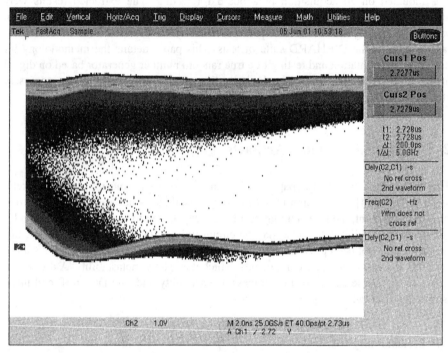

Fig. 1. Recovery from metastable statedisplayed on a TDS7254 Tektronix oscilloscope using the P7240 active probe. The signal resolving from a metastable state was captured using infinite persistence in the Fast Acquisition mode [y-axis: 1volt/division with 8 divisions shown; x-axis: 2ns/division with 10 divisions shown]. Lower intensity (lighter traces) shows more frequent signal traces. Note the single trace where the signal oscillates toward zero but eventually stabilizes at a one

2.2 Oscillator Drift and Jitter

A clock period is never a precise constant, even in highly regulated clock oscillators, such as those that utilize a crystal. Careful observation shows that the oscillating signal has slight changes of phase. Such perturbations of oscillator period are called jitter. Some components of jitter are random [6]. Precise measurement of jitter is more of an art that a science [7].

The design uses two free-running (without a crystal or similar reference) oscillators, which are allowed to drift away from each other. Such oscillators are known to exhibit a great deal of variability even over short periods of time. Thus, the different internal noise of the two similar oscillators (causing phase jitter, accumulated as phase drift) can be utilized as a source of randomness. After some time the instantaneous voltage is latched by a digital bi-stable circuit, capturing the random state.

3 Integrated Circuit Design

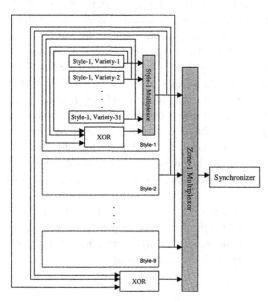

In order to validate these concepts, an integrated circuit containing nine distinct types (or styles) of random number generators was constructed. Each design utilized a bi-stable device, which in most cases contained a pair of gates to form the memory element. Given the obvious dependence on circuit delays each style of random generator was replicated in 15 to 31 different varieties for a total of 247 distinct random number generators. The varieties used different gate sizes, and thus different circuit delays, in pairs of matched or sometimes unmatched gates in an effort to explore the entire problem space for each of the nine styles. All of the gates were drawn from a standard 0.18 micron CMOS

Fig. 2. Zone circuit; Selection of various RNG designs was designated through multiple levels of multiplexers

library and laid out automatically. No effort was made to minimize or match wiring delays, although given the small size of the circuits it is likely that some varieties are very similar.

A multiplexer was utilized to allow for the selection of a particular variety for that style of generator. Every variety from a particular style was also connected to a network of XOR gates thus combining all of the varieties of a style into a single output. This XOR output became one of the inputs to the style multiplexer as well. Any single variety or the XOR of the varieties in the style could be selected via a control word as shown in figure 2.

Each of the styles was then fed to a zone multiplexer. As with the varieties, all of the styles were then XOR'ed together to create another input for the zone multiplexer. A control word was configured to choose any particular style or the XOR combination of the all nine styles (see figure 2). Ultimately 288 variations could be individually selected:

- A specific variety of a specific style of the random number generator
- The XOR value of all varieties of a specific style
- The XOR value of specific varieties combining all styles
- The XOR value of all varieties of all styles.

By including a wide variety of designs, numerous combinations of digital circuit artifacts were tested to determine if a real world random number generator could be created via a single circuit design or a combination of designs.

Once a bi-stable device becomes metastable its non-logic voltages can propagate throughout a circuit and cause other flip-flops in the circuit to become metastable as well. While it is desirable that the random number generator has an unpredictable output the same cannot be said for the typical circuit that requires the random bits. Figure 2, includes a synchronizer circuit that effectively isolates the random number generator from the rest of the circuit by capturing the bits in a series of three flip-flops. Typical flip-flops will not enter a metastable state easily. However, even if the first of these flip-flops were to enter the metastable state, the clock period is sufficiently long so that it is likely that the flip-flop will have left the metastable state and resolved to a zero or one prior to the needed setup time for the second flip-flop. Similarly, the second flip-flop protects the third one. This method is a well-known technique [2] for reducing chances of failure due to metastable behavior. The chances of metastable behavior propagating through the synchronization circuit can be measured in tens or hundreds of years.

In the interest of creating the greatest possible set of variations, the entire zone circuit that is shown in figure 2 was replicated eight times. By purposely replicating the design numerous times, further variations in the circuit layout may have been introduced. The outputs of all zones were directed off chip for analysis.

4   Circuit Description of a Random Number Generator

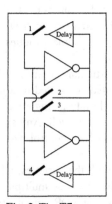

Fig. 3. The T7 concept

Nine different styles or types of the random number generator were implemented in the test chip. Results shown in Appendix A, Table 1 found that six of the styles, designated T1-T6, of the random number generator failed completely by giving only a single binary value of one or zero. These six styles attempted, in different ways, to cause a library flip-flop to become metastable by violating the setup and hold requirements. The failure of these styles to generate truly random sequences can be partially attributed to the fact that modern flip-flops are designed to suppress the metastable artifact. Of the remaining styles, one (T9) gave results that failed to pass all of the DIEHARD Tests. Two styles (T7 and T8) produced results that were random when all of the varieties were XOR'ed together and sometimes from a single random number generator circuit. Since the T7 and T8 designs were quite similar, the focus of the remaining testing effort was on style T7. Figure 3 shows the basic principle behind the design of T7. Later research may explore the usefulness of designs T8 and T9 and discover why T1-T6 failed to produce random results (see appendix A).

4.1 Theory of Operation

The basic concept behind style T7 is relatively simple. The design consists of two inverters and four switches, numbered 1 through 4, as shown in figure 4.

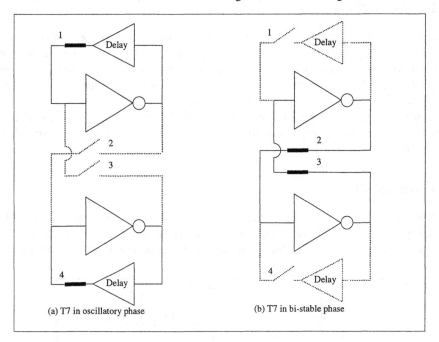

(a) T7 in oscillatory phase   (b) T7 in bi-stable phase

Fig. 4. Operating phases of T7

The switches can be implemented as transition gates or as multiplexers. The actual design was implemented as multiplexers as shown in figure 5. In this case the multiplexers served as the delay elements as well as the switching elements.

Returning to the switch based design in figure 4a, if switches 1 and 4 are closed while switches 2 and 3 are open a configuration is created where the inverters form two independent, free-running ring oscillators, caused by the delay in the negative feedback loop. The inverters can be supplemented by delay elements implemented as a series of buffer gates or an even number of inverters. Ideally each of the oscillators should be sufficiently different so that if the circuit encounters a strong external signal there is little chance that both oscillators would synchronize to it.

If switches 1 and 4 are opened while switches 2 and 3 are closed the connected inverters form a bi-stable memory device as shown in figure 4b. Because of positive feedback the outputs of the inverters eventually resolve, by clipping, to a consistent logic state. This final logic state creates one random bit. The randomness of the bit is derived from the conditions created when the two free-running oscillators are stopped (the oscillator feedback loops are opened and the two inverters get cross-connected). At that point the relative and absolute values of the instantaneous output voltages and the internal noise determine the eventual logic state the circuit will settle to, sometimes even via the artifact of metastability. Thus, the randomness of the circuit exploits two different mechanisms.

4.2 Drift and Jitter

The two oscillators drift apart in different ways because when the oscillators are switched on they don't start immediately, they "hesitate" for a short period of time then the voltage goes either up or down, creating an uncertain starting point. The loop gain of the oscillators should be small otherwise they will start instantly. As the two oscillators continue to oscillate random circuit noise affects the unregulated oscillators so that their clock periods are inconsistent from cycle to cycle. The combined effects ensure that the two oscillators will find themselves in different states each time they are stopped.

4.3 Metastability

When the oscillators are stopped, the gates are cross-connected forming a bi-stable memory element. At this point one or both output voltages may be between logic levels. The bi-stable device, which is formed by the inverters, must settle to a logic

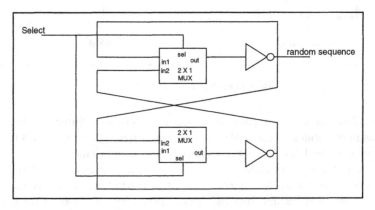

Fig. 5. Actual T7 design based on bi-stable memory element. The random sequence generator is a pair of cross-connected inverters where the multiplexers are used to switch between oscillatory and bi-stable phases

state over time. However, sometimes the bi-stable device will initially find itself in a conflicted state as the two oscillators do not agree or even arrive at a consistent state of disagreement. The bi-stable device will therefore oscillate for a period of time until a final, stable state can be achieved. Thus the final state of the metastable device produces a random bit.

4.5 Circuit Details

There were several varieties, of the design laid out for the test chip. The types of inverters were varied to achieve different delays. One option not explored was to replace a single inverter with a chain of inverters for larger delays. Short (and different) delays are preferred because the ring oscillators will produce smaller amplitude, sinusoid like signals, which should provoke metastability more often.

Very short delays might prevent oscillation if the gain of the circuit is too small at the fundamental frequency of the feedback loop. However, the circuit should work even in this case. The large negative feedback forces the output of the inverters to an intermediate voltage, close to halfway between logic levels. Flipping the switches to activate the bi-stable configuration at an intermediate voltage often forces metastability [10], [11]. The final state achieved under these conditions is shown to be unpredictable if the initial conditions of the bi-stable circuit are at intermediate voltages.

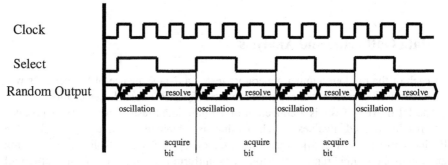

Fig. 6. Random bit acquisition; The 'Select' signal is used to drive the multiplexers that choose between acquisition and oscillation phase of the random number generator.

In the actual design a synchronous circuit was used to collect random numbers. Figure 6 shows how a divided version of the clock was used to switch the multiplexers via the select signal. When the select signal is high the circuit is in the oscillation state and each inverter operates independently. At that point the output is not in any logic state as is indicated in by the diagonal crosshatches. Sufficient time is allowed for the oscillators to diverge in the oscillation state. When the select line goes low the circuit is in the bi-stable configuration and resolves to a single value, either a 1 or 0, via metastable oscillations. The resolution time can be short or long but sufficient time must be allowed in order for the value to be resolved on the vast majority of occasions. Should the random bit be unresolved the synchronization circuit described in section 3 will resolve it. The rising edge of the select signal can be used to acquire the random bit as shown in figure 6.

5 Improving Randomness

Variations within the manufacturing tolerances and unpredictable environmental changes (temperature, supply voltage etc.) will alter the behavior of the randomness circuit. The circuit may fail in an obvious way such as producing all 0 or all 1 output or possibly a heavily biased sequence. The simplest solution to this problem is to lay out many, slightly different versions of the circuit, on the chip. Under different environmental conditions some of the versions will work randomly while others will produce a biased output. If the differences in the circuits are small and there are a large number of varieties of the randomness circuit, with high probability at least one variety will always produce random results.

The randomness of all of the different varieties is collected by simply XOR'ing all of the outputs of the random circuits on the chip [8]. If there is at least one truly random output sequence the final result will still be random, as shown by Matsui [9]. Since each random source is small, only a few hundred standard logic gates will provide very high quality random numbers in real world conditions.

6 Data Gathering and Analysis

For each of the random number generators tested a data sequence of 80 megabits was collected. Each sequence was submitted to the DIEHARD Tests for evaluation.

The DIEHARD Tests is a collection of 16 individual tests that altogether produce 215 results (called Pvalues) Each Pvalue is obtained by applying a function [Pvalue = $F_i(X)$ (i = 1-215)], where the function F_i seeks to establish a distribution function of the sample random variable X as uniform between 0 and 1. In addition all of the functions F_i are an asymptotic approximation, for which the fit will be worst in the tails. Thus only rarely will one find Pvalues near 0 or 1, such as 0.0012 or 0.9983, if the sequence is random. When a sequence is decidedly non-random numerous Pvalues of 0 or 1 to six or more decimal places (i.e. 1.000000) will be present. Thus for a random sequence, only a small number of Pvalues should have a value near zero or one, although the presence of occasional Pvalues of 1.000000 or 0.000000 is insufficient to suggest that a sequence is non-random.

Accordingly, we established a scoring system for the DIEHARD results, where sequences were declared random provided that:

1. There were no more than a single "hard failure" (Pvalue = 1.000000 or 0.000000) for the entire sequence of Pvalues. Two failures were permitted if, and only if, both failures occurred in different varieties of a single DIEHARD test.

2. There were fewer than 5 "near" failure values (where a "near" failure is a Pvalue < 0.000099 or Pvalue > 0.999900)

If a sequence exhibited a small number of "hard failures" or more than 6 "near" values the results were considered as "indeterminate" and a retest was performed with a new dataset. If the retested data produced acceptable results, the generator was considered acceptable. Just to be certain, we also looked for "almost" values (Pvalue < 0.099999 or > 0.900000) and clusters of similar values. In general failed sequences had many "hard failures" to the point where few "near" or "almost" values exist in the entire sequence. Passed sequences almost never had a "hard failure" and very few "near" values.

Ten different chips were received from the CMOS manufacturing facility. Four distinct tests were performed to confirm the results. The first chip was tested twice for each XOR result at each voltage. Additionally, a second chip was tested using another prototype board and the results were compared and verified. While some DIEHARD results produced "hard failures", there were never more than two of such failures in any given sequence. Occasionally, two "hard failures" occurred in the same test of the DIEHARD suite of tests. No repetition of "hard failures", or "near failures" was found among the four test runs. The results obtained from the four distinct test runs show that the same design tested in identical configurations produced different and random results.

7 Results and Interpretation

As explained previously, six of the nine proposed designs failed to produce any non-trivial bits (the output remained at constant 1 or 0 value). One design (T9) produced results that failed to pass DIEHARD Tests. Two designs (T7 and T8) produced results that were random when all of the varieties were XOR'ed together. In addition, T7 produced random results from a single generator circuit at the nominal operating voltage of 1.8 volts (see Appendix A, Table 1). Since designs T7 and T8 were closely related, testing and analysis focused only on the T7 design.

In order to simulate real world conditions where gate delays can vary due to voltage, temperature and processing differences, the circuit was tested at various voltages as is shown in Appendix A, Table 2. In all cases the XOR of the 15 varieties of design T7 produced random sequences. Intuitively this seems to indicate that as some varieties became more biased, other varieties became less biased at different voltage levels. However all varieties produce independent, random results. The XOR appears to represent a collection of the entropy of all varieties, which can be explained as follows.

We define the bias b of a random binary variable X as

$$b = \left| \mathrm{Prob}(X = 1) - \frac{1}{2} \right| = \left| \mathrm{Prob}(X = 0) - \frac{1}{2} \right|$$

The T7 design uses the XOR of 15 binary sequences, each of which is believed to be random, as the output. According to the "Piling-up lemma" [9], if these input sequences are independent, the bias of the output sequence is

$$b = 2^{n-1} \prod_{i=1}^{n} b_i$$

Here n is the number of input sequences and b_i $(1 \leq i \leq n)$ is the bias of each of the input sequences. It can be easily shown that $b \leq b_i$ $(1 \leq i \leq n)$ and the equality holds if and only if $b_j = 0$ or $b_j = \frac{1}{2}$ for all $j \neq i$. In general, the XOR will greatly reduce the collective bias of the independent input sequences. For example, assume $n = 15$ and $b_i = \frac{1}{4}$ for all i, then $b = (\frac{1}{2})^{16} \approx 0.000015$, which is negligible. As a practical matter the experiments show that even when none of the individual sequences passes the DIEHARD test, the XOR of the results is still measured as random. Since the cost of the circuit is quite small, a further reduction in bias can be achieved via XOR'ing two of these circuits together.

Further off-line processing of longer sequences from circuits that did not pass the DIEHARD tests provides an indication of the randomness of those circuits. Specifically, when a simple Von Neumann corrector was applied to long sequences that failed, the resulting shorter sequences passed the DIEHARD suite of tests. The Von Neumann corrector is used in many applications to remove bias [16], [17] at the expense of lower throughput. This shows that aside from bias, the results gathered from different varieties of T7 will provide good sources of random bits when properly debiased. This also seems to suggest that the effects of voltage variations on different varieties of T7 are most pronounced in the bias within the sequence. By XOR'ing different varieties of T7, the effects of bias are reduced to tolerable levels as shown in Appendix A, Table 3. In essence the XOR function has allowed us to trade off area, as more varieties of T7 are needed, for speed since the XOR'ed result does not need to be debiased and higher throughput is possible.

The total gate count for a random number generator based on the T7 design is as follows. Two AND gates, one inverter and one OR gate are used to implement each multiplexer. Thus a single randomness circuit requires four AND gates, four inverters and two OR gates. Since the design incorporates 15 instances of T7 and 14 XOR gates to collect the bits, the total number of gates is 60 AND gates, 60 inverters, 30 OR gates and 14 XOR gates. The small number of gates required to realize this random number generator makes the design suitable for applications that mandate strict power and area constraints.

8 Conclusions

In this paper, we have demonstrated that a practical random number generator can be built using standard digital gates and standard layout tools. The generator is stable even over large changes in operating voltage. This is a strong indication that such a generator will have good characteristics in extreme temperature conditions, such as may be found when a Smartcard is used at an outdoor Automatic Teller Machine (ATM) as well as resistance to attack by variation of voltage or temperature (side channel attacks).

The generator produces random bits by exploiting analog circuit artifacts found in common digital circuits. These artifacts are utilized by creating circuits that are noise sensitive, allowing naturally occurring semiconductor noise to determine the final output. Even though an individual circuit may not be perfectly unbiased, a relatively small number of similar circuits could be XOR'ed to produce a usable random result.

Additionally, we used different varieties of gates (all of which are standard library components) to immunize the overall design against expected environmental and process variations.

8.1 Recommendations for Future Work

Future work will involve the testing of the circuit over temperature extremes. Likewise, the circuit should be tested at greater switching frequencies to determine the maximum usable bit rate. Varying the duty cycle of the switch signal so that different "oscillation" and "resolution" times are applied to the circuit would also be interesting for all of the designs. Combining changes in temperature, voltage, and frequency may also yield interesting results.

Further work using de-biased versions of each generator would help, but not prove the independence of the different varieties. It is possible that even a simple von-Neumann corrector applied to a single variety would produce acceptable results. However, such a design may not have some of voltage immunity a combined design seems to have.

References

1.   T.J. Chaney. Measured flip-flop responses to marginal triggering. IEEE Transactions on Computers, Vol.32 n.12, December 1983, pp. 1207–1209
2.   G.R. Couranz and D.F. Wann. Theoretical and experimental behavior of synchronizers operating in the metastable region. IEEE Transactions on Computers, Vol.24 n.6, June1975, pp. 604–616
3.   S.W. Golomb. Shift register sequences. 1967. Reprinted by Aegean Park Press in 1982
4.   L. Kleeman, A. Cantoni. "Metastable Behavior in Digital Systems", IEEE Design and Test of Computers, vol. 4. Dec. 1987, pp. 4–19
5.   L.R. Marino. "General theory of metastable operation", IEEE Transactions on Computers, Vol.30 n.2, February 1981, pp. 107–115
6.   H. Johnson. Random and deterministic jitter, EDN Magazine, June 27, 2002, pp. 24
7.   M. Rowe, Jitter Discrepancies: not explained, EDN Magazine, February 6, 2003, pp. 48
8.   B. Schneier. Applied Cryptography. Wiley & Sons, 2nd edition, 1995, pp. 425-426
9.   M. Matsui, Linear Cryptanalysis Method for DES cipher, Advances in Cryptology – Eurocrypt'93, Lecture Notes in Computer Science, Springer-Verlag 765, 1993, pp. 386–397
10. M.J. Bellido, A.J. Acosta, et al., A simple binary random number generator: new approaches for CMOS VLSI. 35th MIDWEST Symposium on Circuits and Systems. August 1992

11. M. J. Bellido, A. J. Acosta, M. Valencia, A. Barriga,and J. L. Huertas,"Simple Binary Random Number Generator", Electronics Letters, Vol. 28, No. 7, March 1992, pp. 617–618

12. S. Walker and S. Foo, Evaluating metastability in electronic circuits for random number generation. IEEE Computer Society Workshop on VLSI, April 2001, pp. 99–102

13. L. Kleeman. The jitter model for metastability and its application to redundant synchronizers. IEEE Transactions on Computers, Vol.39. n.7, July 1990, pp. 930–942

14. L.M. Reyneri, L.M. del Corso, B. Sacco. "Oscillatory Metastability in Homogeneous and Inhomogeneous Flip-flops". IEEE J. of Solid-State Circ. Vol.25. n.1. Feb. 1990, pp. 254–264

15. Marsaglia, G. (1996) "DIEHARD: A Battery of Tests of Randomness," http://stat.fsu.edu/~geo

16. Robert Davies: Hardware random number generators, Statistics Research Associates Limited. http://robertnz.net/hwrng.htm

17. Benjamin Jun and Paul Kocher: The Intel® random number generator, Cryptography research, inc. April 22, 1999 http://www.connectedpc.com/design/chipsets/rng/CRIwp.pdf

Appendix A: Tabulated DIEHARD Results

Table 1. DIEHARD results for T1-T9 at 1.8 volts and room temperature

		Zone 1	Zone 2	Zone 3	Zone 4	Zone 5	Zone 6	Zone 7	Zone 8
T1									
T2									
T3		Stable result, non-random sequence							
T4									
T5									
T6									
T7	V1	FAIL	FAIL	FAIL	FAIL	FAIL	FAIL	FAIL	FAIL
	V2	FAIL	FAIL	FAIL	FAIL	FAIL	FAIL	FAIL	FAIL
	V3	FAIL	FAIL	FAIL	FAIL	FAIL	FAIL	FAIL	FAIL
	V4	FAIL	FAIL	FAIL	FAIL	FAIL	FAIL	FAIL	FAIL
	V5	FAIL	FAIL	FAIL	FAIL	FAIL	FAIL	FAIL	FAIL
	V6	FAIL	FAIL	FAIL	FAIL	FAIL	FAIL	FAIL	FAIL
	V7	FAIL	FAIL	FAIL	FAIL	FAIL	FAIL	FAIL	FAIL
	V8	PASS	PASS	PASS	PASS	PASS	PASS	FAIL	PASS
	V9	FAIL	FAIL	FAIL	FAIL	FAIL	FAIL	FAIL	FAIL
	V10	FAIL	FAIL	FAIL	FAIL	FAIL	FAIL	FAIL	FAIL
	V11	FAIL	FAIL	FAIL	FAIL	FAIL	FAIL	FAIL	FAIL
	V12	FAIL	FAIL	FAIL	FAIL	FAIL	FAIL	FAIL	FAIL
	V13	FAIL	FAIL	FAIL	FAIL	FAIL	FAIL	FAIL	FAIL
	V14	FAIL	FAIL	FAIL	FAIL	FAIL	FAIL	FAIL	FAIL
	V15	FAIL	FAIL	FAIL	FAIL	FAIL	FAIL	FAIL	FAIL
	XOR	PASS	PASS	PASS	PASS	PASS	PASS	PASS	PASS
T8		Results similar to T7							
T9		Stable result, non-random sequence							

Table 2. DIEHARD results for T7 at room temperature and varying supply voltage

Variety	Supply Voltage	Zone 1	Zone 2	Zone 3	Zone 4	Zone 5	Zone 6	Zone 7	Zone 8
V7	1.2 volts	FAIL	FAIL	FAIL	FAIL	FAIL	FAIL	FAIL	FAIL
	1.4 volts	FAIL	FAIL	FAIL	FAIL	FAIL	FAIL	FAIL	FAIL
	1.6 volts	FAIL	FAIL	FAIL	**PASS**	FAIL	FAIL	FAIL	FAIL
	1.8 volts	FAIL	FAIL	FAIL	FAIL	FAIL	FAIL	FAIL	FAIL
	2.0 volts	FAIL	FAIL	FAIL	FAIL	FAIL	FAIL	FAIL	FAIL
V8	1.2 volts	FAIL	FAIL	FAIL	FAIL	FAIL	FAIL	FAIL	FAIL
	1.4 volts	FAIL	FAIL	FAIL	FAIL	FAIL	FAIL	FAIL	FAIL
	1.6 volts	FAIL	FAIL	FAIL	FAIL	FAIL	FAIL	FAIL	FAIL
	1.8 volts	**PASS**	**PASS**	**PASS**	**PASS**	**PASS**	**PASS**	FAIL	**PASS**
	2.0 volts	FAIL	FAIL	FAIL	FAIL	FAIL	FAIL	FAIL	FAIL
V9	1.2 volts	FAIL	FAIL	FAIL	FAIL	FAIL	FAIL	FAIL	FAIL
	1.4 volts	FAIL	FAIL	FAIL	FAIL	FAIL	FAIL	FAIL	FAIL
	1.6 volts	FAIL	FAIL	FAIL	FAIL	FAIL	FAIL	FAIL	FAIL
	1.8 volts	FAIL	FAIL	FAIL	FAIL	FAIL	FAIL	FAIL	FAIL
	2.0 volts	FAIL	FAIL	FAIL	FAIL	FAIL	FAIL	FAIL	FAIL
XOR	1.2 volts	**PASS**	**PASS**	**PASS**	**PASS**	**PASS**	**PASS**	**PASS**	**PASS**
	1.4 volts	**PASS**	**PASS**	**PASS**	**PASS**	**PASS**	**PASS**	**PASS**	**PASS**
	1.6 volts	**PASS**	**PASS**	**PASS**	**PASS**	**PASS**	**PASS**	**PASS**	**PASS**
	1.8 volts	**PASS**	**PASS**	**PASS**	**PASS**	**PASS**	**PASS**	**PASS**	**PASS**
	2.0 volts	**PASS**	**PASS**	**PASS**	**PASS**	**PASS**	**PASS**	**PASS**	**PASS**

Table 3. Von Neumann correction of variety 7 and 9 for T7 at 1.8 volts. Table shows the DIEHARD test results for biased and debiased sequence, and the reduction in the size of the debiased sequence after passing the baised sequence through the Von Neumann corrector

		Zone 1	Zone 2	Zone 3	Zone 4	Zone 5	Zone 6	Zone 7	Zone 8
	Biased	FAIL	FAIL	FAIL	FAIL	FAIL	FAIL	FAIL	FAIL
V7	Debiased	**PASS**	**PASS**	**PASS**	**PASS**	**PASS**	FAIL	**PASS**	**PASS**
	Data Reduction	14.87%	11.85%	12.23%	10.64%	24.77%	23.79%	24.39%	10.46%
	Biased	FAIL	FAIL	FAIL	FAIL	FAIL	FAIL	FAIL	FAIL
V9	Debiased	**PASS**	**PASS**	**PASS**	**PASS**	**PASS**	**PASS**	**PASS**	**PASS**
	Data Reduction	22.35%	24.56%	19.12%	24.85%	25.10%	25.08%	21.98%	22.16%

True Random Number Generators Secure in a Changing Environment

Boaz Barak, Ronen Shaltiel, and Eran Tromer

Department of Computer Science and Applied Mathematics
Weizmann Institute of Science , Rehovot, ISRAEL
{boaz,ronens,tromer}@wisdom.weizmann.ac.il

Abstract. A true random number generator (TRNG) usually consists of two components: an "unpredictable" source with high entropy, and a *randomness extractor* — a function which, when applied to the source, produces a result that is statistically close to the uniform distribution. When the output of a TRNG is used for cryptographic needs, it is prudent to assume that an adversary may have some (limited) influence on the distribution of the high-entropy source. In this work:

1. We define a mathematical model for the adversary's influence on the source.
2. We show a simple and efficient randomness extractor and *prove* that it works for *all* sources of sufficiently high-entropy, even if individual bits in the source are correlated.
3. Security is guaranteed even if an adversary has (bounded) influence on the source.

Our approach is based on a related notion of "randomness extraction" which emerged in complexity theory. We stress that the statistical randomness of our extractor's output is *proven*, and is not based on any unproven assumptions, such as the security of cryptographic hash functions.

A sample implementation of our extractor and additional details can be found at a dedicated web page [Web].

1   Introduction

1.1   General Setting

It is well known that randomness is essential for cryptography. Cryptographic schemes are usually designed under the assumption of availability of an endless stream of unbiased and uncorrelated random bits. However, it is not easy to obtain such a stream. If not done properly, this may turn out to be the Achilles heel of an otherwise secure system (e.g., Goldberg and Wagner's attack on the Netscape SSL implementation [GW96]).

In this work we focus on generating a stream of truly random bits. This is the problem of constructing a *true random number generator* (TRNG). The usual way to construct such a generator consists of two components:

C.D. Walter et al. (Eds.): CHES 2003, LNCS 2779, pp. 166–180, 2003.

1. The first component is a device that obtains some digital data that is unpredictable in the sense that it has high entropy.[1] This data might come from various sources, such as hardware devices based on thermal noise or radioactive decay, a user's keyboard typing pattern, or timing data from the hard disk or network. We stress that we only assume that this data has high entropy. In particular, we do *not* assume that it has some nice structure (such as independence between individual bits). We call the distribution that is the result of the first component the *high-entropy source.*
2. The second component is a function, called here a *randomness extractor*, which is applied to the high-entropy source in order to obtain an output string that is shorter, but is random in the sense that it is distributed according to the uniform distribution (or a distribution that is statistically very close to the uniform distribution).

Our focus is on the second component. The goal of this work is to construct a single extractor which can be used with *all* types of high-entropy sources, and that can be *proven* to work, even in a model that allows an adversary some control over the source.

Running a TRNG in adversarial environments. The high entropy source used in a TRNG can usually be influenced by changes in the physical environment of the device. These changes can include changes in the temperature, changes in the voltage or frequency of the power supply, exposure to radiation, etc.. In addition to natural changes in the physical environment, if we are using the output of a TRNG for cryptographic purposes, it is prudent to assume that an adversary may be able to control at least some of these parameters. Of course, if the adversary can have enough control over the source to ensure that it has zero entropy then, regardless of the extractor function used, the TRNG will be completely insecure. However, a reasonable assumption is that the adversary has only partial control over the source in a way that he can influence the source's output, but not remove its entropy completely.

1.2   Our Results

In this paper, we suggest a very general model which captures such adversarial changes in the environment and show how to design a randomness extractor that will be secure even under such attacks.

In all previous designs we are aware of, either there is no mathematical treatment or the source of random noise is assumed to have a nice mathematical structure (such as independence between individual samples). As the nature of cryptanalytic attacks cannot be foreseen in advance, it is hard to be convinced

---

[1] Actually, the correct measure to consider here is not the standard Shannon entropy, but rather the measure of *"Min-Entropy"* (see Remark 1). In the remainder of the paper we will use the word "entropy" loosely as a measure of the amount of randomness in a probability distribution.

of the security of a TRNG based on a set of statistical tests that were performed on a prototype in ideal conditions. We also remark that it may be dangerous to assume that the source of randomness has a nice mathematical structure, especially if the environment in which the TRNG operates may be altered by an adversary.

Our extractor is simple and efficient, and compares well with previous designs. It is based on pairwise-independent hash function [WC81].[2] Our approach is inspired by a somewhat different notion of "randomness extractors" defined in complexity theory (see surveys [NTS99,Sha02] and Section 1.3).

Our design works in two phases:

Preprocessing: In this phase the manufacturer (or the user) chooses a string π which we call a *public parameter*. This string is then hardwired into the implementation and need not be kept secret. The *same* string π can be distributed in all copies of the randomness extractor device, and will be used whenever they are executed. (We discuss this in detail in Section 5).

Runtime: In this phase the randomness extractor gets data from the high-entropy source and its output is a function of this data and the public parameter π.

The analysis guarantees that if π is chosen appropriately in the preprocessing phase and the high-entropy source has sufficient entropy then the output of the TRNG is essentially uniformly distributed even when the environment in which the TRNG operates is altered by an adversary. This guarantee holds as long as the adversary has limited influence on the high-entropy source.

In particular, we make no assumption on the *structure* of the high-entropy distribution except for the necessary assumption that it contains sufficient entropy. Existing designs of high-entropy sources seem to achieve this goal.

1.3   Previous Works

Randomness extractors used in practice. As far as we are aware, all extractors previously used in practice as a component in a TRNG, fall under the following two categories:

Designs assuming mathematical structure. These are extractors that work under the assumption that the physical source has some "nice" mathematical structure.

An example of such an extractor is the *von Neumann extractor* [vN51], used in the design of the Intel TRNG [JK99]. On input a source $X_1, \ldots, X_n$ the von Neumann extractor considers successive pairs of the form X_{2i}, X_{2i+1}; for each pair, if $X_{2i} \neq X_{2i+1}$ then X_{2i} is sent to the output, otherwise nothing it sent. The von Neumann extractor works if one assumes that the all bits in the source are independent and are identically distributed. That is, each

---

[2] Some choices of the parameters require use of ℓ-wise independent hash functions for $\ell > 2$.

bit in the source will be equal to 1 with the same probability p, and this will happen independently of the values of the other bits. However, it may fail if different bits are correlated or have different biases.

Other constructions that are sometimes used have every bit of the output be XOR of bits in the source that are "far from each other". Such constructions assume that these "far away" bits are independent.

RFC 1750 [ErCS94] also suggests some heuristics such as applying a Fast Fourier Transform (FFT) or a compression function to the source. However, we are not aware of any analysis of the conditions on the source under which these heuristic will provide a uniform output.

Applying a cryptographic hash function. Another common approach (e.g., [ErCS94], [Zim95]) is to extract the randomness by a applying a cryptographic hash function (or a block cipher) to the high-entropy source. The result is expected to be a true random (or at least *pseudo*-random) output. As there is no mathematical guarantee of security, confidence that such constructions work comes from the extensive cryptanalytic research that has been done on these hash function. However, this research has mostly been concentrated on specific "pseudorandom" properties (e.g., collision-resistance) of these functions. It is not clear whether this research applies to the behavior of such hash functions on sources where the only guarantee is high entropy, especially when these sources may be influenced by an adversary that knows the exact hash function that is used.

Randomness extractors in complexity theory. The problem of extracting randomness from high-entropy distributions is also considered in complexity theory (for surveys, see [NTS99,Sha02]). However, the model considered there allows the adversary to have full control over the source distribution. The sole restriction is that the source distribution has high entropy. One pays a heavy price for this generality: it is impossible to extract randomness by a *deterministic* randomness extractor.[3] Consequently, this notion of randomness extractors (defined in [NZ96]) allows the extractor to also use few additional truly random bits. The rationale is that the extractor will output many more random bits than initially spent. While this concept proves to be very useful in many areas of computer science, it does not provide a way to generate truly random bits for cryptographic applications.[4]

Nevertheless, our solution uses techniques from this area. For the reader familiar with this area, we remark that our solution builds on observing that a weaker notion of security (the one described in this paper) can be

---

[3] Consider even the simpler task of extracting a single bit. Every candidate randomness extractor $E : \{0,1\}^n \to \{0,1\}$ partitions $\{0,1\}^n$ into two sets B_0, B_1 where B_i is the set of all strings mapped to i by E. Assume w.l.o.g. that $|B_0| \geq |B_1|$. Then the adversary can choose the source distribution X to be the uniform distribution over B_0 and thus, $E(X)$ is always fixed as 0 and is not at all random. Note that if $E(\cdot)$ can be computed efficiently, then this distribution X can also be sampled efficiently.

[4] Some weaker notions of randomness extractors were proposed. These notions usually suggest considering restricted classes of random sources. See [TV02] and the references there.

guaranteed even when the few additional random bits are chosen once and for all by the manufacturer and made public.

1.4   Advantages and Disadvantages of Our Scheme

The main advantage of our scheme is that it is *proven* to work for *every* high-entropy source, provided that the adversary has only limited control on the distribution of the source. By contrast, previous schemes are either known to fail for some very natural high-entropy sources (e.g., the von Neumann's extractor), or lack a relevant formal analysis (see above).

Efficiency. It is natural to measure the performance of a randomness extractor in terms of the cost per output bit. This measure depends on the following factors:

1. *Cost:* The speed and size of the hardware or software implementation of the extractor.
2. *Entropy rate:* The amount of entropy contained in the source.
3. *Entropy loss:* The difference between the amount of entropy that the high-entropy source contains and the number of bits extracted.

Our design allows tuning the running time and entropy loss as a function of the expected entropy rate and the desired resiliency against adversarial effects on the source. This tuning helps to achieve good overall performance in different scenarios. We discuss specific scenarios below.

In general, our approach is quite simple and efficient and is suitable for a hardware implementation. Its cost is comparable to that of cryptographic hash functions, and it can provably achieve low entropy loss and extract more than half of the entropy present in the source (by comparison, the von Neumann extractor extracts at most half of the entropy)[5].

Example: low entropy rate. For example, consider the case where the source is the typing patterns of a user. In this case the speed at which one can sample the high-entropy source is comparatively slow, and furthermore sampling the source may be expensive. It is thus crucial to minimize entropy loss and extract as much as possible from the entropy present in the source. Our design allows extracting 3/4 of the entropy in the source at a slight cost to the running time. In this case, the running time is less significant as the bottleneck is the sampling speed from the random source.

Example: high entropy rate. Consider the case where the source is sampling of thermal noise. Now the running time is important and we can tune our design to work faster at the cost of higher entropy loss.

The existence of a formal proof of security can be helpful when optimizing the implementation. Our proof shows that any implementation of "universal hash

---

[5] The basic von Neumann extractor can be extended to extract more bits at some cost to the algorithm's efficiency [Per92].

functions" (or "ℓ-wise independent hash functions") suffices for our randomness extractor. Thus, a designer can choose the most efficient implementation he finds and optimize it to suit his particular architecture. This is contrast to cryptographic hash functions, which do not have a proof of security and where the effect of changes (e.g., removing a round) is unknown, and thus such optimizations are not recommended in practice.

A public parameter. One *disadvantage* of our scheme is the fact that it uses a *public parameter*.[6] The security of the scheme is proven under the assumption that the parameter is chosen at random. This parameter needs to be chosen only once and the resulting scheme will be secure with extremely high probability.

We stress that we do *not* assume that this parameter is kept secret. Moreover, this parameter can be chosen once and for all by the manufacturer and hardwired into all copies of the device. We also do not assume that the distribution of the high-entropy source is completely independent from the choice of this parameter — our model allows this distribution to be partially controlled by a computationally-unbounded adversary that knows the public parameter.

Note that a public parameter is *necessary* to obtain the security properties that we require.

2 The Formal Model

2.1 Preliminaries

Min-Entropy. The *min-entropy* of the source X, denoted by min-Ent(X), the maximal number k such that for every $x \in X$, $\Pr[X = x] \leq 2^{-k}$.

Remark 1. Min-entropy is a stricter notion than the standard (Shannon) entropy, in the sense that the min-entropy of X is always smaller than or equal to the Shannon entropy of X.

It is easy to see that it is impossible to extract m bits from a distribution X with min-Ent$(X) \leq m - 1$. This is because such a distribution gives probability at least $2^{-(m-1)}$ to some element x. It follows that for any candidate extractor function $E : \{0, 1\}^n \rightarrow \{0, 1\}^m$ the element $y = E(x)$ has probability at least $2^{-(m-1)}$ and thus $E(X)$ is far from being uniformly distributed.

We conclude that having min-entropy larger than m is a *necessary* condition for randomness extraction. In this paper we show that having min-entropy k slightly larger than m is a *sufficient* condition.

Statistical Distance. We use dist(X, Y) to denote the *statistical distance* between X and Y that is: $\frac{1}{2}\sum_a |\Pr[X = a] - \Pr[Y = a]|$. We say that X is ϵ-*close* to Y if dist$(X, Y) < \epsilon$.

---

[6] The description of many hash functions and block ciphers includes various semi-arbitrary constants; arguably these can also be considered public parameters.

Table 1. List of parameters

n	The length (in bits) of a sample from the high-entropy source.
k	The min-entropy of the high-entropy source.
t	The adversary can alter the environment in at most 2^t different ways
m	The length (in bits) of the output of the randomness extractor.
ϵ	The statistical distance between the uniform distribution and the output of the randomness extractor.

Notation for probability distributions. We denote by U_m the uniform distribution on strings of length m. If X is a distribution then by $x \in_R X$ we mean that x is chosen at random according to the distribution X. If X is a set then we mean that x is chosen according to the uniform distribution on the set X. If X is a distribution and $f(\cdot)$ is a function, then we denote by $f(X)$ the random variable that is the result of choosing $x \in_R X$ and computing $f(x)$.

2.2 The Parameters

The parameters for our design are listed in Table 1. We think of samples from the source as coming in blocks of n bits.

The goal is to design an extractor that, given an adversarially-chosen n-bit source with k bits of entropy, is resilient against as much adversary influence as possible (i.e., maximize t), while extracting as many random bits as possible (i.e, maximize m), with negligible statistical distance ϵ. In Section 2.5 we give a sample setting of the parameters.

2.3 Definition of Security

Definition 1 (Extractor). *An* extractor *is a function* $E : \{0,1\}^n \times S \to \{0,1\}^m$ *for some set* S.

Denote by $E^\pi(\cdot) = E(\cdot, \pi)$ the one-input function that is the result of fixing the parameter π to the extractor E. We would like the output $E(X, \pi) = E^\pi(X)$ to be (close to) uniformly distributed, where $X \in \{0,1\}^n$ is the output of the high-entropy source and $\pi \in S$ is the public parameter.

Defining Security. We consider the following ideal setting:

1. An adversary chooses 2^t distributions $\mathcal{D}_1, \ldots, \mathcal{D}_{2^t}$ over $\{0,1\}^n$, such that min-Ent$(\mathcal{D}_i) > k$ for all $i = 1, \ldots, 2^t$.
2. A public parameter π is chosen at random and independently of the choices of $\mathcal{D}_i$.
3. The adversary is given π, and selects $i \in \{1, \ldots, 2^t\}$.
4. The user computes $E^\pi(X)$, where X is drawn from $\mathcal{D}_i$.

Definition 2 (*t*-resilient extractor). *Given n, k, m, ϵ and t, an extractor E is t-resilient if, in the above setting, with probability $1 - \epsilon$ over the choice of the public parameter the statistical distance between $E^\pi(X)$ and U_m is at most ϵ.*

Interpretation. The above ideal setting is intended to capture security in the following scenario. A manufacturer designs a noise generating device whose output is a random variable X. Ideally, we would like the adversary not to be able to influence the distribution of X at all. However, in a realistic setting the adversary has some control over the environment in which the device operates (temperature, voltage, frequency, timing, etc.), and it is possible that that changes in this environment affect the distribution of X. We assume that the adversary can control at most t boolean properties of the environment, and can thus create at most 2^t different environments. The user observes the value of X which, conditioned on the choice of environment being i, is distributed as $\mathcal{D}_i$. The definition of security guarantees that the output of the extractor is close to uniformly distributed.

In fact, the security definition (which our construction fulfills) may be stronger than necessary in several senses:

- We do not assume any computational bound on the adversary.
- We do not assume that the user knows which environment i was chosen.
- More fundamentally, we do not require either the user nor the manufacturer need to knows which properties of the environment are controllable by the adversary. The only limitation is that the adversary can control at most t (boolean) properties, and that the source entropy is at least k.
- We allow adversarial choice of all the source distribution for each of the 2^t environment settings. Thus, even for "normal" environment settings, the source may behave in the worst possible way subject to the above requirements. By contrast, in the real world the source distributions would be determined by the manufacturer, presumably in the most favorable way.
- The behavior of the source may change arbitrarily for different environments (i.e., the distributions $\mathcal{D}_i, \mathcal{D}_j$ for $i \neq j$ need not be related in any way). In the real world, many properties of the source would persist for all but the most extreme environment settings.

Remark 2. One can make a stricter security requirement by allowing the adversary to choose i not only as a single value, but also as a random variable with an arbitrary distribution over $\{1, \ldots, 2^t\}$. The two definitions are equivalent.

Remark 3. Many applications require a long stream of output bits. In such cases, our extractor can simply be applied to successive blocks of inputs, always with the same fixed public parameter. The security is guaranteed as long as each input block contains k bits of *conditional min-entropy* (defined analogously to conditional entropy), conditioned on all previous blocks. Of course, this still requires that the conditional distribution of every input block is one of $\mathcal{D}_1, \ldots, \mathcal{D}_{2^t}$.

2.4   Our Result

Our main result is an efficient design for a t-resilient extractor for a wide range of the parameters. We have the following theorem:

Theorem 1. *For every n, k, m and ϵ there is a t-resilient extractor with a public parameter of length $2n$ such that* [7]

$$t = \frac{k - m}{2} - 2\log(1/\epsilon) - 1$$

We can increase t at the cost of an increase in the running time and the length of the public parameter π, to obtain:

Theorem 2. *For every n, k, m and ϵ and $\ell \geq 2$ there is a t-resilient extractor with a public parameter of length ℓn such that*

$$t = \frac{\ell}{2}(k - m - 2\log(1/\epsilon) - \log \ell + 2) - m - 2 - \log(1/\epsilon)$$

We explain our construction in Section 3 and prove its security in Appendix A. In Section 4 we present potential implementations.

Note that in the above theorems, t does not depend on n. In other words, the resiliency t depends only on the amount of of entropy loss (i.e., $k - m$), the statistical distance (i.e., ϵ) that we are willing to accept and the parameter ℓ.

2.5   Sample Settings of Parameters

As the theorems involve many parameters we give concrete examples for natural choices. In the following examples we will consider extracting $m = 256$ bits which are $\epsilon = 2^{-35}$-close to uniform from a source containing $k = 512$ bits of entropy. The choice of n (the length of the source) should depend on the expected quality of the random sample. For example, if the source is the typing pattern of a user then its entropy rate is low and we may need to set $n \approx 2500$ in order to have 512 bits of entropy, while for dedicated noise-generation hardware we may assume that the entropy rate is very high and set $n = 768$.

Using Theorem 1 we get $t = 57$ for $k = 512$. Using the less efficient design of Theorem 2 we can improve both entropy loss and security: choosing $\ell = 16$ we get $t = 667$ and can even reduce k to $k = 448$. These numbers are just an illustration and different tradeoffs between performance, entropy loss and guarantee of security can be made.

3   Our Design

In this section, we present our construction which is based on the notion of "ℓ-wise independent hash functions". We formally define this notion and describe our construction in these term in Section 3.1 and prove correctness in Appendix A. We discuss implementation of this construction in Section 4.

---

[7] In fact, the constructions described in Section 4 have a shorter public parameter, of length n and $n + m - 1$.

3.1 Randomness Extractors from ℓ-Wise Independence

We start by recalling the notion of ℓ-wise independence.

Definition 3 (ℓ-wise independence). *A collection $Z_1, \cdots, Z_n$ of random variables is called ℓ-wise independent if for every $i_1, \cdots, i_\ell \in \{1, \cdots, n\}$ the random variables $Z_{i_1}, \cdots, Z_{i_\ell}$ are independent.*

A very useful tool is the notion of ℓ-wise independent families of hash functions. Intuitively, such functions have some properties of random functions even though they're much less random.

Definition 4 (ℓ-wise independent families of hash functions). *Given a collection $H = \{h_s\}_{s \in S}$ of functions $h_s : \{0,1\}^n \to \{0,1\}^m$, we consider the probability space of choosing $s \in_R S$. For every $x \in \{0,1\}^n$ we define the random variable $R_x = h_s(x)$ (Note that x is fixed and s is chosen at random). We say that H is an ℓ-wise independent family of hash functions if:*

- *For every x, R_x is uniformly distributed in $\{0,1\}^m$.*
- *The random variables $\{R_x\}_{x \in \{0,1\}^n}$ are ℓ-wise independent.*

The usefulness of this definition stems from the fact that there are such families which are relatively small (of size $2^{\ell n}$) and for which $h_s(x)$ can be efficiently computed given s and x.

In these terms, our construction is described as follows: randomly choose $s \in_R S$ (this is the "public parameter"), and let the randomness extractor be simply

$$E(x) = h_s(x)$$

In Appendix A we show that for appropriate parameters, this yields a t-resilient extractor. The following section describes concrete constructions.

4 Implementation

This section describes several extractor implementations based on known constructions of pairwise-independent hash functions [CW79,WC81]. Recall that an implementation of our randomness extractor is constructed has two phases:

- Preprocessing: Choosing the public parameter π.
- Runtime: Running $E^\pi(x)$ on a given string x.

As the first phase is done once and for all at a preprocessing phase, we focus on optimizing the resources used by the second phase. Also, the known implementations of ℓ-wise independent hash functions (for large ℓ) have higher implementation cost than 2-wise independent hash functions; we thus consider only implementations of 2-wise independent hash functions.

4.1   Linear Functions

Theorem 3. *Let $GF(2^n)$ be the field with 2^n elements, and $S = \{(a,b)|a,b \in GF(2^n)\}$. For $s = (a,b)$ and $m < n$ define: $h_s(x) = (a \cdot x + b)_{1,\dots,m}$ (i.e., the first m bits of $a \cdot x + b$ where arithmetic operations are in $GF(2^n)$). Then the family $H_{n,m} = \{h_s\}_{s \in S}$ is a 2-wise independent family of hash functions.*

The field $GF(2^n)$ can be realized as as the field of polynomials over $GF(2)$ modulo an irredicible polynomial of degree n. When the field element are represented as coefficient vectors, addition is the bitwise XOR operation while multiplication requires a modular reduction operation that is easily implemented in hardware for small n, but grows expensive for larger ones and is not well suited for software implementations.

In Appendix A we show that for every random variable X with min-Ent$(X) \geq k$, for an appropriate choice of m we have that for most pairs $s = (a,b)$, $h_s(X)$ is close to uniform.

We have $s = (a,b)$, $h_s(x) = (ax)_{1,\dots,m} \oplus b_{1,\dots,m}$. Define $g_a(x) = (ax)_{1,\dots,m}$. For every fixed pair (a,b), $b_{1,\dots,m}$ is constant. Thus, $h_{(a,b)}(X)$ is close to uniform if and only if $g_a(X)$ is close to uniform. This yields the following extractor for any $m < n$, where the relation to t and ϵ follows from Theorem 1.

- Preprocessing: choose some irreducible polynomial of degree n. Choose a string $a \in \{0,1\}^n$ at random and set $\pi = a$.
- Runtime: $E^\pi(x) = (a \cdot x)_{1,\dots,m}$, using multiplication in $GF(2^n)$.

Remark 4. A family of ℓ-wise independent hash functions can be constructed in a similar manner by setting $s = (a_1, \cdots, a_\ell)$ and $h_s(x) = \sum_{1 \leq i \leq l} a_i \cdot x^{i-1}$.

4.2   Binary Toeplitz Matrices

Theorem 4. *Let $\mathcal{F}$ be a finite field and let n', m' be integers, $m' < n'$. Let $S = \mathcal{F}^{n'+m'-1}$. For $s \in S$ and $x \in \mathcal{F}^n$, define $h_s(x) \in \mathcal{F}^{m'}$ by $(h_s(x) \in \mathcal{F}^{m'})_i = \sum_{j=0}^n x_i s_{i+j}$. Then the family $H_{n',m'} = \{h_s\}_{s \in S}$ is a 2-wise independent family of hash functions.*

The function $h_s(x)$ can be thought of as multiplication of an $m \times n$ Toeplitz matrix (whose diagonals are given by s) by the vector x. Alternatively, it can be considered a convolution of x and y. For $\mathcal{F} = GF(2)$, $n' = n$, $m' = m$, $h_s(x)$ we get the following extractor any $m < n$, (as before, t and ϵ follows from Corollary 1).

- Preprocessing: Choose a string $\pi \in \{0,1\}^{n+m-1}$ at random.
- Runtime: m bits such that the i-th bit is $\bigoplus_{i=0}^n (x_i \wedge s_{i+j})$.

For reasonable n and m, this can be implemented very efficiently in hardware, though in software the bit operations are somewhat inconvenient. To evaluate this, we tested a software implementation of this construction, written in plain C. For the parameters $n = 768$, $m = 256$ of Section 2.5, this implementation had a throughput of 36Mbit/sec (measured at the input) when executed on a 1.7GHz Pentium Xeon processor (cf. [Web]).

4.3 Toeplitz Matrices over $GF(2^k)$

Since Theorem 4 applies to any finite field, we may benefit from using $GF(2^k)$ for $k > 2$. For example, consider $\mathcal{F} = \mathcal{F}'[x]/(x^2 + rx + 1) \cong GF(2^{16})$ where $\mathcal{F}' = GF(2)[x']/(x'^8 + x'^4 + x'^3 + x' + 1) \cong GF(2^8)$ and $r = x' \in \mathcal{F}'$ (both modulus polynomials are irreducible over the respective fields; the latter is taken from the Rijndael cipher.) Using this field, a direct implementation of the convolution performs just $n/16$ field multiplications for every 16 output bits.

The fields $\mathcal{F}, \mathcal{F}'$ are very suitable for software implementation, as follows. For $a, b \in \mathcal{F}' \setminus \{0\}$, multiplication in $\mathcal{F}'$ can be realized as $ab = \exp(\log_z a + \log_z b)$, where z is some generator of $\mathcal{F}'$; $\exp_z$ and $\log_z$ can be implemented via two small lookup tables [Gla02]. Multiplication in $\mathcal{F}$ can then realized by 5 multiplications in $\mathcal{F}'$:

$$(a_1 x + a_0) \cdot (b_1 + x b_0) \equiv a_0 b_0 - c + (a_0 b_1 + a_1 b_0 - rc)x \quad \text{over } \mathcal{F}$$

where $c = a_1 b_1$ and $r = x' \in \mathcal{F}'$ is a constant (02_h as a bit vector). All additions can be done as bitwise XOR, so overall each multiplication over $\mathcal{F}$ requires 4 XOR operations, 5 integer additions, a few shifts, $5 \cdot 3$ table lookups (into the L1 cache) and $5 \cdot 2$ tests whether $a, b = 0$. [8] A straightforward C implementation of the above description achieved a throughput of 56Mbit/sec in the same settings as above (cf. [Web]).

4.4 Randomness Tests

We subjected the above implementations to the DIEHARD suite of statistical tests [Mar95]. For the seed, we used 1023 truly random bit generated by the /dev/random TRNG of Linux 2.4.20. For the source, we generated a 90MB file of English text by retrieving a large number of e-texts from Project Guttenberg, discarding the first 1000 lines of each file (this contains a common header) and eliminating all whitespaces. Thus the source data included the texts of Moby Dick and Frankenstein, the complete works of Shakespeare, and other such "random" data. We then executed the extractor on successive blocks of n bits, with $n = 768$ and $m = 256$, to get 30MB of output bits. The DIEHARD tests did not detect any anomaly in this output.[9] The test report is available at [Web].

5 Conclusions

In this work, we provide an extractor function that is *proven* to work in a model that allows for some adversarial influence on the high-entropy source. The most obvious question is whether the real world conditions satisfy the assumptions

---

[8] The multiplication of c by the constant r can be computed by a single lookup; this eliminates 1 addition, 2 lookups and 2 tests.

[9] In fact, on first attempt DIEHARD did report certain anomalies in one test. A careful inspection revealed that our source data accidentally included several nearly-identical versions of some literary works.

of our model. For example, suppose that a manufacturer constructs a device that outputs a distribution with min entropy at least k in the "benign" (i.e., non-adversarial) settings. Suppose now that he wants to apply our extractor to the output of this device, in an environment that may be somewhat influenced by the adversary.

One concern the manufacturer may have is how to ensure that under all possible adversarial influences, the entropy of the source will remain sufficient? This is indeed a valid concern, but if this is not met then the result will be insecure regardless of the extractor function used (since the adversary will be able to reduce the source's entropy, and no extractor function can *add* entropy). Therefore, it is the responsibility of the manufacturer to make sure that the device satisfies this condition. When this is fulfilled, our construction gives explicit guarantees on the quality of the extractor's output.

We stress that while the manufacturer still needs to carefully design the high-entropy source to be as independent as possible from environmental influence, the overall scheme will work even if design is not perfect and the adversary can affect the source in unpredictable ways, subject to the constraints assumed in our security model.

Acknowledgements. We thank Moni Naor and Adi Shamir for helpful discussions, and the anonymous referees for their constructive comments.

References

[BR94]   M. Bellare and J. Rompel. Randomness-efficient oblivious sampling. In *35th Annual Symposium on Foundations of Computer Science*, 1994.

[CW79]   L. Carter and M. Wegman. Universal hash functions. *JCSS: Journal of Computer and System Sciences*, 18:143–154, 1979.

[ErCS94]   D. Eastlake, 3rd, S. Crocker, and J. Schiller. RFC 1750: Randomness recommendations for security, December 1994.

[Gla02]   Brian Gladman. A specification for Rijndael, the AES algorithm. Available from http://fp.gladman.plus.com/cryptography_technology/rijndael/aesspec.pdf , 2002.

[GW96]   Ian Goldberg and David Wagner. Randomness and the netscape browser. *Dr. Dobb's Journal*, pages 66–70, 1996.

[ILL89]   R. Impagliazzo, L.A. Levin, and M. Luby. Pseudorandom generation from one-way functions. In *Proceedings of the 21st ACM Symposium on Theory of Computing*, 1989.

[JK99]   Benjamin Jun and Paul Kocher. The Intel random number generator. Technical report, Cryptography Research Inc., 1999. Available from http://www.intel.com/design/security/rng/rngppr.htm .

[Mar95]   George Marsaglia. DIEHARD, a battery of tests for random number generators. Available from http://stat.fsu.edu/~geo/diehard.html , 1995.

[NTS99]   Nisan and Ta-Shma. Extracting randomness: A survey and new constructions. *JCSS: Journal of Computer and System Sciences*, 58, 1999.

[NZ96]   Noam Nisan and David Zuckerman. Randomness is linear in space. *Journal of Computer and System Sciences*, 52(1):43–52, February 1996.

[Per92]   Yuval Peres. Iterating von Neumann's procedure for extracting random bits. *Ann. Statist.*, 20(1):590–597, 1992.

[Sha02]   Ronen Shaltiel. Recent developments in extractors. *Bulletin of the European Association for Theoretical Computer Science*, 77, 2002.

[TV02]   L. Trevisan and L. Vadhan. Pseudorandomness and average-case complexity via uniform reductions. In *Proceedings of the 17th Annual Conference on Computational Complexity*, 2002.

[vN51]   John von Neumann. Various techniques used in connection with random digits. *Applied Math Series*, 12:36–38, 1951.

[WC81]   M. N. Wegman and J. L. Carter. New hash functions and their use in authentication and set equality. *JCSS: Journal of Computer and System Sciences*, 22, 1981.

[Web]   Web page for this paper. Available from
          http://www.wisdom.weizmann.ac.il/~tromer/trng/ .

[Zim95]   Philip R. Zimmermann. *PGP: Source Code and Internals*. MIT Press, Cambridge, MA, USA, 1995.

A   Proof of the Main Theorems

We begin by showing that if $H = \{h_s\}_{s \in S}$ is an ℓ-wise independent family of hash function for sufficiently large ℓ, then for any fixed distribution X with sufficiently large min-Ent(X), for most choices of $s \in S$, $h_s(X)$ is close to the uniform distribution. The interpretation is that for most choices of s, h_s is a good randomness extractor for X. This is formally stated in the next lemma.

Lemma 1. *Let X be an n-bit random variable with* min-Ent$(X) \geq k$. *Let $H = \{h_s\}_{s \in S}$ be a family of ℓ-wise independent hash functions from n bits to m bits, $\ell \geq 2$. For at least a $1 - 2^{-u}$ fraction of $s \in S$, $h_s(X)$ is ϵ-close to uniform for*

$$u = \frac{\ell}{2}\left(k - m - 2\log(1/\epsilon) - \log\ell + 2\right) - m - 2$$

The proof of uses standard arguments on ℓ-wise independent hash functions (this technique was used in a very related context in [TV02]). We will need the following tail inequality for ℓ-wise independent distributions.

Theorem 5. [BR94] *Let $A_1, \cdots, A_n$ be ℓ-wise independent random variables in the interval $[0,1]$. Let $A = \sum_{i=1}^{n} A_i$ and $\mu = \mathrm{E}(A)$ and $\delta < 1$. Then,*

$$\Pr[|A - \mu| \geq \delta\mu] \leq c_\ell \left(\frac{\ell}{\delta^2\mu}\right)^{\lfloor \ell/2 \rfloor}$$

Where $c_\ell < 3$ and $c_\ell < 1$ for $\ell \geq 8$.

Proof (of Lemma 1). For $x \in \{0,1\}^n$ let $p_x = \Pr[\mathbf{X} = x]$. We consider the probability space of choosing $s \in_R S$. For every $x \in \{0,1\}^n$ and $y \in \{0,1\}^m$ we define the following random variable:

$$Z_{x,y} = \begin{cases} p_x & h(x) = y \\ 0 & \text{otherwise} \end{cases}$$

We also define $A_{x,y} = Z_{x,y}2^k$. Recall that for every x, $h_s(x)$ is uniformly distributed, and therefore for every x, y, $\mathrm{E}(Z_{x,y}) = p_x2^{-m}$. Let $Z_y = \sum_{x \in \{0,1\}^n} Z_{x,y}$ and $A_y = \sum_{x \in \{0,1\}^n} A_{x,y}$. It follows that for every $y \in \{0,1\}^m$, $\mathrm{E}(Z_y) = 2^{-m}$ and thus, $\mathrm{E}(A_y) = 2^{k-m}$. Note that for every x, y, $A_{x,y}$ lies in the interval $[0,1]$ and that for every y, the variables $A_{x,y}$ are ℓ-wise independent. Applying Theorem 5 we obtain that for every y and $\delta < 1$

$$\Pr[|A_y - 2^{k-m}| \geq \delta 2^{k-m}] \leq c_\ell \left(\frac{\ell}{\delta^2 2^{k-m}} \right)^{\ell/2}$$

Substituting Z_y for A_y and choosing $\delta = 2\epsilon$, we get that for every $\epsilon < 1/2$

$$\Pr[|Z_y - 2^{-m}| \geq 2\epsilon 2^{-m}] \leq c_\ell \left(\frac{\ell}{4\epsilon^2 2^{k-m}} \right)^{\ell/2}$$

By a union bound, it follows that with probability $1 - 2^m c_\ell \left(\frac{\ell}{4\epsilon^2 2^{k-m}} \right)^{\ell/2} > 1 - 2^{-u}$ over $s \in_R S$, for all $y \in \{0,1\}^m$ $|Z_y - 2^{-m}| < 2\epsilon 2^{-m}$. We now argue that for such s, $h_s(X)$ is ϵ-close to uniform. Observe that Z_y is the probability that $h_s(X) = y$ (we think of s as fixed, with x chosen according to X). The statistical distance between $h_s(X)$ and the uniform distribution is given by:

$$1/2 \sum_{y \in \{0,1\}^m} |Z_y - 2^{-m}| < 1/2 \sum_{y \in \{0,1\}^m} 2\epsilon 2^{-m} \leq \epsilon$$

∎

When applying Lemma 1 with $\ell = 2$, one must set $m < k/2$. This can be avoided as shown by the following lemma, for the special case $\ell = 2$.

Lemma 2. *Let X be an n-bit random variable with min-Ent$(X) \geq k$. Let $H = \{h_s\}_{s \in S}$ be a family of 2-wise independent hash functions from n bits to m. For at least a $1 - 2^{-u}$ fraction of $s \in S$, $h_s(X)$ is ϵ-close to uniform for*

$$u = \frac{k - m}{2} - \log(1/\epsilon) - 1$$

The proof is based on a technique introduced in [ILL89], and will appear in the full version of this paper. The next corollary follows easily, by a union bound.

Corollary 1. *Let $X_1, \cdots, X_{2^t}$ be random variables with values in $\{0,1\}^n$ such that for each $1 \leq i \leq 2^t$, min-Ent$(X_i) \geq k$. Let H and u be as in Lemma 1 (or Lemma 2). For at least a $1 - 2^{t-u}$ fraction of $s \in S$ it holds that for all i, $h_s(X_i)$ is ϵ-close to uniform.*

Theorems 1 and 2 follow by setting $u = log(1/\epsilon)$.

How to Predict the Output of a Hardware Random Number Generator

Markus Dichtl

Siemens AG, Corporate Technology
Markus.Dichtl@siemens.com

Abstract. A hardware random number generator was described at CHES 2002 in [Tka03]. In this paper, we analyze its method of generating randomness and, as a consequence of the analysis, we describe how, in principle, an attack on the generator can be executed.

Keywords: Random Number Generator, Entropy, Linear Feedback Shift Register, Cellular Automaton

1 Introduction

Both designs for hardware random number generators and the evaluation of hardware random number generators have not been treated often in publications. This means a serious contrast between – on the one hand – the importance of hardware random number generation and their evaluation for the security of applications and – on the other hand – the little attention this topic has found in the published literature.

One case where the absence of a suitable physical random number generator received considerable public attention were the defects found in the random number generation for SSL implemented in an early version of the Netscape browser. The time of the day used as the only source of true randomness did not provide enough entropy. This lack of entropy could be used for a spectacular attack [GW96].

Physical random number generators deriving their randomness from a physical random process, are also called true random number generators (TRNGs). TRNGS have to be distinguished from pseudo random number generators (PRNGs). PRNGs derive their output algorithmically from a secret initial state. The unpredictability of PRNGs relies on the computational infeasibilty of trying all possible initial states, and on some assumptions on the algorithm used.

We will see that the hardware random number generator described at CHES 2002 in [Tka03] is a combination of TRNG and PRNG elements. Therefore we just call it the RNG (random number generator) in this paper.

We will show that the RNG in some cases produces very little entropy, so that its output can be predicted. This is in contrast with one of the design requirements for the generator cited in [Tka03].

This paper provides theoretical analysis based on the properties of the RNG described in [Tka03]; no experiments on a real chip were made.

C.D. Walter et al. (Eds.): CHES 2003, LNCS 2779, pp. 181–188, 2003.

2   A Hardware Random Number Generator

The RNG described in [Tka03] uses two free running oscillators, implemented as ring oscillators, to clock two deterministic finite state machines.

One of the finite state machines is a binary linear feedback shift register (LFSR) of length 47. The feedback polynomial of the LFSR is primitive.

The other finite state machine is a one dimensional binary cellular automaton (CA) with a neighbourhood of 3. The CA consists of 37 cells. All cells except one follow one rule to derive their next state: The new value is the XOR of the old values of the two neighbouring cells. Only cell 28 is subject to another, albeit similar, rule: Its new value is the XOR sum of its old value and the old values of its two neighbours. For the two cells at the border of the CA, null boundary conditions are used; that is those neighbouring cells which are required by the CA rules, but which are beyond the limits of the CA, are assumed to have the fixed value 0. If started from a not all zeros state, this CA has a cycle length of $2^{37} - 1$.

When the RNG has to produce a random number, 32 bits of the LFSR and 32 bits of the CA are taken from fixed positions of those finite state machines. The 32 bits from the LFSR are XORed with the 32 bits from the CA in order to receive the 32 bit output word.

3   Where Does the Randomness in the RNG Come From?

The main elements of the RNG are two free running oscillators, a linear feedback shift register, and a cellular automaton. Free running oscillators can be a basis of TRNGs, whereas linear feedback shift registers and cellular automata, which are deterministic, are frequently used in the construction of PRNGs. In order to get a clear understanding of the origin of randomness in the RNG, the TRNG parts and the PRNG parts have to be separated mentally.

The linear feedback shift register used in the RNG can be seen as a special counter with a period length of $2^{43} - 1$. The counter states are not represented as the familiar binary numbers, but are encoded as subsequent shift register states. Clearly the conversion between the representation of a counter state as a binary number and a shift register state is completely deterministic.

Analogously, the cellular automaton used in the RNG can be seen as a special counter with a period length of $2^{37} - 1$. Here, the counter state is represented as a cellular automaton state, but again, the conversion between these cellular automaton states and the familiar binary numbers is completely deterministic.

Hence, the only source of entropy in the RNG are the initial states of the registers in the linear feedback shift register and the cellular automaton, and the number of clocks that occurred for the linear feedback shift register and for the cellular automaton. The number of clocks for the linear feedback shift register is only relevant modulo $2^{43} - 1$, the number of clocks for the cellular automaton modulo $2^{37} - 1$.

4   How Random Is This?

As we have identified the sources of entropy in the RNG, the question arises how much entropy they provide.

The first source of randomness are the undetermined initial states of the registers. Even if each register assumed the 0 and 1 state with probability 1/2 each time the RNG is initialized, and even if there were no dependencies between the states of the registers at different initializations, this would not help the RNG much on the long run, because a good TRNG has to must produce continually new entropy as it runs, and not rely on an initial stock of entropy. However, the initial states of flip flops turn out to be no reliable source of entropy. Due to manufactoring variations, they are not completely symmetrical and as a consequence, most flip flops have an initial state which they initially take with a probability close to 1.

The other sources of randomness are the number of clocks occurring for the cellular automaton (modulo $2^{37} - 1$) and for the linear feedback shift register (modulo $2^{43} - 1$) since the initialization.

When an attacker Alice knows the frequencies of the free running oscillators clocking the LFSR and the CA only with limited precision, the RNG becomes completely unpredictable after a sufficiently long waiting time.

Let us assume Alice knows the frequencies of the free running oscillators with a precision of 10 percent, and that she also knows the initial state of the CA. Roughly speaking she looses all information about the state of the CA after about $10 \cdot (2^{37} - 1) \approx 10^{12}$ clocks of the CA. Even if the CA were clocked with 1 GHz, this would mean a waiting time of about 22 minutes. And in order to achieve completely independent CA states, one would also need waiting times of about 22 minutes between each 32 bit block of random values generated.

In [Tka03], too, a minimum sampling period for the subsequent generation of 32 bit blocks of random values is given. It is considerably smaller than 10^{12}, namely 86 cyles of the oscillator clocking the LFSR. Subsequently, we shall study the predictability of the RNG when this minimum sampling period is used.

5   How to Predict the RNG Bits

5.1   How Well Does an Attacker Know the Frequencies of the Free Running Clocking Oscillators?

Evidently, the better the attacker Alice knows the frequencies of the free running oscillators clocking the LFSR and the CA, the better she can predict the numbers of clocks occurring for the LSFR and the CA.

The knowledge of these frequencies depends heavily on the circumstances of the attack. The main environmental parameters influencing the frequencies of free running oscillators are the temperature and the supply voltage. Sometimes these parameters are difficult to predict for Alice. In other applications, she may know these parameters precisely , or may even choose them. For example, professionally run trust centers tend to have their computers in stable air conditioned environments without much variation in temperature or supply voltage. On the other hand, smartcards will encounter enormous variations in environmental conditions, but when the user of a smartcard wants to attack the physical random number generator of the smartcard, she may choose the environment temperature and the supply voltage at her will.

But even when Alice knows the operating conditions of the oscillators perfectly, their frequencies cannot be predicted perfectly, because of non-deterministic effects in the oscillators. For example, there are physically unavoidable noise voltages in the transistors of the oscillator. This noise influences to some degree the exact moments when the transistors switch.

In order to infer the clocking frequencies from the environmental data, the attacker can either perform experiments on a chip with the hardware random number generator, or she must know the design details of the oscillators. The manufacturer of the chip of course knows these design details, and the outcome of a good hardware random number generator should be unpredictable even for the manufacturer of the hardware.

We will not elaborate a statistical model for the attacker's knowledge of the clocking frequencies, because this is not crucial for the attack. As we will see later on, we can easily increase the number of tries to guess the number of clocks occurring for the LFSR and the CA if our assummptions about the knowledge of the clocking frequencies are wrong.

In order to make the attack as efficient as possible, we concentrate on the case where the RNG is subsequently sampled as fast as possible. In [Tka03], the minimum time between the sampling of two output words is defined by the requirement that both state machines (CA and LFSR) clock at least twice their length.

For our attack we also need to assume an upper bound on the ratio of the frequency of the faster free running oscillator and the slower oscillator. This is not an arbitrary restriction, but performance and power consumption considerations make it advisable to choose the clocking frequencies of both oscillators in the same order of magnitude. If a very fast oscillator is used for one finite state machine and a slow oscillator for the other, one gets a RNG with a low data rate but high power consumption. The low data rate is caused by the slow oscillator and the design rule that the finite state machine must be clocked twice its length before it can be sampled again. The high power consumption is due to the fast oscillator, because power consumption and clocking frequency of the state machine are roughly proportional. Subsequently, we assume an upper bound of 3 for the frequency ratio of the oscillators. If the frequency ratio were higher, more guesses would be needed to find the correct number of clocks. A lower bound would speed up the attack.

In the scenario where the attacker defines the environmental conditions, she should be able to know the clock frequencies with a precision of 1 percent. If the attacker does not control the environmental conditions, she might be able to determine the clocking frequencies with a precision of 10 percent.

5.2   Guessing the Number of Clocks

Subsequently, we will consider three 32 bit words sampled from the RNG at top speed. This means that we have to consider the number of clocks occurring between the first and second sample, and between the second and third sample, for each oscillator. We use f_{CA} and f_{LFSR} to denote the clock frequencies of the CA and the LFSR, respectively.

Case A. Here we consider the case $f_{LFSR} \leq f_{CA}$. In this case, the maximum sampling frequency is limited by the rule that the LFSR must clock twice its length before it can be sampled again. This means that at top speed the LFSR clocks 86 times, or, since Alice knows the frequency only with a precision of one percent in the scenario of an environment controlled by her, there may also occur 85 or 87 clocks. By our bound of 3 on the frequency ratio between the oscillators, the number of CA clocks is bounded by 258. With an error of one percent in Alices knowledge of the frequency, this leads to at most 7 possible numbers of CA clocks. Analogously, with a 10 percent insecurity for the frequencies, there are 19 possibilities for the number of LFSR clocks, and at most 53 for the number of CA clocks.

Case B. Here we consider the case $f_{LFSR} > f_{CA}$. We have to distinguish two subcases.

Case B1. When $37 f_{LFSR} \leq 43 f_{CA}$ holds, that is f_{LFSR} is only slightly larger than f_{CA}, the maximum sampling rate allowed for the RNG is still determined by the LFSR frequency. As in case A, we have 3 possibilities for the number of LFSR clocks, if we know the frequencies with a precision of 1 percent. Since the CA is clocked at a lower rate, at most 3 numbers of CA clocks are possible. For the scenario of a 10 percent precision in the knowledge of the frequencies, we get 19 possible numbers of LFSR clocks and also 19 possible numbers of CA clocks.

Case B2. When $37 f_{LFSR} > 43 f_{CA}$ holds, the maximum sampling rate is determined by the CA. If the attacker knows the frequencies with a precision of 1 percent, this leads to 3 possible numbers of CA clocks, and to at most 7 possible numbers of LFSR clocks. With a 10 percent accuracy in the frequencies, there are 19 possible numbers of CA clocks, and 46 possible numbers of LFSR clocks.

In the case of frequencies known with a precision of 1 percent, the worst case is that we have a total of 21 possibilities for the numbers of clocks for both finite state machines. With a precsision of 10 percent, the worst case are 1007 possibilities.

Since we need the numbers of clocks occurring between the first and second sample, and between the second and third sample, we get a total of 441 cases (1 percent case) or 1014049 (10 percent case). This numbers are just a very coarse upper bound on the number of cases to consider, because the numbers of clocks between the different samples are strongly dependend. If, for example, the attacker knows that the number of LFSR clocks is between 200 and 240, she should not begin with 200 for the number of clocks between the first and second sample, and 240 for the number of clocks between the second and third sample. This combination is quite improbable to occur, because the frequencies of the oscillators do not change suddenly from very low to very high. Instead, the best strategy for the attacker is to assume that the clock frequency changed only very little from the second to the third sample. So, combinations of numbers of clocks with small differences should be tried first.

5.3   Determining the Internal States of the CA and the LFSR

In this section we assume that we have correctly guessed the number of clocks of both the CA and the LFSR occurring between three top speed samplings of the RNG. We try to find out the internal states of the machines from the three 32 bit output words.

Since we assume that we know the number of clocks occurring we could try a brute force approach. The almost 2^{43+37} possible intial states make this quite impractical. An efficient solution must rely on the properties of the state machines.

A closer inspection of the two finite state machines makes the solution very easy: both are linear in GF(2). The function combining bits from each finite state machine to compute the RNG output is also linear. We have to solve a system of 96 linear equations in order to determine the 80 bits of the states of the CA and the LFSR.

The fact that the number of equations exceeds the number of variables by 16, helps to eliminate wrong guesses of the number of clocks of the finite state machines. With a probability of $1 - 1/2^{16}$, a wrong guess results in a system of linear equations without a solution.

When one tries to write down the linear equations, one encounters a minor problem: [Tka03] does not specify which 32 bits from each finite state machine are used and how they are permuted. An attacker could reverse engineer the chip in order to receive this information. The information is also known to the manufacturer of the chip. And, as already mentioned above, the output of a good RNG should be unpredictable even for the manufacturer of the chip. In our further analysis, we assume that the attacker knows which bits of the finite state machines are used for the output, and how they are permuted.

To determine the time required to find the solution a system of equations as described above, fixed random choices of bits and fixed random permutations were used. Clearly these choices do not have essential influence on the complexity of solving the system of linear equations.

On a 400 MHz Pentium II, Mathematica 4.2 solved the system of equations in 0.06 seconds using the function LinearSolve[] . This time can definitively be improved significantly by using a faster PC or dedicated software for solving systems of linear equations over GF(2). But even when it takes 0.06 seconds to solve the system of linear equations, in the scenario of clock frequencies known with a precision of 1 percent, all 441 possible systems can be solved in 27 seconds. In the 10 percent scenario, it takes 17 hours to try all 1014049 possibilities. But as pointed out above, many combinations of numbers of clocks are quite improbable, so a good strategy for ordering the tries will enable the attacker to find the solution much faster. If the attacker tries all 1014049 possibilties, she will find about 15 solutions not corresponding to the internal states of the finite machines. The reason is that wrong guesses lead to a solvable system of linear equations with a probabilty of $1/2^{16}$. The attacker should prefer solutions for which the differences in the number of clocks are small.

5.4   Predicting Bits

Once the attacker knows the internal states of the finite machines, she is well off. In order to predict the next output bits, she just has to guess the numbers of clocks of the finite state machines until the time the next random sample was generated. We have seen above that the number of cases to consider is quite small. But now the task of finding the right number of clocks is easier in two ways, compared to finding the correct number of clocks to determine the state. To be able to get the equations, Alice had to guess the right number of clocks for two samples. Here the number of clocks for one sample is sufficient. Alice can also profit from knowledge aquired when finding out the internal states of the finite machines. She may have started with little knowledge of the oscillator frequencies, but now she knows them with high precision, because she knows for which numbers of clocks the system of linear equations could be solved. This good knowledge of the oscillator frequencies leads to very few possibilities for the numbers of clocks for the finite state machines. Alice applies these numbers of clocks to a simulation of the finite state machines in order to compute the next output of the RNG.

6   Is the Described Attack Practically Relevant?

The attack described above enables an attacker to predict output bits from the RNG after having seen some earlier output bits. The question is whether there are practical security applications where such an attack could be applied.

One straight forward application of cryptographic RNGs is the generation of keys for symmetric cryptography. When a number of keys is generated subsequently for different users, the recipient Alice of a key could find out the key generated for the next user by applying the technique described above. Today, symmetric keys usually have 128 bits or more, so Alice can use her own key to determine the state of the RNG and only has to try a very small number of possible keys for the next user. Of course she does not have to stop there, she can continue with the next user but one, and so on.

In the scenario just described, the attacker had to participate actively in a protocol in order to get her own key, from which she could derive they keys of other users. Can the attack of section 5 also be used by a passive attacker? For such an attack we need a protocol which generates and communicates random numbers in plaintext, and subsequently uses the RNG to generate a secret. This turns out to occur very often, namely the generation of a random challenge for challenge and response authentication, and subsequently the generation of a session key.

7   Conclusion

We showed that the random number generator described in [Tka03] is a combination of TRNG and PRNG elements. The TRNG elements produce little entropy when the random number generator is sampled at top rates. The output of the device can be predicted by taking into account both the small amount of entropy generated and the linearity of the PRNG elements.

How can these problems be overcome? Obviously by strengthening the TRNG elements and/or the PRNG elements.

The problem with the TRNG elements is that at top sampling rates the amount of state information it outputs largely exceeds the amount of entropy it generates. This can be cured by sampling less frequently, or by sampling less bits each time. We have seen in section 4 that the required reduction of the sampling frequency is rather impractical. To sample less bits each time the RNG is invoked, is more efficient. For example, if only one bit of output is generated in each output of the RNG, the data rate drops to 1/32 of the original design. But attacks like the one described above are impossible, because the device produces more entropy than it outputs state information.

Concerning the PRNG elements, non-linear components could be used to prevent attacks like the one described above. The disadvantage of only fixing the PRNG parts of the RNG is that this provides only computational security. Attacks are in principle still possible but require a large – hopefully too large for practical application – computational effort. In contrast, TRNGs provide information theoretical security.

References

[GW96]   I. Goldberg and I. Wagner, *Randomness and the Netscape browser*, Dr. Dobb's Journal (1996) 66–70

[Tkac]   T.E. Tkacik, *A hardware random number generator*, Cryptographic Hardware and Embedded Systems – CHES 2002 (B.S. Kaliski Jr., C.K. Koç, and Chr. Paar, eds.), Lecture Notes in Computer Science, vol. 2523, Springer-Verlag, 2003, pp. 450–453

On Low Complexity Bit Parallel Polynomial Basis Multipliers

Arash Reyhani-Masoleh and M. Anwar Hasan

Centre for Applied Cryptographic Research,
University of Waterloo, Waterloo, Ontario, Canada N2L 3G1.
{areyhani, ahasan}@uwaterloo.ca

Abstract. Representing finite field elements with respect to the polynomial (or standard) basis, we consider a bit parallel multiplier architecture for the finite field $GF(2^m)$. Time and space complexities of such a multiplier heavily depend on the field defining irreducible polynomials. Based on a number of important classes of irreducible polynomials, we give exact complexity analyses of the multiplier gate count and time delay. In general, our results match or outperform the previously known best results in similar classes. We also present exact formulations for the coordinates of the multiplier output. Such formulations are expected to be useful to efficiently implement the multiplier using hardware description languages, such as VHDL and Verilog, without having much knowledge of finite field arithmetic.

Keywords: Finite or Galois field, Mastrovito multiplier, pentanomial, polynomial basis, trinomial and equally-spaced polynomial.

1 Introduction

With the rapid expansion of the Internet and wireless communications, more and more digital systems are becoming increasingly equipped with some form of cryptosystems to provide various kinds of data security. Many such cryptosystems rely on computations in very large finite fields and require fast computations in the fields [5,1]. Among the basic arithmetic operations over finite field $GF(2^m)$, addition is easily realized using m two-input XOR gates while multiplication is costly in terms of gate count and time delay.

In the past, many bit parallel multipliers were proposed (see for example [3, 9,2,11,6,10]). In [4,3], Mastrovito proposed an algorithm along with its hardware architecture for polynomial (PB) basis multiplication. In his scheme, first a binary matrix is formed which is then multiplied with a binary vector to obtain the required result. Halbutogullari and Koc have given a method for constructing the Mastrovito multiplier for arbitrary irreducible polynomials [2]. This method considers general as well as special classes of irreducible polynomials such as *trinomials, all-one polynomials* (AOPs) and *equally-spaced polynomials* (ESPs). So far, for these special polynomials, the XOR gate count and time delay of the Halbutogullari-Koc algorithm appear to be the lowest. In [11], Zhang and

C.D. Walter et al. (Eds.): CHES 2003, LNCS 2779, pp. 189–202, 2003.

Parhi give a systematic method to design the Mastrovito multiplier. Moreover, in [11], the method is extended to design the modified Mastrovito multiplication scheme proposed in [8]. They also present new results on the complexities of the Mastrovito multiplier for two classes of irreducible *pentanomials*. Recently, Rodriguez-Henriquez and Koc in [7] have proposed a PB multiplier for special case of pentanomials and have given its time and gate complexities.

In this article, first we review the multiplication scheme and its bit-parallel architecture presented in [6]. Then, using the *reduction* matrix $\mathbf{Q}$, the complexities of the multiplier based on a number of irreducible polynomials are obtained. We also present explicit formulations for the output coordinates of the multiplier in terms of its inputs. Such formulations can be directly coded using VHDL or Verilog languages to implement an efficient multiplier by someone who is not that familiar with finite field arithmetic. It is shown that for general irreducible polynomials, the space and time complexities of the proposed structure are lower than those available in the literature in terms of combined gate count and time delay. Furthermore, this architecture has fewer signals to be routed which is advantageous for VLSI implementation.

2   Polynomial Basis Multiplications over $GF(2^m)$

Let $P(x) = x^m + \sum_{i=0}^{m-1} p_i x^i$ be a monic irreducible polynomial over $GF(2)$ of degree m, where $p_i \in GF(2)$ for $i = 0, 1, \cdots, m-1$. Let $\alpha \in GF(2^m)$ be a root of $P(x)$, *i.e.*, $P(\alpha) = 0$. Then the set $\{1, \alpha, \alpha^2, \cdots, \alpha^{m-1}\}$ is referred to as the polynomial or standard basis and each element of $GF(2^m)$ can be written with respect to (w.r.t.) the polynomial basis (PB). Let A be an element in $GF(2^m)$, then the representation of A w.r.t. the PB is $A = \sum_{i=0}^{m-1} a_i \alpha^i$, $a_i \in \{0, 1\}$, where a_i's are the coordinates. For convenience, these coordinates will be denoted in vector notation[1] as $\mathbf{a} = [a_0, a_1, a_2, \cdots, a_{m-1}]^T$, where T denotes the transposition. Using this vector notation, the representation of A can be written as $A = \boldsymbol{\alpha}^T \mathbf{a}$, where $\boldsymbol{\alpha} = [1, \alpha, \alpha^2, \cdots, \alpha^{m-1}]^T$. Let S be the binary polynomial of degree not more than $2m - 2$ obtained by the direct multiplication of the PB representations of any two elements A and B of $GF(2^m)$, i.e.,

$$S = \left(\sum_{i=0}^{m-1} a_i \alpha^i \right) \cdot \left(\sum_{j=0}^{m-1} b_j \alpha^j \right) = \sum_{k=0}^{m-1} d_k \alpha^k + \sum_{k=0}^{m-2} e_k \alpha^{m+k}, \qquad (1)$$

where

$$\mathbf{d} = [d_0, d_1, \cdots, d_{m-1}]^T = \mathbf{Lb}, \qquad (2)$$

$$\mathbf{e} = [e_0, e_1, \cdots, e_{m-1}]^T = \mathbf{Ub}, \qquad (3)$$

---

[1] In this paper, vectors and matrices are shown with small and capital bold faces, respectively.

$$
\mathbf{L} \triangleq
\begin{bmatrix}
a_0 & 0 & 0 & 0 & \cdots & 0 \\
a_1 & a_0 & 0 & 0 & \cdots & 0 \\
a_2 & a_1 & a_0 & 0 & \cdots & 0 \\
\vdots & \vdots & & \ddots & \ddots & \vdots \\
a_{m-2} & a_{m-3} & \cdots & a_1 & a_0 & 0 \\
a_{m-1} & a_{m-2} & \cdots & a_2 & a_1 & a_0
\end{bmatrix}
, \mathbf{U} \triangleq
\begin{bmatrix}
0 & a_{m-1} & a_{m-2} & \cdots & a_2 & a_1 \\
0 & 0 & a_{m-1} & \cdots & a_3 & a_2 \\
\vdots & \vdots & & \ddots & \ddots & \vdots \\
0 & 0 & \cdots & 0 & a_{m-1} & a_{m-2} \\
0 & 0 & \cdots & 0 & 0 & a_{m-1}
\end{bmatrix}.
\tag{4}
$$

Then, the product $C = A \cdot B$ can be obtained by the following modulo reduction.

$$
C \triangleq \sum_{i=0}^{m-1} c_i \alpha^i \equiv S \bmod P(\alpha).
\tag{5}
$$

Definition 1. *[3] The reduction matrix* $\mathbf{Q}$ *is an* $m - 1$ *by* m *binary matrix which is obtained from*

$$
\alpha^\uparrow \equiv \mathbf{Q}\alpha \ (\bmod \ P(\alpha)),
\tag{6}
$$

where $\alpha^\uparrow = [\alpha^m, \ \alpha^{m+1}, \ \cdots, \ \alpha^{2m-2}]^T$.

Theorem 1. *[6] Let* C *be the product of* A *and* $B \in GF(2^m)$. *Then,*

$$
\mathbf{c} = [c_0, \ c_1, \ \cdots, \ c_{m-1}]^T = \mathbf{d} + \mathbf{Q}^T \mathbf{e},
\tag{7}
$$

where $\mathbf{d}$, $\mathbf{e}$ *and* $\mathbf{Q}$ *are defined in (2), (3), and (6) respectively.*

The corresponding architecture for polynomial basis multiplication over $GF(2^m)$ is shown in Figure 1. This structure is divided into two parts: IP-network and Q-network. The IP-network has m blocks (denoted as $I_0, I_1, \cdots, I_{m-1}$) which generates vectors $\mathbf{d}$ and $\mathbf{e}$ in accordance with (2) and (3), using m^2 AND gates and $(m-1)^2$ XOR gates. Using (2) and (3), the delay for d_j, $0 \le j \le m - 1$, and e_i, $0 \le i \le m - 2$, can be calculated from

$$
T(d_j) = T_A + \lceil \log_2(j+1) \rceil T_X, \qquad 0 \le j \le m - 1,
\tag{8}
$$
$$
T(e_i) = T_A + \lceil \log_2(m-i-1) \rceil T_X, \qquad 0 \le i \le m - 2.
\tag{9}
$$

In Figure 1, the Q-network takes $\mathbf{d}$ and $\mathbf{e}$ as inputs and generates $\mathbf{c}$. It is noted that the number of lines on the interconnection bus IB is fixed and is equal to the number of e_j's, i.e., $m - 1$. In Figure 1, there are three buses, A, B and IB, and the number of lines on the buses is $3m - 1$.

In the following sections, we attempt to minimise the number of XOR gates of the Q-network for special irreducible polynomials, namely equally-spaced polynomials, trinomials, and pentanomials. We start with equally-spaced polynomials which are very structured and will help us present the remaining special cases with less difficulties.

3 Multipliers Using Equally-Spaced Polynomials

Definition 2. *A polynomial* $P(x) = x^{ns} + x^{(n-1)s} + \cdots + x^s + 1$, *over* $GF(2)$, *with* $ns = m$ *and* $1 \le s \le \lfloor \frac{m}{2} \rfloor$, *is called an equally-spaced polynomial (denoted as s-ESP) of degree* m.

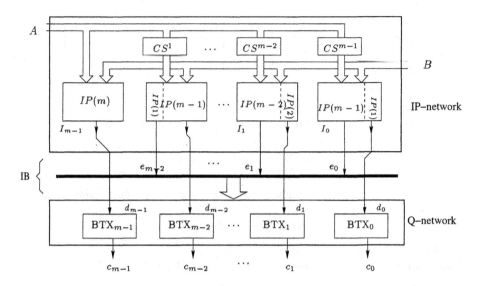

Fig. 1. Architecture of the multiplier over $GF(2^m)$, where CS^i represents an i- fold cyclic shift.

When $s = 1$, we have 1-ESP which is the same as the all-one polynomial (AOP) which has the highest Hamming weight among all polynomials of degree m. On the other hand, $s = \lfloor \frac{m}{2} \rfloor$ results in the least Hamming weight irreducible polynomial (i.e., trinomial) of degree m. It is easy to check that for an equally spaced trinomial m is even and $s = \frac{m}{2}$.

Theorem 2. *For an s-ESP based multiplier over $GF(2^m)$, the number of AND gates (N_A), the number of XOR gates (N_X) and time delay (T_C) are $N_A = m^2$, $N_X = m^2 - s$, and $T_C = T_A + (1 + \lceil \log_2 m \rceil) T_X$, respectively.*

Proof. When α is a root of the s-ESP of degree m as defined above, we have

$$\alpha^{m+i} = \begin{cases} \alpha^i + \alpha^{s+i} + \cdots + \alpha^{(n-1)s+i}, & 0 \leq i < s, \\ \alpha^{i-s}, & s \leq i \leq m-2. \end{cases} \quad (10)$$

Using (10), the reduction matrix $\mathbf{Q}$ is obtained as

$$\mathbf{Q} = \begin{bmatrix} \mathbf{I_s} \, \mathbf{I_s} \, \cdots \, \mathbf{I_s} \\ \mathbf{I_{m-s-1}} \, \mathbf{0_{s+1}} \end{bmatrix}, \quad (11)$$

where $\mathbf{I_j}$ is the $j \times j$ unity matrix and $\mathbf{0_{s+1}}$ is a zero matrix which has $m - s - 1$ rows and $s + 1$ columns. The graphical representations of $\mathbf{Q}$ in (11) for different values of s are shown in Figure 2. In this figure, non-zero entries of $\mathbf{Q}$ are shown with the small squares.

In order to obtain exact expressions for N_X and T_C, first we obtain the coordinates of C. To this end, from Theorem 1 and (11), one can write

$$c_j = d'_j + e_{j \bmod s}, \quad 0 \leq j \leq m-1, \quad (12)$$

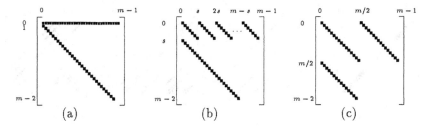

Fig. 2. Graphical representations of the locations of non-zeros entries of $\mathbf{Q}$ for s-ESP $P(x) = x^{ns} + x^{(n-1)s} + \cdots + x^s + 1$, $m = ns$. (a) $s = 1$ (AOP), (b) $1 < s < \frac{m}{2}$, (c) $s = \frac{m}{2}$ (trinomial).

where

$$d'_j = \begin{cases} d_j + e_{j+s} & 0 \leq j \leq m - s - 2, \\ d_j & m - s - 1 \leq j \leq m - 1. \end{cases} \tag{13}$$

Thus, using (12) and (13), the exact XOR gate count for an s-ESP based multiplier is $N_X = m^2 - s$. Also, by using (8) and (9), d'_j of (13) can be generated with a maximum gate delay of $T_A + (1 + \lceil \log_2 m \rceil) T_X$.

It is worth mentioning that the resultant number of signal lines on IB reduces from $m - 1$ to s, which is considerably lower than the s-ESP based Mastrovito multiplier which has $\frac{m(m-s)}{2s} + m$ signal lines [4]. Thus, the total number of lines on the buses of the multiplier is $2m + s$.

4   Extension to More Generic Polynomials

Here we consider irreducible polynomials of the form $P(x) = x^m + x^{k_t} + \cdots + x^{k_2} + x^{k_1} + 1$, where $1 \leq k_1 < k_2 < \cdots < k_t \leq \frac{m}{2}$. The Hamming weight of $P(x)$ is $t + 2$ and the degree of the second leading term is less than or equal to $\frac{m}{2}$. All five binary fields recommended by NIST for ECDSA can be constructed by such irreducible polynomials.

In order to apply the general formulation stated in Section 2 to these polynomials, first we obtain the corresponding $\mathbf{Q}$ matrix. Note that all the rows of the $\mathbf{Q}$ matrix are the PB representations of α^{m+i}, $0 \leq i \leq m - 2$, where α is a root of $P(x)$. Since $P(\alpha) = 0$, then $\alpha^m = 1 + \alpha^{k_1} + \alpha^{k_2} + \cdots + \alpha^{k_t}$. Thus, the 0-th row, i.e., $i = 0$, has only ones in these $t + 1$ columns of $\mathbf{Q}$: $0, k_1, k_2, \cdots, k_t$. The consecutive rows of this matrix can be obtained by using a linear feedback shift register (LFSR). As a result, the rows with $i = 0$ to $m - k_t - 1$ of $\mathbf{Q}$ have $t + 1$ ones.

The $\mathbf{Q}$ matrix for $t = 1$ and $t = 3$ (i.e., trinomials and pentanomials, respectively) are shown in Figure 3. As shown in this figure, row i, $0 \leq i \leq m - k_t - 1$ of $\mathbf{Q}$ has $t + 1$ ones corresponding to the $t + 1$ segmented lines. When the last column of $\mathbf{Q}$ contains one which takes place in row $i = m - k_j - 1$, $j = t, \cdots, 2, 1$, the next row originates new $t + 1$ lines in columns: $0, k_1, k_2$, up to k_t provided

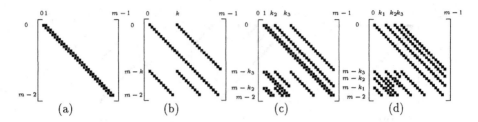

Fig. 3. Graphical representations of the reduction matrix $\mathbf{Q}$ for trinomials: (a) $k = k_1 = 1$ (b) $1 < k < \frac{m}{2}$ (see Figure 2(c) for $k_1 = \frac{m}{2}$); and for pentanomials: (c) $k_1 = 1$ (d) $1 < k_1 \leq \frac{m}{2}$.

that there is no previous lines that pass these columns. If there exists a previous line that passes the column k_j, $1 \leq j \leq t$, then the previous line terminates in column $k_j - 1$ and no new line originates from column k_j due to XORing of two lines. This happens in row $\frac{m}{2}$ and column $\frac{m}{2}$ in Figure 2(c) for trinomials when $k_1 = \frac{m}{2}$. This is also the case for pentanomials where $t = 3$ and it is shown in Figures 3(c) and 3(d) for $k_1 = 1$ and $1 < k_1 \leq \frac{m}{2}$, respectively.

We divide the lines of $\mathbf{Q}$ into $t + 1$ sets (see Figure 4 for $t = 3$) such that $\mathbf{Q} = \mathbf{Q}_0 + \mathbf{Q}_1 + \mathbf{Q}_2 + \cdots + \mathbf{Q}_t$ where non-zero entries of $\mathbf{Q}_i$, $0 \leq i \leq t$ start from the column k_i (assume that $k_0 = 0$). It is noted that the last non-zero entry of sub-matrix $\mathbf{Q}_i$, $1 \leq i \leq t$ is in column $m - 1$, whereas the one in $\mathbf{Q}_0$ is in column $m - 2$. Moreover, the number of ones in each column of $\mathbf{Q}_i$, $0 \leq i \leq t$ is at most $t + 1$ if $k_1 > 1$, and t if $k_1 = 1$.

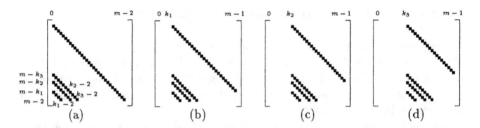

Fig. 4. Graphical representations of submatrices of $\mathbf{Q} = \mathbf{Q}_0 + \mathbf{Q}_1 + \mathbf{Q}_2 + \mathbf{Q}_3$ for pentanomials $P(x) = x^m + x^{k_3} + x^{k_2} + x^{k_1} + 1$, where $1 < k_1 < k_2 < k_3 \leq \frac{m}{2}$, (see Figure 3(d) for $\mathbf{Q}$). (a) $\mathbf{Q}_0$, (b) $\mathbf{Q}_1$, (c) $\mathbf{Q}_2$, (d) $\mathbf{Q}_3$.

Theorem 3. *The number of XOR gates and the time delay of the multiplier based on the irreducible polynomial* $P(x) = x^m + x^{k_t} + \cdots + x^{k_2} + x^{k_1} + 1$, $1 \leq k_1 < k_2 < \cdots < k_t \leq \frac{m}{2}$ *are*

$$N_X = (m + t)(m - 1)$$

and

$$T_C = T_A + \left(\lceil \log_2(t+1) \rceil + \left\lceil \log_2(\left\lceil \frac{t}{2} \right\rceil + 1) \right\rceil + \lceil \log_2(m-1) \rceil \right) T_X.$$

Proof. Let us denote $\mathbf{e}^{(i)} = [e_0^{(i)}, e_1^{(i)}, \cdots, e_{m-1}^{(i)}]^T = \mathbf{Q}_i^T \mathbf{e}$, $0 \leq i \leq t$, then using Theorem 1, we can obtain the coordinates of the pentanomial based multiplication as

$$\mathbf{c} = \mathbf{d} + \mathbf{e}^{(0)} + \mathbf{e}^{(1)} + \mathbf{e}^{(2)} + \cdots + \mathbf{e}^{(t)}. \tag{14}$$

First, let us assume $k_1 \neq 1$. Using $\mathbf{Q}_0$ (see Figure 4(a) for $t = 3$), the elements of $\mathbf{e}^{(0)}$ are as follows:

$$e_j^{(0)} = \begin{cases} e_j + e_{j+m-k_t} + \cdots + e_{j+m-k_2} + e_{j+m-k_1}, & \text{if } 0 \leq j \leq k_1 - 2 \\ e_j + e_{j+m-k_t} + \cdots + e_{j+m-k_2} & \text{if } k_1 - 1 \leq j \leq k_2 - 2 \\ \vdots & \vdots \\ e_j + e_{j+m-k_t} & \text{if } k_{t-1} - 1 \leq j \leq k_t - 2 \\ e_j & \text{if } k_t - 1 \leq j \leq m - 2 \\ 0 & \text{if } j = m - 1. \end{cases} \tag{15}$$

The total number of XOR gates to form $e_j^{(0)}$'s, $0 \leq j \leq k_t - 2$, is $N_1 = t(k_1 - 1) + (t-1)(k_2 - k_1) + \cdots + k_t - k_{t-1} = \sum_{i=1}^{t} k_i - t$. Let $T(e_j^{(0)})$ denote the time delay due to gates to find $e_j^{(0)}$. As seen in (15), the longest path delay is to obtain $e_0^{(0)} = e_0 + e_{m-k_t} + \cdots + e_{m-k_2} + e_{m-k_1}$, i.e., $T(e_j^{(0)}) \leq T(e_0^{(0)})$. In order to reduce this delay, we first add any two terms except c_0, e.g., $e_{m-k_j} + e_{m-k_i}$, $1 \leq i, j \leq t$, $i \neq j$. Then add these $\lceil \frac{t}{2} \rceil$ signals to c_0 using a binary tree of XOR gates. Since $T(e_j) = T_A + \lceil \log_2(m - j - 1) \rceil T_X$, then $T(e_{m-k_j} + e_{m-k_i}) \leq T_X + T(e_{m-k_t}) = T_A + (1 + \lceil \log_2(k_t - 1) \rceil) T_X \leq T_A + \lceil \log_2(m-1) \rceil T_X$, where the last inequality is due to $k_t \leq \frac{m}{2}$. Thus, we have

$$T(e_j^{(0)}) \leq \begin{cases} T_A + \left(\lceil \log_2(\lceil \frac{t}{2} \rceil + 1) \rceil + \lceil \log_2(m-1) \rceil \right) T_X, & \text{if } 0 \leq j \leq k_t - 2 \\ T_A + \lceil \log_2(m-1) \rceil T_X & \text{if } k_t - 1 \leq j \leq m - 2. \end{cases} \tag{16}$$

By reusing the signals of $e_j^{(0)}$'s, the coordinates of $\mathbf{e}^{(i)}$, for $1 \leq i \leq t$, can be obtained as

$$e_j^{(i)} = \begin{cases} 0, & if \ 0 \leq j \leq k_i - 1 \\ e_{j-k_i}^{(0)} & \text{otherwise.} \end{cases} \tag{17}$$

This results in the coordinates of $C = AB$ as

$$c_j = d_j + \begin{cases} e_j^{(0)} & \text{if } 0 \leq j \leq k_1 - 1 \\ e_j^{(0)} + e_j^{(1)} & \text{if } k_1 \leq j \leq k_2 - 1 \\ \vdots & \vdots \\ e_j^{(0)} + e_j^{(1)} + \cdots + e_j^{(t-1)} & \text{if } k_{t-1} \leq j \leq k_t - 1 \\ e_j^{(0)} + e_j^{(1)} + \cdots + e_j^{(t)} & \text{if } k_t \leq j \leq m - 2 \\ e_j^{(1)} + e_j^{(2)} + \cdots + e_j^{(t)} & \text{if } j = m - 1 \end{cases} \tag{18}$$

by using (14). To realize (18) in hardware, one requires $N_2 = m + (k_2 - k_1) + 2(k_3 - k_2) + \cdots + (t-1)(k_t - k_{t-1}) + t(m - k_3 - 1) + t - 1 = (t+1)m - \sum_{i=1}^{t} k_i - 1$ XOR gates. Thus, the total XOR gates needed for the multiplier is $(m-1)^2 + N_1 + N_2 = (m+t)(m-1)$.

To obtain the time delay of the proposed multiplier, we use a binary tree for each coordinate in (18). For $j \notin [k_t, m-2]$, it is seen in (18) that $T_C \leq \lceil \log_2(t+1) \rceil T_X + T(e_0^{(0)})$ and the proof is complete by using (16). Now, we need only to obtain the time delay of $c_j' s$ for $k_t \leq j \leq m-2$. For $j \in [k_t, m-2]$, if we form $c_j = (d_j + e_j^{(0)}) + e_j^{(1)} + e_j^{(2)} + \cdots + e_j^{(t)}$ such that $d_j + e_j^{(0)}$ is calculated first, then

$$T(d_j + e_j^{(0)}) \leq T_A + (1 + \lceil \log_2(m-1) \rceil) T_X$$
$$\leq T_A + \left(\left\lceil \log_2\left(\left\lceil \frac{t}{2} \right\rceil + 1 \right) \right\rceil + \lceil \log_2(m-1) \rceil \right) T_X.$$

Also, using (17) and (16), one can see

$$T(e_j^{(t)}) \leq T_A + \left(\left\lceil \log_2\left(\left\lceil \frac{t}{2} \right\rceil + 1 \right) \right\rceil + \lceil \log_2(m-1) \rceil \right) T_X$$

which implies that

$$T_C \leq T_A + \left(\lceil \log_2(t+1) \rceil + \left\lceil \log_2\left(\left\lceil \frac{t}{2} \right\rceil + 1 \right) \right\rceil + \lceil \log_2(m-1) \rceil \right) T_X$$

and the proof is complete.

In addition to the three buses shown in Figure 1 now, there will be another bus in the middle of the **Q**-network for signals $e_j^{(0)}$ for $0 \leq j \leq k_t - 2$. Thus, the total number of lines on the buses is $3m + k_t - 2$.

Corollary 1. *For $k_1 = 1$ and $t > 1$, the time delay would reduce to*

$$T_A + \left(\lceil \log_2(t+1) \rceil + \left\lceil \log_2 \left\lceil \frac{t}{2} \right\rceil \right\rceil + \lceil \log_2(m-1) \rceil \right) T_X.$$

Based on the above results, one can obtain the time delay and the number of XOR gates for the trinomial based multiplier by substituting $t = 1$ in Theorem 3, for $k_1 \neq \frac{m}{2}$ and $s = \frac{m}{2}$ in Theorem 2 for $k_1 = \frac{m}{2}$. Note that the results for $k_1 = \frac{m}{2}$ are obtained using the implementation of the $\frac{m}{2}$-ESP based multiplier.

5   Special Classes of Pentanomials

A polynomial with five non-zero coefficients, i.e., $P(x) = x^m + x^{k_3} + x^{k_2} + x^{k_1} + 1$, where $1 \leq k_1 < k_2 < k_3 \leq m-1$, is called a *pentanomial* of degree m. The non-zero constant term is due to the irreducibility properly needed to define the field. In terms of the values of k_is, the pentanomials can be divided into a number of different classes. Below we consider two special classes of irreducible pentanomials as proposed in [11].

5.1 Class 1: $k_3 \leq \frac{m}{2}$

For this class of irreducible pentanomial where $k_3 \leq \frac{m}{2}$, one can apply $t = 3$ to the complexity results we have presented in Section 4. This yields the following.

Corollary 2. *The gate counts and time delay of the multiplier for the the pentanomial* $P(x) = x^m + x^{k_3} + x^{k_2} + x^{k_1} + 1$, *where* $1 \leq k_1 < k_2 < k_3 \leq \frac{m}{2}$, *are*

$$N_A = m^2,$$
$$N_X = m^2 + 2m - 3,$$
$$T_C = \begin{cases} T_A + (3 + \lceil \log_2(m - 1) \rceil) T_X, & \text{if } k_1 = 1 \\ T_A + (4 + \lceil \log_2(m - 1) \rceil) T_X, & \text{otherwise,} \end{cases}$$

and the number of lines on the buses is $3m + k_3 - 2$.

The number of XOR gates can be reduced if we choose a pentanomial such that $k_1 = k_3 - k_2$. Towards this, let us introduce the following set of new signals

$$e'_j = e_{j+m-k_3} + e_{j+m-k_2}, 0 \leq j \leq k_2 - 2. \tag{19}$$

Equation (19) can be used to generate $e_j^{(0)}$, $0 \leq j \leq k_2 - 2$, by substituting $t = 3$ in (15) as follows

$$e_j^{(0)} = \begin{cases} e_j + e'_j + e_{j+m-k_1}, & \text{if } 0 \leq j \leq k_1 - 2 \\ e_j + e'_j & \text{if } k_1 - 1 \leq j \leq k_2 - 2 \\ e_j + e_{j+m-k_3} & \text{if } k_2 - 1 \leq j \leq k_3 - 2 \\ e_j & \text{if } k_3 - 1 \leq j \leq m - 2 \\ 0 & \text{if } j = m - 1. \end{cases} \tag{20}$$

The total number of XOR gates needed to generate $e_j^{(0)}$'s (see (20)) is $N_1 = k_1 + k_2 + k_3 - 3$ where $k_2 - 1$ of which is due to (19). Also, the maximum delay due to gates in (20) is

$$T(e_j^{(0)}) \leq \begin{cases} T_A + (2 + \lceil \log_2(m - 1) \rceil) T_X & \text{if } 0 \leq j \leq k_1 - 2 \\ T_A + (1 + \lceil \log_2(m - 1) \rceil) T_X & \text{if } k_1 - 1 \leq j \leq k_3 - 2 \\ T_A + \lceil \log_2(m - 1) \rceil T_X & \text{if } k_3 - 1 \leq j \leq m - 1. \end{cases} \tag{21}$$

Lemma 1. *With symbols defined as above, one has*

$$e_j^{(0)} + e_j^{(1)} = e'_{j+k_2-m}, \text{ for } m - k_2 \leq j \leq m - 2,$$
$$e_j^{(2)} + e_j^{(3)} = e_{j-k_2}^{(0)} + e_{j-k_2}^{(1)}, \text{ for } k_3 \leq j \leq m - 1.$$

Let us represent $e_j^{(01)}$, $0 \leq j \leq m - 1$, as the elements of $(\mathbf{Q}_0 + \mathbf{Q}_1)^T \mathbf{e}$, where $\mathbf{Q}_0$ and $\mathbf{Q}_1$ are shown in Figure 4(a) and Figure 4(b), respectively. Then, substituting $t = 3$ in the general case given in (18) and using the above lemma, we can obtain the coordinates of $C = AB$ as follows:

$$c_j = d_j + e_j^{(01)} + e_{j-k_2}^{(01)}, 0 \leq j \leq m - 1, \tag{22}$$

where $e_{j-k_2}^{(01)} = 0$ for $j < k_2$, and

$$
e_j^{(01)} = \begin{cases} e_j^{(0)} & \text{if } 0 \le j \le k_1 - 1 \\ e_j^{(0)} + e_j^{(1)} & \text{if } k_1 \le j \le m - k_2 - 1 \\ e'_{j+k_2-m} & \text{if } m - k_2 \le j \le m - 2 \\ e_j^{(1)} & \text{if } j = m - 1. \end{cases} \tag{23}
$$

As seen in (23), one has to realize $e_j^{(0)} + e_j^{(1)}$ for all $k_1 \le j \le m - k_2 - 1$ which requires $m - k_2 - k_1$ XOR gates. Once $e_j^{(01)}$'s are obtained, then equation (22) requires $2m - k_2$ XOR gates. Thus, the total number of XOR gates needed for the multiplier is $(m-1)^2 + N_1 + m - k_2 - k_1 + 2m - k_2 = m^2 + m + k_1 - 2$. Due to the reuse of terms e'_j, $0 \le j \le k_2 - 1$, and $e_j^{(0)} + e_j^{(1)}$, $k_1 \le j \le m - k_2 - 1$, additional lines needed on the bus in the **Q**-network are $(k_2 - 1)$ and $(m - k_1 - k_2)$, respectively. Thus, the total number of lines on the buses is increased to $4m + k_2 - 3$.

To obtain the time delay of the proposed multiplier, we use Table 1 which shows the maximum delay of the used signals in (22) for the given ranges of j in each row. In this figure i, $0 \le i \le 4$, represents the time delay of $T_A + (i + \lceil \log_2(m-1) \rceil) T_X$, and the numbers inside brackets are for $k_1 = 1$. Also, x determines either $e_j^{(01)}$ or $e_{j-k_2}^{(01)}$ to be added with d_j first to obtain c_j. In each row of this table, the delays are obtained for the first digit of the given range. This is because as j increases, the time delays of the used signals in each row of this table decreases. As seen in this table, the maximum delay of the multiplier is $T_A + (4 + \lceil \log_2(m-1) \rceil) T_X$. For $k_1 = 1$, only one signal, i.e., c_{k_3}, has the delay of $T_A + (4 + \lceil \log_2(m-1) \rceil) T_X$. One can reduce this delay to $T_A + (3 + \lceil \log_2(m-1) \rceil) T_X$ if only c_{k_3} is realized as $c_{k_3} = ((d_{k_3} + e_j^{(0)}) + e_j^{(1)}) + e_{k_3-k_2}^{(01)}$ by using one extra XOR gate.

Table 1. Maximum time delays of the signals, where i, $0 \le i \le 4$, represents the time delay of $T_A + (i + \lceil \log_2(m-1) \rceil) T_X$, numbers inside brackets are for $k_1 = 1$, and x determines either $e_j^{(01)}$ or $e_{j-k_2}^{(01)}$ to be added first with d_j.

j	$e_j^{(0)}$	$e_j^{(1)}$	$e_j^{(01)}$	$e_{j-k_2}^{(01)}$	$d_j + x$	c_j
$0 \le j \le k_1 - 1$	2(1)	-	2(1), x	-	3	3
$k_1 \le j \le k_2 - 1$	1	2(1)	3(2), x	-	4(3)	4(3)
$k_2 \le j \le k_3 - 1$	1	2(1)	3(2)	2(1), x	3(2)	4(3)
$k_3 \le j \le k_3 + k_1 - 1$	0	1	2, x	3(2)	3	4
$k_3 + k_1 \le j \le m - k_2 - 1$	0	0	1, x	3(2)	2	4(3)
$m - k_2 \le j \le m - 1$	0	0	1, x	3(2)	2	4(3)
$j = m - 1$	-	0	1, x	1	2	3

Based on the above results, we can state the following.

Theorem 4. *The gate counts and time delay of the multiplier based on the pentanomial* $P(x) = x^m + x^{k_3} + x^{k_2} + x^{k_1} + 1$, *where* $1 \le k_1 < k_2 < k_3 \le \frac{m}{2}$, *and* $k_3 - k_2 = k_1$ *are*

$$
N_A = m^2,
$$

$$N_X = \begin{cases} m^2 + m & if\, k_1 = 1 \\ m^2 + m + k_1 - 2\ otherwise, \end{cases}$$

$$T_C = \begin{cases} T_A + (3 + \lceil \log_2(m-1) \rceil)\, T_X, & if\, k_1 = 1 \\ T_A + (4 + \lceil \log_2(m-1) \rceil)\, T_X, & otherwise, \end{cases}$$

and the number of lines on the buses is $4m + k_2 - 3$.

Remark 1. To verify that class 1 irreducible pentanomials exist, we have used a Maple^{TM} program for $m \in [160, 600]$ and have found that at least one irreducible pentanomial exists for each m in the range of 160 to 600. This is of interest to elliptic curve cryptosystem designers. In order to minimise the number of XOR gates of the multiplier, we have obtained irreducible pentanomials such that k_1 is minimum. We have also observed that, k_1 is less than or equal six for all m in the above mentioned range.

It is noted that the pentanomial presented in [7] is a special case when $k_1 = 1$.

5.2   Class 2: $m - k_3 = k_3 - k_2 = k_2 - k_1 = s$, $\frac{m-1}{8} \leq s \leq \frac{m-1}{3}$

We refer to polynomials $P(x) = x^m + x^{k_3} + x^{k_2} + x^{k_1} + 1$, where $1 \leq k_1 < k_2 < k_3 \leq m - 1$, and $m - k_3 = k_3 - k_2 = k_2 - k_1 = s$ as class 2 type. Similar to the other special irreducible polynomials, here we first obtain the corresponding reduction matrix. Then the coordinates and complexities of the multiplier can be obtained. Based on the values of s (or $k_1 = m - 3s$), we can divide the reduction matrix into different forms. Because of lack of space, only three of them are presented here. These $\mathbf{Q}$ matrices for $\frac{m-1}{8} \leq s \leq \frac{m-1}{3}$ (or $1 \leq k_1 \leq 5s + 1$) are shown in Figure 5. Based on this figure, we can state the following theorem.

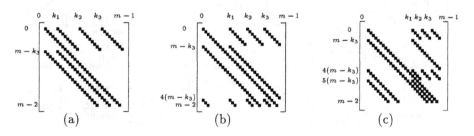

Fig. 5. Graphical representations of the reduction matrix $\mathbf{Q}$ for class 2 pentanomials $P(x) = x^m + x^{k_3} + x^{k_2} + x^{k_1} + 1$, where $m - k_3 = k_3 - k_2 = k_2 - k_1 = s$. (a) $\frac{m-1}{4} \leq s \leq \frac{m-1}{3}$ or $1 \leq k_1 \leq s + 1$ (see Figure 2(a) for $k_1 = s$), (b) $\frac{m-1}{5} \leq s \leq \frac{m-1}{4}$ or $s + 1 < k_1 \leq 2s + 1$, (c) $\frac{m-1}{8} \leq s < \frac{m-1}{5}$ or $2s + 1 < k_1 \leq 5s + 1$.

Theorem 5. *The gate counts and the time delay of the multiplier for the pentanomial* $P(x) = x^m + x^{m-s} + x^{m-2s} + x^{m-3s} + 1$, *for* $\frac{m-1}{8} \leq s \leq \frac{m-1}{3}$ *are* $N_A = m^2,$

$$N_X = \begin{cases} m^2 + m - s - 1, & if \frac{m-1}{4} \leq s \leq \frac{m-1}{3} \\ m^2 + 2m - 5s - 2 & if \frac{m-1}{5} \leq s < \frac{m-1}{4} \\ m^2 + m - 2 & if \frac{m-1}{8} \leq s < \frac{m-1}{5} \end{cases}$$

$$T_C = \begin{cases} T_A + (3 + \lceil \log_2(m-1) \rceil) T_X, & if \frac{m-1}{5} \leq s \leq \frac{m-1}{3} \\ T_A + (4 + \lceil \log_2(m-1) \rceil) T_X, & otherwise. \end{cases}$$

Remark 2. Using Maple^{TM}, we have found that there exists 147 values of m, where $m \in [160, 600]$ such that polynomial $P(x) = x^m + x^{m-s} + x^{m-2s} + x^{m-3s} + 1$, $1 \leq s \leq \frac{m-1}{3}$ is irreducible. Among them only 23 have $1 \leq s < \frac{m-1}{8}$.

Table 2. Comparison of related polynomial basis multipliers.

Reference	Special Case	#XOR	Time delay
$P(x) = x^{ns} + x^{(n-1)s} + \cdots + x^s + 1, m = ns$			
This paper,[2,11]	-	$m^2 - s$	$T_A + (1 + \lceil \log_2 m \rceil) T_X$
$P(x) = x^m + x^k + 1$			
[9,2,11]	$k = 1$	$m^2 - 1$	$T_A + (1 + \lceil \log_2 m \rceil) T_X$
[9,2,11]	$1 < k \leq \frac{m}{2}$	$m^2 - 1$	$T_A + (2 + \lceil \log_2 m \rceil) T_X$
This paper,[10]	$1 \leq k \leq \frac{m}{2}$	$m^2 - 1$	$T_A + (2 + \lceil \log_2(m-1) \rceil) T_X$
$P(x) = x^m + x^{k_t} + \cdots + x^{k_2} + x^{k_1} + 1, 1 \leq k_1 < k_2 < \cdots < k_t \leq \frac{m}{2}$			
[11]	$t > 1$	$(m+t)(m-1)$	$T_A + (2t + \lceil \log_2 m \rceil) T_X$
This paper	$t > 1$	$(m+t)(m-1)$	$T_A + (\lceil \log_2(\lceil \frac{t}{2} \rceil + 1) \rceil + \lceil \log_2(t+1) \rceil + \lceil \log_2(m-1) \rceil) T_X$
$P(x) = x^m + x^{k_3} + x^{k_2} + x^{k_1} + 1, 1 < k_1 < k_2 < k_3 \leq \frac{m}{2}$			
[11]	$k_1 \geq 1$	$m^2 + 2m - 3$	$T_A + (6 + \lceil \log_2 m \rceil) T_X$
This paper	$k_1 > 1$	$m^2 + 2m - 3$	$T_A + (4 + \lceil \log_2(m-1) \rceil) T_X$
This paper	$k_1 = 1$	$m^2 + 2m - 3$	$T_A + (3 + \lceil \log_2(m-1) \rceil) T_X$
This paper	$k_3 - k_2 = k_1$	$m^2 + m + k_1 - 2$	$T_A + (4 + \lceil \log_2(m-1) \rceil) T_X$
[7]	$k_3 - k_2 = k_1 = 1$	$m^2 + m + 2k_2$	$T_A + (3 + \lceil \log_2(m-1) \rceil) T_X$
This paper	$k_3 - k_2 = k_1 = 1$	$m^2 + m$	$T_A + (3 + \lceil \log_2(m-1) \rceil) T_X$
This paper,[7]	$k_i = i$	$m^2 + m$	$T_A + (3 + \lceil \log_2(m-1) \rceil) T_X$
$P(x) = x^m + x^{m-s} + x^{m-2s} + x^{m-3s} + 1$			
[11]	$1 \leq s \leq \frac{m-1}{3}$	$m^2 + 4m - 5s - 5$	$T_A + (\lfloor \frac{d}{4} \rfloor + 4 + \lceil \log_2(m-1) \rceil) T_X$
[11]	$s \leq \frac{m-1}{3}$	$\geq m^2 + 2.33m - 7$	$\geq T_A + (4 + \lceil \log_2(m-1) \rceil) T_X$
This paper	$\frac{m-1}{8} \leq s \leq \frac{m-1}{3}$	$\leq m^2 + m$	$\leq T_A + (4 + \lceil \log_2(m-1) \rceil) T_X$

6   Complexity Results and Concluding Remarks

In this article, time and space complexities of bit parallel multipliers for $GF(2^m)$ have been considered. A comparison of our newly derived gate counts and delays

Table 3. Comparison of the structure of Figure 1 with the Mastrovito multiplier in terms of number of number of lines on the buses.

Multipliers	# Lines on the buses			
	trinomial	s-ESP	pentanomial	generic
Mastrovito [4]	$3m - 1$	$\frac{m(m-s)}{2s} + 2m$	$5m - 3$	$(t + 2)(m - 1) + 2$
This paper	$3m - 1$	$2m + s$	$\leq 4m + k_2$	$3m + k_t - 2$

with those of existing ones is shown in Table 2. As seen in this table, for trinomial $x^m + x + 1$, the multiplier of Figure 1 has one additional XOR gate delay compared to the best one available in the literature, i.e., [2,11]. However, our results for the ESPs and trinomials ($k \neq 1$) match the corresponding best results available ([2, 11] and [9]). Also, the resultant gate and time complexities for trinomials match those presented in [10].

For a more generic irreducible polynomial as discussed in Section 4, the multiplier in Figure 1 has the same gate count but a shorter time delay compared to [11]. For class 1 pentanomials, this multiplier is faster than [11] and has fewer XOR gates if the special case of $k_3 - k_2 = k_1$ is used. This proposed special case of class 1 covers the case of pentanomials reported in [7], where $k_1 = 1$. Compared to the multiplier proposed in [7], the multiplier discussed in this paper for the special case of $k_1 = k_3 - k_2 = 1$ has $2k_2$ fewer XOR gates and match the ones proposed in [7] for $k_1 = 1$ and $k_2 = 2$. Also, for class 2 pentanomials, our multiplier is either faster or has the same gate delay and has at least $1.33m - 7$ fewer XOR gates than the multiplier reported in [11].

In VLSI implementation, in addition to the gate counts, the number of lines on the buses is also an important parameter which determines the space complexity and consequently its actual time delay. Table 3 compares this metric of the proposed architecture with that of Mastrovito multiplier [4]. As shown in this table, the architectures discussed here have a fewer number of lines on the buses compared to the well known Mastrovito multiplier.

Acknowledgements. This work has been supported in part by an NSERC postdoctoral fellowship awarded to A. Reyhani-Masoleh and in part by an NSERC grant awarded to M. A. Hasan.

References

1. G. B. Agnew, R. C. Mullin, and S. A. Vanstone. "An Implementation of Elliptic Curve Cryptosystems Over $F_{2^{155}}$". *IEEE J. Selected Areas in Communications*, 11(5):804–813, June 1993.
2. A. Halbutogullari and C. K. Koc. "Mastrovito Multiplier for General Irreducible Polynomials". *IEEE Transactions on Computers*, 49(5):503–518, May 2000.
3. E. D. Mastrovito. "VLSI Designs for Multiplication over Finite Fields $GF(2^m)$". In *LNCS-357, Proc. AAECC-6*, pages 297–309, Rome, July 1988. Springer-Verlag.
4. E. D. Mastrovito. *VLSI Architectures for Computation in Galois Fields*. PhD thesis, Linkoping Univ., Linkoping Sweden, 1991.

5. A.J. Menezes, I.F. Blake, X. Gao, R.C. Mullin, S.A. Vanstone, and T. Yaghoobian. *Applications of Finite Fields*. Kluwer Academic Publishers, 1993.

6. A. Reyhani-Masoleh and M. A. Hasan. "A New Efficient Architecture of Mastrovito Multiplier over $GF(2^m)$". In 20th *Biennial Symposium on Communications*, pages 59–63, Kingston, Ontario, Canada, May 2000.

7. F. Rodriguez-Henriquez and C. K. Koc. "Parallel Multipliers Based on Special Irreducible Pentanomials". *IEEE Transactions on Computers*, to appear, 2003, available at http://islab.oregonstate.edu/koc/Publications.html.

8. L. Song and K. K. Parhi. "Low Complexity Modified Mastrovito Multipliers over Finite Fields $GF(2^M)$". In *ISCAS-99, Proc. IEEE International Symposium on Circuits and Systems*, pages 508–512, 1999.

9. B. Sunar and C. K. Koc. "Mastrovito Multiplier for All Trinomials". *IEEE Transactions on Computers*, 48(5):522–527, May 1999.

10. H. Wu. "Bit-Parallel Finite Field Multiplier and Squarer Using Polynomial Basis". *IEEE Transactions on Computers*, 51(7):750–758, July 2002.

11. T. Zhang and K. K. Parhi. "Systematic Design of Original and Modified Mastrovito Multipliers for General Irreducible Polynomials". *IEEE Transactions on Computers*, 50(7):734–748, July 2001.

Efficient Modular Reduction Algorithm in $\mathbb{F}_q[x]$ and Its Application to "Left to Right" Modular Multiplication in $\mathbb{F}_2[x]$

Jean-François Dhem

Gemplus Corporate Product R&D
Security Technologies Department
Card Security Group (STD/CSG/CHA)
Avenue du Jujubier – ZI Athelia IV
F-13705 La Ciotat Cedex
France
jf.dhem@ieee.org

Abstract. This paper describes a new efficient method of modular reduction in $\mathbb{F}_q[x]$ suited for both software and hardware implementations. This method is particularly well adapted to smart card implementations of elliptic curve cryptography over $\mathsf{GF}(2^p)$ using a polynomial representation. Many publications use the equivalent in $\mathbb{F}_2[x]$ of Montgomery's modular multiplication over integers. We show here an equivalent in $\mathbb{F}_q[x]$ to the generalized Barrett's modular reduction over integers. The attractive properties of the last method in $\mathbb{F}_2[x]$ allow nearly ideal implementations in hardware as well as in software with minimum additional resources as compared to what is available on usual processor architecture.

An implementation minimizing the memory accesses is described for both Montgomery's implementation and ours. This shows identical computing and memory access resources for both methods. The new method also avoids the need for the bulky normalization (denormalization) which is required by Montgomery's method to obtain a correct result.

Keywords: Smart card, cryptography, modular multiplication, quotient evaluation, elliptic curves, ECDSA, Montgomery, Barrett, multiply and add without carries, multiplications in $\mathbb{F}_2[x]$.

1 Introduction

Montgomery's multiplication in $\mathbb{F}_2[x]$ is a well known "right to left" modular reduction (see for example [1]) and directly derives from Montgomery's modular reduction over integers [2,3]. It is often used for efficient software implementations of elliptic curves crypto-systems like ECDSA [4], using polynomial representations. The main disadvantage of Montgomery's method is the bulky normalization - denormalization phase required to obtain a correct modular reduction. In the set of integers, the corresponding "left to right" methods [5,6,

C.D. Walter et al. (Eds.): CHES 2003, LNCS 2779, pp. 203–213, 2003.

7] are less efficient in software because they require additional non standard (on general purpose microprocessors) resources (special multiplier) to obtain similar performances. We will show that the generalized Barrett's method can be made more efficient in $\mathbb{F}_2[x]$ so that it can compete directly with the Montgomery's method. Even more, because of the absence of normalization, the memory requirements for data and ROM code of our method are smaller. This could be of main importance in constrained implementations (e.g. smart cards).

The easiest "left to right" method to compute a quotient and a remainder is the one taught at school: the quotient is calculated by first zeroing the upper (most significant) part of the numerator by adding / subtracting a multiple of the denominator. Montgomery's method, however, zeroes the least significant bits of the number to reduce.

In section 2, the way how the quotient is computed in order to perform an improved "left to right" modular reduction in $\mathbb{F}_q[x]$ is described.

In section 3, the particular case of the modular multiplication in $\mathbb{F}_2[x]$ is studied.

2   Quotient Evaluation in $\mathbb{F}_q[x]$

In order to compute $S(x) = U(x) \bmod N(x)$ we could first evaluate, in a scholastic way, the quotient $Q(x)$ defined by the equation $U(x) = Q(x)N(x) + S(x)$ where $S(x)$ and $Q(x)$ are respectively the remainder and the quotient of the Euclidean division [8] of $U(x)$ by $N(x)$. By similarity to computations over integers we define $Q(x) = \left\lfloor \frac{U(x)}{N(x)} \right\rfloor$. The degree p of the polynomial $N(x)$ is noted $\deg(N)$. We also define $\alpha = \deg(U) - \deg(N)$.

In most applications, like in elliptic curve crypto-systems, $N(x)$ is fixed (when working on a given elliptic curve). To speed up the computations, we can precompute $\left\lfloor \frac{x^{p+\beta}}{N(x)} \right\rfloor$ $(= R(x))$, for some value of β defined hereafter. The quotient's evaluation may then be reduced to a multiplication and an appropriate shift (division by a power of x) as shown in equation 1 (see [5] and [7] when working over integers).

$$\hat{Q}(x) = \left\lfloor \frac{\left\lfloor \frac{U(x)}{x^p} \right\rfloor \left\lfloor \frac{x^{p+\beta}}{N(x)} \right\rfloor}{x^\beta} \right\rfloor = \left\lfloor \frac{T(x)R(x)}{x^\beta} \right\rfloor \tag{1}$$

For the comprehension of equation 1, $\lfloor A(x)/B(x) \rfloor$ represents the quotient of the polynomial division of $A(x)$ by $B(x)$, discarding the remainder ($A(x)$ and $B(x)$ are some polynomials). In the next sections we will demonstrate the equivalence of equation 1 with the real quotient $Q(x) = \lfloor U(x)/N(x) \rfloor$.

2.1   Equivalence between $\hat{Q}(x)$ and $Q(x)$

In the present section, $A_k(x)$ represents the polynomial $A(x)$ of degree k (where $A(x)$ is some polynomial).

We can write:

$$\frac{U(x)}{x^p} = \Phi_\alpha(x) + \frac{\varphi_{p-1}(x)}{x^p}$$

where $\Phi_\alpha(x)$ is the quotient and $\varphi_{p-1}(x)$ the remainder of the division. Similarly, we can write:

$$\frac{x^{p+\beta}}{N(x)} = \Lambda_\beta(x) + \frac{\lambda_{p-1}(x)}{N_p(x)}$$

Where $\Lambda_\beta(x)$ is the quotient and $\lambda_{p-1}(x)$ the remainder of the division. We now can write:

$$Q(x) = \left\lfloor \frac{\left(\frac{U(x)}{x^p}\right)\left(\frac{x^{p+\beta}}{N(x)}\right)}{x^\beta} \right\rfloor \tag{2}$$

$$= \left\lfloor \frac{\left(\Phi_\alpha(x) + \frac{\varphi_{p-1}(x)}{x^p}\right)\left(\Lambda_\beta(x) + \frac{\lambda_{p-1}(x)}{N_p(x)}\right)}{x^\beta} \right\rfloor \tag{3}$$

$$= \left\lfloor \frac{\Phi_\alpha(x)\Lambda_\beta(x) + \left(\Phi_\alpha(x)\frac{\lambda_{p-1}(x)}{N_p(x)} + \Lambda_\beta(x)\frac{\varphi_{p-1}(x)}{x^p} + \frac{\varphi_{p-1}(x)}{x^p}\frac{\lambda_{p-1}(x)}{N_p(x)}\right)}{x^\beta} \right\rfloor \tag{4}$$

The second term of equation 4 is nullified only if $\beta \geq \alpha$. In that case only we can write:

$$Q(x) = \left\lfloor \frac{\Phi_\alpha(x)\Lambda_\beta(x)}{x^\beta} \right\rfloor = \left\lfloor \frac{\lfloor\Phi_\alpha(x)\rfloor\lfloor\Lambda_\beta(x)\rfloor}{x^\beta} \right\rfloor \tag{5}$$

$$= \left\lfloor \frac{\left\lfloor\Phi_\alpha(x) + \frac{\varphi_{p-1}(x)}{x^p}\right\rfloor\left\lfloor\Lambda_\beta(x) + \frac{\lambda_{p-1}(x)}{N_p(x)}\right\rfloor}{x^\beta} \right\rfloor \tag{6}$$

$$= \left\lfloor \frac{\left\lfloor\frac{U(x)}{x^p}\right\rfloor\left\lfloor\frac{x^{p+\beta}}{N(x)}\right\rfloor}{x^\beta} \right\rfloor = \hat{Q}(x) \tag{7}$$

$\square$

There is no need to chose $\beta > \alpha$ because this would require more computations than when choosing $\beta = \alpha$ as $R(x)$ will then be larger.

The previous quotient computation is thus valid for $\mathbb{F}_q[x]$.

In the next section, we will use our method in $\mathbb{F}_2[x]$ which will show improvements of its software (and hardware) implementations. $Q(x)$ is then a binary polynomial of degree α requiring $\alpha + 1$ bits for its binary representation.

3   Modular Multiplication in $\mathbb{F}_2[x]$

In $\mathbb{F}_2[x]$, the coefficients n_i of the polynomial $N(x) = n_p x^p + n_{p-1}x^{p-1} + \ldots + n_1 x + n_0$ are either '0' or '1'. This gives a binary representation of the polynomials in $\mathbb{F}_2[x]$, the upper bits of the representation being the upper coefficients of the polynomial. For example, the polynomial $x^5 + x^3 + 1$ can be represented as the binary number '101001'.

The modular multiplication in $\mathbb{F}_2[x]$ is one of the most important operation in elliptic curves cryptography over $\mathsf{GF}(2^p)$. We will show that one can obtain, with the method described in the previous section, similar performances to the one using Montgomery's modular multiplication in $\mathsf{GF}(2^p)$ [1].

From now on, we will suppose that the polynomial modular multiplication is implemented on a t-bit architecture (e.g. 32-bit). The modular multiplication $A(x)B(x) \bmod N(x)$ can be written as a sum of modular products of words of the first operand by the second operand:

$$\sum_{i=0}^{p_A-1} A_i(x)B(x)x^{it} \bmod N(x) = U(x) \bmod N(x) \qquad (8)$$

where, $A_i(x)$ is a polynomial of degree $t-1$ such that $A(x) = \sum_{i=0}^{p_A-1} A_i(x)x^{it}$ and where $p_A = \lceil \frac{p_a}{t} \rceil$ with p_a, degree of $A(x)$.

Let's first recall some characteristics on the polynomial computations in $\mathbb{F}_2[x]$:

- The product of a polynomial of degree $t-1$ (which can be represented as a t-bit binary vector) by a polynomial of degree $n-1$ is a polynomial of degree $n+t-2$ represented as a $(n+t-1)$-bit vector. Over integers however, the result of a product of a t-bit by an n-bit integer is a $(n+t)$-bit number.
- The result of a polynomial addition of two polynomials of degree p is a polynomial of degree p (same number of bits in its binary representation). Over integers, the result may have one more bit in its binary representation because of carry propagations.
- The "modulo" operation (remainder of the division between two polynomials) gives a polynomial which is one degree smaller than the modulus. This means that the binary vector representing the remainder has always one bit less than the one of the modulus. Over integers, the remainder is smaller than the modulus but can still have the same number of bits in its binary representation.

Given equation 8, we can now evaluate how the size of quotient $Q(x)$ (equation 1) changes. To reduce the required memory and minimize accesses to it, equation 8 can be computed by interleaving the multiplication (from the highest index of A_i to the smallest one) with the reduction (by $N(x)$) as shown in figure 1.

Using the above characteristics of computations in $\mathbb{F}_2[x]$, the temporary polynomial $U(x)$ is always, at every stage as shown in figure 1, at a maximum degree of $t + p_N - 1$. This means that when computing the quotient $Q(x)$ as in equation 1, α has to be replaced by $t-1$. The corresponding $T(x)$ in equation 1 will then be of degree $t-1$ and $R(x)$ in equation 1 of degree $t-1$. A binary

```
1 :    U(x) = 0
2 :    for i = p_A − 1 down to 0
3 :        U(x) = U(x)x^t ⊕ A_i(x)B(x)
4 :        Q(x) = ⌊U(x)/N(x)⌋
5 :        U(x) = U(x) ⊕ Q(x)N(x)    [U(x) = U(x) mod N(x)]
6 :    end for i
```

Fig. 1. Modular multiplication in $\mathbb{F}_2[x]$.

vector representation in t-bit will thus perfectly match the computations on a t-bit architecture (CPU).

This also means that our method would require "standard" computations as in Montgomery's method described in [1]. In Montgomery's multiplication, a t-bit polynomial multiplication (with polynomials of degree $t-1$) and a division by x^t is required. In the present case, a t-bit polynomial multiplication and a division by x^{t-1} is needed. The remaining part of the computations can be the same in both methods but the way the computations are made is different (we start here from the most significant word A_{p_A-1} instead of A_0 in the Montgomery's method).

```
1 :    U(x) = A_{p_A−1}(x)B(x)
2 :    for i = p_A − 2 down to 0
3 :        Q(x) = ⌊(T(x)R(x))/x^{t−1}⌋
4 :        for j = 0 to p_N − 1
5 :            U(x) = (U(x) ⊕ Q(x)N_j(x))x^{t(j+1)} ⊕ A_i(x)B_j(x)x^{tj}
6 :        end for j
7 :    end for i
8 :    Q = ⌊U(x)/N(x)⌋
9 :    U(x) = U(x) ⊕ Q(x)N(x)
```

Fig. 2. Interleaved modular multiplication in $\mathbb{F}_2[x]$.

It is possible to further reduce the memory accesses on $U(x)$. This is very important since memory accesses are often an important limiting factor in terms of execution time, namely in smart cards. To do so, we take the first computation $A_{p_A-1}(x)B(x)$ out of the i loop as shown in figure 2 allowing the i loop to start with the quotient computation $(Q(x))$ and the two computations of $U(x)$ in line (3) and (5) as shown in figure 1 to be merged into line (5) as shown in figure 2. Such a computation requires a final reduction outside the i loop (lines (8) and (9) in figure 2). The only disadvantage of interleaving the multiplication and the reduction phase using a unique j loop is that the numbers of $N_{j(x)}$ and $B_j(x)$ are identical ($p_B = p_N$), meaning that if, for example, the degree of $B(x)$ is smaller than the one of $N(x)$ it should be padded left with zeroes when storing it in $B[j]$'s. This does nevertheless not influence the speed of practical implementations since $B(x)$ is normally considered with an identical size to $N(x)$.

3.1   Software Implementations on a t-Bit Processor Architecture

For the sake of clearness, when comparing with existing implementations, we will suppose that $p = p_A = p_N$. The detailed implementation of the modular multiplication is shown in figure 4.

In this figure, a "(Hi, Lo)" is a $2t$-bit register (such a register is common on RISC architectures providing a $t \times t$-bit multiplication with a $2t$-bit result) which is the concatenation of registers Hi and Lo (both t-bit registers), where Hi is the upper part (most significant bits) of that register and Lo, the lower part. The expression "(Hi, Lo) $\gg t$" means that this virtual register is shifted left of t-bit. In other words, the result is Hi $= 0$ and Lo $=$ Hi, the old value of Lo being discarded.

In figure 4, a '$\oplus$' represents a bitwise XOR operation and a '$\otimes$' means a polynomial multiplication in $\mathbb{F}_2[x]$ where the polynomials have at most a degree of $t-1$. In other words, a computation like (Hi, Lo)$\oplus = A \otimes B$ is simply a multiply and accumulate calculation, just like the one present on most RISC processors and DSP's, but with the internal carries in multiplications and additions being disabled. An algorithmic representation of this computation is shown in figure 3. Such a calculation is already implemented as an instruction in some high-end smart card microprocessors to improve elliptic curve computations in $\mathsf{GF}(2^p)$.

$$
\begin{array}{ll}
1: & \textbf{for } i = 0 \textbf{ to } t-1 \\
2: & \quad (\mathsf{Hi}, \mathsf{Lo}) = (\mathsf{Hi}, \mathsf{Lo}) \oplus ((A \cdot \underbrace{((B \gg i) \text{ AND } 1)}_{i\text{th bit of } B}) \ll i) \\
3: & \textbf{end for } i
\end{array}
$$

Fig. 3. Simple program simulating the computation of $(\mathsf{Hi}, \mathsf{Lo})\oplus = A \otimes B$.

Ideally, U_{sup} in lines (0) and (0) of figure 4, can be replaced by Lo, only if the Most Significant Bit (MSB) of $N[p-1]$ corresponds to the upper most significant coefficient of $N(x)$. Otherwise $U_{sup} = (Lo \ll k) \oplus (Rs \gg (t-k))$, where k is the shift value that would be needed to align the most significant coefficient of $N(x)$ stored in $N[p-1]$ with the MSB of $N[p-1]$.

Figure 4 also shows the number of multiply and accumulate instructions without carries (column '#$\oplus$') and the number of memory accesses. In comparison with the paper of Koç and Acar [1] we have exactly the same number of multiplications without carries but without the additional XOR operations. To be correct, most of the XOR are included in our multiply and accumulate operation.

Improved implementation. The pseudo-code in figure 4 can still be compacted by computing $A_i B$ interleaved with QN in the reverse order as shown in figure 5 (computing first $A_i B[p-1]$ and $QN[p-1]$). This was made possible since there is no carry propagation when working in $\mathbb{F}_2[x]$ (compared with the computations over integers). Aligning the upper coefficient of $N(x)$ with the upper

		#⊗	#LOAD	#STORE
1:	$Hi = 0$			
2:	$Lo = 0$			
3:	$A_{p-1} = A[p-1]$		1	
4:	**for** $j = 0$ **to** $p-1$			
5:	$(Hi, Lo)\oplus = A_{p-1} \otimes B[j]$	p	p	
6:	$Rs = Lo; U[j] = Rs$			p
7:	$(Hi, Lo) \gg t$			
8:	**end for** j			
9:	**for** $i = p-2$ **down to** 0			
10:	$Q = (U_{sup} \otimes R) \gg (t-1)$	$p-1$		
11:	$A_i = A[i]$		$p-1$	
12:	$Hi = U[0]$		$p-1$	
13:	$Lo = 0$			
14:	$(Hi, Lo)\oplus = A_i \otimes B[0]$	$p-1$	$p-1$	
15:	$U[0] = Lo$			$p-1$
16:	$(Hi, Lo) \gg t$			
17:	**for** $j = 1$ **to** $p-1$			
18:	$Hi = U[j]$		$(p-1)^2$	
19:	$(Hi, Lo)\oplus = A_i \otimes B[j]$	$(p-1)^2$	$(p-1)^2$	
20:	$(Hi, Lo)\oplus = Q \otimes N[j-1]$	$(p-1)^2$	$(p-1)^2$	
21:	$Rs = Lo; U[j] = Rs$			$(p-1)^2$
22:	$(Hi, Lo) \gg t$			
23:	**end for** j			
24:	$(Hi, Lo)\oplus = Q \otimes N[p-1]$	$p-1$	$p-1$	
25:	**end for** i			
26:	$Q = (U_{sup} \otimes R) \gg (t-1)$	1		
27:	$Lo = U[0]$		1	
28:	**for** $j = 0$ **to** $p-2$			
29:	$Hi = U[j+1]$		$p-1$	
30:	$(Hi, Lo)\oplus = Q \otimes N[j]$	$p-1$	$p-1$	
31:	$U[j] = Lo$			$p-1$
32:	$(Hi, Lo) \gg t$			
33:	**end for** j			
34:	$(Hi, Lo)\oplus = Q \otimes N[p-1]$	1	1	
35:	$U[p-1] = Lo$			1
	TOTAL	$2p^2 + p$	$3p^2 + p$	$p^2 + p$

Fig. 4. Interleaved modular multiplication in $\mathbb{F}_2[x]$.

bit of $N[p-1]$, when storing $N(x)$ in the $N[j]$'s (before staring the computations), also simplifies the computations as shown in figure 5. This only requires one additional adaptation (right final shift) to the final result if the modulus is constant during a whole set of modular multiplications. This is exactly the case with most cryptographic algorithms (e.g. ECDSA over $GF(2^p)$ [4]).

		$\#\otimes$	$\#$LOAD	$\#$STORE
1 :	**for** $j = 0$ **to** $p - 1$			
2 :	$U[j] = 0$			p
3 :	**for** $i = p - 1$ **down to** 0			
4 :	$\text{Hi} = U[p - 1]$		p	
5 :	$\text{Lo} = U[p - 2]$		p	
6 :	$A_i = A[i]$		p	
7 :	$(\text{Hi}, \text{Lo})\oplus = A_i \otimes B[p - 1]$	p	p	
8 :	$Q = ((\text{Hi}, \text{Lo})_{sup} \otimes R) \gg (t - 1)$	p		
9 :	$(\text{Hi}, \text{Lo})\oplus = Q \otimes N[p - 1]$	p	p	
10 :	**for** $j = p - 2$ **down to** 1			
11 :	$(\text{Hi}, \text{Lo}) \ll t$			
12 :	$\text{Lo} = U[j - 1]$		$p(p - 2)$	
13 :	$(\text{Hi}, \text{Lo})\oplus = A_i \otimes B[j]$	$p(p - 2)$	$p(p - 2)$	
14 :	$(\text{Hi}, \text{Lo})\oplus = Q \otimes N[j]$	$p(p - 2)$	$p(p - 2)$	
15 :	$U[j + 1] = \text{Hi}$			$p(p - 2)$
16 :	**end for** j			
17 :	$(\text{Hi}, \text{Lo}) \ll t$			
18 :	$(\text{Hi}, \text{Lo})\oplus = A_i \otimes B[0]$	p	p	
19 :	$(\text{Hi}, \text{Lo})\oplus = Q \otimes N[0]$	p	p	
20 :	$U[1] = \text{Hi}$			p
21 :	$U[0] = \text{Lo}$			p
22 :	**end for** i			
	TOTAL	$2p^2 + p$	$3p^2 + p$	$p^2 + p$

Fig. 5. Interleaved modular multiplication with internal loop starting in the reverse order.

In this case (figure 5), U_{sup} is simply $((U[p-1], U[p-2]) \oplus A[i] \otimes B[p-1]) \gg (t-1)$. Indeed, there is no influence of $A[i] \otimes B[p-2]$ on the required upper part of $U(x)$ as this only influences the $(t-1)$ first bits of $(U[p-1], U[p-2]) \oplus A[i] \otimes B[p-1])$ and there is no carry propagation. Another consequence and advantage of such a computation is that there is no more need to compute $A_{p-1}(x)B(x)$ in advance. No additional final reduction by $Q(x)N(x)$ is necessary such that the lines (1) to (0) and (0) to (0) in figure 4 are no more necessary in figure 5.

As shown in figure 5, the global number of operations is identical to the one in figure 4, but the code size is smaller. Except for the last reason, the choice between the two implementations will only depend on the (processor's) architecture.

3.2 Comparison with Montgomery's Modular Multiplication

In figure 6, Montgomery's modular multiplication is implemented in the same way as our method. The main difference in our implementation compared to Koç and Acar's one [1] is the merging between the multiplication and the reduction phase of the algorithm to reduce the memory accesses. The number of memory accesses is smaller in our case as compared to the one required by the Montgomery's method described by Koç and Acar. Their method needs $(6s^2 - s)$ load and $(3s^2 + 2s + 1)$ store operations as given in table 2 of their paper.

As shown in figures 4, 5 and 6, our method is similar to Montgomery's one in terms of the number of multiply and accumulate without carries and the number of memory accesses.

		#⊗	#LOAD	#STORE
1 :	**for** $j = 0$ **to** $p - 1$			
2 :	$U[j] = 0$			p
3 :	**end for** j			
4 :	**for** $i = 0$ **to** $p - 1$			
5 :	$A_i = A[i]$		p	
6 :	$\mathsf{Lo} = U[0]$		p	
7 :	$\mathsf{Hi} = U[1]$		p	
8 :	$(\mathsf{Rt}, \mathsf{Lo})\oplus = A_i \otimes B[0]$	p	p	
9 :	$(\mathsf{Hi}, \mathsf{Q}) = \mathsf{Lo} \otimes N_0'$	p		
10 :	$(\mathsf{Rt}, \mathsf{Lo})\oplus = Q \otimes N[0]$	p	p	
11 :	$(\mathsf{Hi}, \mathsf{Lo}) \gg t$			
12 :	**for** $j = 1$ **to** $p - 2$			
13 :	$\mathsf{Hi} = U[j + 1]$		$p(p - 2)$	
14 :	$(\mathsf{Hi}, \mathsf{Lo})\oplus = A_i \otimes B[j]$	$p(p - 2)$	$p(p - 2)$	
15 :	$(\mathsf{Hi}, \mathsf{Lo})\oplus = Q \otimes N[j]$	$p(p - 2)$	$p(p - 2)$	
16 :	$U[j - 1] = \mathsf{Lo}$			$p(p - 2)$
17 :	$(\mathsf{Hi}, \mathsf{Lo}) \gg t$			
18 :	**end for** j			
19 :	$(\mathsf{Hi}, \mathsf{Lo})\oplus = A_i \otimes B[p - 1]$	p	p	
20 :	$(\mathsf{Hi}, \mathsf{Lo})\oplus = Q \otimes N[p - 1]$	p	p	
21 :	$U[p - 2] = \mathsf{Lo}$			p
22 :	$U[p - 1] = \mathsf{Hi}$			p
23 :	**end for** i			
	TOTAL	$2p^2 + p$	$3p^2 + p$	$p^2 + p$

Fig. 6. Interleaved Montgomery's modular multiplication in $\mathbb{F}_2[x]$.

However, a possible inconvenience of our method (for a software implementation on a general purpose processor), as compared to Montgomery's one, would be the slower "extraction" of U_{sup} from the intermediate values of $U(x)$ as well as the right shift by $(t-1)$ when computing Q. This disadvantage can be simply taken into account, with a very minimal cost, in a hardware implementation.

The main disadvantage of Montgomery's method is not visible in figure 6. Indeed, our method exactly computes $A(x)B(x) \bmod N(x)$ which is not the case for Montgomery's one which computes $A(x)B(x)x^{-p} \bmod N(x)$ [1]. This last computation is mostly not desired. This is why Montgomery's method often requires substantial additional code to deal with that computation (e.g replace $A(x)$ by $A(x)x^p \bmod N(x)$ and $B(x)$ by $B(x)x^p \bmod N(x)$ before the computations and then finish the computations by multiplying the final result by '1' using Montgomery's method). These computations penalize Montgomery's method in terms of code size (it can be critical in smart card's context) and may complicate the use of the Montgomery's method outside the scope of modular exponentiations computations (e.g. for a single modular multiplication).

3.3   Speed Comparisons on a Real Implementation

Table 1 shows the results, in terms of clock cycles, obtained by both Montgomery's multiplication (figure 6) and the two versions of our method described in figures 4 and 5.

	256-bit multiplication	512-bit multiplication
Algorithm in fig. 4	910	3230
Compact Algorithm in fig. 5	812	3028
Montgomery's method (fig. 6)	756	2916

Table 1. Algorithms' speed in clock cycles on a Montgomery's optimized processor.

Measurements were done on a 32-bit processor's (usable in smart cards) simulator using the modified multiply accumulate instruction (without internal carries) as described in section 3.1. This processor was designed to speed up Montgomery's multiplications. This explains why, in table 1, the Montgomery's method has still an advantage. As explained before, this is only due to the small additional computations required for computing $Q(x)$ in our implementation. Indeed, 7 additional clock cycles for each $Q(x)$ computation (equivalent to line (8) in figure 5) are required as compared to what is done in the Montgomery's implementation (line (9) in figure 6). However, similar results for both implementations can be obtained by very slightly modifying a few processor's instructions.

The comparison made here only involves the "core's" modular multiplication itself. As explained in section 3.2, the fact that Montgomery's method computes $A(x)B(x)x^{-p} \bmod N(x)$ in place of the exact value $(A(x)B(x) \bmod N(x))$ for our method, can also deeply influence the choice between one algorithm and the other.

4 Conclusions

We have first extended the generalized Barrett's modular reduction to $\mathbb{F}_q[x]$. We then described an efficient way to implement fast "left to right" modular multiplication in $\mathbb{F}_2[x]$ which is at least as efficient as the best known methods, namely Montgomery's multiplication [1]. Furthermore, our method has the advantage of computing the modular multiplication without the inconvenience of a normalization as needed in the Montgomery's one. This makes this method particularly attractive for smart cards and hardware implementations. A way of reducing the memory accesses in both methods has been described. Both methods can be efficiently implemented by having a multiply and accumulate instruction without internal carries to perform fast competitive software implementations of elliptic curve crypto-systems in $GF(2^p)$.

Acknowledgments. The author would like to thank Jacques Fournier, Marc Joye and the anonymous referees for their useful comments.

References

1. Koç, C., Acar, T.: Montgomery multiplication in $GF(2^k)$. In Publishers, K.A., ed.: Designs, Codes and Cryptography. Volume 14., Boston (1998) 57–69
2. Menezes, A., van Oorschot, P., Vanstone, S.: Handbook of Applied Cryptography. CRC Press (1997)
3. Montgomery, P.: Modular multiplication without trial division. Mathematics of Computation **44** (1985) 519–521
4. IEEE: Std 1363-2000. IEEE standard specifications for public-key cryptography, New York, USA (2000) Informations available at
   http://grouper.ieee.org/groups/1363/.
5. Barrett, P.: Implementing the Rivest Shamir and Adleman public key encryption algorithm on a standard digital signal processor. In Odlyzko, A., ed.: Advances in Cryptology - CRYPTO '86, Santa Barbara, California. Volume 263 of Lecture Notes in Computer Science., Springer-Verlag (1987) 311–323
6. Quisquater, J.J.: Encoding system according to the so-called RSA method, by means of a microcontroller and arrangement implementing this system. U.S. Patent # 5,166,978 (1992)
7. Dhem, J.F., Quisquater, J.J.: Recent results on modular multiplications for smart cards. In Quisquater, J.J., Schneier, B., eds.: Proc. CARDIS'98, Smart Card Research and Applications, Louvain-la-Neuve, Belgium. Volume 1820 of Lecture Notes in Computer Science., Springer-Verlag (2000) 336–352
8. Cohen, H.: A Course in Computational Algebraic Number Theory. 2nd edn. Graduate Texts in Mathematics. Springer (1995)
9. De Win, E., Bosselaers, A., Vandenberghe, S., De Gersem, P., Vandewalle, J.: A fast software implementation for arithmetic operations in $GF(2^n)$. In Kim, K., Matsumoto, T., eds.: Advances in Cryptology - ASIACRYPT '96, Kyongju, Korea. Volume 1163 of Lecture Notes in Computer Science., Springer (1996) 65–76

Faster Double-Size Modular Multiplication from Euclidean Multipliers

Benoît Chevallier-Mames, Marc Joye, and Pascal Paillier

[1] Gemplus, Card Security Group
La Vigie, Avenue du Jujubier, ZI Athélia IV, 13705 La Ciotat Cedex, France
{benoit.chevallier-mames, marc.joye}@gemplus.com —
http://www.geocities.com/MarcJoye/
[2] Gemplus, Cryptography Group
34 rue Guynemer, 92447 Issy-les-Moulineaux, France
pascal.paillier@gemplus.com
http://www.gemplus.com/smart/

Abstract. A novel technique for computing a $2n$-bit modular multiplication using n-bit arithmetic was introduced at CHES 2002 by Fischer and Seifert. Their technique makes use of an Euclidean division based instruction returning not only the remainder but also the integer quotient resulting from a modular multiplication, i.e., on input x, y and z, both $\lfloor xy/z \rfloor$ and $xy \bmod z$ are returned. A second algorithm making use of a special modular 'multiply-and-accumulate' instruction was also proposed.

In this paper, we improve on these algorithms and propose more advanced computational strategies with fewer calls to these basic operations, bringing in a speed-up factor up to 57%. Besides, when Euclidean multiplications themselves have to be emulated in software, we propose a specific modular multiplication based algorithm which surpasses original algorithms in performance by 71%.

Keywords: Modular multiplication, crypto-processors, embedded cryptographic software, efficient implementations, RSA.

1 Introduction

When a cryptographic coprocessor is inherently limited to handle numbers of a specific bitsize n, performing modular arithmetic operations over larger operands turns out to be an intricate implementation problem. One may think of natural and simple solutions like programming multi-precision algorithms such as those of Montgomery [5], Barrett [3], Quisquater [8] or Walter [10]. These algorithms as well as others, however, require processing data blocks via smaller operations that may not be supported by the underlying hardware architecture. A typical example lies in the regular $(n \times n)$-bit integer multiplication (with a $2n$-bit result), which may not be directly available on a crypto-processor like Infineon's ACE where only n-bit modular operations —for n up to 1100— are programmable. To

C.D. Walter et al. (Eds.): CHES 2003, LNCS 2779, pp. 214–227, 2003.
© Springer-Verlag Berlin Heidelberg 2003

remedy this, a conventional trick tells us to take a blocksize $n \leq 1100/2$ (say $n = 512$), so that integer multiplications result from multiplications modulo 2^{1100}. Adopting this strategy, a Montgomery-based implementation for a single 2048-bit multiplication would cost no less than forty 512-bit integer multiplications, an unacceptable performance. So implementing a 2048-bit RSA while sustaining higher expectations in terms of execution speed is not a straightforward task on this platform. Given that context, one has to devise more specific techniques to emulate modular arithmetic operations over operands of larger sizes.

Surprisingly enough, little is known about software strategies that would overcome this hardware-originated length limitation in a very efficient way. Two different techniques, however, have appeared in the literature recently. In [6], Paillier presents a $2n$-bit modular multiplication emulated with $8 \rightarrow 6$ calls to a modular multiplier of bitsize n (plus other, negligible operations). Paillier's algorithm, inspired by Montgomery's technique [5], strongly relies on Residue Number Systems (RNS) [7,9] for representing data and performing partial operations on them. It simplifies earlier, more intricate approaches making use of mixed base representations [1,2]. The efficiency of this system is due to the use of fast base extensions in connection with a specific choice for the RNS base, a choice which also ensures that the result of the double-size multiplication is returned under a representation compatible with the input operands themselves, thereby allowing repeated invocations of the algorithm. Unfortunately, its Montgomery-like style forces one to precompute a modulus-dependent constant prior to multiplying any data.

More recently, in [4], Fischer and Seifert suppress the need for precomputed constants: $2n$-bit operands are handled through a classical radix representation with base 2^n, the new technique outputting a result under the same representation. Independently, in this work, Fischer and Seifert replace the basic operation with an Euclidean multiplication, i.e., an operation that simultaneously returns both the quotient and remainder of the division $xy \div z$, given arbitrary n-bit integers x, y and z. The motivation for this stems from the ease of integrating such an operation in a hardware architecture which already supports modular multiplications. On most architectures indeed, the arithmetic units involved in the execution of a modular reduction could be easily enriched to simultaneously output quotient bits with extremely moderate extra cost.

In this paper, we improve on Fischer and Seifert's algorithms and propose more advanced computational strategies with fewer calls to the Euclidean multiplication. Our improved algorithms use 2^n-radix representations of numbers in the spirit of [4]. We also show that adapting the choice of the radix base according to the modulus may further speed up our technique. This modification can be carried out while maintaining inputs and outputs under the same arithmetic format, thereby making it possible to iterate executions. In the most favorable case, we emulate a double-size modular multiplication with no more than 3 Euclidean multiplications, which leads to a speedup factor of $\frac{7-3}{7} \approx 57\%$. In addition, when Euclidean multiplications themselves must be emulated in software from modular multiplications, we show how to use these directly without referring to RNS-based approaches [1,2,7,9]. More precisely, we propose a simple alternative to these works which keeps numbers under a radix representation and runs as fast as 2 Euclidean multiplications. This accelerates original algorithms by $\frac{14-4}{14} \approx 71\%$. We remind that, whatever the computational strategy, it is agreed that n-bit linear operations such as signed additions, subtractions, xors, conditional branchings and so forth, are always available and that their respective running times remain negligible in comparison with operations of multiplicative nature. As usual, we consider these as being virtually free operations throughout the paper.

The rest of this paper is organized as follows. In the next section, we review the technique introduced by Fischer and Seifert for emulating a $2n$-bit modular multiplication and show how to improve it in Section 3. Then, in Section 4, we investigate the influence of data representations on the performances of our algorithms. In Section 5, we detail an implementation of an Euclidean multiplication. Section 6 describes a specific strategy for cases when Euclidean multiplications are emulated in software. Finally, we summarize and compare our results in Section 7.

2   Fischer and Seifert's Algorithms

Fischer and Seifert's technique [4] relies on the two basic instructions

$$\texttt{MultModDiv}(x, y, z) \overset{\Delta}{=} (\lfloor (x \cdot y)/z \rfloor, (x \cdot y) \bmod z) \tag{1}$$

and

$$\texttt{MultModDivInit}(x, y, t, z) \overset{\Delta}{=} (\lfloor (x \cdot y + t \cdot 2^n)/z \rfloor, (x \cdot y + t \cdot 2^n) \bmod z), \tag{2}$$

where x, y, t, z are n-bit integers. It is implicitly required in [4] that operands x, y and t can be negative, i.e., that the processor is able to handle them whatever their sign through a signed representation without affecting computation results. In fact, these two instructions should, by extension, work for any non-reduced inputs x, y, t, namely whenever $|x| > z$ for instance, provided that $\lceil |x|/z \rceil$ remains an extremely small value. Subsequent hardware or software corrections are neglected in the description of all algorithms, as proposed in [4].

The algorithms originally proposed by Fischer and Seifert, which we denote by **FS1** and **FS2**, are depicted on Fig. 1 and Fig. 2, respectively. We refer the reader to [4] for proofs of correctness.

3   Improved Algorithms

Our idea consists in rewriting the modular multiplication in terms of manipulations over half-size operands. This is reminiscent of Karatsuba's famous method which we recall here for the sake of completeness.

Input:	$2n$-bit integers $A = A_1 2^n + A_0, B = B_1 2^n + B_0, N = N_1 2^n + N_0$
Output:	$AB \pmod{N}$
Cost:	7 MultModDiv

$(Q^{(1)}, R^{(1)}) = \text{MultModDiv}(B_1, 2^n, N_1)$
$(Q^{(2)}, R^{(2)}) = \text{MultModDiv}(Q^{(1)}, N_0, 2^n)$
$(Q^{(3)}, R^{(3)}) = \text{MultModDiv}(A_1, R^{(1)} - Q^{(2)} + B_0, N_1)$
$(Q^{(4)}, R^{(4)}) = \text{MultModDiv}(A_0, B_1, N_1)$
$(Q^{(5)}, R^{(5)}) = \text{MultModDiv}(Q^{(3)} + Q^{(4)}, N_0, 2^n)$
$(Q^{(6)}, R^{(6)}) = \text{MultModDiv}(A_1, R^{(2)}, 2^n)$
$(Q^{(7)}, R^{(7)}) = \text{MultModDiv}(A_0, B_0, 2^n)$
Return $(R^{(3)} + R^{(4)} - Q^{(5)} - Q^{(6)} + Q^{(7)})2^n + (R^{(7)} - R^{(6)} - R^{(5)})$

Fig. 1. Fischer-Seifert's modular multiplication algorithm FS1

Input:	$2n$-bit integers $A = A_1 2^n + A_0, B = B_1 2^n + B_0, N = N_1 2^n + N_0$
Output:	$AB \pmod{N}$
Cost:	5 MultModDiv + 1 MultModDivInit

$(Q^{(1)}, R^{(1)}) = \text{MultModDiv}(A_1, B_1, N_1)$
$(Q^{(2)}, R^{(2)}) = \text{MultModDivInit}(N_0, -Q^{(1)}, R^{(1)}, N_1)$
$(Q^{(3)}, R^{(3)}) = \text{MultModDiv}(A_1, B_0, N_1)$
$(Q^{(4)}, R^{(4)}) = \text{MultModDiv}(A_0, B_1, N_1)$
$(Q^{(5)}, R^{(5)}) = \text{MultModDiv}(A_0, B_0, 2^n)$
$(Q^{(6)}, R^{(6)}) = \text{MultModDiv}(Q_{(2)} + Q^{(3)} + Q^{(4)}, N_0, 2^n)$
Return $(R^{(2)} + R^{(3)} + R^{(4)} + Q^{(5)} - Q^{(6)})2^n + (R^{(5)} - R^{(6)})$

Fig. 2. Fischer-Seifert's modular multiplication algorithm FS2

Lemma 1 (Karatsuba). *If $A = A_1 2^n + A_0$ and $B = B_1 2^n + B_0$ then*

$$AB = 2^n(2^n - 1)A_1 B_1 + 2^n(A_1 + A_0)(B_1 + B_0) - (2^n - 1)A_0 B_0 .$$

Our first algorithm only makes use of Fischer and Seifert's MultModDiv instruction while our second algorithm also employs the MultModDivInit instruction.

3.1 Using MultModDiv Instructions Only

In this section, we eliminate a MultModDiv instruction in Fischer-Seifert's technique. We state:

Theorem 1. *Given arbitrary $2n$-bit integers N and $A, B \leq N$, the $2n$-bit integer AB mod N can be computed with at most <u>six</u> n-bit MultModDiv instructions.*

Input:	$2n$-bit integers $A = A_1 2^n + A_0, B = B_1 2^n + B_0, N = N_1 2^n + N_0$
Output:	$AB \pmod{N}$
Cost:	6 MultModDiv

$(Q^{(1)}, R^{(1)}) = \text{MultModDiv}(A_1, B_1, N_1)$
$(Q^{(2)}, R^{(2)}) = \text{MultModDiv}(Q^{(1)}, N_0, 2^n)$
$(Q^{(3)}, R^{(3)}) = \text{MultModDiv}(A_1 + A_0, B_1 + B_0, 2^n - 1)$
$(Q^{(4)}, R^{(4)}) = \text{MultModDiv}(A_0, B_0, 2^n)$
$(Q^{(5)}, R^{(5)}) = \text{MultModDiv}(2^n - 1, R^{(1)} + Q^{(3)} - Q^{(2)} - Q^{(4)}, N_1)$
$(Q^{(6)}, R^{(6)}) = \text{MultModDiv}(Q^{(5)}, N_0, 2^n)$
Return $(R^{(3)} + R^{(5)} - Q^{(6)} - R^{(2)} - R^{(4)})2^n + (R^{(2)} + R^{(4)} - R^{(6)})$

Fig. 3. Our improved algorithm A1 for double-size modular multiplication

Our algorithm, denoted A1, is described hereafter on Fig. 3.

Proof (of correctness for A1). For convenience, we write $Z = 2^n$ and denote by $\equiv_N$ the equivalences modulo N. Then, rewriting Lemma 1 gives

$$AB = Z(Z-1)A_1 B_1 + Z(A_1 + A_0)(B_1 + B_0) - (Z-1)A_0 B_0 .$$

Moreover, noticing that $N_1 Z \equiv_N -N_0$, we get

$$\begin{aligned}
Z(Z-1)A_1 B_1 &\equiv_N Z(Z-1)(Q^{(1)} N_1 + R^{(1)}) \\
&\equiv_N -(Z-1)(Q^{(1)} N_0) + Z(Z-1)R^{(1)} \\
&\equiv_N -(Z-1)(Q^{(2)} Z + R^{(2)}) + Z(Z-1)R^{(1)} \\
&\equiv_N Z(Z-1)(R^{(1)} - Q^{(2)}) - (Z-1)R^{(2)} ,
\end{aligned}$$

$$Z(A_1 + A_0)(B_1 + B_0) = Z((Z-1)Q^{(3)} + R^{(3)}) = Z(Z-1)Q^{(3)} + ZR^{(3)}$$

and

$$(Z-1)A_0 B_0 = (Z-1)(ZQ^{(4)} + R^{(4)}) = Z(Z-1)Q^{(4)} + (Z-1)R^{(4)} .$$

Hence, we have

$$\begin{aligned}
AB &\equiv_N Z(Z-1)(R^{(1)} + Q^{(3)} - Q^{(2)} - Q^{(4)}) + ZR^{(3)} - (Z-1)(R^{(2)} + R^{(4)}) \\
&\equiv_N Z(Q^{(5)} N_1 + R^{(5)}) + ZR^{(3)} - (Z-1)(R^{(2)} + R^{(4)}) \\
&\equiv_N -Q^{(5)} N_0 + Z(R^{(3)} + R^{(5)}) - (Z-1)(R^{(2)} + R^{(4)}) \\
&\equiv_N -(Q^{(6)} Z + R^{(6)}) + Z(R^{(3)} + R^{(5)}) - (Z-1)(R^{(2)} + R^{(4)}) \\
&\equiv_N (R^{(3)} + R^{(5)} - Q^{(6)} - R^{(2)} - R^{(4)})Z + (R^{(2)} + R^{(4)} - R^{(6)}) ,
\end{aligned}$$

which proves the correctness of Algorithm A1 and of Theorem 1. $\square$

3.2   Using `MultModDiv` and `MultModDivInit` Instructions

Here again, we invoke Lemma 1 and improve Fischer and Seifert's double size multiplier FS2. Formally, we state:

Theorem 2. *Given arbitrary $2n$-bit integers N and $A, B \leq N$, the $2n$-bit integer $AB \bmod N$ can be computed with at most <u>four</u> n-bit `MultModDiv` instructions and <u>one</u> n-bit `MultModDivInit` instruction.*

Our algorithm, denoted A2, is described below on Fig. 4.

Input:	$2n$-bit integers $A = A_1 2^n + A_0, B = B_1 2^n + B_0, N = N_1 2^n + N_0$
Output:	$AB \pmod{N}$
Cost:	4 `MultModDiv` $+$ 1 `MultModDivInit`

$(Q^{(1)}, R^{(1)}) = \texttt{MultModDiv}(A_1, B_1, N_1)$
$(Q^{(2)}, R^{(2)}) = \texttt{MultModDiv}(A_1 + A_0, B_1 + B_0, 2^n - 1)$
$(Q^{(3)}, R^{(3)}) = \texttt{MultModDiv}(A_0, B_0, 2^n)$
$(Q^{(4)}, R^{(4)}) = \texttt{MultModDivInit}(Q^{(1)}, N_0, Q^{(3)} - R^{(1)} - Q^{(2)}, N_1)$
$(Q^{(5)}, R^{(5)}) = \texttt{MultModDiv}(N_0 + N_1, Q^{(4)}, 2^n)$
Return $(R^{(2)} + Q^{(5)} - R^{(3)} - R^{(4)})2^n + (R^{(3)} + R^{(4)} + R^{(5)})$

Fig. 4. Improved algorithm A2 for double-size modular multiplication

Proof (of correctness for A2*).* As before, we set $Z = 2^n$. We have

$$Z(Z-1)A_1 B_1 \equiv_N Z(Z-1)(Q^{(1)} N_1 + R^{(1)}) \equiv_N (Z-1)(-Q^{(1)} N_0 + R^{(1)} Z),$$
$$Z(A_1 + A_0)(B_1 + B_0) = Z((Z-1)Q^{(2)} + R^{(2)}) = (Z-1)Q^{(2)} Z + ZR^{(2)},$$
$$(Z-1)A_0 B_0 = (Z-1)(ZQ^{(3)} + R^{(3)}) = (Z-1)Q^{(3)} Z + (Z-1)R^{(3)}$$

so that

$$\begin{aligned}
AB &\equiv_N -(Z-1)\left(Q^{(1)} N_0 + (Q^{(3)} - R^{(1)} - Q^{(2)})Z\right) + ZR^{(2)} - (Z-1)R^{(3)} \\
&\equiv_N -(Z-1)(Q^{(4)} N_1 + R^{(4)}) + ZR^{(2)} - (Z-1)R^{(3)} \\
&\equiv_N (N_0 + N_1)Q^{(4)} + ZR^{(2)} - (Z-1)(R^{(3)} + R^{(4)}) \\
&\equiv_N (Q^{(5)} Z + R^{(5)}) + ZR^{(2)} - (Z-1)(R^{(3)} + R^{(4)}) \\
&\equiv_N (R^{(2)} + Q^{(5)} - R^{(3)} - R^{(4)})Z + (R^{(3)} + R^{(4)} + R^{(5)})
\end{aligned}$$

since $N_1(Z-1) \equiv_N -N_0 - N_1$. $\square$

4   Further Improvements Using Specific Representations

All algorithms considered so far manipulate integers in radix representation with base 2^n. We now show how changing that representation may lead to further cost savings in our algorithms. Although we explicitly describe only a couple of (modulus-dependent) representations in what follows, there might exist other ones which would reveal quite as efficient. In both cases, the idea is simply to employ a clever representation base derived from modulus N. This computation is performed prior to the execution of the corresponding double-size modular multiplication and can be executed once and for all, especially when the multiplication is invoked repeatedly.

4.1 Down to 5 `MultModDiv`

Let $X = \lceil \sqrt{N} \rceil$. Then setting $\alpha = X^2 \bmod N$, we have $\alpha < 2X$. We state:

Theorem 3. *Given arbitrary 2n-bit integers N and $A, B \leq N$, the 2n-bit integer $AB \bmod N$ can be computed with at most five n-bit* `MultModDiv` *instructions.*

We denote our new algorithm by A3 and describe it on Fig. 5.

Input:	radix base X, 2n-bit integers $A = A_1 X + A_0, B = B_1 X + B_0$
Output:	$AB \pmod N$
Cost:	5 `MultModDiv`

$(Q^{(1)}, R^{(1)}) = \texttt{MultModDiv}(A_0, B_0, X)$
$(Q^{(2)}, R^{(2)}) = \texttt{MultModDiv}(A_1 + A_0, B_1 + B_0, X)$
$(Q^{(3)}, R^{(3)}) = \texttt{MultModDiv}(A_1, B_1, X)$
$(Q^{(4)}, R^{(4)}) = \texttt{MultModDiv}(\alpha, Q^{(3)}, X)$
$(Q^{(5)}, R^{(5)}) = \texttt{MultModDiv}(\alpha, -Q^{(1)} + Q^{(2)} - Q^{(3)} + Q^{(4)} + R^{(3)}, X)$
Return $(R^{(5)} + R^{(1)}) + (R^{(4)} - R^{(1)} + Q^{(1)} + R^{(2)} - R^{(3)} + Q^{(5)})X$

Fig. 5. Double-size modular multiplication algorithm A3

Proof (of Algorithm A3*).* Using $X^2 \equiv_N \alpha$ and

$$AB = X(X-1)A_1 B_1 + X(A_1 + A_0)(B_1 + B_0) - (X-1)A_0 B_0 \ ,$$

we get

$$(X-1)A_0 B_0 \equiv_N (X-1)(Q^{(1)}X + R^{(1)}) \equiv_N Q^{(1)}\alpha - R^{(1)} + (R^{(1)} - Q^{(1)})X \ ,$$
$$X(A_1 + A_0)(B_1 + B_0) \equiv_N X(Q^{(2)}X + R^{(2)}) \equiv_N Q^{(2)}\alpha + R^{(2)}X \ ,$$

and

$$X A_1 B_1 \equiv_N X(Q^{(3)}X + R^{(3)}) \equiv_N Q^{(3)}\alpha + R^{(3)}X \ ,$$
$$X^2 A_1 B_1 \equiv_N X(R^{(3)}X + Q^{(3)}\alpha) \equiv_N R^{(3)}\alpha + Q^{(3)}\alpha X \ ,$$
$$X(X-1)A_1 B_1 \equiv_N (-Q^{(3)} + R^{(3)})\alpha + (-R^{(3)} + Q^{(3)}\alpha)X \ ,$$

where-from

$$AB \equiv_N \alpha(-Q^{(1)} + Q^{(2)} - Q^{(3)} + R^{(3)}) + R^{(1)}$$
$$+ X(-R^{(1)} + Q^{(1)} + R^{(2)} - R^{(3)} + Q^{(3)}\alpha)$$
$$\equiv_N \alpha(-Q^{(1)} + Q^{(2)} - Q^{(3)} + R^{(3)}) + R^{(1)}$$
$$+ X(Q^{(4)}X + R^{(4)} - R^{(1)} + Q^{(1)} + R^{(2)} - R^{(3)})$$
$$\equiv_N \alpha(-Q^{(1)} + Q^{(2)} - Q^{(3)} + Q^{(4)} + R^{(3)}) + R^{(1)}$$
$$+ X(R^{(4)} - R^{(1)} + Q^{(1)} + R^{(2)} - R^{(3)})$$
$$\equiv_N (R^{(5)} + R^{(1)}) + (R^{(4)} - R^{(1)} + Q^{(1)} + R^{(2)} - R^{(3)} + Q^{(5)})X \,,$$

which proves the correctness of A3. $\square$

Again, this algorithm uses only 5 `MultModDiv` instructions. But if we take a careful look at its description, we observe that a couple of `MultModDiv` instructions are performed directly with operand α. Therefore, having a small value for α would render these two `MultModDiv` instructions significantly faster. Suppose for example that an n-bit X can be found given N such that $\alpha \leq 2^{n/2}$. Assuming that the execution time of `MultModDiv` is essentially linear in the bitsize of its first operand, then Algorithm A3 would have a time consumption close to 4 `MultModDiv`, resulting in an additional speedup of 20%.

4.2   Extreme Cases: Down to 3 `MultModDiv`

Optimal performances are reached when $\alpha = -1, 2, 3$ for instance, in which cases the computational cost of our algorithm reduces to 3 `MultModDiv` instructions. One may of course ask under which circumstances there exists an n-bit integer X with such a trivial square modulo a $2n$-bit RSA modulus N. A practical way to ensure this consists in modifying the RSA key generation. We believe that simple algebraic techniques allow to do that while preserving the security of RSA moduli.

Other choices for the representation base may also present interesting properties, as we now illustrate. Assume for instance that for a given N, there exists an n-bit $Y \geq \lceil\sqrt{N}\rceil$ such that

$$Y^2 \equiv \alpha + \delta Y \pmod{N},$$

where we try to make α and δ as trivial as possible. If α and δ are simple numbers (ideally $\delta = 1$), A3 simplifies into Algorithm A4 depicted on Fig. 6.

Proof (of correctness for A4). Using $Y^2 \equiv_N \alpha + \delta Y$ and Lemma 1, one gets

$$(Y - 1)A_0B_0 \equiv_N (Y - 1)(Q^{(1)}Y + R^{(1)})$$
$$\equiv_N -Q^{(1)}Y - R^{(1)} + R^{(1)}Y + Q^{(1)}(\alpha + \delta Y)$$
$$\equiv_N Q^{(1)}\alpha - R^{(1)} + (R^{(1)} - Q^{(1)} + Q^{(1)}\delta)Y \,,$$

$$Y(A_1 + A_0)(B_1 + B_0) \equiv_N Y(Q^{(2)}Y + R^{(2)})$$
$$\equiv_N R^{(2)}Y + Q^{(2)}(\alpha + \delta Y)$$
$$\equiv_N Q^{(2)}\alpha + (R^{(2)} + Q^{(2)}\delta)Y \,,$$

Input:	radix base Y, $2n$-bit integers $A = A_1Y + A_0, B = B_1Y + B_0$
Output:	$AB \pmod N$
Cost:	3 `MultModDiv`

$(Q^{(1)}, R^{(1)}) = $ `MultModDiv`(A_0, B_0, Y)
$(Q^{(2)}, R^{(2)}) = $ `MultModDiv`$(A_1 + A_0, B_1 + B_0, Y)$
$(Q^{(3)}, R^{(3)}) = $ `MultModDiv`(A_1, B_1, Y)

Return $\alpha(-Q^{(1)} + Q^{(2)} - Q^{(3)} + R^{(3)} + Q^{(3)}\delta) + R^{(1)}$
$+ Y(-R^{(1)} - R^{(3)} + Q^{(1)} + R^{(2)} + Q^{(3)}(\alpha + \delta^2) + (-Q^{(3)} + R^{(3)} - Q^{(1)} + Q^{(2)})\delta)$

Fig. 6. Double-size modular multiplication algorithm A4

$$YA_1B_1 \equiv_N Y(Q^{(3)}Y + R^{(3)}) \equiv_N Q^{(3)}\alpha + (R^{(3)} + Q^{(3)}\delta)Y$$
$$Y^2A_1B_1 \equiv_N Y(Q^{(3)}\alpha + (R^{(3)} + Q^{(3)}\delta)Y)$$
$$\equiv_N Q^{(3)}\alpha Y + (R^{(3)} + Q^{(3)}\delta)(\alpha + \delta Y)$$
$$\equiv_N (R^{(3)} + Q^{(3)}\delta)\alpha + ((R^{(3)} + Q^{(3)}\delta)\delta + Q^{(3)}\alpha)Y \; ,$$

$$Y(Y - 1)A_1B_1 \equiv_N (R^{(3)} - Q^{(3)} + Q^{(3)}\delta)\alpha$$
$$+ ((R^{(3)} + Q^{(3)}\delta)(\delta - 1) + Q^{(3)}\alpha)Y$$
$$\equiv_N (R^{(3)} - Q^{(3)} + Q^{(3)}\delta)\alpha + (-R^{(3)} + (-Q^{(3)} + R^{(3)})\delta$$
$$+ Q^{(3)}(\delta^2 + \alpha))Y \; ,$$

so that

$$AB \equiv_N \alpha(-Q^{(1)} + Q^{(2)} - Q^{(3)} + R^{(3)} + Q^{(3)}\delta) + R^{(1)}$$
$$+ Y(-R^{(1)} - R^{(3)} + Q^{(1)} + R^{(2)} + Q^{(3)}(\alpha + \delta^2)$$
$$+ (-Q^{(3)} + R^{(3)} - Q^{(1)} + Q^{(2)})\delta) \; ,$$

thereby proving Algorithm A4.   □

Again, this algorithm has a cost of 3 `MultModDiv` instructions provided that the values for α, δ and $\alpha + \delta^2$ are simple constant numbers. This could be ensured by properly adapting the RSA key generation algorithm.

5   Emulating Euclidean Multiplications

When the Euclidean multiplication itself is not directly available in hardware, it can be emulated easily with a couple of modular multiplications. The quotient of `MultModDiv` in [4] is calculated from the remainders of $x \cdot y$ modulo z and modulo $(z + 1)$. However, such a situation is most unfortunate for fast modular multiplication algorithms based on Montgomery's technique as either z or $z + 1$ is even. Although extensions of Montgomery to even moduli exist, we suggest a simple alternative hereafter. Our method is based on the next lemma.

Lemma 2. *If $0 < xy \le (z-1)^2$ then*

$$\left\lfloor \frac{xy}{z+\beta} \right\rfloor \le \left\lfloor \frac{xy}{z} \right\rfloor \le \left\lfloor \frac{xy}{z+\beta} \right\rfloor + \beta$$

for any nonnegative β.

Proof. Since $z < z + \beta$, it follows that $xy/(z+\beta) < xy/z$ and consequently $\lfloor xy/(z+\beta) \rfloor \le \lfloor xy/z \rfloor$. For the second inequality, we observe that

$$\frac{xy}{z} = \frac{xy}{z+\beta}\left(1 + \frac{\beta}{z}\right) \le \frac{xy}{z+\beta} + \frac{(z-1)^2\beta}{(z+\beta)z} < \frac{xy}{z+\beta} + \beta \; .$$

Therefore, we get $\lfloor xy/z \rfloor \le \lfloor xy/(z+\beta) \rfloor + \lceil \beta \rceil = \lfloor xy/(z+\beta) \rfloor + \beta$. □

So, letting

$$\Delta_\beta = \left\lfloor \frac{xy}{z} \right\rfloor - \left\lfloor \frac{xy}{z+\beta} \right\rfloor \quad \text{and} \quad C_\beta = xy \bmod (z+\beta) \, ,$$

(with $0 \le \Delta_\beta \le \beta$ by Lemma 1), one expresses the integer quotient resulting from the modular multiplication, $C = xy \bmod z$, as

$$\left\lfloor \frac{xy}{z} \right\rfloor = \frac{C - C_\beta - \Delta_\beta(z+\beta)}{\beta} \; . \tag{3}$$

Proof. By definition, we have $xy = \lfloor xy/z \rfloor z + C = \lfloor xy/(z+\beta) \rfloor(z+\beta) + C_\beta = (\lfloor xy/z \rfloor + \Delta_\beta)(z+\beta) + C_\beta$, which implies $C = \lfloor xy/z \rfloor \beta + \Delta_\beta(z+\beta) + C_\beta$. □

In particular, the value $\beta = 2$ yields the integer quotient from two modular reductions with moduli having the same parity as z. Carrying out a division by β is inexpensive as it amounts to a shift of a single bit to the right. Finally, since $\Delta_2 \le 2$, there are (at most) only two negligible corrections to make to get the exact value of the quotient.

Remark 1. This method readily extends for any value of β; the powers of 2 are of particular interest. Note also that a way to lower the expected error (cf. Δ_β) consists in increasing the numerator in $\lfloor xy/(z+\beta) \rfloor$.

6   A Modular Multiplication Based Algorithm

When Euclidean multiplications are emulated in software from modular multiplications, one may wonder if using these directly could yield faster algorithms without necessarily coming back to RNS-based approaches [1,2,7,9]. In this section, we propose a simple alternative to these works that keeps numbers under a radix representation. We rely on the following lemma.

Lemma 3. *Let X be an n-bit odd integer not divisible by 3 and N an arbitrary integer such that $X > \lceil \sqrt{N} \rceil$. There exists an algorithm which, given any $A = A_1 X + A_0$ and $B = B_1 X + B_0$ such that $A, B < N$ outputs the representation*

$$AB = C_3 X^3 + C_2 X^2 + C_1 X + C_0$$

in at most <u>four</u> n-bit modular multiplications.

We refer the reader to Appendix A for a description of such an algorithm, which we denote by Coefficients in the sequel.

Now, very much in the spirit of Section 4.1, we precompute X such that $X > \lceil \sqrt{N} \rceil$ and set $\alpha = X^2 \bmod N$. Here however, as we need $\gcd(X, 6) = 1$, we try out $X = \lceil \sqrt{kN} \rceil$ for increasing values of $k = 1, \ldots$, until X is found odd and coprime to 3. Even if $|X|$ exceeds n, the difference $|X| - n$ will be a very small value in any case, and we refer to the fact that we are able to work with non-reduced numbers when they do not exceed their range too much (see Section 2). Relying on Lemma 3, we devise Algorithm A5 as shown on Fig. 7.

Input:	radix base X, $2n$-bit integers $A = A_1 X + A_0, B = B_1 X + B_0$
Output:	$AB \pmod N$
Cost:	2 MultModDiv + 1 Coefficients

$(U^{(1)}, V^{(1)}, W^{(1)}, R^{(1)}) = \text{Coefficients}(A, B, X)$
$(Q^{(2)}, R^{(2)}) = \text{MultModDiv}(\alpha, U^{(1)}, X)$
$(Q^{(3)}, R^{(3)}) = \text{MultModDiv}(\alpha, V^{(1)} + Q^{(2)}, X)$
Return $R^{(1)} + R^{(3)} + (R^{(2)} + W^{(1)} + Q^{(3)}) X$

Fig. 7. Double-size modular multiplication algorithm A5

Proof (of correctness for A5). By definition,

$$\begin{aligned}
AB &= U^{(1)} X^3 + V^{(1)} X^2 + W^{(1)} X + R^{(1)} \\
&\equiv_N R^{(1)} + V^{(1)} \alpha + (U^{(1)} \alpha + W^{(1)}) X \\
&\equiv_N R^{(1)} + V^{(1)} \alpha + (Q^{(2)} X + R^{(2)} + W^{(1)}) X \\
&\equiv_N R^{(1)} + (V^{(1)} + Q^{(2)}) \alpha + (R^{(2)} + W^{(1)}) X \\
&\equiv_N R^{(1)} + R^{(3)} + (R^{(2)} + W^{(1)} + Q^{(3)}) X ,
\end{aligned}$$

thereby validating A5.   □

As indicated, our algorithm runs two MultModDiv and one Coefficients operations, which (relying on Section 5 or [4]) yields 8 modular multiplications among which 4 are executed with operand α. Then, we can combine A5 with a proper modification of the RSA key generator to ensure that α is some small (absolute) constant. In this context of use, the cost of a double size modular multiplication by A5 reduces to four n-bit multiplications only, *i.e.*, becomes computationally equivalent to 2 calls to MultModDiv thereby yielding a speedup factor of $(14 - 4)/14 \approx 71\%$ in comparison with FS1.

7   Conclusion

In this paper, we showed how to optimally reduce the cost of Fischer and Seifert double-size modular multiplications, provided that the same basic operation (Euclidean multiplication) is available. We highlighted the role of the data representation towards the performance of emulated multiplications and proposed new ones featuring dramatic cost savings.

Table 1. Number of calls in double-size modular multiplication algorithms. The last line displays the number of equivalent n-bit modular multiplications

Calls	Fischer-Seifert		Our algorithms				
	FS1	FS2	A1	A2	A3	A4	A5
MultModDiv	7	5	6	4	$5 \to 3$	3	$2 \to 0$
MultModDivInit	0	1	0	1	0	0	0
Coefficients	0	0	0	0	0	0	1
Equiv. MultMod	14	12	12	10	$10 \to 6$	6	$8 \to 4$

We stress that in each and every of our algorithms, modifications of the radix base can be carried out while maintaining inputs and outputs under the same arithmetic format, which allows repeated executions with the same modulus. Naturally, the same algorithms may readily be used to perform double-size modular squarings. In the most favorable case, we emulate a double-size modular multiplication with no more than 3 Euclidean multiplications, resulting in a speedup factor of 57% in comparison with Fischer and Seifert's original procedures, as indicated in Table 1. When Euclidean multiplications cannot be carried out in hardware, we provide a variation based on modular multiplications only which surpasses original algorithms in performance by 71%. Although we doubt the existence of more advanced yet simple techniques, we challenge the cryptographic community for better results.

References

1. J.-C. Bajard, L.-S. Didier, and P. Kornerup. An RNS Montgomery multiplication algorithm. In *13th IEEE Symposium on Computer Arithmetic (ARITH 13)*, pp. 234–239, IEEE Press, 1997.
2. J.-C. Bajard, L.-S. Didier, and P. Kornerup. An RNS Montgomery multiplication algorithm. *IEEE Transactions on Computers*, vol. 47, no. 7, pp. 766–776, 1998.
3. P. Barrett. Implementing the Rivest Shamir and Adleman public key encryption algorithm on a standard digital signal processing. In A.M. Odlyzko, Ed., *Advances in Cryptology – CRYPTO '86*, vol. 263 of *Lecture Notes in Computer Science*, pp. 311–323, Springer-Verlag, 1987.

4. W. Fischer and J.-P. Seifert. Increasing the bitlength of crypto-coprocessors via smart hardware/software co-design. In B.S. Kaliski Jr., Ç.K. Koç, and C. Paar, Eds., *Cryptographic Hardware and Embedded Systems – CHES 2002*, vol. 2523 of *Lecture Notes in Computer Science*, pp. 71–81, Springer-Verlag, 2003.
5. P.L. Montgomery. Modular multiplication without trial divisions. *Mathematics of Computations*, vol. 44, no. 170, pp. 519–521, 1985.
6. P. Paillier. Low-cost double-size modular exponentiation or how to stretch your cryptoprocessor. In H. Imai and Y. Zheng, Eds., *Public-Key Cryptography*, vol. 1560 of *Lecture Notes in Computer Science*, pp. 223–234, Springer-Verlag, 1999.
7. K.C. Posh and R. Posh. Modulo reduction in Residue Number Systems. *IEEE Transactions on Parallel and Distributed Systems*, vol. 6, no. 5, pp. 449–454, 1995.
8. J.-J. Quisquater. Fast modular exponentiation without division. Rump session of EUROCRYPT '90, Århus, Denmark, 1990.
9. J. Schwemmlein, K.C. Posh and R. Posh. RNS modulo reduction upon a restricted base value set and its applicability to RSA cryptography. *Computer & Security*, vol. 17, no. 7, pp. 637–650, 1998.
10. C.D. Walter. Faster modular multiplication by operand scaling. In J. Feigenbaum, Ed., *Advances in Cryptology – CRYPTO '91*, vol. 576 of *Lecture Notes in Computer Science*, pp. 313–323, Springer-Verlag, 1992.

A   Proof of Lemma 3

Our four modular multiplications will be

$$\begin{cases} R_0 & = (A \bmod X)(B \bmod X) \bmod X \ , \\ R_1 & = (A \bmod (X+1))(B \bmod (X+1)) \bmod (X+1) \ , \\ R_2 & = (A \bmod (X+2))(B \bmod (X+2)) \bmod (X+2) \ , \\ R_3 & = (A \bmod (2X+3))(B \bmod (2X+3)) \bmod (2X+3) \ . \end{cases}$$

In what follows, we use the notations

$$k_0 = AB \operatorname{div} X \ , \qquad\qquad k_1 = AB \operatorname{div} (X+1) \ ,$$
$$k_2 = AB \operatorname{div} (X+2) \ , \qquad k_3 = AB \operatorname{div} (2X+3) \ ,$$

and we have by definition

$$AB = k_0 X + R_0 = k_1(X+1) + R_1 = k_0(X+1) + (R_0 - k_0)$$
$$= k_2(X+2) + R_2 = k_0(X+2) + (R_0 - 2k_0) \ ,$$
$$2AB = 2k_0 X + 2R_0 = 2k_3(2X+3) + 2R_3 = k_0(2X+3) - 3k_0 + 2R_0 \ ,$$

so that

$$k_0 \equiv R_0 - R_1 \pmod{(X+1)}$$
$$2k_0 \equiv R_0 - R_2 \pmod{(X+2)}$$
$$3k_0 \equiv 2(R_0 - R_3) \pmod{(2X+3)} \ .$$

Since X is coprime to 3, if we call $a = (R_0 - R_2 + ((R_0 - R_2) \bmod 2)(X + 2))/2 \bmod X + 2$, $b = R_0 - R_1 \bmod (X + 1)$ and $c = (2(R_0 - R_3) + (2(R_0 - R_3) \bmod 3)(2X + 3))/3 \bmod (2X + 3)$, we get that $k_0 \equiv b \pmod{(X + 1)}$, $k_0 \equiv a \pmod{(X + 2)}$ and $k_0 \equiv c \pmod{(2X + 3)}$. Starting from these equations, we can perform Chinese remaindering, because $X + 1$, $X + 2$ and $2X + 3$ are pairwise relatively prime:

$$k_0 \bmod ((X + 1)(X + 2)) = ((b - a) \bmod (X + 1))(X + 2) + a .$$

Letting $d = ((b - a) \bmod (X + 1))$ and $e = a + 2d$, we have

$$k_0 \bmod ((X + 1)(X + 2)) = dX + e .$$

Moreover, remarking that $(X + 1)(X + 2)(-4) \equiv 1 \pmod{(2X + 3)}$ and letting $f = -6d + 4e - 4c \bmod (2X + 3)$, we notice that the second CRT recombination

$$
\begin{aligned}
k_0 &= [-4(c - dX - e) \bmod (2X + 3)](X + 1)(X + 2) + dX + e \\
&= [2d(2X + 3) - 6d + 4e - 4c) \bmod (2X + 3)](X + 1)(X + 2) + dX + e \\
&= f(X + 1)(X + 2) + dX + e
\end{aligned}
$$

is easily rewritten as $k_0 = fX^2 + (d + 3f)X + (e + 2f)$. Consequently, $C = AB$ is computed in 4 modular multiplications as $C = C_3X^3 + C_2X^2 + C_1X + C_0$ with

$$
\begin{aligned}
C_3 &= f , \\
C_2 &= d + 3f , \\
C_1 &= e + 2f , \\
C_0 &= R_0 .
\end{aligned}
$$

Note that these operations are not of size $2n \times 2n$ modulo n but of size $n \times n$ modulo n, because, from $A = A_1X + A_0$ and $B = B_1X + B_0$, R_0, R_1, R_2 and R_3 can be computed as

$$
\begin{aligned}
R_0 &= A_0B_0 \bmod X , \\
R_1 &= (A_0 - A_1)(B_0 - B_1) \bmod (X + 1) , \\
R_2 &= (A_0 - 2A_1)(B_0 - 2B_1) \bmod (X + 2) , \\
R_3 &= (A_0 + (A_1 \bmod 2)X - 3(A_1 \operatorname{div} 2)) \\
&\quad \times (B_0 + (B_1 \bmod 2)X - 3(B_1 \operatorname{div} 2)) \bmod (2X + 3) .
\end{aligned}
$$

As before, the cost of auxiliary operations (additions, subtractions, parity bits, etc.) is neglected. □

Efficient Exponentiation for a Class of Finite Fields $GF(2^n)$ Determined by Gauss Periods

Soonhak Kwon[1], Chang Hoon Kim[2], and Chun Pyo Hong[2]

[1] Inst. of Basic Science and Dept. of Mathematics, Sungkyunkwan University,
Suwon 440-746, Korea
shkwon@math.skku.ac.kr
[2] Dept. of Computer and Information Engineering, Daegu University,
Kyungsan 712-714, Korea
chkim@dsp.taegu.ac.kr, cphong@daegu.ac.kr

Abstract. We present a fast and compact hardware architecture of exponentiation in a finite field $GF(2^n)$ determined by a Gauss period of type (n, k) with $k \geq 2$. Our construction is based on the ideas of Gao et al. and on the computational evidence that a Gauss period of type (n, k) over $GF(2)$ is very often primitive when $k \geq 2$. Also in the case of a Gauss period of type $(n, 1)$, i.e. a type I optimal normal element, we find a primitive element in $GF(2^n)$ which is a sparse polynomial of a type I optimal normal element and we propose a fast exponentiation algorithm which is applicable for both software and hardware purposes. We give an explicit hardware design using the algorithm.

Keywords: Finite field, Gauss period, primitive element, exponentiation, optimal normal basis

1 Introduction

Arithmetic of finite fields finds various applications in many cryptographic areas these days. Especially, fast exponentiation is very important in such applications as Diffie-Hellman key exchange and pseudo random bit generators. Though exponentiation is the most time consuming and complex arithmetic operation, in some situations such as Diffie-Hellman key exchange, one can devise an efficient exponentiation algorithm since a fixed (primitive) element is raised to many different powers. Let $GF(q^n)$ be a finite field with q^n element where q is a power of a prime and let $g \in GF(q^n)$ be a primitive element (or an element of high multiplicative order). Roughly speaking, the computation of g^s for arbitrary values of s is studied from two different directions. One is the use of precomputation with vector addition chains such as BGMW method [1] and its improvements by Lim and Lee [6] and also by Rooij [7]. The other approach is suggested by Gao et al. [4,5] and it uses a special primitive element called a Gauss period which generates a normal basis for $GF(q^n)$ over $GF(q)$. The BGMW method and its improvements are applicable to arbitrary finite field $GF(q^n)$ and very flexible. On the other hand, an ideal version of BGMW method requires a memory of order $O(n \log q / \log(n \log q))$ values in $GF(q^n)$ and multiplications of order

C.D. Walter et al. (Eds.): CHES 2003, LNCS 2779, pp. 228–242, 2003.

$O(\log(n \log q))$ which amounts to an order of $O(n^2 \log^2 q \log(n \log q))$ bit additions. An algorithm proposed by Gao et al. is not applicable to all finite fields. However, it does not need a precomputation and the complexity of the algorithm is $O(kqn^2)$ additions. Therefore if q is small and if there is a Gauss period of high order of type (n, k) for a small value of k, then the method of Gao et al. outperforms the precomputation methods. In this paper, we will discuss an improved algorithm of Gao et al. for a hardware arrangement. We will present a compact and fast hardware architecture for exponentiation using Gauss periods of type (n, k) in $GF(2^n)$ where $k \geq 2$, and detailed explanations will be given for $k = 2, 3$. Also we will give an algorithm for efficient exponentiation in the field determined by an irreducible all one polynomial (AOP). This is possible since we may successfully find a primitive element which is a trinomial of a root of an AOP for most of the cases. Since none of the papers in [1,4,5,6,7] mentions an explicit hardware architecture for exponentiation and since our construction of the circuit has the features of regularity and modularity for VLSI implementation, our result may have possible applications such as smart card purposes.

2 Gauss Periods of Type (n, k) in $GF(q^n)$

We will briefly review the theory of Gauss periods and the method of Gao et al.. Let n, k be positive integers such that $p = nk + 1$ is a prime not dividing q. Let $K = \langle \tau \rangle$ be a unique subgroup of order k in $GF(p)^\times$. Let $ord_p q$ be the order of q modulo p and assume $gcd(nk/ord_p q, n) = 1$. Let β be a primitive pth root of unity in $GF(q^{nk})$. Then the the following element

$$\alpha = \sum_{j=0}^{k-1} \beta^{\tau^j} \tag{1}$$

is called a Gauss period of type (n, k) over $GF(q)$. It is well known that α is a normal element in $GF(q^n)$. That is, letting $\alpha_i = \alpha^{q^i}$ for $0 \leq i \leq n - 1$, $\{\alpha_0, \alpha_1, \alpha_2, \cdots, \alpha_{n-1}\}$ is a basis for $GF(q^n)$ over $GF(q)$. Since $K = \langle \tau \rangle$ is a subgroup of order k in $GF(p)^\times$, a cyclic group of order $p - 1 = nk$, the quotient group $GF(p)^\times / K$ is also a cyclic group of order n and the generator of the group is qK. Therefore we have a coset decomposition of $GF(p)^\times$ as a disjoint union,

$$GF(p)^\times = K_0 \cup K_1 \cup K_2 \cdots \cup K_{n-1}, \tag{2}$$

where $K_i = q^i K, 0 \leq i \leq n - 1$. Note that any element in $GF(p)^\times$ is uniquely written as $\tau^s q^t$ for some $0 \leq s \leq k - 1$ and $0 \leq t \leq n - 1$. Now for each $0 \leq i \leq n - 1$, we have

$$\alpha \alpha_i = \sum_{s=0}^{k-1} \beta^{\tau^s} \sum_{t=0}^{k-1} \beta^{\tau^t q^i}$$

$$= \sum_{s=0}^{k-1} \sum_{t=0}^{k-1} \beta^{\tau^s (1 + \tau^{t-s} q^i)} = \sum_{s=0}^{k-1} \sum_{t=0}^{k-1} \beta^{\tau^s (1 + \tau^t q^i)}. \tag{3}$$

Notice that there is unique $0 \le u \le k-1$ and $0 \le v \le n-1$ such that $1+\tau^u q^v = 0 \in GF(p)$. If $t \ne u$ or $i \ne v$, then we have $1+\tau^t q^i \in K_{\sigma(t,i)}$ for some $0 \le \sigma(t,i) \le n-1$ depending on t and i. Thus we may write $1+\tau^t q^i = \tau^{t'} q^{\sigma(t,i)}$ for some t'. Now when $i \ne v$,

$$
\begin{aligned}
\alpha\alpha_i = \sum_{s=0}^{k-1}\sum_{t=0}^{k-1} \beta^{\tau^s(1+\tau^t q^i)} &= \sum_{s=0}^{k-1}\sum_{t=0}^{k-1} \beta^{\tau^s(\tau^{t'} q^{\sigma(t,i)})} \\
&= \sum_{t=0}^{k-1}\sum_{s=0}^{k-1} \beta^{\tau^{s+t'} q^{\sigma(t,i)}} = \sum_{t=0}^{k-1} \alpha^{q^{\sigma(t,i)}} = \sum_{t=0}^{k-1} \alpha_{\sigma(t,i)}.
\end{aligned}
\tag{4}
$$

Also when $i = v$,

$$
\begin{aligned}
\alpha\alpha_v = \sum_{s=0}^{k-1}\sum_{t=0}^{k-1} \beta^{\tau^s(1+\tau^t q^v)} &= \sum_{t \ne u}^{k-1}\sum_{s=0}^{k-1} \beta^{\tau^s(\tau^{t'} q^{\sigma(t,v)})} + \sum_{s=0}^{k-1} \beta^{\tau^s(1+\tau^u q^v)} \\
&= \sum_{t \ne u}^{k-1}\sum_{s=0}^{k-1} \beta^{\tau^{s+t'} q^{\sigma(t,v)}} + \sum_{s=0}^{k-1} 1 = \sum_{t \ne u} \alpha^{q^{\sigma(t,v)}} + k = \sum_{t \ne u} \alpha_{\sigma(t,v)} + k.
\end{aligned}
\tag{5}
$$

Therefore $\alpha\alpha_i$ is computed by the sum of at most k basis elements in $\{\alpha_0, \alpha_1, \cdots, \alpha_{n-1}\}$ for $i \ne v$ and $\alpha\alpha_v$ is computed by the sum of at most $k-1$ basis elements and the constant term $k \in GF(q)$. Using these ideas, Gao et al. [4] showed the following.

Theorem 1. *Let α be a Gauss period of type (n,k) over $GF(q)$, with k and q bounded. For any $0 \le r < q^n$, α^r can be computed in $O(n^2)$ additions in $GF(q)$.*

Sketch of Proof. Write $r = \sum_{j=0}^{n-1} r_j q^j$ with $0 \le r_j < q$. Then the following algorithm gives an output α^r.

Table 1. An exponentiation algorithm in [4]

Input: $r = \sum_{j=0}^{n-1} r_j q^j$
Output: $\alpha^r = \prod_{0 \le i \le n-1} \alpha_i^{r_i}$
$A \leftarrow 1$
for $(i = 0$ to $n-1$; $i++)$
   if $r_i \ne 0$
      for $(j = 1$ to r_i ; $j++)$
         $A \leftarrow A\alpha_i$
      end for
   end if
end for

Assuming that qth Frobenius map $\alpha \to \alpha^q$ is almost free, $A\alpha_i$ is computed by $O(nk)$ additions in $GF(q)$ in a redundant basis $\{\alpha_0, \alpha_1, \cdots, \alpha_{n-1}, 1\}$. For each i, the inner loop $A \leftarrow A\alpha_i$ runs r_i times. Therefore the total number of multiplications $A \leftarrow A\alpha_i$ is $\sum_{i=0}^{n-1} r_i \leq (q-1)n$. Since $A\alpha_i$ is computed by $O(nk)$ additions, one can compute α^r by $O(kqn^2)$ additions in $GF(q)$.   □

If above theorem should have any application, it must be guaranteed that the Gauss period α is a primitive element in $GF(q^n)$, or at least is of high order. This is not always satisfied. For example, a Gauss period α of type $(n, 1)$ is never a primitive element since $\alpha^{n+1} = 1$ and $n + 1 << q^n$. However, various computational results imply that the Gauss period α of type (n, k), $k \geq 2$, over $GF(2)$ is very often primitive, and even in the cases that α is not primitive, it usually has a very high multiplicative order. For example, it is known [4] that, among the 177 values of $n \leq 1000$ for which a Gauss period α of type $(n, 2)$ over $GF(2)$ exists, α is a primitive element for 146 values of n. Moreover, when α is not primitive, it is usually of very high order. The same table in [4] implies that a Gauss period of type (n, k) over $GF(2)$ is also very often primitive for $k \geq 3$. In the table, it is shown that for approximately 1050 values of $2 \leq n \leq 1200$, there is a primitive Gauss period of type (n, k) for some k, and in many cases, one can choose $k < 20$. A theorem supporting this experimental evidence is obtained by Gathen and Shparlinski [17], where it is shown that a Gauss period of type $(n, 2)$ in $GF(q^n)$ has order at least $2^{\sqrt{2n}-2}$ for infinitely many n.

3   Hardware Arrangements for Exponentiation Using Gauss Periods of Type (n, k) in $GF(2^n)$ for $k \geq 2$

Throughout this section, let us assume that $q = 2$. The algorithm in section 2 is not suitable for a hardware arrangement since one has to multiply different α_j for each step and since the exact number of additions in the coefficients of the expression $A\alpha_j$ is unclear. From now on, instead of using a redundant basis as in section 2, we will always use a normal basis $\{\alpha_0, \alpha_1, \cdots, \alpha_{n-1}\}$ because our approach is more suitable for a unified and simple hardware architecture. Let $A = \sum_{i=0}^{n-1} a_i\alpha_i$ with $a_i \in GF(2)$. Notice that there exist unique $0 \leq u \leq k - 1$ and $0 \leq v \leq n - 1$ such that $1 + \tau^u 2^v \equiv 0 \pmod{p}$. In this case there is no $0 \leq \sigma(u, v) \leq n - 1$ satisfying $0 = 1 + \tau^u 2^v \in K_{\sigma(u,v)}$. Therefore from the equations (4) and (5),

$$A\alpha = \sum_{i=0}^{n-1} a_i\alpha_i\alpha = a_v k + \sum_{\substack{i=0 \\ (t,i)\neq(u,v)}}^{n-1}\sum_{t=0}^{k-1} a_i\alpha_{\sigma(t,i)}. \tag{6}$$

For each $0 \leq i \leq n - 1$, letting $t_{ij} = |\{0 \leq t \leq k - 1 | 1 + \tau^t q^i \in K_j\}| = |\{0 \leq t \leq k - 1 | \sigma(t, i) = j\}|$, we have

$$A\alpha = a_v k + \sum_{i=0}^{n-1} a_i \sum_{j=0}^{n-1} t_{ij}\alpha_j = a_v k + \sum_{j=0}^{n-1}\left(\sum_{i=0}^{n-1} a_i t_{ij}\right)\alpha_j. \tag{7}$$

Lemma 1. *If k is even, each coefficient of $A\alpha$ is computed by the sum of at most k $a_i s$, $0 \leq i \leq n-1$, and if k is odd, it is computed by the sum of at most $k+1$ $a_i s$.*

Proof. For each j, it is almost clear that the number of $0 \leq i \leq n-1$ such that $t_{ij} \neq 0$ is at most k. If not, there are $k+1$ different i_s with $s = 1, 2, \cdots, k+1$ such that $1 + \tau^{t_s} 2^{i_s} \in K_j$ for some t_s, $s = 1, 2, \cdots, k+1$. Since the coset $K_j = 2^j K$ is a set with k elements, there exist $1 \leq s \neq l \leq k+1$ such that $1 + \tau^{t_s} 2^{i_s} = 1 + \tau^{t_l} 2^{i_l}$, which implies $i_s = i_l$. Therefore $\sum_{i=0}^{n-1} a_i t_{ij}$ is the sum of at most k $a_i s$. Because we have the field $GF(2^n)$ of characteristic two, $a_v k = 0$ if k is even and $a_v k = a_v$ if k is odd. Since the coefficient of α_j in $A\alpha$ is $\sum_{i=0}^{n-1} a_i t_{ij}$ if k is even and $a_v + \sum_{i=0}^{n-1} a_i t_{ij}$ if k is odd, our assertion is verified. $\square$

Now we are ready to give a modified algorithm which is easily applicable to a hardware arrangement.

Table 2. A modified exponentiation algorithm for a hardware purpose

---

Input: $r = \sum_{j=0}^{n-1} r_j 2^j$
Output: α^r
$A \leftarrow 1$
for $(i = n-1$ to 0 ; $i--)$
    $A \leftarrow A^2 \alpha^{r_i}$
end for

Above algorithm is just a simple form of binary window method which computes

$$\alpha^r = \alpha^{\sum_{i=0}^{n-1} r_i 2^i} = (\cdots(((\alpha^{r_{n-1}})^2 \alpha^{r_{n-2}})^2 \alpha^{r_{n-3}})^2 \cdots)^2 \alpha^{r_0}. \qquad (8)$$

Notice that by lemma 1, the operation $A \leftarrow A^2 \alpha$ in our algorithm needs at most $k-1$ or k additions under the normal basis expression. One may realize above exponentiation in a linear array circuit consisting of n flip-flops, n 2-1 MUXs and at most $n(k-1)$ or nk (depending on the parity of k) XOR gates. The initial value $A = 1$ is loaded in n flip-flops, i.e. we have $a_0 = a_1 = \cdots = a_{n-1} = 1$ initially. The signal of $r = \sum_{j=0}^{n-1} r_j 2^j$ is loaded serially in descending order. That is, $r_0, \cdots, r_{n-2}, r_{n-1} \longrightarrow$. Since $A \leftarrow A^2$ is free in a hardware arrangement (just a rewiring), $A \leftarrow A^2 \alpha^{r_i}$ is computed at most $k-1$ or k additions for each coefficient. This operation can be done in one clock cycle. Namely, at ith clock cycle, all the coefficients of A^2 and $A^2 \alpha$ are loaded as input values of the MUXs where the control signal is r_{n-i}. Therefore if $r_{n-i} = 0$, then A^2 is selected, and if $r_{n-i} = 1$, then $A^2 \alpha$ is selected. Let us remind that XOR is a 2-input XOR gate and MUX is a 2-1 multiplexer. Also D_X is the delay time of a XOR and D_M is the delay time of a MUX.

Proposition 1. *Let α be a Gauss period of type (n, k), $k \geq 2$, in $GF(2^n)$. Let $r = \sum_{i=0}^{n-1} r_i 2^i$ with $r_i = 0, 1$. Then,*

(a) we can construct a linear array which computes α^r using n flip-flops, n 2-1 MUXs, and at most $n(k - 1)$ XOR gates if k is even, at most nk XOR gates if k is odd.

(b) Each coefficient of α_j consists of an XOR tree with at most $k - 1$ or k XOR gates. Thus the depth of each XOR tree is at most $\lceil \log_2 k \rceil$ and the critical path delay of our architecture is $\lceil \log_2 k \rceil D_X + D_M$.

We present the design of the circuit in Fig. 1. Note that we get the result α^r after n clock cycles and at each ith clock cycle, A^2 and $A^2 \alpha$ are simultaneously computed and pass through MUX to get the correct value $A \leftarrow A^2 \alpha^{n-i}$.

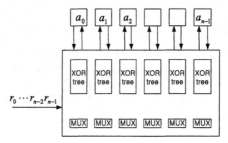

Fig. 1. A circuit for exponentiation α^r using a type (n, k) Gauss period in $GF(2^n)$

To show the power of our architecture, which is a linear array but involves many parallel computations, let us think of a finite field $GF(2^{1188})$. It is known [4] that the lowest complexity primitive Gauss period in $GF(2^{1188})$ is of type $(1188, 19)$, i.e. $k = 19$. In this case, our architecture needs 1188 flip-flops and MUXs, and at most 21384 XOR gates. But the critical path delay is only $\lceil \log_2 k \rceil D_X + D_M = 5D_X + D_M$. When $n = 1194$, we have a primitive Gauss period of type $(1194, 2)$ in $GF(2^{1194})$. Thus we need only 1194 XOR gates and the critical path delay is $D_X + D_M$. It should be mentioned that a linear array for exponentiation is proposed by Wu and Hasan [15] using a polynomial basis. Though their method is quite efficient, the complexity and the structure of the design heavily depends on the choice of primitive irreducible polynomial. However our array provides high flexibility and modularity with respect to field size n. In the following subsections, we will discuss the circuits of Gauss periods of type $(n, 2)$ and $(n, 3)$ which have low computational complexity. In these cases, the exact number of necessary gates will be determined rather easily.

3.1 Optimal Normal Basis of Type II over $GF(2)$

Let $\alpha = \beta + \beta^{-1}$ be a Gauss period of type $(n, 2)$ in $GF(2^n)$, where $2n + 1 = p$ is a prime and β is a primitive pth root of unity in $GF(2^{2n})$. It is also called an optimal normal element of type II and $\{\alpha_0, \alpha_1, \cdots, \alpha_{n-1}\}$ is called an optimal normal basis of type II over $GF(2)$. It has the lowest complexity in the sense that the sum of the number of nonzero terms in the expression of $\alpha\alpha_i$ for $0 \leq i \leq n-1$ is minimal, which is $2n - 1$. Since $\{1, 2, \cdots, 2n\}$ and $\{\pm1, \pm2, \cdots, \pm n\}$ are the same reduced residue system (mod p), we easily find $\{\alpha_0, \alpha_1, \cdots, \alpha_{n-1}\}$ and $\{\beta + \beta^{-1}, \beta^2 + \beta^{-2}, \cdots, \beta^n + \beta^{-n}\}$ are same sets. Letting $\alpha'_s = \beta^s + \beta^{-s}, 1 \leq s \leq n$, it is clear that $\alpha\alpha'_i = (\beta + \beta^{-1})(\beta^i + \beta^{-i}) = \alpha'_{i-1} + \alpha'_{i+1}$. A multiplication table can be constructed easily using above property. Or we may use the self dual property of a Gauss period of type (n, k) for even k. We say that two bases $\{\beta_1, \beta_2, \cdots, \beta_n\}$ and $\{\gamma_1, \gamma_2, \cdots, \gamma_n\}$ of $GF(2^n)$ are dual if the trace map, $Tr : GF(2^n) \to GF(2)$, with $Tr(\beta) = \beta + \beta^2 + \cdots + \beta^{2^{n-1}}$, satisfies $Tr(\beta_i \gamma_j) = \delta_{ij}$ for all $1 \leq i, j \leq n$, where $\delta_{ij} = 1$ if $i = j$, zero if $i \neq j$. A basis $\{\beta_1, \beta_2, \cdots, \beta_n\}$ is said to be self dual if $Tr(\beta_i \beta_j) = \delta_{ij}$. One can directly prove that the Gauss period of type $(n, 2)$ in $GF(2^n)$ generates a self dual normal basis or more generally, one may refer the result in [3] which says that a normal basis of Gauss period of type (n, k) in $GF(2^n)$ is self dual if and only if k is even. Using this self duality, or by a straightforward computation, one can show [20] that

Lemma 2. *Let β is a primitive pth root of unity in $GF(2^{2n})$ where $p = 2n + 1$ is a prime, and let $\alpha = \beta + \beta^{-1}$ be an optimal normal element of type II in $GF(2^n)$. Let $\alpha'_i = \beta^i + \beta^{-i}$ for all $1 \leq i \leq n$. Let $A = \sum_{i=1}^{n} a_i \alpha'_i$ and $B = \sum_{i=1}^{n} b_i \alpha'_i$ be elements in $GF(2^n)$. Then we have $AB = \sum_{j=1}^{n}(AB)_j \alpha'_j$, where the jth coefficient $(AB)_j$ satisfies*

$$(AB)_j = \sum_{i=1}^{n} b_i(a_{j-i} + a_{j+i}),$$

where it is defined that $a_0 = 0$, $a_s = a_{-s}$ if s is negative, and $a_s = a_{2n+1-s}$ if $s > n$.

For our purpose, we only need to know the formula of $A\alpha$ with respect to the basis $\{\alpha'_1, \alpha'_2, \cdots, \alpha'_n\}$. Letting $B = \alpha = \alpha'_1$ in above lemma, we get $b_1 = 1$ and b_i is zero if $i \neq 1$. Thus $(A\alpha)_j = a_{j-1} + a_{j+1}$ for all $1 \leq j \leq n$. That is,

$$A\alpha = a_2\alpha'_1 + (a_1 + a_3)\alpha'_2 + (a_2 + a_4)\alpha'_3 + \cdots + (a_{n-2} + a_n)\alpha'_{n-1} + (a_{n-1} + a_n)\alpha'_n. \tag{9}$$

Using this formula and since $\{\alpha_0, \alpha_1, \cdots, \alpha_{n-1}\}$ and $\{\alpha'_1, \alpha'_2, \cdots, \alpha'_n\}$ are same sets where $\alpha_i = \alpha^{2^{i-1}}$ and $\alpha'_i = \beta^i + \beta^{-i}$, we find that the circuit for exponentiation needs exactly n flip-flops, n 2-1 MUXs, and $n - 1$ XOR gates.

Proposition 2. *Let α be a type II optimal normal element in $GF(2^n)$. Then we can construct a linear array which computes α^r for any $r = \sum_{i=0}^{n-1} r_i 2^i$ with $r_i = 0, 1$ using n flip-flops, n 2-1 MUXs and $n - 1$ XOR gates. The critical path delay of our architecture is $D_X + D_M$ and the latency is n.*

Example 1. Let $p = 11$ and $n = 5$ where the existence of a type II optimal normal element $\alpha = \beta + \beta^{-1}$ is well known. Notice that α is a primitive element. Also note the following correspondence,

$$\alpha_0 = \alpha'_1, \alpha_1 = \alpha'_2, \alpha_2 = \alpha'_4, \alpha_3 = \beta^8 + \beta^{-8} = \alpha'_3, \alpha_4 = \beta^{16} + \beta^{-16} = \alpha'_5, \tag{10}$$

where $\beta^{11} = 1$ is used. Now let $A = a_0\alpha_0 + a_1\alpha_1 + a_2\alpha_2 + a_3\alpha_3 + a_4\alpha_4$. Then $A^2 = a_4\alpha_0 + a_0\alpha_1 + a_1\alpha_2 + a_2\alpha_3 + a_3\alpha_4$. From the correspondence (10), we get $A^2 = a_4\alpha_1' + a_0\alpha_2' + a_2\alpha_3' + a_1\alpha_4' + a_3\alpha_5'$. Thus from the formula (9),

$$A^2\alpha = A^2\alpha_1'$$
$$= a_0\alpha_1' + (a_2 + a_4)\alpha_2' + (a_0 + a_1)\alpha_3' + (a_2 + a_3)\alpha_4' + (a_1 + a_3)\alpha_5' \quad (11)$$
$$= a_0\alpha_0 + (a_2 + a_4)\alpha_1 + (a_2 + a_3)\alpha_2 + (a_0 + a_1)\alpha_3 + (a_1 + a_3)\alpha_4.$$

The basis of our circuit is $\{\alpha_0, \alpha_1, \cdots, \alpha_{n-1}\}$, and at each ith clock cycle, the serial input r_{n-i} selects via MUX one of the two values, A^2 or $A^2\alpha$. This is realized in the following circuit shown in Fig. 2.

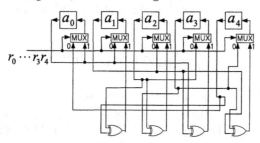

Fig. 2. A circuit for exponentiation α^r using a type II optimal normal element in $GF(2^n)$ for $n = 5$

3.2 Gauss Period of Type $(n, 3)$ over $GF(2)$

Let $3n + 1 = p$ is a prime and β is a primitive pth root of unity in $GF(2^{3n})$. Let $\alpha = \beta + \beta^\tau + \beta^{\tau^2} \in GF(2^n)$ be a Gauss period of type $(n, 3)$ where τ is a generator of the unique cyclic subgroup K of order 3 in $GF(p)^\times$. Note that there is unique u and v such that

$$1 + \tau^u 2^v = 0 \in GF(p). \quad (12)$$

Also notice $v \neq 0$ because $-1 \notin K = \langle\tau\rangle$. We claim that v is a unique integer satisfying

$$1 + \tau, 1 + \tau^2 \in K_v = 2^v K. \quad (13)$$

Since τ is an element of order 3 in $GF(p)^\times$, we get

$$\tau^2 + \tau + 1 = 0. \quad (14)$$

Thus by (12) and (14),

$$\tau + \tau^2 = -1 = \tau^u 2^v, \quad (15)$$

which implies

$$1 + \tau = \tau^{u-1} 2^v, \text{ and } 1 + \tau^2 = \tau^2(1 + \tau) = \tau^{u+1} 2^v. \quad (16)$$

Therefore the equation (13) is verified. Now from the equation (5),

$$\alpha\alpha_v = \sum_{t \neq u} \alpha^{2^{\sigma(t,v)}} + 3 = \alpha_{\sigma(t_1,v)} + \alpha_{\sigma(t_2,v)} + 1, \qquad (17)$$

where $t_1, t_2 \neq u$. We claim that $\sigma(t_1, v) \neq \sigma(t_2, v)$. In fact, more generally we have

Lemma 3. *If $i \neq 0, v$, then $\sigma(0, i), \sigma(1, i), \sigma(2, i)$ are all different. If $i = 0$, then $\sigma(1, 0) = \sigma(2, 0) = v$ and $\sigma(0, 0) = 1$. If $i = v$, then $\sigma(t_1, v) \neq \sigma(t_2, v)$.*

Proof. The second statement is already proved in view of the equation (13). Now suppose $i \neq 0, v$. To prove the first statement, we have to show that $1 + 2^i, 1 + \tau 2^i, 1 + \tau^2 2^i$ are in all different cosets of K in $GF(p)^\times$. Suppose on the contrary that there exist $s \neq t$ such that $1 + \tau^s 2^i$ and $1 + \tau^t 2^i$ belong to the same coset. Then we have

$$\frac{1 + \tau^t 2^i}{1 + \tau^s 2^i} \in K = \{1, \tau, \tau^2\}. \qquad (18)$$

Since $s \neq t$, $\frac{1+\tau^t 2^i}{1+\tau^s 2^i} = 1$ is impossible. Suppose $\frac{1+\tau^t 2^i}{1+\tau^s 2^i} = \tau$. Then $1 + \tau^t 2^i = \tau + \tau^{s+1} 2^i$. If $t \equiv s + 1 \pmod 3$, we get $\tau = 1$ which is absurd. Therefore $t \equiv s + 2 \pmod 3$ and we get $1 + \tau^{s+2} 2^i = \tau + \tau^{s+1} 2^i$. Thus we get $2^i = \frac{\tau - 1}{\tau^{s+2} - \tau^{s+1}} = \tau^{-s-1} \in K$, which is a contradiction since $0 < i \leq n - 1$ and n is the least positive integer satisfying $2^n \in K$. Now suppose $\frac{1+\tau^t 2^i}{1+\tau^s 2^i} = \tau^2$. Then $\frac{1+\tau^s 2^i}{1+\tau^t 2^i} = \tau^{-2} = \tau$ and the same technique can be applied. The proof of the last statement is also same. $\qquad \square$

From lemma 3, the multiplication structure of $\alpha\alpha_i$ is completely determined. That is, when $i = 0$, $\alpha\alpha_0 = \alpha_1$ consists of one basis element. For $i = v$, we have $\alpha\alpha_v = \alpha_{\sigma(t_1,v)} + \alpha_{\sigma(t_2,v)} + 1$ with $\sigma(t_1, v) \neq \sigma(t_2, v)$. And for $i \neq 0, v$, we get $\alpha\alpha_i = \alpha_{\sigma(0,i)} + \alpha_{\sigma(1,i)} + \alpha_{\sigma(2,i)}$ where all the summands are different. Therefore, except for the constant term 1 in the expression $\alpha\alpha_v$, the number of elements which are in the summands of $\alpha\alpha_i$, $0 \leq i \leq n - 1$ is exactly $3(n - 1)$. On the other hand, α_v appears only once as a summand of $\alpha\alpha_i$ for some i since two α_v in the expression of $\alpha\alpha_0$ are cancelled each other. Moreover α_0 appears twice as a summand of $\alpha\alpha_i$ for two different values of i. This is because we have only two different pairs of (t, i) satisfying $1 + \tau^t 2^i \in K$, i.e. $1 + \tau^t 2^i = 1$ is never satisfied. Since the proof of lemma 1 says that α_j appears at most 3 times as a summand of $\alpha\alpha_i$, $0 \leq i \leq n - 1$, we conclude that α_j $(j \neq 0, v)$ appears exactly 3 times as a summand of $\alpha\alpha_i$. Letting $A = \sum_{i=0}^{n-1} a_i \alpha_i$, the multiplication structure of $A\alpha$ in the equation (7) says,

$$A\alpha = \sum_{j=0}^{n-1} \left(a_v + \sum_{i=0}^{n-1} a_i t_{ij} \right) \alpha_j. \qquad (19)$$

From the observations on the number of basis element as a summand of $\alpha\alpha_i$, we conclude that $a_v + \sum_{i=0}^{n-1} a_i t_{ij}$ needs 2 additions if $j = 0$, one addition if $j = v, \sigma(t_1, v), \sigma(t_2, v)$ and 3 additions if $j \neq 0, v, \sigma(t_1, v), \sigma(t_2, v)$.

Proposition 3. *Let α be a Gauss period of type $(n, 3)$ in $GF(2^n)$. Then we can construct a linear array which computes α^r for any $r = \sum_{i=0}^{n-1} r_i 2^i$ with $r_i = 0, 1$ using n flip-flops, n 2-1 MUXs and $3n - 7$ XOR gates. Each coefficient of α_j $(j \neq 0, v, \sigma(t_1, v), \sigma(t_2, v))$ consists of 3 XOR gates, For $j = v, \sigma(t_1, v), \sigma(t_2, v)$, each coefficient consists of one XOR gate, and the coefficient of α_0 needs 2 XOR gates. Thus the critical path delay of our architecture is $2D_X + D_M$ and the latency is n.*

Example 2. Let $p = 19$ and $n = 6$ where a Gauss period α of type $(6, 3)$ in $GF(2^6)$ exists and is primitive. In this case, the unique cyclic subgroup of order 3 in $GF(19)^\times$ is $K = \{1, 7, 11\}$. Let β be a primitive 19th root of unity in $GF(2^{18})$. Thus letting $\tau = 7$, α is written as $\alpha = \beta + \beta^7 + \beta^{11}$. The computations of $\alpha\alpha_i$, $0 \leq i \leq 5$ is easily done from the following table. For each block regarding K and K', (s, t) entry with $0 \leq s \leq 2$ and $0 \leq t \leq 5$ denotes $\tau^s 2^t$ and $1 + \tau^s 2^t$ respectively.

Table 3. Computation of K_i and K_i'

K_0	K_1	K_2	K_3	K_4	K_5	K_0'	K_1'	K_2'	K_3'	K_4'	K_5'
1	2	4	8	16	13	2	3	5	9	17	14
7	14	9	18	17	15	8	15	10	0	18	16
11	3	6	12	5	10	12	4	7	13	6	11

From above table, we easily deduce

$$\alpha\alpha = \alpha_1, \qquad \alpha\alpha_1 = \alpha_1 + \alpha_2 + \alpha_5, \qquad \alpha\alpha_2 = \alpha_0 + \alpha_4 + \alpha_5, \qquad (20)$$
$$\alpha\alpha_3 = \alpha_2 + \alpha_5 + 1, \qquad \alpha\alpha_4 = \alpha_2 + \alpha_3 + \alpha_4, \qquad \alpha\alpha_5 = \alpha_0 + \alpha_1 + \alpha_4. \qquad (21)$$

For example, see the block K_2' for the expression of $\alpha\alpha_2$. The entries of K_2' are 5, 10, 7. Now see the blocks of $K_i s$ and find $5 \in K_4, 10 \in K_5, 7 \in K_0$. Thus we get $\alpha\alpha_2 = \alpha_4 + \alpha_5 + \alpha_0$. Note that $v = 3$ and $\sigma(t_1, v), \sigma(t_2, v) = 2, 5$ in our example. Let $A = \sum_{i=0}^{5} a_i \alpha_i$ be an element in $GF(2^6)$. Then $A^2 = \sum_{i=0}^{5} a_i \alpha_{i+1}$ where α_6 is understood as α_0. Thus

$$
\begin{aligned}
A^2\alpha &= \sum_{i=0}^{5} a_i \alpha_{i+1}\alpha \\
&= a_0(\alpha_1 + \alpha_2 + \alpha_5) + a_1(\alpha_0 + \alpha_4 + \alpha_5) + a_2(\alpha_2 + \alpha_5 + 1) \\
&\quad + a_3(\alpha_2 + \alpha_3 + \alpha_4) + a_4(\alpha_0 + \alpha_1 + \alpha_4) + a_5\alpha_1 \\
&= a_2 + (a_1 + a_4)\alpha_0 + (a_0 + a_4 + a_5)\alpha_1 + (a_0 + a_2 + a_3)\alpha_2 \\
&\quad + a_3\alpha_3 + (a_1 + a_3 + a_4)\alpha_4 + (a_0 + a_1 + a_2)\alpha_5 \\
&= (a_1 + a_2 + a_4)\alpha_0 + (a_0 + a_2 + a_4 + a_5)\alpha_1 + (a_0 + a_3)\alpha_2 \\
&\quad + (a_2 + a_3)\alpha_3 + (a_1 + a_2 + a_3 + a_4)\alpha_4 + (a_0 + a_1)\alpha_5.
\end{aligned}
\tag{22}
$$

Thus the exponentiation algorithm is realized in the following circuit.

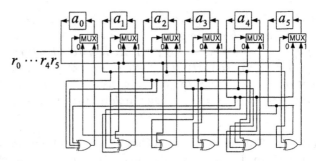

Fig. 3. A circuit for exponentiation α^r using Gauss period of type $(n, 3)$ in $GF(2^n)$ for $n = 6$

4   Primitive Elements Spanned by an Optimal Normal Basis of Type I over $GF(2)$

Let α be a Gauss period of type $(n, 1)$ over $GF(2)$, where $n + 1 = p$ is a prime. α is also called a type I optimal normal element and the corresponding normal basis is called a type I optimal normal basis. Note that α is never primitive and has a very low order, i.e. $\alpha^{n+1} = 1$ where $n + 1 << 2^n - 1$. Therefore one cannot use the algorithms in Table 1, 2 for exponentiation for a practical purpose. On the other hand, it should be noticed that there are not so few primitive elements in arbitrary finite field $GF(2^n)$. That is, the number of primitive elements in $GF(2^n)$ is $\phi(2^n - 1)$, where $\phi(x)$ is Euler's *phi*-function. Thus, the probability for a randomly chosen element $\alpha \in GF(2^n)^\times$ to be a primitive element is

$$\phi(2^n - 1)/(2^n - 1) = \prod_{q | 2^n - 1} \left(1 - \frac{1}{q} \right), \tag{23}$$

where the product runs through all primes q dividing $2^n - 1$. As long as $2^n - 1$ is not a product of many small prime factors, which is a necessary condition to avoid the Pohlig-Hellman attack for discrete logarithm problem, the probability is not so small. In fact, the following formula for average value of the probability is well known [25],

$$\sum_{n=1}^{N} \phi(n)/n = \frac{6}{\pi^2} N + O(\log N). \tag{24}$$

Of course, our choice of α is not a randomly chosen element. Though α is not a primitive element, we may ask a natural question whether there exists a primitive element which is a sparse polynomial of α, for example, a binomial of the form $\alpha^s + \alpha^t$. However it turns out that they are never primitive if $n > 4$. To show this, note that $\alpha^s + \alpha^t = \alpha^s(1 + \alpha^{t-s}) = \alpha^s(1 + \alpha)^{2^j}$ for some j. Also from the observation, $(1 + \alpha)^{2^{n/2}} = 1 + \alpha^{2^{n/2}} = 1 + \alpha^{-1} = (1 + \alpha)/\alpha$, we get $\alpha = (1 + \alpha)^{-2^{n/2}+1}$. Therefore, neither $1 + \alpha$ nor $\alpha^s + \alpha^t$ is a primitive element if

$n+1 < 2^{n/2}+1$, i.e. if $n > 4$. The next possible choice is the elements of the form $\alpha^s + \alpha^t + \alpha^l$, or more simply, $1 + \alpha^s + \alpha^t$ because the multiplication by α contributes a negligible order $n+1$. For this type of elements, we could not use the same technique as proving $1 + \alpha$ is of low order. In fact we found, by a computation, that a trinomial $1 + \alpha + \alpha^s$ of a type I optimal normal element α in $GF(2^n)$ is always a primitive element for some s with only one exception among all $n \le 550$.

Table 4. List of $n \le 550$ for which a type I optimal normal element α exists and its corresponding primitive element

n	primitive element	n	primitive element	n	primitive element
4	$1+\alpha+\alpha^3$	138	$1+\alpha+\alpha^4$	372	$1+\alpha+\alpha^6$
10	$1+\alpha+\alpha^3$	148	$1+\alpha+\alpha^5$	378	$1+\alpha+\alpha^3$
12	$1+\alpha+\alpha^3$	162	$1+\alpha+\alpha^3$	388	$1+\alpha+\alpha^7$
18	$1+\alpha+\alpha^4$	172	$1+\alpha+\alpha^3$	418	$1+\alpha+\alpha^5$
28	$1+\alpha+\alpha^2+\alpha^6$	178	$1+\alpha+\alpha^3$	420	$1+\alpha+\alpha^8$
36	$1+\alpha+\alpha^6$	180	$1+\alpha+\alpha^5$	442	$1+\alpha+\alpha^3$
52	$1+\alpha+\alpha^4$	196	$1+\alpha+\alpha^9$	460	$1+\alpha+\alpha^7$
58	$1+\alpha+\alpha^3$	210	$1+\alpha+\alpha^3$	466	$1+\alpha+\alpha^3$
60	$1+\alpha+\alpha^3$	226	$1+\alpha+\alpha^3$	490	$1+\alpha+\alpha^7$
66	$1+\alpha+\alpha^7$	268	$1+\alpha+\alpha^8$	508	$1+\alpha+\alpha^3$
82	$1+\alpha+\alpha^3$	292	$1+\alpha+\alpha^3$	522	$1+\alpha+\alpha^3$
100	$1+\alpha+\alpha^7$	316	$1+\alpha+\alpha^7$	540	$1+\alpha+\alpha^7$
106	$1+\alpha+\alpha^3$	346	$1+\alpha+\alpha^7$	546	$1+\alpha+\alpha^3$
130	$1+\alpha+\alpha^3$	348	$1+\alpha+\alpha^6$		

We used MAPLE for above computation. In the case of $n = 28$, there was no primitive element which is a trinomial of α, so we chose the next simple expression. Let $\gamma = 1 + \alpha^s + \alpha^t$ be a fixed primitive element in $GF(2^n)$ where α is a type I optimal normal element. Let $\{1, \alpha, \alpha^2, \cdots, \alpha^n\}$ be an extended AOP (all one polynomial) basis. Then we have the following algorithm which computes γ^r using the basis $\{1, \alpha, \alpha^2, \cdots, \alpha^n\}$.

Table 5. Exponentiation using $\gamma = 1 + \alpha^s + \alpha^t$ under the extended AOP basis

---

Input: $r = \sum_{j=0}^{n-1} r_j 2^j$
Output: γ^r
$A \leftarrow 1$
for $(i = n-1$ to 0 ; $i--)$
    $A \leftarrow A^2 \gamma^{r_i}$
end for

Note that above algorithm is applicable for both software and hardware purposes. Though the case $q = 2$ is dealt in above algorithm, one may also use other small primes $q = 3, 5, \cdots$ for efficient exponentiation. The operation $A \leftarrow A^2$ is free in our basis because $\{1, \alpha, \alpha^2, \cdots, \alpha^n\} = \{1, \alpha, \alpha^2, \alpha^{2^2}, \cdots, \alpha^{2^{n-1}}\}$. Now letting $A = \sum_{i=0}^{n} a_i \alpha^i$, we get

$$
\begin{aligned}
A\gamma &= A(1 + \alpha^s + \alpha^t) = A + A\alpha^s + A\alpha^t \\
&= \sum_{i=0}^{n} a_i \alpha^i + \sum_{i=0}^{n} a_{i-s} \alpha^i + \sum_{i=0}^{n} a_{i-t} \alpha^i = \sum_{i=0}^{n} (a_i + a_{i-s} + a_{i-t}) \alpha^i,
\end{aligned}
\tag{25}
$$

where the coefficients a_i, a_j are understood as $a_i = a_j$ if $i \equiv j \pmod{n+1}$ since $\alpha^{n+1} = 1$. Therefore the computation $A \leftarrow A\gamma$ needs 2 additions for each coefficient of α^i ($0 \le i \le n$) and the total number of bit additions needed to compute γ^r is $2n(n+1)$ which is of $O(n^2)$. By following the same ideas of previous section, we find that

Proposition 4. *Let α be a type I optimal normal element in $GF(2^n)$ and assume that $\gamma = 1 + \alpha^s + \alpha^t$ is a primitive element for some s and t. Then we can construct a linear array which computes γ^r for any $r = \sum_{i=0}^{n-1} r_i 2^i$ with $r_i = 0, 1$ using $n+1$ flip-flops, $n+1$ 2-1 MUXs and $2n+2$ XOR gates. The critical path delay of our architecture is $2D_X + D_M$ and the latency is n.*

Example 3. Let $n = 4$ and let α be a type one optimal normal element, i.e. α is a 5th root of unity over $GF(2)$. It is trivial to show that $\gamma = 1 + \alpha + \alpha^3$ is a primitive element in $GF(2^4)$. Let $A = a_0 + a_1\alpha + a_2\alpha^2 + a_3\alpha^3 + a_4\alpha^4$ be an element in $GF(2^4)$ with respect to the extended AOP basis. From

$$
\begin{aligned}
A^2 &= a_0 + a_3\alpha + a_1\alpha^2 + a_4\alpha^3 + a_2\alpha^4, \\
A^2\alpha &= a_2 + a_0\alpha + a_3\alpha^2 + a_1\alpha^3 + a_4\alpha^4, \\
A^2\alpha^3 &= a_1 + a_4\alpha + a_2\alpha^2 + a_0\alpha^3 + a_3\alpha^4,
\end{aligned}
\tag{26}
$$

we find

$$
\begin{aligned}
A^2\gamma &= A^2(1 + \alpha + \alpha^3) \\
&= (a_0 + a_1 + a_2) + (a_0 + a_3 + a_4)\alpha \\
&\quad + (a_1 + a_2 + a_3)\alpha^2 + (a_0 + a_1 + a_4)\alpha^3 + (a_2 + a_3 + a_4)\alpha^4.
\end{aligned}
\tag{27}
$$

From this information, the computation γ^r, $r = \sum_{i=0}^{n-1} r_i 2^i$ is easily realized in the following circuit.

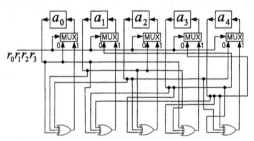

Fig. 4. A circuit for exponentiation γ^r in $GF(2^n)$ for $n = 4$, where γ is a trinomial of a type I optimal normal element α

Table 6. Comparison with previously proposed exponentiation architectures

	[10]	[12]	[13]	Fig. 1
latency	$2n^2 + 2n$	$n(n-1) + \lfloor \frac{n}{2} \rfloor + 1$	$n(n-1)$	n
critical path delay	$2D_{AND} + 2D_{XOR}$	$D_{AND} + 2D_{XOR}$	$2D_{AND} + 2D_{XOR}$	$\lceil \log_2 k \rceil D_{XOR} + D_{MUX}$
complexity				
AND	$4n^2(n-1)$	$3n^2(n-1)$	$3n(n-1)$	0
XOR	$4n^2(n-1)$	$3n^2(n-1)$	$3n(n-1)$	kn
MUX	0	0	0	n
flip-flop	$14n^2(n-1)$	$\frac{9}{2}n^2(n-1)$	$4n(n-1)$	n

5   Conclusions

We proposed a compact and fast exponentiation architecture using a Gauss period of type (n, k) in $GF(2^n)$ for all k. Using the computational evidence that a Gauss period of type (n, k), $k \geq 2$, is very often primitive and by modifying the multiplication algorithm given by Gao et al., we successfully constructed low complexity arithmetic circuits which have possible applications such as smart cards. Also for the case of a type I optimal normal element, i.e. a Gauss period of type $(n, 1)$, we found primitive elements which yield low complexity multiplication structure and we gave an exponentiation algorithm which is applicable for both software and hardware purposes. We presented explicit designs of the circuits for Gauss periods of type (n, k) when $k = 1, 2, 3$. Table 6 implies that our linear array has many superior properties in terms of latency and gate complexity compared with other existing exponentiation architectures, though our circuit works only for a fixed primitive element. The critical path delay $\lceil \log_2 k \rceil D_{XOR} + D_{MUX}$ in our architecture is not so long since in most of the cases of $n \leq 1200$, we could choose a Gauss period of type (n, k) where $k \leq 32$. That is $\lceil \log_2 k \rceil \leq 5$.

Acknowledgements. This work was supported by grant No. R05-2003-000-11325-0 from the Basic Research Program of the Korea Science & Engineering Foundation.

References

1. E.F. Brickell, D.M. Gordon, K.S. McCurley, and D.B. Wilson, "Fast exponentiation with precomputation," *Eurocrypt 92, Lecture Notes in Computer Science*, vol. 658, pp. 200–207, 1992.
2. T. Beth, B.M. Cook, and D. Gollman, "Architectures for exponentiation in $GF(2^n)$," *Crypto 86, Lecture Notes in Computer Science*, vol. 263, pp. 302–310, 1986.
3. S. Gao, J. von zur Gathen, and D. Panario, "Gauss periods and fast exponentiation in finite fields," *Latin 95, Lecture Notes in Computer Science*, vol. 911, pp. 311–322, 1995.

4. S. Gao, J. von zur Gathen, and D. Panario, "Orders and cryptographical applications," *Math. Comp.*, vol. 67, pp. 343–352, 1998.
5. S. Gao and S. Vanstone, "On orders of optimal normal basis generators," *Math. Comp.*, vol. 64, pp. 1227–1233, 1995.
6. C.H. Lim and P.J. Lee, "More flexible exponentiation with precomputation," *Crypto 94, Lecture Notes in Computer Science*, vol. 839, pp. 95–107, 1994.
7. P. de Rooij, "Efficient exponentiation using precomputation and vector addition chains," *Eurocrypt 94, Lecture Notes in Computer Science*, vol. 950, pp. 389–399, 1994.
8. B. Sunar and Ç.K. Koç, "An efficient optimal normal basis type II multiplier," *IEEE Trans. Computers*, vol 50, pp. 83–87, 2001.
9. A.J. Menezes, I.F. Blake, S. Gao, R.C. Mullin, S.A. Vanstone, and T. Yaghoobian, *Applications of finite fields*, Kluwer Academic Publisher, 1993.
10. C.L. Wang, "Bit level systolic array for fast exponenetiation in $GF(2^m)$," *IEEE Trans. Computers*, vol. 43, pp. 838–841, 1994.
11. P.A. Scott, S.J. Simmons, S.E. Tavares, and L.E. Peppard, "Architectures for exponentiation in $GF(2^m)$," *IEEE J. on Selected Areas in Communications*, vol. 6, pp. 578–586, 1988.
12. S.K. Jain, L. Song, and K.K. Parhi, "Efficient semisystolic architectures for finite field arithmetic," *IEEE Trans. VLSI Syst.*, vol. 6, pp. 101–113, 1998.
13. S.W. Wei, "VLSI architectures for computing exponentiations, multiplicative inverses, and divisions in $GF(2^m)$," *IEEE Trans. Circuits Syst. II*, vol. 44, pp. 847–855, 1997.
14. H. Wu, M.A. Hasan, I.F. Blake, and S. Gao, "Finite field multiplier using redundant representation," *IEEE Trans. Computers*, vol. 51, pp. 1306–1316, 2002.
15. H. Wu and M.A. Hasan, "Efficient exponentiation of a primitive root in $GF(2^m)$," *IEEE Trans. Computers*, vol. 46, pp. 162–172, 1997.
16. J. von zur Gathen and M.J. Nöcker, "Exponentiation in finite fields: Theory and Practice," *AAECC 97, Lecture Notes in Computer Science*, vol. 1255, pp. 88–133, 1997.
17. J. von zur Gathen and I. Shparlinski, "Orders of Gauss periods in finite fields," *ISAAC 95, Lecture Notes in Computer Science*, vol. 1004, pp. 208–215, 1995.
18. G.B. Agnew, R.C. Mullin, I. Onyszchuk, and S.A. Vanstone, "An implementation for a fast public key cryptosystem," *J. Cryptology*, vol. 3, pp. 63–79, 1991.
19. G.B. Agnew, R.C. Mullin, and S.A. Vanstone, "Fast exponentiation in $GF(2^n)$," *Eurocrypt 88, Lecture Notes in Computer Science*, vol. 330, pp. 251–255, 1988.
20. S. Kwon and H. Ryu, "Efficient bit serial multiplication using optimal normal bases of type II in $GF(2^m)$," *ISC 02, Lecture Notes in Computer Science*, vol. 2433, pp. 300–308, 2002.
21. D.M. Gordon, "A survey of fast exponentiation methods," *J. of Algorithms*, vol. 27, pp. 129–146, 1998.
22. W. Geiselmann and D. Gollmann, "VLSI design for exponentiation in $GF(2^n)$," *Auscrypt 90, Lecture Notes in Computer Science*, vol. 453, pp. 398–405, 1990.
23. Ç.K. Koç and B. Sunar, "Low complexity bit parallel canonical and normal basis multipliers for a class of finite fields," *IEEE Trans. Computers*, vol. 47, pp. 353–356, 1998.
24. C. Paar, P. Fleischmann, and P. Roelse, "Efficient multiplier archtectures for Galois fields $GF(2^{4n})$," *IEEE Trans. Computers*, vol. 47, pp. 162–170, 1998.
25. G. Tenenbaum, "*Introduction to analytic and probabilistic number theory*," Cambridge Univ. Press, 1995.

GCD-Free Algorithms for Computing Modular Inverses

Marc Joye[1] and Pascal Paillier[2]

[1] Gemplus, Card Security Group
La Vigie, Avenue du Jujubier, ZI Athélia IV, 13705 La Ciotat Cedex, France
marc.joye@gemplus.com − http://www.geocities.com/MarcJoye/
[2] Gemplus, Cryptography Group
34 rue Guynemer, 92447 Issy-les-Moulineaux, France
pascal.paillier@gemplus.com
http://www.gemplus.com/smart/

Abstract. This paper describes new algorithms for computing a modular inverse $e^{-1} \bmod f$ given coprime integers e and f. Contrary to previously reported methods, we neither rely on the extended Euclidean algorithm, nor impose conditions on e or f. The main application of our gcd-free technique is the computation of an RSA private key in both standard and CRT modes based on simple modular arithmetic operations, thus boosting real-life implementations on crypto-accelerated devices.

Keywords: Modular inverses, RSA key generation, prime numbers, efficient implementations, embedded software, GCD algorithms.

1 Introduction

The usual way one computes a modular inverse is by applying the extended Euclidean algorithm [8, Algorithm X, p. 325]. Given e and f on input, this algorithm returns integers α and β such that $\alpha e + \beta f = \gcd(e, f)$. Assuming that e is relatively prime to f, and therefore that e has an inverse modulo f, it follows that $\alpha e \equiv 1 \pmod{f}$ meaning that $\alpha \equiv e^{-1} \pmod{f}$.

Unfortunately, for code-optimization reasons, this algorithm is hardly available on embedded platforms. Instead, algebraic tricks based on simple modular arithmetic are highly preferred because gcd-type calculations may be too intricate to handle on cryptoprocessors compared to modular operations. As an example, executing an extended binary gcd may require much less arithmetic or logic operations on large numbers than *glue instructions* such as register switches, loop control, pointer management, etc., rendering this approach comparatively prohibitive to straightforward, arithmetic-only implementations. But then, one requires that one of the two input values, e or f, is prime. Indeed, when f is prime, the inverse of e modulo f is given by Fermat Little Theorem stat-

C.D. Walter et al. (Eds.): CHES 2003, LNCS 2779, pp. 243–253, 2003.
© Springer-Verlag Berlin Heidelberg 2003

ing that $e^{-1} \equiv e^{f-2} \pmod{f}$; when e is prime, $e^{-1} \bmod f$ is given by Arazi's well-known inversion formula.[1] Little is known about the other cases.

This paper presents simple ways for computing $e^{-1} \bmod f$ without the (extended) Euclidean algorithm (or variants thereof) and *without any restrictions* on e or f. Our techniques only invoke usual basic operations like (possibly modular) additions, multiplications and exponentiations. Since these operations are optimized on devices supporting public-key cryptography, the technique we propose is especially well-suited for smart-card on-board computation of an RSA private key, in both standard and CRT modes.

2   Arazi's Inversion Formula

When f is prime, the inverse of e modulo f is given by Fermat Little Theorem because $d = e^{-1} \bmod f = e^{f-2} \bmod f$. When f is not prime and *provided that e is prime*, the usual trick consists in applying Arazi's inversion formula, which expresses $e^{-1} \bmod f$ in terms of $f^{-1} \bmod e$.

Lemma 1 (Arazi). *Let e and f be two positive integers. If* $\gcd(e, f) = 1$ *then*

$$d = e^{-1} \bmod f = \frac{1 + f(-f^{-1} \bmod e)}{e} . \tag{1}$$

Proof. Define $U = e(e^{-1} \bmod f) + f(f^{-1} \bmod e)$. Since $U \equiv 1 \pmod{e}$ and $U \equiv 1 \pmod{f}$, it follows that $U \equiv 1 \pmod{ef}$. Hence, noting that $1 < e + f \leq U < 2ef$, this implies that $U = 1 + ef$, or equivalently that $e^{-1} \bmod f = [1 + f\{e - (f^{-1} \bmod e)\}]/e$ as desired. $\qquad\square$

Hence if e is prime, its inverse d modulo f can easily be computed as

$$d = \frac{1 + f(-f^{e-2} \bmod e)}{e} .$$

This formula is limited to prime values for e, but is easily extended to

$$d = \frac{1 + f(-f^{\lambda(e)-1} \bmod e)}{e} , \tag{2}$$

whenever $\lambda(e)$ is known. We recall that computing Carmichæl's function $\lambda(e)$ from e requires to factor e, a task which imposes a very strong computational requirement. So, the extended technique given by Eq. (2) is of no interest if the inversion algorithm is not given $\lambda(e)$ as an input. The same remarks are independently stated in [6].

---

[1] Named after Arazi who was the first to take advantage of this folklore theorem to implement fast modular inversions of RSA exponents on a crypto-processor.

2.1   Implementing Arazi's Formula with Modular Operations

Equation (2) requires an integer division (i.e., over $\mathbb{Z}$), which is performed either directly by the cryptoprocessor (some of them may provide this functionality) or in the following way. We compute $d = d \bmod 2^{|f|}$ where $|f|$ stands for the binary length of f. The multiplication and incrementation in Eq. (2) are done modulo $2^{|f|}$, and then the division is replaced by a multiplication by $e^{-1} \bmod 2^{|f|}$. This value may be hard-coded in the program when this is possible, or dynamically computed as depicted on the algorithm of Fig. 1. Another algorithm for performing an integer division can be found in [5, p. 235].

Input: e (odd), $|f|$
Output: $e^{-1} \bmod 2^{|f|}$

$T \leftarrow \lceil \log_2(|f|) \rceil$; $y \leftarrow 1$
for $i = 1$ to T do
  $y \leftarrow y(2 - ey) \bmod 2^i$
endfor

return $y \bmod 2^{|f|}$

Fig. 1. Inversion $e \mapsto e^{-1} \bmod 2^{|f|}$

Note that all operations involved here are modular. Besides, this technique turns out to be extremely fast (only 10 iterations for a typical size $|f| = 1024$), especially for small values of e.

2.2   The Case of Composite Numbers

In the sequel, Π will always denote the product of small primes $\Pi = \prod_{i \in I} p_i$ for $I \subset \mathbb{N}$ and where p_i is the ith prime (i.e., $p_1 = 2$, $p_2 = 3$, ...). Unless stated otherwise, we assume that $I = [1, k]$ for a certain bound k depending on the context of use for Π. We also assume that the choice for Π has been done once and for all, and that Π and $\lambda(\Pi)$ are absolute constants coded in our algorithms.

Now suppose that f is some composite number with unknown factorization. We consider different scenarios depending on the information we have about the operand e:

1. e is known at compile time. We thus have access to $\lambda(e)$ which may be written or coded in the program itself;
2. e is an input data and is provided along with $\lambda(e)$;
3. e is provided alone, but is known to be prime (and thus $\lambda(e) = e - 1$);
4. e is given (e.g., dynamically loaded and provided) but nothing else is known about e.

The first three situations lead to the implementation given below. The fourth context of use is somewhat more intricate and requires a specific treatment as shown later in Section 3.

We adopt the following twofold approach:

- one attempts to retrieve $\lambda(e)$, or a multiple thereof, in order to invoke the previous, very efficient technique;
- if unsuccessful, one computes d without that knowledge but in an heavier, somewhat pathological way.

In some cases, retrieving $\lambda(e)$ may be quite efficient. When e is smooth enough so that $e \mid \Pi$, then a multiple of $\lambda(e)$ is simply $\lambda(\Pi)$. This may also hold without necessarily having $e \mid \Pi$. In many situations, the following will be sufficient. Set

$$\widehat{\lambda} = e(e-1)\lambda(\Pi)\Pi \ , \tag{3}$$

and execute the inversion algorithm of the previous section by replacing $\lambda(e) - 1$ by $\widehat{\lambda} - 1$. Get the result $\widehat{d}$ and test whether $e\,\widehat{d} \equiv 1 \pmod{f}$. We output $\widehat{d}$ if this equality holds. Otherwise, we know that the structure of e is less simple than originally thought.[2]

The next section describes new efficient approaches that always return the value of $d = e^{-1} \bmod f$ *whatever* the conditions on e and/or f.

3   Extended Algorithms for Composite Integers

3.1   Algorithm 1

Our idea is fairly simple. It is based on the somewhat obvious observation that

$$e^{-1} \equiv (e + Cf)^{-1} \pmod{f}$$

for any integer C. Therefore we can add an appropriate multiple of f to e so that the result is prime and then apply Arazi's inversion formula directly to it. Define

$$\hat{e} = e + Cf \ .$$

We require $\hat{e}$ to be a prime. A naive way to find such an $\hat{e}$ consists in trying $C = 1, 2, \dots$ and so on until $e + Cf$ is prime [4]. We can however do much better.

Proposition 1. *Let e and f be two positive integers with $\gcd(e, f) = 1$. Let also $\Pi = \prod p_i$ be a product of (small) primes. Define*

$$\hat{e} = e + Cf \quad with \quad C = \left[1 - e^{\lambda(\Pi)}\right] \bmod \Pi \ . \tag{4}$$

Then we have $\gcd(\hat{e}, p_i) = 1$ for all primes p_i dividing Π. Moreover, we have $\gcd(\hat{e}, f) = 1$.

---

[2] In this event, e is a composite number with at least one prime factor e_i with $e_i \nmid \Pi$ such that $e_i - 1$ has some large prime factor ...

Proof. (i) Consider first the case $\gcd(f, p_i) = p_i$. Then from Eq. (4) we have $\hat{e} \equiv e \pmod{p_i}$. Moreover, since by definition $\gcd(e, f) = 1$, it follows that $\gcd(\hat{e}, p_i) = \gcd(e, p_i) = 1$.

(ii) Suppose now that $\gcd(f, p_i) = 1$. If $\gcd(e, p_i) = p_i$ then Eq. (4) yields $C \equiv 1 \pmod{p_i}$, which implies $\hat{e} \equiv f \pmod{p_i}$ and consequently $\gcd(\hat{e}, p_i) = \gcd(f, p_i) = 1$. Conversely, assuming $\gcd(e, p_i) = 1$ forces $C \equiv 0 \pmod{p_i}$ and then $\hat{e} \equiv e \pmod{p_i}$ which, again, leads to $\gcd(\hat{e}, p_i) = \gcd(e, p_i) = 1$.

Finally, as $\hat{e} \equiv e \pmod{f}$, it follows that $\gcd(\hat{e}, f) = \gcd(e, f) = 1$. □

As $\hat{e}$ is co-prime to all small primes $p_i \mid \Pi$, it is likely to be a prime number which we can test using some primality test.[3] If the test is unsuccessful, we re-iterate the process with another candidate $\hat{e}^{(\text{new})} = \hat{e}^{(\text{old})} + f\Pi$, and so forth until $\hat{e}$ is a prime number. Remark that the updated $\hat{e}$, $\hat{e}^{(\text{new})}$, satisfies $\hat{e}^{(\text{new})} \equiv \hat{e}^{(\text{old})}$ $(\bmod\ \{p_i, f\})$ and so also verifies Proposition 1.

We note that our technique differs from the one described in [7] in several ways, and in particular, the building of $\hat{e}$ from e is not probabilistic. The resulting algorithm is detailed on Fig. 2.

Input: e, f with $\gcd(e, f) = 1$
$\quad\quad\quad \Pi = \prod p_i,\ \lambda(\Pi)$
Output: $d = e^{-1} \bmod f$

$\quad C \leftarrow \left[1 - e^{\lambda(\Pi)}\right] \pmod{\Pi};\ \hat{e} \leftarrow e + Cf$
$\quad$ while ($\hat{e}$ is not prime) do
$\quad\quad\quad \hat{e} \leftarrow \hat{e} + f\Pi$
$\quad$ endwhile
$\quad u \leftarrow f^{\hat{e}-2} \bmod \hat{e};\ d \leftarrow [1 + f(\hat{e} - u)]/\hat{e}$
$\quad$ return d

Fig. 2. A first inverting algorithm (Algorithm 1)

There may exist variations of Algorithm 1. To illustrate the diversity of our technique, we provide here another alternative. We state:

Proposition 2. *Let e and f be two positive integers with $\gcd(e, f) = 1$. Let also $\Pi = \prod p_i$ be a product of (small) primes, and $c \in \mathbb{Z}_{\Pi}^*$. Define*

$$\hat{e} = e + Cf \quad \text{with} \quad C = \left[(c - e)f^{\lambda(\Pi)-1}\right] \bmod \Pi . \tag{5}$$

Then $\gcd(\hat{e}, p_i) = 1$ for all p_i dividing Π.

---

[3] Popular primality tests like Fermat's test or Miller-Rabin's test are easy to implement with basic modular operations and hence are quite fast in practice.

Proof. (i) Consider first the case $\gcd(f, p_i) = p_i$. Then from Eq. (5) we have $\widehat{e} \equiv e \pmod{p_i}$. Moreover, since by definition $\gcd(e, f) = 1$, it follows that $\gcd(\widehat{e}, p_i) = 1$.

(ii) Suppose now that $\gcd(f, p_i) = 1$. Then Eq. (5) yields $\widehat{e} \equiv c \pmod{p_i}$ and thus we find again $\gcd(\widehat{e}, p_i) = 1$ since $c \in \mathbb{Z}_\Pi^*$. □

As $\widehat{e}$ is co-prime to all small primes $p_i \mid \Pi$, it is likely a prime number. Otherwise we re-iterate the process with another candidate $c \in \mathbb{Z}_\Pi^*$.

```
Input:   f, e, Π, and a ∈ Z*_Π \ {1}
Output:  d_f
```

```
1. Compute U ← f^(λ(Π)−1) mod Π
2. Set C ← [(c − e)U mod Π] and ê ← e + Cf
3. If (T(ê) = false) then
     a) Set c ← ac (mod Π)
     b) Go to Step 2
4. Compute F ← −f^(ê−2) mod ê
5. Output d_f = (Ff + 1)/e
```

Fig. 3. An alternative algorithm (Algorithm 1')

Again, note that all operations in the above algorithm exclusively rely on basic modular arithmetic. If the cryptoprocessor cannot handle integer divisions directly, the division by e in the last step can be computed with the algorithm described in Fig. 1.

3.2   Algorithm 2

A second algorithm can be derived by exchanging the roles of e and f in Proposition 1. Doing so, we obtain a prime $\widehat{f}$. Two applications of Arazi's formula will give thus the expected result.

Since $\widehat{f}$ is prime, the inverse of e modulo $\widehat{f}$ is given by $u = e^{\widehat{f}-2} \bmod \widehat{f}$. Noting that $f \equiv \widehat{f} \pmod{e}$, a first application of Arazi's formula enables to recover the value of $f^{-1} \bmod e$ as

$$v := f^{-1} \bmod e = \frac{1 + e(\widehat{f} - u)}{\widehat{f}}, \tag{6}$$

and a second application yields

$$d = e^{-1} \bmod f = \frac{1 + f(e - v)}{e}.$$

In many cases, the value of e is small compared to that of f. Moreover, using the fact that $\gcd(e, f) = \gcd(e \bmod f, e)$, we obtain the following corollary of Proposition 1.

Corollary 1. *With the notations of Proposition 1, define $\bar{e} = e \bmod f$ and*

$$\hat{\bar{e}} = \bar{e} + Cf \quad \text{with} \quad C = \left[1 - \bar{e}^{\lambda(\Pi)}\right] \bmod \Pi .$$

Then we have $\gcd(\hat{\bar{e}}, p_i) = 1$ for all primes p_i dividing Π. Moreover, we have $\gcd(\hat{\bar{e}}, f) = 1$.

Proof. Straightforward by replacing e with $\bar{e}$ in Proposition 1.   □

Therefore, we can advantageously consider $\hat{f}$ instead of f (remember that for our second algorithm the roles of e and f are exchanged in Proposition 1) and so evaluate v in Eq. (6) as

$$v = \bar{f}^{-1} \bmod e = \frac{1 + e(\hat{\bar{f}} - u)}{\hat{\bar{f}}}$$

where $\bar{f} = f \bmod e$.

Putting all together, we obtain a second algorithm for computing modular inverses.

Input: e, f with $\gcd(e, f) = 1$
$\Pi = \prod p_i, \ \lambda(\Pi)$
Output: $d = e^{-1} \bmod f$

$\hat{\bar{f}} \leftarrow f \bmod e$
if ($\hat{\bar{f}}$ is not prime) then
$\quad \bar{C} \leftarrow \left[1 - \hat{\bar{f}}^{\lambda(\Pi)}\right] \ (\bmod \ \Pi); \ \hat{\bar{f}} \leftarrow \hat{\bar{f}} + \bar{C}e$
$\quad$ while ($\hat{\bar{f}}$ is not prime) do
$\quad\quad \hat{\bar{f}} \leftarrow \hat{\bar{f}} + e\Pi$ $\hspace{3cm}$ [i]
$\quad$ endwhile
endif
$u \leftarrow e^{\hat{\bar{f}}-2} \ (\bmod \ \hat{\bar{f}})$
$d \leftarrow [\hat{\bar{f}} + f(e\,u - 1)]/(e\hat{\bar{f}})$ $\hspace{2cm}$ [ii]
return d

Fig. 4. Our second algorithm (Algorithm 2)

This second algorithm is particularly efficient when e is small since then Π may be chosen smaller, which in turn implies smaller values for $\bar{C}$ and for $\hat{\bar{f}}$. On the contrary, the first algorithm (Fig. 2) is more suitable when the size of e is sensibly the same as that of f.

As easily seen, the choice for parameter Π remains completely free. We now discuss the best way to choose Π in practice. Primality checks executed in our 'while' loop involve integers of bitsize close to $|e\Pi|$. As most practical implementations for primality testing are of cubic complexity, a single test has a cost $\propto |e\Pi|^3$. Moreover, the average number of tests amounts to

$$|e\Pi| \cdot \ln 2 \cdot \frac{\phi(\mathrm{lcm}(e, \Pi))}{\mathrm{lcm}(e, \Pi)} .$$

Totalling these two facts, and upper bounding the ratio $\phi(\mathrm{lcm}(e, \Pi))/\mathrm{lcm}(e, \Pi)$ by $\phi(\Pi)/\Pi$, the average workfactor for finding $\hat{\bar{f}}$ is bounded by a function proportional to $(|e\Pi|)^4 \phi(\Pi)/\Pi$. Therefore, provided that $\Pi = \prod_1^k p_i$, an optimal choice for k with respect to a given $|e|$ is easily found. Interestingly, for small parameter lengths such as $|e| = 32$ or 64, the optimum is obtained for $k = 3$ ($\Pi = 2 \cdot 3 \cdot 5$), i.e., for an extremely small value of Π. Algorithm 2 then performs only a few primality checks over integers of size close to the one of e, and is therefore very fast.

Remark 1. Certain hardware implementations return the value of f div e together with the value of f mod e when computing the remainder of an integer division. In this case, the division by $e\hat{\bar{f}}$ in the expression of d (cf. [‡] in Fig. 3) can be reduced to a division by $\hat{\bar{f}}$. Initializing $\bar{C}$ to 0 and keeping track of its accumulated value, we can replace Line [i] by

$$\bar{C} \leftarrow \bar{C} + \Pi; \hat{\bar{f}} \leftarrow \hat{\bar{f}} + e\Pi \qquad\qquad [i']$$

and Line [ii] by

$$d \leftarrow [fu - (f \text{ div } e) + \bar{C}]/\hat{\bar{f}} . \qquad\qquad [ii']$$

4   Application to RSA

RSA [11], named after its inventors Rivest, Shamir, and Adleman, is undoubtedly the most widely used cryptosystem. We give hereafter a short description and refer the reader to the original paper or any textbook on cryptography for further details.

Let $n = pq$ be the product of two large primes. We let e and d denote a matching pair of public exponent/private exponent, according to

$$e\,d \equiv 1 \pmod{\lambda(n)} , \qquad\qquad (7)$$

where λ is Carmichæl function. In particular, for the RSA, we have $n = pq$ and $\lambda(n) = \mathrm{lcm}(p - 1, q - 1)$.

Given $x \in]0, n[$, the public operation (e.g., encryption of a message or verification of a signature) consists in raising x to the power e, modulo n, i.e., in computing $c = x^e \bmod n$. Next, from c, the corresponding private operation (e.g., decryption of a ciphertext or a signature generation) is $c^d \bmod n$. From the definition of e and d, we obviously have that $c^d \equiv x \pmod{n}$.

4.1   Standard Mode

In standard mode, on input p, q and e, one has to compute the private exponent d satisfying Eq. (7). We assume that we are given Π, the product of small primes, along with $\lambda(\Pi)$. These numbers are pure constants, and are thus easily hard-coded into the implementation.

When e (or its factorization) cannot be determined in advance, a direct application of Algorithm 1 (or Algorithm 2) with e and $f = \text{lcm}(p - 1, q - 1)$ on input will output the corresponding secret key d.

If the value of $\text{lcm}(p - 1, q - 1)$ cannot be computed, one can replace the Carmichæl function of n by the Euler totient function and take $f = (p-1)(q-1)$. This, however, results in a larger yet valid value for d.

From a computational viewpoint, taking $|f| = 1024$ and $|e| = 32$ for instance, a typical implementation of Algorithm 2 would use the specific choice $\Pi = 2 \cdot 3 \cdot 5$. Thus, around 6.83 primality tests over 37-bit numbers are required, on average. When $|f| = 1024$ and $e = 64$, the same choice for Π yields an average of 12.75 primality tests over 69-bit numbers. In addition to that, operations starting and ending the algorithm are almost negligible: computing $\bar{C}$ amounts to a few squares modulo $2 \cdot 3 \cdot 5$; u requires an exponentiation of size close to $|e|$; and the computation of d (thanks to our technique on Fig. 1) boils down to a few multiplications carried out modulo 2^{1024}.

4.2   CRT Mode

The private operation can be speeded up through Chinese remaindering (CRT mode) [9]. The computations are performed modulo p and q and then recombined. The private parameters are (p, q, d_p, d_q, i_q) with $d_p = d \mod (p - 1)$, $d_q = d \mod (q - 1)$ and $i_q = q^{-1} \mod p$. We then obtain $c^d \mod n$ as

$$\text{CRT}(x_p, x_q) = x_q + q[i_q(x_p - x_q) \mod p] \, ,$$

where $x_p = c^{d_p} \mod p$ and $x_q = c^{d_q} \mod q$. The expected speed-up factor is 4, compared to the standard (i.e., non-CRT) mode.

In CRT mode, the procedure is readily the same. We apply Algorithm 1 or 2 where inputs e and f are initialized to $e \pmod{(p - 1)}$ and $p - 1$, respectively. This yields the value of the private exponent d_p. Similarly, the exponent d_q is obtained by initializing e and f to $e \pmod{(q - 1)}$ and $q - 1$, respectively.

4.3   Standard Mode (II)

There is another way to compute the private key in standard mode. We first compute d_p and d_q as described in the previous section. Next, letting $\mathcal{Q} := q - 1$ and $\Lambda := \lambda(n)/(q - 1)$, we compute the inverse of $\mathcal{Q}$ modulo Λ, say $\mathcal{I}_{\mathcal{Q}}$, thanks to the algorithm of Fig. 2 as[4]

---

[4] Remark here that we have to compute the inverse of $(q - 1)$ modulo $\frac{p-1}{\gcd(p-1,q-1)} = \frac{\text{lcm}(p-1,q-1)}{q-1}$ as $e^{-1} \mod f$ exists if and only if $\gcd(e, f) = 1$.

$$\mathcal{I}_{\mathcal{Q}} = \frac{1 + \Lambda\left[-\Lambda^{\widehat{\mathcal{Q}}-2} \bmod \widehat{\mathcal{Q}}\right]}{\widehat{\mathcal{Q}}} \tag{8}$$

where

$$\widehat{\mathcal{Q}} = q - 1 + C_{\mathcal{Q}}\,\Lambda \quad \text{with} \quad C_{\mathcal{Q}} = \left[1 - (q-1)^{\lambda(\Pi)}\right] \bmod \Pi + \mu \cdot \Pi$$

for some $\mu \geq 0$ such that $\widehat{\mathcal{Q}}$ is prime. Therefore, Chinese remaindering on d_p and d_q finally gives $d = d_q + (q-1)\left[\mathcal{I}_{\mathcal{Q}}(d_p - d_q) \bmod \Lambda\right]$.

5   Conclusion

We devised new algorithms for computing modular inverses in a gcd-free manner. We stress that, implementing our techniques, an RSA key generation process can be executed on any given crypto-enhanced embedded processor in almost every circumstances.

Acknowledgements. We are grateful to Jean-François Dhem for pointing out reference [4] and to Karine Villegas for her careful reading of an earlier version of this paper.

References

1. A.O.L. Atkin and F. Morain. Elliptic curves and primality proving. *Mathematics of Computation* **61**:29–68, 1993.
2. D. Boneh and M. Franklin. Efficient generation of shared RSA keys. In *Advances in Cryptology – CRYPTO '97*, vol. 1294 of Lecture Notes in Computer Science, pp. 425–439, Springer-Verlag, 1997.
3. W. Bosma and M.-P. van der Hulst. Faster primality testing. In *Advances in Cryptology – CRYPTO '89*, vol. 435 of Lecture Notes in Computer Science, pp. 652–656, Springer-Verlag, 1990.
4. M.F.A. Derôme. Generating RSA keys without the Euclid algorithm. *Electronics Letters* **29**(1):19–21, 1993.
5. S.R. Dussé and B.S. Kaliski Jr. A cryptographic library for the Motorola DSP 56000. In *Advances in Cryptology – EUROCRYPT '90*, vol. 473 of *Lecture Notes in Computer Science*, pp. 230–234, Springer-Verlag, 1991.
6. W. Fischer and J.-P. Seifert. Note on fast computation of secret RSA exponents. In *Information Security and Privacy (ACISP 2002)*, vol. 2384 of *Lecture Notes in Computer Science*, pp. 136–143, Springer-Verlag, 2002.
7. M. Joye and P. Paillier. Constructive methods for the generation of prime numbers. *Proc. of the 2nd NESSIE Workshop*, Egham, UK, September 12–13, 2001.
8. D.E. Knuth. *The Art of Computer Programming, v. 2. Seminumerical Algorithms.* Addison-Wesley, 2nd edition, 1981.
9. J.-J. Quisquater and C. Couvreur. Fast decipherment algorithm for RSA public-key cryptosystem. *Electronics Letters* **18**:905–907, 1982.
10. H. Riesel. *Prime Numbers and Computer Methods for Factorization*, Birkhäuser, 1985.

11. R.L. Rivest, Adi Shamir, and L.M. Adleman. A method for obtaining digital signatures and public-key cryptosystems. *Communications of the ACM* **21**(2):120–126, 1978.

12. R. Solovay and V. Strassen. A fast Monte-Carlo test for primality. *SIAM Journal on Computing* **6**: 84–85, 1977.

Attacking Unbalanced RSA-CRT Using SPA

Pierre-Alain Fouque, Gwenaëlle Martinet, and Guillaume Poupard

DCSSI Crypto Lab
51, Boulevard de Latour-Maubourg
75700 Paris 07 SP, France
Pierre-Alain.Fouque@ens.fr
Gwenaelle.Martinet@worldonline.fr
Guillaume.Poupard@m4x.org

Abstract. Efficient implementations of RSA on computationally limited devices, such as smartcards, often use the CRT technique in combination with Garner's algorithm in order to make the computation of
modular exponentiation as fast as possible. At PKC 2001, Novak has proposed to use some information that may be obtained by simple power
analysis on the execution of Garner's algorithm to recover the factorization of the RSA modulus. The drawback of this approach is that it
requires chosen messages; in the context of RSA decryption it can be realistic but if we consider RSA signature, standardized padding schemes
make impossible adaptive choice of message representative.

In this paper, we use the same basic idea than Novak but we focus
on the use of known messages. Consequently, our attack applies to
RSA signature scheme, whatever the padding may be. However, our
new technique based on SPA and lattice reduction, requires a small
difference, say 10 bits, between the bit lengths of modulus prime factors.

Keywords: Simple Power Analysis, RSA signature, factorization, LLL
algorithm.

1 Introduction

Since the introduction in 1996 of the timing attacks by Kocher [5], many papers
have considered various side channel attacks and the potential countermeasures.
Side channels attacks allow to extract some information on the manipulated data
which can be used to recover secret data. This general kind of attacks can be
divided into several different techniques: timings attacks [5], or Simple Power
Analysis (SPA) and Differential Power Analysis (DPA), both introduced in 1999
by Kocher, Jaffe and Jun [6]. Lots of countermeasures have been proposed against
such attacks but addressing all weaknesses when implementing an algorithm is
a hard task. Power attacks are very difficult to prevent, and thus, most of the
time, countermeasures do not suffice to thwart all of them.

In the public key setting, many papers have focused on the security of cryptosystems against such side channel attacks [5,11,10]. In particular, the RSA
signature and encryption schemes have been mainly studied. To sign a message

C.D. Walter et al. (Eds.): CHES 2003, LNCS 2779, pp. 254–268, 2003.
© Springer-Verlag Berlin Heidelberg 2003

M with RSA, M is first transformed using appropriate padding scheme and hash function into a representative $m \in \mathbb{Z}_N$, where N is the product of two primes p and q. Then $m^d \bmod N$ is computed, where d is the secret key of the signer. This yields the signature for the message M.

Side channel attacks on RSA extract information from the exponentiation step. For example, by precisely measuring the time it takes for the cryptographic device to perform the RSA signature, an attacker can recover the secret key, as shown by Kocher in [5]. This timing attack can be mounted against naive implementations of RSA using the repeated square and multiply algorithm. The attack recovers the secret exponent d, one bit at a time. Indeed, for each non zero bit on the secret exponent d, an additional multiplication is performed. Analyzing differences between running time for various input values reveals the secret key. Another classical way to attack basic RSA implementation is to use SPA technique that consists in measuring the power consumption during exponentiation [3]. Since power consumption also allows to determine if the additional multiplication is done, all bits of the secret exponent can be recovered by monitoring only one exponentiation.

Optimized implementations are also subject to attacks. The Chinese Remainder Theorem is a well-known technique to optimize RSA exponentiation. In a CRT implementation, the signer first computes separately the signature modulo each prime factors p and q. He then uses the Chinese Remainder Theorem to compute the signature $S \bmod N$. Since the size of p and q is about half the size of N, CRT exponentiation is about four times faster than direct exponentiation. The first attack on RSA-CRT has been presented in 1997 by Boneh, DeMillo and Lipton [1]. It is based on fault injection during computation. By using a valid signature for a message and a faulty one, the modulus N can be efficiently factored. A timing attacks against RSA with the Chinese Remainder Theorem (CRT) is also possible [11], when the Montgomery algorithm is used for squaring and multiplication operations. Recently, Novak [10] has described an adaptive chosen message attack against smart cards implementations of RSA decryption when the Garner's algorithm implements the CRT. This attack is based on a simple power analysis (SPA). The power consumption of the card leaks information on the secret manipulated data. The cryptanalyst goal is to relate such information to the bits of the secret key. Although this attack can be mounted against RSA decryption scheme, it is not realistic in practice against the RSA signature scheme. Indeed, a padding scheme is used in practical implementations and then chosen inputs attacks cannot be made.

In this paper, we show how to extend this attack to the case of RSA signature based on any encoding scheme, such as PKCS#1 [7]. In particular we show that with a simple power analysis, if the RSA modulus $N = pq$ is such that $q < p/2^\ell$, the RSA factors p and q can be recovered by performing $60 \times 2^\ell$ signatures on average. The value ℓ should be larger than an explicit bound we precise in this paper. These signatures can be computed on any messages, not necessarily chosen by the adversary.

- Input: A message M to sign, the private key (p, q, d),
  with $p > q$, the pre-calculated values $d_p = d \bmod p - 1$,
  $d_q = d \bmod q - 1$, and $u = q^{-1} \bmod p$.
- Output: a valid signature S for the message M
  1. Encode the message M in $m \in \mathbb{Z}_N$ (with PKCS#1)
  2. Compute $s_p = m^{d_p} \bmod p$
  3. Compute $s_q = m^{d_q} \bmod q$
  4. Set $t = s_p - s_q$
  5. If $t < 0$ then $t \leftarrow t + p$
  6. Compute $S = s_q + ((t \cdot u) \bmod p) \cdot q$
  7. Return S as a signature for the message M

Fig. 1. The RSA-CRT signature generation with Garner's algorithm

In the next section, we briefly describe the RSA-CRT signature scheme implemented with the Garner's algorithm. Then, we develop a new technique to factor N when having access to a set of special form integers modulo N. In section 4, we precisely describe the attack and how to collect such special form RSA signature. We also give practical results on experiments. Finally, we propose classical countermeasures to thwart this attack.

2   RSA Signature Scheme

Let $N = pq$ an n-bit RSA modulus. The public key of the signer is denoted by (N, e) and the private key by (p, q, d), where e and d are such that $e \cdot d = 1 \bmod (p-1)(q-1)$. Let M be a message to sign. A signature for M is $S = m^d \bmod N$, where m is deduced from M by an encoding scheme, randomized or not, such as PKCS#1 for example [7]. To check if a signature S is valid for M, a verifier simply computes m and checks if the equality $m = S^e \bmod N$ holds. Note that the encoding step is mainly used in practice to avoid some basic attacks on RSA. The requirement on the encoding scheme is that the outputs are uniformly distributed in $\mathbb{Z}_N$.

Smart cards implementations of RSA frequently use the Chinese Remainder Theorem to speed up the computation of $S = m^d \bmod N$. The Garner's algorithm is an efficient method to determine the signature S from $s_p = S \bmod p$ and $s_q = S \bmod q$. This algorithm does not require any reduction modulo N but uses instead reductions modulo the factors p and q. It is thus more efficient than the classical implementation of the CRT. A detailed description of this algorithm can be found in [9] and in [4]. In figure 1, we describe the RSA signature generation using the Garner's algorithm.

Step 5 of this algorithm needs some explanation: we first remark that the value t computed at the previous step may be negative, since, as we assume

$q < p$, it lies in the range $[-q, p]$. However, the modular multiplication $tu \bmod p$ has to be performed in the next step. Since inputs for modular multiplications have to be already reduced in the range $[0, p-1]$, if t is negative, p should be added so that $t > 0$. Therefore step 5 consists in computing $t \bmod p$ before the modular multiplication with u.

In [10], Novak has described a method to factor an RSA modulus when the Garner's algorithm is used for CRT. This attack applies to the RSA encryption schemes. It is based on the observation that, for a message m encrypted into c, if $m_p = c^{d_p} \bmod p$ is smaller than $m_q = c^{d_q} \bmod q$, step 5 of the decryption algorithm (similar to 5 of the signature generation in 1) is performed. Otherwise, no addition is made in this step. The analysis of the power trace gives the information on the execution of such a conditional step. In other words, using a SPA analysis, an adversary is able to detect whether the addition $t \leftarrow t + p$ is performed, and then to deduce if $m_p < m_q$. This information allows a binary search to recover the factor p: an attacker searches for a plaintext m such that $m \bmod p < m \bmod q$ and $(m-1) \bmod p \geq (m-1) \bmod q$. Such a plaintext can be efficiently found with a binary search combined with a simple power analysis. Once m is found, Novak has remarked that m is in fact a multiple of the factor p that can then be deduced as the GCD of m and the modulus N. However, this attack is only possible in a chosen-plaintext scenario. Thus it cannot be made in practical implementations of the RSA signature scheme due to the encoding step. Indeed, an adversary is still supposed to choose the message M to sign but does not have enough control over the encoding m of M, particularly when randomization techniques are used.

In the following we show how to recover the factor q using this leaked information even if a padding scheme is used. However, we require the prime factors of the modulus to be slightly unbalanced.

3 Lattice Based Techniques

3.1 Preliminaries on Lattices

We denote by $\|\mathbf{x}\|$ the Euclidean norm of the vector $\mathbf{x} = (x_1, \ldots, x_{d+1})$, defined by $\|\mathbf{x}\| = \sqrt{\sum_{i=1}^{d+1} x_i^2}$. Let $\mathbf{v}_1, \ldots, \mathbf{v_d}$, be d linearly independent vectors such that for $1 \leq i \leq d$, $\mathbf{v}_i \in \mathbb{Z}^{d+1}$. We denote by L, the lattice spanned by the matrix V whose rows are $\mathbf{v}_1, \ldots, \mathbf{v_d}$. L is the set of all integer linear combinations of $\mathbf{v}_1, \ldots, \mathbf{v_d}$:

$$L = \left\{ \sum_{i=1}^{d} c_i \mathbf{v_i}, \ c_i \in \mathbb{Z} \right\}$$

Geometrically, $\det(L)$ is the volume of the parallelepiped spanned by $\mathbf{v}_1, \ldots, \mathbf{v_d}$. The Hadamard's inequality says that $\det(L) \leq \|\mathbf{v_1}\| \times \ldots \times \|\mathbf{v_d}\|$.

Given $\langle \mathbf{v_1}, \ldots, \mathbf{v_d} \rangle$ the LLL algorithm [8] will produce a so called "reduced" basis $\langle \mathbf{b_1}, \ldots, \mathbf{b_d} \rangle$ of L such that

$$\|\mathbf{b_1}\| \leq 2^{(d-1)/2} \det(L)^{1/d} \tag{1}$$

in time $O(d^4 \log(M))$ where $M = \max_{1 \le i \le d} \|\mathbf{v}_i\|$. Consequently, given a basis of a lattice, the LLL algorithm finds a short vector $\mathbf{b_1}$ of L satisfying equation (1). Moreover, we assume in the following that the new basis vectors are of the same length and also have all their coordinates of approximatively the same length. Indeed, a basis for a random lattice can be reduced into an almost orthonormal basis. Therefore, $\|b_i\| \approx \|b_1\|$ for $1 \le i \le d$, and so $\|b_i\|^d \approx \det(L)$.

3.2   Factoring Using LLL

In the following we describe a new method, based on lattice reduction, to factor a modulus N given some special form integers s_i in $\mathbb{Z}_N$.

Let $N = p \times q$ be an RSA modulus such that p and q are two prime integers. Let s_1, s_2, $\ldots$, s_d be d integers from $\mathbb{Z}_N$. For each s_i, we consider its euclidian division by p. $\forall i \in [1, d]$, we can write $s_i = r_i + u_i \times p$ with $r_i \in \mathbb{Z}_p$ and $u_i \in \mathbb{Z}_q$. Let us assume that, instead of being distributed all over the set $\mathbb{Z}_p$, the r_is values are smaller than a bound $A < p/2$. We further consider the lattice L spanned by the $d + 1$ rows of the following matrix:

$$\begin{pmatrix} N & 0 & \ldots & \ldots & 0 \\ 0 & N & \ddots & & \vdots \\ \vdots & \ddots & \ddots & \ddots & \vdots \\ 0 & \ldots & 0 & N & 0 \\ -s_1 & -s_2 & \ldots & -s_d & A \end{pmatrix}$$

Theorem 1. *Assuming the LLL algorithm returns the shortest vector of a lattice, the reduction of lattice L computes the factorization of modulus N with probability $> 1 - \varepsilon_0$ if the bit-length difference between p and A is such that*

$$\log p - \log A > \max\left(\frac{\log q - \log \varepsilon_0 - 0.105}{d} + 2.047, \log\left(2\sqrt{d+1}\right) \right)$$

As an example, for a 512-bit prime factor q and a probability of success of the algorithm $> 1 - 2^{-10}$, we obtain a minimum $\log p - \log A \approx 10.4$ for $d = 60$. Note that this minimum does not strongly depend on the size of q and is about 5 bits for any cryptographic size of this factor.

Sketch of proof. [A complete proof is proposed in appendix A]
By definition of the lattice L, a vector of L is an integer combination of the rows of the matrix. In other words, we may define the lattice in the following way :

$$L = \left\{ (c_1 N - cs_1, c_2 N - cs_2, \ldots, c_d N - cs_d, Ac) ; (c_1, c_2, \ldots, c_d, c) \in \mathbb{Z}^{d+1} \right\}$$

For a fixed choice of the integer coefficients $(c_1, c_2, \ldots, c_d, c)$, we note

$$b(c_1, c_2, \ldots, c_d, c) = (c_1 N - cs_1, c_2 N - cs_2, \ldots, c_d N - cs_d, Ac) \in L$$

In other words, the lattice L is the set of all the vectors $b(c_1, c_2, \ldots, c_d, c)$ for $(c_1, c_2, \ldots, c_d, c) \in \mathbb{Z}^{d+1}$.

A special vector of the lattice, strongly related with the q prime factor of N, is "abnormally" short. The consequence is that we can expect the LLL lattice reduction algorithm to compute this short vector.

This special vector is $b^* = b(u_1, u_2, \ldots, u_d, q)$, where u_i is defined by division of the s_i by p, and q is the other factor of the modulus $N = p \times q$. Note that the knowledge of b^* immediately reveals q since its last coordinate is Aq and A is known. The size of b^*, i.e. its euclidian norm, can be easily estimated and we obtain $\|b^*\| < \sqrt{d+1} Aq$.

Then, in order to prove that the vector b^* is the shortest one, we study the Euclidean norm of the vectors $b(c_1, c_2, \ldots, c_n, c)$, and we prove that, if $c \neq q$, those vectors are larger that b^*, whatever the c_is may be, for all the s_i but only a very small fraction. A precise analysis, described in appendix A, leads to the result of theorem 1.

$\square$

Therefore the knowledge of d values s in $\mathbb{Z}_N$ such that $|s \bmod p| < A$ allows us to factor N. In the following, we describe how SPA can be used to find such d values in the context of RSA signature generation, with "slightly" unbalanced modulus.

4 Application to RSA-CRT Signature Scheme

In this section, we use the results presented above to extend the chosen ciphertext attack described by Novak in [10]. In particular, we show that if the factors p and q of the modulus are such that $|p| - |q| > \ell$, for a given bound ℓ, then they can be recovered with a known message attack combined with a simple power analysis. In the following we suppose that a SPA attack allows us to detect if the addition of step 5 is performed during a signature generation. Such an assumption is realistic in practice if no countermeasure is implemented, and a detailed way to extract this information can be found in [10] and in [5].

Attack. In the previous section we have shown how the prime factors p and q of a modulus N can be recovered, given a set of integers s in $\mathbb{Z}_N$ such that $s \bmod p$ is less than a given bound. We apply this result by using a simple power analysis on the RSA signature scheme in order to find these integers.

We assume that the prime factors p and q are such that $|p| - |q| > \ell$, for ℓ a small integer. Such an assumption is realistic in many actual implementations since, in many descriptions of the RSA algorithm, we can find that p and q have to be of "roughly" the same length, or "about the same bit-length". Here, we consider that p and q have a very small bit-length difference, about 10 bits for an 1024-bit modulus. This does not constitute a contradiction with the usual description of RSA, that can be interpreted in many different ways.

Let S be a signature for a random message M, computed by using the algorithm described in figure 1. We suppose that the step 5 of the algorithm has been performed to generate the signature S. Otherwise, we choose another random message until this step is executed. Since the optional addition has been made, we know that $s_p - s_q < 0$. Thus, since we assume that $q < p$, we simply have that $s_p < s_q < q < p$. By definition of $s_p = S \bmod p$, S can always be written as $S = s_p + u \times p$ for an integer $u < q$. Consequently S is a candidate input to the factoring algorithm, given in section 3.2, for the upperbound $A = q$ on $s \bmod p$. The problem here is that clearly, this bound is not known to the attacker. Thus, the last entry A of the matrix cannot be explicitly given. However, A is an upperbound on the $s_i \bmod p$ where s_i is the ith input of the last row. Thus choosing an integer $A > q$ is a correct choice and we choose in practice A to be the largest integer such that $|A| = |q|$.

To run the lattice reduction described in the previous section, we have to find d signatures s_i such that $s_i \bmod p$ is less than the bound A we choose. We thus query the signature of messages and we perform a SPA attack on each generation. Each signature verifying $s_p < s_q$ is kept as input to the matrix. We query the signing card until d valid candidate signatures have been found.

The number d of required signatures has to be sufficiently large, according to the bit-length difference ℓ between the factors p and q, and according to the modulus length. To estimate the average number of queries made to the signing card, we compute the probability that $s_p = m^{d_p} \bmod p$ is less than $s_q = m^{d_q} \bmod q$, for $q < p/2^\ell$, and for random integers $s \in \mathbb{Z}_N$. We suppose that the values s are uniformly and independently distributed in $\mathbb{Z}_N$. This is verified in practice when an appropriate hash function, such as SHA-1, is used. In this case we assume that the output m of the encoding scheme is uniformly distributed in $\mathbb{Z}_N$ so that $s = m^d \bmod N$ is uniformly distributed. Let s_p and s_q the values computed during the signature generation in steps 2 and 3 respectively. We have:

$$\Pr\{s_p < s_q\} = \sum_{B=0}^{q-1} \Pr\{s_p < B | s_q = B\} \cdot \Pr\{s_q = B\}$$

$$= \sum_{B=0}^{q-1} \frac{B}{p} \cdot \frac{1}{q} = \frac{q(q-1)}{2pq} < \frac{1}{2^{\ell+1}}$$

where the last inequality comes from the fact that $q < p/2^\ell$.

Thus in a set of 2^ℓ signatures, there is a probability of one half that at least one of them is such that $s_p < s_q$. Detecting such a signature is possible with a SPA attack during the signature generation: when the step 5 is performed, we know that the signature s_i is a good candidate input for our algorithm, if we write it as $s_i = (s_i \bmod p) + u_i p$ where $s_i \bmod p < p/2^\ell$. Otherwise, we query another signature until we find a good candidate. On average, 2^ℓ trials are needed. To have d such signatures, a set of $d \cdot 2^\ell$ signatures is required. The algorithm described in section 3.2 to factor a modulus N can then be used with the candidate signatures as inputs. Following the analysis made above, the number d of required signatures must be such that $\ell \geq \frac{\log q - 10}{d} + 2$ so that the algorithm successfully ends with probability greater than $1 - 2^{-10}$. However, we

modulus length in bits	$\|p\| - \|q\|$	lattice dimension $d+1$	average number of required signatures	time to factor
512	8	41	2^{13}	30 s
	7	51	2^{13}	2 min
	6	61	2^{12}	14 min
768	10	43	2^{15}	50 min
	9	51	2^{15}	2 min
	8	61	2^{14}	15 min
1024	12	46	2^{17}	2 min
	11	56	2^{17}	6 min
	10	61	2^{16}	16 min
1536	16	53	2^{22}	7 min
	15	56	2^{21}	10 min
	14	61	2^{20}	32 min
2048	20	53	2^{26}	11 min
	19	56	2^{25}	20 min
	18	61	2^{24}	32 min

Fig. 2. Experimental results on the RSA-CRT with unbalanced modulus.

assume here that $q < p/2^\ell$, and thus, $\log q = \frac{\log N - \ell}{2}$. Thus, after some simple computations we obtain:

$$\ell \geq \frac{\log N - 20 + 4d}{2d + 1}$$

Thus, if p and q are such that $q < p/2^\ell$ for ℓ greater than $\frac{\log N - 20 + 4d}{2d+1}$, we can factor N from $d \times 2^\ell$ signatures on average and with probability at least $1 - 2^{-10}$.

Experimental results. In practice, the lattice dimension depends on the value ℓ, and on the number of available signatures. However, if the lattice dimension is too large, then the LLL algorithm fails. Particularly, under a reasonable time, it is not possible to run the LLL algorithm on a 100×100 dimensional matrix where the entries are 1024-bit numbers.

For each modulus length, we give in figure 2 the integer ℓ such that $q < p/2^\ell$, the dimension $d+1$ of the lattice, the average number of required signatures and the time needed to recover p and q. The number of required signatures, equal to $d \times 2^{\|p\|-\|q\|}$, is upperbounded by $2^{6+\|p\|-\|q\|}$ since we always have $d < 2^6$. The tests have been run on an Intel Pentium IV, XEON 1.5 GHz, with the Victor Shoup's library NTL ([12]).

From this previous table, we show that the LLL algorithm works better than the theoretical results indicated in section 3.2. For 1024-bit modulus, the expected result gives $\ell > 10$. These values are realistic since a difference of 10 bits between p and q for a 1024-bit modulus means that p is a 517-bit prime and q is a 507-bit prime number. This distance between p and q is not ruled out by the specifications of the key generation of the RSA algorithm. It is worth noticing that this attack can be extended to more "secure" moduli, say 2048 bits long. Moreover, this attack can be mounted in practice since the number of required signatures for known messages is not large, one million if $n = 1536$.

5   Countermeasures

Such an attack proves that an implementation of the Garner's algorithm should be carefully checked so that SPA or other side channels attacks should not be possible. We now propose some countermeasures to avoid this attack.

In order to balance time execution and power consumption, dummy operations can be added. This can be done by modifying the step 4 of the algorithm as follows: first, t is computed as $s_p - s_q$. Then, new variables t' and t'' are respectively set as $t+p$ and t. The step 5 is then defined as: 5. If $t < 0$ then $t \leftarrow t'$, else, $t \leftarrow t''$. In this case, the implementation does not leak any information about the difference $s_p - s_q$ since the addition is always performed. The crucial remark here is that this implementation should use a probabilistic encoding step (Step 1 of figure 1). If not, another attack is possible: suppose for example that PKCS#1 v1.5 is used. Thus the encoding of a message is always the same. In this case, this countermeasure can be broken by using safe errors attacks [13]. Such attacks use fault injection at particular computational step to produce an error during operations, possibly unused depending on some secret data. Here a fault can be performed during the computation of t'. If the resulting signature is not valid (the card outputs a failed error), we learn that for this signature, t' has been used, that is $t < 0$. In this case, we ask again the card with the same message without producing any error during the generation. We know that the resulting signature is such that $t < 0$, since it has been computed on the same input m. We can then use this signature as input for our algorithm.

Another classical countermeasure [2] is based on the randomization of m: a signature s is computed on $r^e \times m$. The signature for m is given as $s/r \bmod N$. In this case, since r is kept secret, there is no relationship between the information leaked by the card on the value t and the output signature $s/r \bmod N$. Thus, in this case, our attack is no longer feasible. Note that it is also possible to fully randomize all the parameters of the signature generation, as done by many actual implementations of RSA signatures: the factors p and q are randomized as $p' = r_1 \times p$ and $q' = r_2 \times q$ and the signature s is deduced from $s \bmod p'$ and $s \bmod q'$ respectively computed as $((m \bmod p) + r_1' \times p)^{d_p + r_1'' \times (p-1)}$ and $((m \bmod q) + r_2' \times q)^{d_q + r_2'' \times (q-1)}$. The signature is finally given by $s \bmod N$. Due to a full randomization of each step of the signature generation, such an implementation is not subject to our attack.

Acknowledgments. We wish to thank anonymous referees for their constructive remarks and suggestions of countermeasures.

References

1. D. Boneh, R. A. DeMillo, and R. J. Lipton. On the Importance of Checking Cryptographic Protocols for Faults. In W. Fumy, editor, *Advances in Cryptology – Eurocrypt'97*, volume 1233 of *LNCS*, pages 37–51. Springer-Verlag, Berlin, 1997.
2. D. Chaum. Blind signatures for untraceable payments. In D. Chaum, R.L. Rivest, and A.T. Sherman, editors, *Advances in Cryptology – Crypto'82*, pages 199–204. Plenum Publishing, 1982.
3. B. den Boer, K. Lemke, and G. Wicke. A DPA Attack Against the Modular Reduction within a CRT Implemenataion of RSA. In B. S. Kaliski Jr, Ç. K. Koç, and C. Paar, editors, *Cryptographic Hardware and Embedded Systems – CHES 2002*, volume 2523 of *LNCS*, pages 228–243. Springer-Verlag, 2002.
4. D. E. Knuth. *The Art of Computer Programming, Vol 2: Semi Numerical Algorithms*. Addison Wesley, 1969.
5. P. C. Kocher. Timing Attacks on Implementations of Diffie-Hellman, RSA, DSS, and Others Systems. In N. Koblitz, editor, *Advances in Cryptology – Crypto '96*, volume 1109 of *LNCS*, pages 104–113. Springer-Verlag, 1996.
6. P. C. Kocher, J. Jaffe, and B. Jun. Differential Power Analysis. In M. Wiener, editor, *Advances in Cryptology – Crypto '99*, volume 1666 of *LNCS*, pages 388–397. Springer-Verlag, 1999.
7. RSA Laboratories. PKCS #1 v2.1 : RSA Cryptography Standard, 2002. Available at http://www.rsalabs.com/pkcs/pkcs-1.
8. A. K. Lenstra, H. W. Lenstra, and L. Lovász. Factoring Polynomials with Rational Coefficients. *Mathematische Annalen*, 261(4):515–534, 1982.
9. A. Menezes, P. van Oorschot, and S. Vanstone. *Handbook of Applied Cryptography*. CRC Press, 1996.
10. R. Novak. SPA-Based Adaptive Chosen-Ciphertext Attack on RSA Implementation. In D. Naccache and P. Paillier, editors, *PKC 2002*, volume 2274 of *LNCS*, pages 252–261. Springer-Verlag, 2002.
11. W. Schindler. A Timing Attack against RSA with the Chinese Remainder Theorem. In Ç. K. Koç and C. Paar, editors, *Cryptographic Hardware and Embedded Systems 2000*, volume 1965 of *LNCS*, pages 109–124. Springer-Verlag, 2000.
12. V. Shoup. Number Theory C++ Library (NTL), version 5.0b. Available at http://www.shoup.net.
13. S.M. Yen and M. Joye. Checking before output may not be enough against fault-based cryptanalysis. *IEEE Trans. on Computers*, 49:967–970, 2000.

A   Proof of Theorem 1

By definition of the lattice L, a vector of L is an integer combination of the rows of the matrix. In other words, we may define the lattice in the following way:

$$L = \left\{ (c_1 N - c s_1, c_2 N - c s_2, \dots, c_d N - c s_d, Ac) \; ; \; (c_1, c_2, \dots, c_d, c) \in \mathbb{Z}^{d+1} \right\}$$

For a fixed choice of the integer coefficients $(c_1, c_2, \dots, c_d, c)$, we note

$$b(c_1, c_2, \dots, c_d, c) = (c_1 N - c s_1, c_2 N - c s_2, \dots, c_d N - c s_d, Ac) \in L$$

In other words, the lattice L is the set of all the vectors $b(c_1, c_2, \dots, c_d, c)$ for $(c_1, c_2, \dots, c_d, c) \in \mathbb{Z}^{d+1}$.

We now show that a special vector of the lattice, strongly related with the q prime factor of N, is "abnormally" short. The consequence is that we can expect the LLL lattice reduction algorithm to compute this short vector.

This special vector is $b^* = b(u_1, u_2, \ldots, u_d, q)$, where u_i is defined by division of the s_i by p, and q is the other factor of the modulus $N = p \times q$. Note that the knowledge of b^* immediately reveals q since its last coordinate is Aq and A is known. Let us evaluate the size of b^*, i.e. its euclidian norm

$$
\begin{aligned}
||b^*||^2 &= ||b(u_1, u_2, \ldots, u_d, q)||^2 \\
&= ||(u_1 N - qs_1, u_2 N - qs_2, \ldots, u_d N - qs_d, Aq)||^2 \\
&= \sum_{i=1}^{d} (u_i N - q(r_i + u_i p))^2 + A^2 q^2 \\
&= \sum_{i=1}^{d} q^2 r_i^2 + A^2 q^2
\end{aligned}
$$

Since $0 \le r_i < A$, we immediately obtain that $||b^*||^2 < dq^2 A^2 + A^2 q^2$, and so $||b^*|| < \sqrt{d+1} Aq$.

In order to prove that the vector b^* is abnormally short, we now show that the other elements of the lattice L are, with overwhelming probability, larger. With this aim in view, we define, for any integer c, the function

$$
\mathcal{F}(c) = \min_{(c_1, c_2, \ldots, c_d) \in \mathbb{Z}^d} ||b(c_1, c_2, \ldots, c_d, c)||
$$

i.e., $\mathcal{F}(c)$ is the size of the shortest vector in L whose last coordinate is equal to Ac. From the definition of $\mathcal{F}(c)$, we can derive the following expression

$$
\mathcal{F}(c) = \min_{(c_1, c_2, \ldots, c_d) \in \mathbb{Z}^d} \sqrt{\sum_{i=1}^{d} (c_i N - cs_i)^2 + A^2 c^2}
$$

Then, it is easy, for a fixed c, to find the c_is that reach the minimum since $\mathcal{F}(c)$ is the minimum of a sum of independent squares. As a consequence, the minimum is reached when each term is as small as possible. This means that c_i is the nearest integer of cs_i/N; we note $\lfloor \frac{cs_i}{N} \rceil$ this integer. We finally obtain

$$
\mathcal{F}(c) = \sqrt{\sum_{i=1}^{d} \left(\left\lfloor \frac{cs_i}{N} \right\rceil N - cs_i \right)^2 + A^2 c^2}
$$

We can now notice that, by definition of $\mathcal{F}(c)$ and $b^* = b(u_1, u_2, \ldots, u_d, q)$, we have $\mathcal{F}(q) \le ||b^*|| < \sqrt{d+1} Aq$. In fact, b^* is exactly the smallest vector with last coordinate equal to Aq because

$$\mathcal{F}(q)^2 = \sum_{i=1}^{d} \left(\left\lceil \frac{qs_i}{N} \right\rceil N - qs_i \right)^2 + A^2 q^2$$

$$= q^2 \times \left(\sum_{i=1}^{d} \left(\left\lceil \frac{r_i + u_i \times p}{p} \right\rceil p - r_i - u_i \times p \right)^2 + A^2 \right)$$

$$= q^2 \times \left(\sum_{i=1}^{d} \left(\left\lceil \frac{r_i}{p} \right\rceil p - r_i \right)^2 + A^2 \right) = q^2 \times \left(\sum_{i=1}^{d} r_i^2 + A^2 \right)$$

since, for $0 \le r_i < A < \frac{p}{2}$, $\left\lceil \frac{r_i}{p} \right\rceil = 0$. This finally proves that $\mathcal{F}(q) = ||b^*||$.

Then, we can further notice that if $c > \sqrt{d+1} \times q$, $\mathcal{F}(c)$ is obviously greater than Ac so $\mathcal{F}(c) > \sqrt{d+1} \times q \times A > \mathcal{F}(q)$. We finally need to evaluate, for a fixed $c \neq q$, the probability that $\mathcal{F}(c) < \mathcal{F}(q)$ where the probabilities are computed when the s_i are uniformly distributed in

$$\mathcal{S} = \{r + u \times p \,;\, 0 \le r < A \,,\, 0 \le u < q\} \subset \mathbb{Z}_N$$

If this probability is negligible, we can conclude that b^* is, with overwhelming probability, the shortest vector of the lattice.

We first notice that the distribution of the $\left\lceil \frac{c \times s_i}{N} \right\rceil \times N - c \times s_i$, when s_i is uniformly distributed in $\mathbb{Z}_N$, is uniform between $\frac{-(N-1)}{2}$ and $\frac{N-1}{2}$. This is an obvious consequence of the fact that $\left\lceil \frac{c \times s_i}{N} \right\rceil - \frac{c \times s_i}{N} \in [-1/2; 1/2]$ and that if $\left\lceil \frac{c \times s_i}{N} \right\rceil \times N - c \times s_i = \alpha$ for an integer α, we obtain by modular reduction that $-c \times s_i = \alpha \bmod N$ and thus that $s_i = -\alpha \times c^{-1} \bmod N$ if c is prime with N.

If we restrict the possible values of s_i to the set $\mathcal{S}$, the previous result implies

$$\mathcal{D}_c = \left\{ \left\lceil \frac{cs_i}{N} \right\rceil \times N - cs_i \,;\, s_i = r_i + u_i \times p \in \mathcal{S} \right\} \subset \left[\frac{-(N-1)}{2}, \frac{N-1}{2} \right]$$

We can further write

$$\mathcal{D}_c = \left\{ \left\lceil \frac{cs_i}{N} \right\rceil \times N - cs_i \,;\, s_i = r_i + u_i p \,,\, 0 \le r_i < A \,,\, 0 \le u_i < q \right\}$$

$$= \left\{ \left\lceil \frac{cr_i}{N} + \frac{cu_i}{q} \right\rceil \times N - cr_i - cu_i p \,;\, 0 \le r_i < A \,,\, 0 \le u_i < q \right\}$$

$$= \left\{ \left\lceil \frac{cr_i}{N} + \frac{cu_i \bmod q}{q} \right\rceil \times N - cr_i - (cu_i \bmod q) \times p \,;\, 0 \le r_i < A \,,\, u_i \in \mathbb{Z}_q \right\}$$

$$= \left\{ \left\lceil \frac{cr_i}{N} + \frac{v_i}{q} \right\rceil \times N - cr_i - v_i p \,;\, 0 \le r_i < A \,,\, 0 \le v_i < q \right\}$$

Then, for any $\alpha \in [-(N-1)/2, (N-1)/2]$, if $\left\lceil \frac{c \times r_i}{N} + \frac{v_i}{q} \right\rceil \times N - c \times r_i - v_i \times p = \alpha$ we obtain that $\alpha = -cr_i - v_i p \bmod N$. Since v_i is uniformly distributed in $\mathbb{Z}_q$ and $\alpha + p = -cr_i - (v_i - 1 \bmod q) \times p \bmod N$, we conclude that, if α is an element of $\mathcal{D}_c$, $\alpha + p$ (or $\alpha + p - N$ if $\alpha + p > N/2$) is also an element of $\mathcal{D}_c$. So, if $\gcd(c, N) = 1$, the set $\mathcal{D}_c$ is a subset of $\left[\frac{-(N-1)}{2}, \frac{N-1}{2} \right]$ that is invariant by (circular) translation of length p.

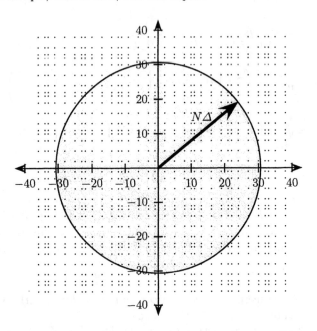

Fig. 3. Graphical representation of the pairs (s_1, s_2) for the parameters $d = 2$, $p = 11$, $q = 7$, $N = 77$ and $A = 5$. The disk of radius $N\Delta$ covers a ratio of those points that is illustrated in figure 4.

In other words, this formalizes the idea that, if c is prime with N, the elements of $\mathcal{D}_c$ are well distributed in $\left[\frac{-(N-1)}{2}, \frac{N-1}{2}\right]$. As a toy example, figure 3 represents the elements of $\mathcal{D}_3 \times \mathcal{D}_3$ for $N = 7 \times 11$ and $A = 5$.

Always with the aim of computing the probability for $\mathcal{F}(c)$ to be less that $\mathcal{F}(q)$, if $\gcd(c, N) = 1$, we can state that

$$\Pr_{(s_1, s_2, \dots, s_d) \in \mathcal{S}^d} \{\mathcal{F}(c) < \mathcal{F}(q)\} < \Pr_{(s_1, s_2, \dots, s_d) \in \mathcal{S}^d} \left\{\mathcal{F}(c) < \sqrt{d+1} Aq\right\}$$

$$< \Pr_{(s_1, s_2, \dots, s_d) \in \mathcal{S}^d} \left\{\sum_{i=1}^{d} \left(\left\lceil \frac{cs_i}{N} \right\rceil N - cs_i\right)^2 < (d+1)A^2 q^2 - A^2 c^2\right\}$$

$$< \Pr_{(s_1, s_2, \dots, s_d) \in \mathcal{S}^d} \left\{\sum_{i=1}^{d} \left(\left\lfloor \frac{cs_i}{N} \right\rceil N - cs_i\right)^2 < (d+1)A^2 q^2\right\}$$

Let $\Delta = \sqrt{d+1} A/p$; we need to evaluate the probability

$$\Pr_{(s_1, s_2, \dots, s_d) \in \mathcal{S}^d} \left\{\sum_{i=1}^{d} \left(\left\lfloor \frac{cs_i}{N} \right\rceil N - cs_i\right)^2 < N^2 \Delta^2\right\}$$

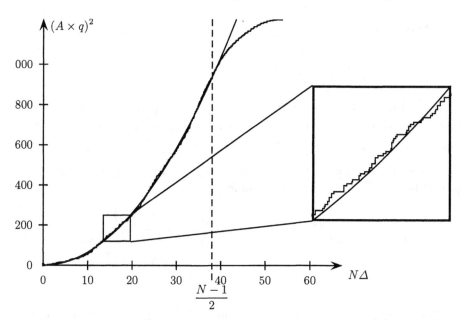

Fig. 4. Ration of points in figure 3 covered by a disk of radius $N\Delta$. The irregular experimental curve represents the number of points covered by the disk and the smooth curve is based on the approximation by the surface of this disk.

From the previous result on the distribution of $X_i = \left\lfloor \frac{cs_i}{N} \right\rceil - \frac{cs_i}{N}$ in $\left[-\frac{1}{2}, \frac{1}{2} \right]$, this probability can be approximated, for large N, by

$$\Pr_{(X_1, X_2, \ldots, X_d) \in_u \left[-\frac{1}{2}, \frac{1}{2} \right]} \left\{ \sum_{i=1}^{d} X_i^2 < \Delta^2 \right\}$$

where the X_is are independent and uniformly distributed over $\left[-\frac{1}{2}, \frac{1}{2} \right]$. If $0 \le \Delta < \frac{1}{2}$, this probability is equal to the volume of the d-dimensional ball of radius Δ. Figure 4 provides a graphical illustration of this fact, using the toy example of figure 3. If d is even, this volume is equal to $\pi^{d/2} \Delta^d / (d/2)!$.

Note that this approximation of the number of points in the ball of radius $N\Delta$ (see figure 3) using the volume of this ball is very good for values of c that are relatively prime with q, even if the repartition of the points is not perfectly uniform. This is mainly due to the compensation of local errors of estimation. When $c = q$, such a compensation does not apply and the approximation can no longer be used. Then, using the well-known Stirling formula, we obtain the upper-bound

$$\Pr_{(s_1, s_2, \ldots, s_d) \in S^d} \left\{ \sum_{i=1}^{d} \left(\left\lfloor \frac{cs_i}{N} \right\rceil N - cs_i \right)^2 < N^2 \Delta^2 \right\} < \left(\frac{2e\pi\Delta^2}{d} \right)^{\frac{d}{2}} \times \frac{1}{\sqrt{d\pi}}$$

We finally obtain

$$\Pr_{(s_1,s_2,\ldots,s_d)\in\mathcal{S}^d} \{\mathcal{F}(c) < \mathcal{F}(q)\} < \left(\frac{2e\pi\Delta^2}{d}\right)^{\frac{d}{2}} \times \frac{1}{\sqrt{d\pi}}$$

Note that this is true only if $\Delta < \frac{1}{2}$, i.e., if $A < \frac{p}{2\sqrt{d+1}}$. In other words, in order to make the proof correct, this means that the difference $-\log(A/p)$ of bit-length of p and A must fulfill the inequality $-\log\left(\frac{A}{p}\right) > \log\left(2\sqrt{d+1}\right)$

Using the fact that $\left(\frac{d+1}{d}\right)^d < e$, we finally obtain an upper bound of the probability for $\mathcal{F}(c)$ to be smaller than $\mathcal{F}(q)$

$$\Pr_{(s_1,s_2,\ldots,s_d)\in\mathcal{S}^d} \{\mathcal{F}(c) < \mathcal{F}(q)\} < \left(\frac{2e\pi A^2}{p^2} \times \frac{d+1}{d}\right)^{\frac{d}{2}} \times \frac{1}{\sqrt{d\pi}}$$

$$< \sqrt{\frac{e}{d\pi}}\left(\sqrt{2\pi e}\right)^d \left(\frac{A}{p}\right)^d$$

The last step is to estimate for which values of the parameters we can consider that $\mathcal{F}(c) > \mathcal{F}(q) = \|b^*\|$ for any $c \neq q$. Let $P_0 = 1 - \varepsilon_0$ be a lower bound of the probability for b^* to be the smallest vector of the lattice. From the previous results, using an approximative argument of independence of the probabilities for different values of c, we deduce

$$\Pr_{(s_1,s_2,\ldots,s_d)\in\mathcal{S}^d} \{\forall c \neq q\ \mathcal{F}(c) > \mathcal{F}(q)\} > \left(1 - \sqrt{\frac{e}{d\pi}}\left(\sqrt{2\pi e}\right)^d \left(\frac{A}{p}\right)^d\right)^{\sqrt{d+1}\times q}$$

If $\varepsilon_0 > (\sqrt{d+1} \times q) \times \sqrt{\frac{e}{d\pi}}(\sqrt{2\pi e})^d\left(\frac{A}{p}\right)^d$, the last expression is greater than P_0. Thus, the inequality can be reworded as follows

$$-\log\left(\frac{A}{p}\right) > \frac{1}{d}\left(\frac{1}{2}\log\left(\frac{d+1}{d}\right) - \log\varepsilon_0 + \log q + \frac{1}{2}\log\left(\frac{e}{\pi}\right) + d\log\left(\sqrt{2\pi e}\right)\right)$$

$$> \frac{\log q - \log\varepsilon_0 + \frac{1}{2}\log\left(\frac{e}{\pi}\right)}{d} + \log\left(\sqrt{2\pi e}\right)$$

In other words, this means that, if the difference $-\log(A/p)$ of bit-length of p and A is larger than $\frac{\log q - \log\varepsilon_0 - 0.105}{d} + 2.047$, the algorithm finds q with probability $> 1 - \varepsilon_0$, assuming that LLL returns the shortest vector of the lattice. $\qquad\square$

The Doubling Attack – *Why Upwards Is Better than Downwards*

Pierre-Alain Fouque and Frederic Valette

DCSSI Crypto Lab
51, Boulevard de Latour-Maubourg
75700 Paris 07 SP
France
Pierre-Alain.Fouque@ens.fr
Frederic.Valette@m4x.org

Abstract. The recent developments of side channel attacks have lead implementers to use more and more sophisticated countermeasures in critical operations such as modular exponentiation, or scalar multiplication in the elliptic curve setting. In this paper, we propose a new attack against a classical implementation of these operations that only requires two queries to the device. The complexity of this so-called "doubling attack" is much smaller than previously known ones. Furthermore, this approach defeats two of the three countermeasures proposed by Coron at CHES '99.

Keywords. SPA-based analysis, modular exponentiation, scalar multiplication, DPA countermeasures, multiple exponent single data attack.

1 Introduction

Modular exponentiation or scalar multiplication are the main parts of the most popular public key cryptosystems such as RSA [15] or DSA [13]. This very sensitive operation can be efficiently implemented in smart cards products. However, data manipulated during this computation should often be kept secret, so the implementation of such algorithms must be protected against side channel attacks. For example, during the generation of an RSA signature by a device, the secret exponent is used to transform a message related data into a digital signature via modular exponentiation.

Timings and power attacks, initially presented by Kocher [9,10] are now well studied and various countermeasures have been proposed. Those attacks represent a real threat when we consider operations that both involve secret data and require a long computation time. The consequence is that naive implementation of RSA based or discrete log based cryptosystems usually leak information about the secret key.

In this paper we present a new side channel attack, that we called "doubling attack", which allows to recover the secret scalar used in the binary scalar

C.D. Walter et al. (Eds.): CHES 2003, LNCS 2779, pp. 269–280, 2003.
© Springer-Verlag Berlin Heidelberg 2003

multiplication or the secret exponent used in the binary exponentiation algorithm. It is worth to notice that contrary to previous attacks which work for the "Left-to-Right" and the "Right-to-Left" implementations of the binary modular exponentiation, the doubling attack only works for the "Left-to-Right" implementation. Furthermore, the "Left-to-Right" implementation is often used since it requires only one variable. The new attack enables to recover the secret key decryption of RSA [15] or the key decryption of ElGamal [4]. It can also be used to obtain the secret key of the Diffie-Hellman authentication system. We only focus on the decryption cases. In this attack we assume that the adversary mounts a chosen ciphertext attack. This is a valid assumption in a side channel scenario, since randomized paddings avoiding chosen ciphertext attack, such as OAEP [1], are checked after the running of the decryption process. As a consequence, the binary exponentiation or multiplication is always performed and side channel attacks can be mounted on these algorithms. The attack on the RSA cryptosystem is a direct application of the doubling attack. On discrete-log based cryptosystems, we describe the attack in the elliptic curve setting, since the doubling attack allows to defeat classical countermeasures which are mainly proposed to elliptic curve systems.

In this paper, we first remind classical binary scalar multiplication algorithms. Then, we shortly describe different types of side channel attacks such as simple power analysis and differential power analysis but also the attack of Messerges, Dabbish and Sloan [11] in order to motivate the most frequently used countermeasures.

Then, we present an improvement of Messerges *et al* attack that applies when so-called downward algorithms are used. It has a much smaller complexity since it only requires two queries to the device in order to recover all the secret data. This new attack is called *"doubling attack"* since it is based on the doubling operation in the elliptic curve setting. We also explain how to use this new attack to defeat Coron's countermeasures [3].

2   Binary Scalar Multiplication Algorithms

In classical cryptosystems based on the RSA or on the discrete logarithm problem, the main operation is modular exponentiation. In the elliptic curve setting, the corresponding operation is the scalar multiplication. From an algorithmic point of view, those two operations are very similar; the only difference is the underlying group structure. In this paper, we consider operations over a generic group, without using any additional property. The consequence is an immediate application to the elliptic curve setting but it should be clear that all what we state can be easily transposed to modular exponentiation.

Scalar multiplication is usually performed using the "double-and-add" method that computes $d \times P$ using the binary representation of the scalar $d = \sum_{i=0}^{n} d_i \times 2^i$:

$$d \times P = \sum_{i=0}^{n} d_i \times \left(2^i \times P\right)$$

Two versions of the double-and-add algorithm are usually considered, according to the order of the terms in the previous sum. The first routine starts from the most significant bit and works downward. This method is usually called "Left-to-Right" (see figure 1).

$$S = 0$$
for i from n down to 0
$\quad S = 2.S$
$\quad$ if $d_i = 1$ then $S = S + P$
return S

Fig. 1. Downward "Left-to-Right" double-and-add(P,d)

The second routine starts from the least significant bit and works upward. This method is also known as "Right-to-Left" (see figure 2).

$$S = 0$$
$$T = P$$
for i from 0 to n
$\quad$ if $d_i = 1$ then $S = S + T$
$\quad T = 2.T$
return S

Fig. 2. Upward "Right-to-Left" double-and-add(P,d)

The first implementation is the most frequently used since it requires less memory. Up to now, no distinction was made on the security of those routines since all proposed attacks can be adapted to both implementations. In the following sections, we focus on the downward implementations and we show that it may be much more easily attacked than upward versions.

3   Power Analysis Attacks

It is well known that naive double-and-add algorithms are subject to power attacks introduced by Kocher *et al* [10]. More precisely, they introduced two types of power attacks : Simple Power Analysis (SPA) and Differential Power Analysis (DPA) we now shortly remind.

3.1   Simple Power Analysis

The first type of attack consists in observing the power consumption in order to guess which instruction is executed. For example, in the previous algorithm, one can easily recover the exponent $d = \sum_{i=0}^{n} d_i 2^i$ by distinguishing the doubling from the addition instruction. To avoid this attack, downward double-and-add algorithm is usually modified using so-called "dummy" instructions (see figure 3).

$$
\begin{aligned}
&S[0] = 0 \\
&\text{for } i \text{ from } n \text{ down to } 0 \\
&\quad S[0] = 2.S[0] \\
&\quad S[1] = S[0] + P \\
&\quad S[0] = S[d_i] \\
&\text{return } S[0]
\end{aligned}
$$

Fig. 3. Downward double-and-add(P,d) resistant against SPA

Although this new algorithm is immune to SPA, a more sophisticated treatment of power consumption measures can still enable to recover the secret scalar d.

3.2 Differential Power Analysis

DPA uses power consumption to retrieve information on the operand of the instruction. More precisely, it no longer focuses on which instruction is executed but on the Hamming weight of the operands used by the instruction. Such an attack has been described in the elliptic curve setting in [3,14].

This technique can also be used in a different way. Messerges, Dabbish and Sloan introduced "Multiple Exponent Single Data" attack [11]. Note that, for our purpose, a better name would be "Multiple Scalar Single Data". We first assume that we have two identical equipments available with the same implementation of algorithm 3, one with an unknown scalar d and another one with a chosen scalar e. In order to discover the value of d, using correlation between power consumption and operand value, we can apply the following algorithm. We guess the bit d_n of d which is first used in the double-and-add algorithm and we set e_n to this guessed value. Then, we compare the power consumption of the two equipments doing the scalar multiplication of the same message. If the consumption is similar during the two first steps of the inner loop, it means that we have guessed the correct bit d_n. Otherwise, if the consumption differs in the second step, it means that the values are different and that we have guessed the wrong bit. So, after this measure, we know the most significant bit of d. Then, we can improve our knowledge on d by iterating this attack to find all bits as it is illustrated in the algorithm of figure 4.

This kind of attack is well known and some classical countermeasures are often implemented. For example, the Chaum's blinding technique [2] can be used to protect an RSA implementation since it prevents an attacker from knowing the data used in the exponentiation. This method cannot be applied directly for

```
for i from 0 to n
    e_i = 0
for i from 0 to n
    e_{n-i} = 1
    choose M randomly
    double-and-add(P,d) on equipment 1
    double-and-add(P,e) on equipment 2
    if no correlation at step (i + 1) e_{n-i} = 0
return e
```

Fig. 4. MESD attack to find secret scalar d

computations based on the Discrete Logarithm problem as there is no public exponent associated with the secret exponent, as in RSA.

4 Usual DPA Countermeasures for Double-and-Add Algorithm

The most well know countermeasures for scalar multiplication on elliptic curve have been published by Coron [3]. In this paper, the author describes three different countermeasures which are respectively based on the blinding of the scalar, on the blinding of the point or on the blinding of the multiplication. We now recall the description of the first two countermeasures. Then we explain, in section 5, how to defeat them.

4.1 Coron's First Countermeasure

During the computation of a scalar multiplication, this scalar can be blinded by adding a multiple of the number $\mathcal{E}$ of points of the curve. For this purpose, the algorithm needs a random value r which length is fixed to 20 bits in [3]. Then, the algorithm computes $(d + r\mathcal{E})P$ which is obviously equal to dP. This countermeasure, depicted in figure 5, is very efficient since the scalar value changes for each computation.

```
pick random value r
d' = d + rE
return double-and-add(P,d')
```

Fig. 5. Implementation 1 secure against DPA

4.2   Coron's Second Countermeasure

The second solution is based on the same idea as Chaum's blind RSA signature scheme. However, since we use a discrete log based problem, applying this method requires twice the time needed for a single scalar multiplication. To be more efficient, it is proposed in [3] to store a secret point R and the associated value $S' = dR$. The multiplication of P by d is performed by computing $d(P+R)$ and then subtracting S' to the result. The variability is obtained by doubling R and S' at each execution as shown in figure 6.

$$
\begin{aligned}
&\text{pick } b \in \{0, 1\} \text{ at random} \\
&R = (-1)^b.2.R \\
&S' = (-1)^b.2.S' \\
&S = \text{double-and-add}(P + R, d) \\
&\text{return } S - S'
\end{aligned}
$$

Fig. 6. Implementation 2 secure against DPA

In this paper we will not focus on the third countermeasure proposed by Coron since it has been partially broken by Goubin in [5]. Our attack does not work on this countermeasure and does not allow to enhance Goubin's attack.

These two countermeasures are well admitted to be efficient against power attacks. Other recently proposed countermeasures, such as randomized NAF [6] or [7,17,8] mainly focus on improving efficiency in terms of speed.

5   The New Attack

We introduce a new attack mainly based on two reasonable assumptions. This attack is able to recover the secret scalar with a few requests to the card. The adversary needs to send chosen messages directly to the double-and-add algorithm. Indeed, when considering decryption of the ElGamal cryptosystem or of the RSA cryptosystem for instance, the padding can only be verified at the end of the computation.

The idea of the attack is based on the fact that, even if an adversary is not able to tell which computation is done by the card, he can at least detect when the card does twice the same operation. More precisely, if the card computes $2.A$ and $2.B$, the attacker is not able to guess the value of A nor B but he is able to check if $A = B$. Such an assumption is reasonable since this kind of computation usually takes many clock cycles and depends greatly on the value of the operand. This assumption has been used in a stronger variant and validated by Schramm et al. in [16]. Indeed, they are able to distinguish collisions during one DES round computation which is much more difficult than distinguishing collisions during a doubling operation. If the noise is negligible, a simple comparison of the two power consumption curves during the doubling will be sufficient to detect this equality.

If the noise is more important we propose two solutions to detect equalities. The first and easiest one is to compare the average of several consumptions curves with A and B. However asking twice the same computation may be impossible. In that case, a better solution is to use the tiny differences on many clock cycles since a point doubling usually takes a few thousand cycles. By summing the square of the differences between these curves on each clock cycle, we can decrease the influence of noise. This approach is precised in appendix A.

5.1   Description of the Doubling Attack

The so-called *"doubling attack"* is based on the fact that similar intermediate values may be manipulated when working with points P and $2P$. However this idea only works when using the downward routine.

Let us first consider an example. Let $d = 78 = 64 + 8 + 4 + 2$, i.e. $n = 6$ and

$$(d_0, d_1, d_2, d_3, d_4, d_5, d_6) = (0, 1, 1, 1, 0, 0, 1)$$

Then we compare the sequence of operations when the downward binary scalar multiplication algorithm of figure 3 is used to compute $d \times P$ (on the left) and $d \times (2P)$ (on the right) :

i	d_i	comput. of dP	comput. of $d(2P)$
6	1	2×0	2×0
		$0 + P$	$0 + 2P$
5	0	$2 \times P$	$2 \times 2P$
		$2P + P$	$4P + 2P$
4	0	$2 \times 2P$	$2 \times 4P$
		$4P + P$	$8P + 2P$
3	1	$2 \times 4P$	$2 \times 8P$
		$8P + P$	$16P + 2P$
2	1	$2 \times 9P$	$2 \times 18P$
		$18P + P$	$36P + 2P$
1	1	$2 \times 19P$	$2 \times 38P$
		$38P + P$	$76P + 2P$
0	0	$2 \times 39P$	$2 \times 78P$
		$78P + P$	$156P + 2P$
		return $78P$	return $156P$

If we focus on the doubling operations, we notice that some of them manipulate the same operand. More precisely, we observe that the doubling operation at rank i in the computation of dP is the same as the doubling operation at rank $i - 1$ in the computation of $d(2P)$ if and only if $d_{i-1} = 0$. Consequently, all the bits (except the least significant one) can be deduced from the SPA analysis of only two power consumption curves, the first one obtained during the computation of dP and the second one with $d(2P)$.

More formally, let us denote the partial sums $S_k(P) = \sum_{i=0}^{i=k} d_{n-i} 2^{k-i} \times P$. This value is the content of $S[0]$ in the algorithm 3 after $k+1$ iterations. Therefore we also refer to $S_k(P)$ as an intermediate result of the binary scalar multipliation algorithm. So this value is used in the next doubling operation in any case. Besides

$$S_k(P) = \sum_{i=0}^{k} d_{n-i} 2^{k-i} \times P$$

$$= \sum_{i=0}^{k-1} d_{n-i} 2^{k-1-i} \times (2P) + d_{n-k} \times P$$

$$= S_{k-1}(2P) + d_{n-k} \times P$$

Thus the intermediate result of the downward double-and-add algorithm with P at step k will be equal to the intermediate result with $2P$ at step $k-1$ if and only if d_{n-k} is null.

Using the same example as before, we obtain :

value of step k	0	1	2	3	4	5	6
value of d_{n-k}	1	0	0	1	1	1	0
value of $S_k(P)$	P	$2P$	$4P$	$9P$	$19P$	$39P$	$78P$
value of $S_k(2P)$	$2P$	$4P$	$8P$	$18P$	$38P$	$78P$	$156P$

In conclusion, we just need to compare the doubling computation at step $k+1$ for P and at step k for $2P$ to recover the bit d_{n-k}. If both computations are identical, d_{n-k} is equal to 0 otherwise d_{n-k} is equal to 1. This can also be observed by shifting the second measurement curve by one step to the right and comparing it to the first curve. Therefore, with only two requests to the card, it is possible to recover all the bits of the secret scalar.

Note that this attack also works with addition-subtraction chains such as Non Adjacent Form representation [12]. It allows to recover all the zeros in the NAF coding which represent roughly two third of the bits according to the paper of Morain and Olivos [12]. The missing information can be recovered by exhaustive search or by a more efficient method such as an adaptation of the baby step giant step algorithm for short Diffie-Hellman exponent. Indeed, if the prime group order is a 160-bit prime number, then only 54 bits remain to be discovered. Moreover, the Baby Step Giant Step algorithm can be used to reduce the complexity of the discovery of the discrete logarithm to 2^{27} in time and in memory.

5.2   Application of the Doubling Attack to Coron's Countermeasures

The first countermeasure of Coron uses a 20-bit random value to blind the secret scalar at each request to the card. This size of random value is sufficient to resist usual DPA attacks. However it is not enough to resist our new doubling attack. Indeed, due to the birthday paradox, after 2^{10} requests with P and 2^{10} requests with $2P$, there should exist a common scalar with high probability. In order to recover this collision on the scalar, the attacker needs to compare each curve obtained with P to each curve obtained with $2P$. With the method described before, assuming the same scalar is used, it is possible to find the position of zeroes in this scalar. The right pair will be distinguished as it will give a scalar with enough zeroes. Indeed, if the corresponding scalars are different, it is unlikely that common intermediate values appear on both computations. Hence identifying the right pair is quite easy and requires only 2^{20} comparisons. In case several bits of the scalar cannot be clearly identified due to the noise, they can be recovered by exhaustive search. The attack is summarized in figure 7

---

while no correct pair is found
> request computation with P and store measurement $C(P)$ in set $\mathcal{A}$
> request computation with $2P$ and store measurement $C(2P)$ in set $\mathcal{B}$
> compare $C(P)$ with all set $\mathcal{B}$
> compare $C(2P)$ with all set $\mathcal{A}$
> if two measurements have many common intermediate squaring, a correct pair is found

exhaustive search for undefined bits of the scalar recovered with the correct pair.

Fig. 7. Attack of Coron's first countermeasure

If the number of undefined bits is too large, one can notice that the number of correct pairs increases as the square of the number of extra requests. With about 2^{15} requests in each group, the number of correct pairs will be approximately 2^{10} which may help to decrease the work of the exhaustive search.

The second countermeasure is even more vulnerable to the doubling attack. Indeed, the random value which blinds P is itself doubled at each execution. The attack goes as follows : a point P is first sent as the first request. Then the card executes its routine with the value $P + R$. The adversary then requests the computation with the point $2P$. With probability $\frac{1}{2}$, the card will use the point $2P + 2R = 2(P + R)$. So the attacker is then able to compare the two measurements and to recover the secret scalar. If the noise is too important, the adversary can use a statistical approach. He can choose a random point Q, send Q and $2Q$ to the card and make the difference between the first curve and the second one shifted by one step to the right. By summing the square of those differences, we can recover all the bits of the secret scalar. Indeed, when the bit at step i is equal to 0, half of the differences at step $i + 1$ are null. So the curve representing the sum of the differences will be flatter at positions corresponding to a zero than at positions corresponding to a one.

6 Conclusion

A new powerful attack on scalar multiplication and modular exponentiation has been presented which takes advantage of some implementation choices that were not considered as a security concern up to now. As regards to this attack, it appears that the "bad" choice was the most commonly used, due to efficiency criteria. This vulnerability considerably weakens usual countermeasures used to defeat power attacks.

Since no attack as efficient as the doubling attack is known on the upward double-and-add algorithm (from the least to the most significant bit), we recommend to use this routine to compute scalar multiplication combined with the appropriate countermeasures.

It is an open problem to study whether our attack and Goubin's attack can be combined in order to defeat the combination of Coron countermeasures.

References

1. M. Bellare and P. Rogaway. Optimal Asymetric Encryption - How to Encrypt with RSA. In *Eurocrypt '94*, LNCS 950, pages 92–111. Springer-Verlag, 1994.
2. D. Chaum. Blind Signatures for Untraceable Payments. In *Crypto '82*, pages 199–203. Plenum, NY, 1983.
3. J.S. Coron. Resistance against differential power analysis for elliptic curve. In *CHES '99*, LNCS 1717, pages 292–302. Springer-Verlag, 1999.
4. T. El Gamal. A Public Key Cryptosystem and a Signature Scheme Based on Discrete Logarithms. In *IEEE Transactions on Information Theory*, volume IT–31, no. 4, pages 469–472, july 1985.
5. L. Goubin. A Refined Power-Analysis Attack on Elliptic Curve Cryptosystems. In *PKC '2003*, LNCS 2567, pages 199–210. Springer-Verlag, 2003.
6. J. Ha and S. Moon. Randomized signed-scalar Multiplication of ECC to resist Power Attacks. In *Pre-Proceeding CHES'02*, pages 553–565. Springer Verlag, 2002.
7. K. Itoh, J. Yajima, M. Takenaka, and N. Torii. DPA Countermeasures by Improving the Window Method. In *Pre-Proceeding CHES'02*, pages 304–319. Springer Verlag, 2002.
8. T. Izu, B. Moller and T. Takagi. Improved Elliptic Curve Multiplication Methods Resistant against Side Channel Attacks. In *IndoCrypt '2002*, LNCS 2551, pages 296–313. Springer-Verlag, 2002.
9. P.C. Kocher. Timing Attacks on Implementations of Diffie-Hellman, RSA, DSS, and Others Systems. In *Crypto '96*, LNCS 1109, pages 104–113. Springer-Verlag, 1996.
10. P.C. Kocher, J. Jaffe, and B. Jun. Differential Power Analysis. In *Crypto '99*, LNCS 1666, pages 388–397. Springer-Verlag, 1999.
11. T. S. Messerges, E. A. Dabbish, and R. H. Sloan. Power Analysis Attack of Modular Exponentiation in Smartcards. In *CHES '99*, LNCS 1717. Springer-Verlag, 1999.
12. F. Morain and J. Olivos. Speeding up the computation on an elliptic curve using addition-substraction chains. *Inform Theory Appl.*, 24:531–543, 1990.
13. NIST. Digital Signature Standard (DSS). Federal Information Processing Standards PUBlication 186–2, february 2000.

14. K. Okeya and K. Sakurai. A second-Order DPA Attack Breaks a Window-Method Based Countermeasure against Side Channel Attacks. In *ISC '2002*, LNCS 2433, pages 389–401. Springer-Verlag, 2002.
15. R.L. Rivest, A. Shamir, and L.M. Adleman. A method for obtaining digital signatures and public-key cryptosystem. *Communications of the ACM*, 21(2):120–126, 1978.
16. K. Schramm, T. Wollinger, and C. Paar. A New Class of Collision Attacks and its Application to DES. In *FSE '2003*, LNCS, pages –. Springer-Verlag, 2003.
17. E. Trichina and A. Bellezza. Implementation of Elliptic Curve Cryptography with Built-in Counter Measures against Side Channel Attacks. In *Pre-Proceeding CHES'02*, pages 98–113. Springer Verlag, 2002.

A   Statistical Approach of Noise Reduction

Let c denotes the number of cycles of a point doubling operation. More precisely, we only consider the cycles that are data dependant, i.e., for which the power consumption differs when operands are changed. The power consumption (without noise) of the computation with A and B at cycle i are respectively named $C_A(i)$ and $C_B(i)$. The noise N can be modeled by random independent variables $N_A(i)$, $N_B(i)$ with mean μ and variance σ. The power consumption observed on cycle i are then equal to $C_A(i) + N_A(i)$ and $C_B(i) + N_B(i)$. The indicator I is defined as follow:

$$I = \frac{1}{c} \sum_{i=1}^{c} (C_A(i) + N_A(i) - C_B(i) - N_B(i))^2$$

$$= \frac{1}{c} \sum_{i=1}^{c} (C_A(i) - C_B(i))^2 + \frac{1}{c} \sum_{i=1}^{c} (N_A(i) - N_B(i))^2$$

$$+ \frac{2}{c} \sum_{i=1}^{c} (N_A(i) - N_B(i))(C_A(i) - C_B(i))$$

If c is large enough, we can evaluate the mean of the indicator when $A = B$ and $A \neq B$. In the first case, $I = \frac{1}{c} \sum_{i=1}^{c} (N_A(i) - N_B(i))^2$, which is a sum of assumed independent variables $Y(i) = (N_A(i) - N_B(i))^2$. The mean of $Y(i)$ is

$$\begin{aligned} E(Y) &= E((N_A(i) - N_B(i))^2) \\ &= Var((N_A(i) - N_B(i)) - [E(N_A(i) - N_B(i))]^2 \\ &= Var(N_A(i)) + Var(N_B(i)) - 0 = 2\sigma^2 \end{aligned}$$

So the mean of the indicator, if $A = B$, is $E(I(A = B)) = 2\sigma^2$.

In the second case ($A \neq B$), assuming that

$$\forall i \leq c, \varepsilon_1 \leq |C_A(i) - C_B(i)| \leq \varepsilon_2$$

the mean of the indicator can be bounded as follow

$$2\sigma^2 + \varepsilon_1^2 \leq E(I(A \neq B)) \leq 2\sigma^2 + \varepsilon_2^2$$

With this bound in mind, it appears that the indicator can be used to distinguish if the manipulated data are equal or not. The confidence on this indicator relies on its variance and the number of clock cycles c to perform the operation.

An Analysis of Goubin's Refined Power Analysis Attack

Nigel P. Smart

Dept. Computer Science,
University of Bristol,
Merchant Venturers Building,
Woodland Road,
Bristol, BS8 1UB,
United Kingdom.
nigel@cs.bris.ac.uk

Abstract. Power analysis attacks on elliptic curve based systems work by analysing the point multiplication algorithm. Recently Goubin observed that if an attacker can choose the point P to enter into the point multiplication algorithm then none of the standard three randomizations can fully defend against a DPA attack. In this paper we examine Goubin's attack in more detail and completely discount its effectiveness when the attacker chooses a point of finite order, for the remaining cases we propose a defence based on using isogenies of small degree.

1 Introduction

Elliptic curves were first introduced into cryptography by Koblitz [9] and Miller [14] in 1985. Since that time, due to their perceived advantages in bandwidth and required computing resources, there has been increasing interest in using them in low-cost cryptographically enabled devices such as smart cards.

Smart cards are a particularly interesting environment due to the ability for the attacker to mount side-channel attacks based on, for example, power analysis [10] and [11]. The idea behind these attacks is to measure the power consumption of the card and use this to derive information about the underlying secret key contained in the card. Such power attacks come in two variants; Simple Power Analysis, or SPA, uses only a single observation of the power to obtain information. Differential Power Analysis, or DPA, makes many measurements and then uses a statistical technique to deduce information about the underlying secret.

In the context of elliptic curve cryptography, power analysis is applied to determine the multiplier used in a point multiplication. In other words for public $P \in E(K)$ and a private $d \in \mathbb{Z}$ one uses power analysis to determine the value of d from the power consumed in computing

$$Q = [d]P.$$

C.D. Walter et al. (Eds.): CHES 2003, LNCS 2779, pp. 281–290, 2003.

Since DPA requires multiple measurements this means one can only apply DPA to protocols in which one applies the same private multiplier d over multiple protocol runs, with possibly different values of P in each protocol run. Hence DPA can not be applied to ECDSA, two pass ECDH or two pass ECMQV. It can however be applied to ECIES, single pass ECDH or single pass ECMQV, where one of the "ephemeral" Diffie-Hellman/MQV keys is kept constant (i.e. it is a long term static public key). On the other hand, SPA can be applied to any algorithm in which one needs to keep the multiplier secret. Hence, SPA applies to all elliptic curve protocols.

A number of ways of defending against SPA have been proposed in the literature, for example the "double and add always" method of Coron [5], or the use of the Montgomery form [15] which helps prevent both SPA and timing attacks, [16], [17]. These defences try to prevent information leaking because of the different power profile of the addition and doubling formulae for the elliptic curve.

An approach attracting increasing interest is to use group formulae which are identical for both addition and doubling. This idea was introduced in the context of the Jacobi form of an elliptic curve by Liardet and Smart [12]. This was extended to cover the Hessian form of a curve, see [7] and [19],

$$x^3 + y^3 + 1 = Dxy.$$

Note that the Hessian form curves are particularly efficient in characteristic three [20], yet this advantage can only be exploited at the expense of having different routines for addition and doubling. Finally Brier and Joye have given a single formula for both the addition and doubling law for elliptic curves in standard form [4]. To recap, the standard form for a curve in characteristic two is given by

$$y^2 + xy = x^3 + ax^2 + b,$$

whilst in large prime characteristic it is given by

$$y^2 = x^3 + ax + b.$$

For efficiency reasons it is common to select $a = 1$ in characteristic two and $a = -3$ in large prime characteristic, see [3] for the reasons, and many curves recommended, or mandated, in standards documents satisfy these extra conditions on a.

Yet SPA defences are not enough to prevent DPA attacks. Coron [5] proposed three possible DPA defences namely; randomizing the secret exponent d, adding random points to P to randomize the base point, using a randomized projective representation. Only the third of these can be done with minimal cost, whilst the other two are not as effective and add additional computational costs into the point multiplication algorithm. In a similar vein to randomized projective coordinates Joye and Tymen introduced two other cheap randomizations, namely random curve isomorphisms and random field isomorphisms [8].

So a combined approach of using indistinguishable group laws and a randomization of (at least one of) the projective point representation, the curve representation or the field representation; would appear to offer a defence against power analysis for elliptic curve systems. However, recently Goubin [6] observed that if the attacker can choose the point P to enter into the point multiplication algorithm then none of these three randomizations can fully defend against a DPA attack.

In this paper we examine Goubin's attack in more detail and discount its effectiveness in a large number of cases, for the remaining cases we propose a defence based on using isogenies of small degree. The paper is organized as follows: In Section 2 we describe Goubin's attack and his notion of "Special Points" and examine the three anti-DPA randomizations mentioned above. We divide the special points into two types, those of small order and those of large order. Then in Section 3 we explain how careful implementation of existing standards definitions means we need not worry about special points of small order. Then in Section 4 we recap on various aspects of isogenies on elliptic curves over finite fields. We then use these isogenies to propose a defence for special points of large order in Section 5. In addition we examine whether our proposed defence works for the elliptic curves recommended or mandated in various standards. Finally we end in Section 7 with some conclusions.

2 "Special Points"

Before presenting Goubin's refined power analysis attack we present the three standard randomized DPA defences mentioned in the introduction.

Let $C(X, Y, Z)$ denote a projective representation of the affine elliptic curve we are using in our cryptosystem, whose affine form we shall assume is monic in Y. There is a map from affine coordinates to projective coordinates

$$(x, y) \longmapsto (x, y, 1)$$

and a similar reverse one

$$(X, Y, Z) \longmapsto (X/Z^s, Y/Z^t)$$

where s and t are the "weights" of the projective representation. Note: As above, we shall use lower case letters to denote variables on the affine form of the curve and capital letters for the projective form.

The three proposed randomization defences against DPA are as follows:

2.1 Randomized Projective Coordinates

Here one takes the affine point $P = (x, y)$ and before we apply d to it we first map it into a projective representation, using a random $r \in K^*$,

$$(x, y) \longmapsto (xr^s, yr^t, r).$$

One then performs the point multiplication in projective coordinates. Since multiple runs of the protocol will result in different values of r we see that each run will be uncorrelated with other runs and so a DPA attack seems impossible to mount.

2.2   Randomized Curve Isomorphism

Here we have $P \in C$ given to us and we then define $P' = (r^s x, r^t y)$ for some random $r \in K^*$. We then consider P' as a point on C' where if C is given by

$$C = \sum a_{i,j}\, x^i y^j$$

then C' is given by

$$C' = \theta^v \sum a'_{i,j}\, x^i y^j$$

with

$$a'_{i,j} = a_{i,j} \cdot r^{-(si+jt)},$$

and v chosen so as to make C' monic in the y. The curves C and C' are isomorphic. In our cryptographic operation we now compute $Q' = (X', Y') = [d]P'$ on C' and then map this back to C via $Q = (X, Y) = (X'/r^s, Y'/r^t)$.

2.3   Randomized Field Isomorphism

Here we take $P \in C$ and apply a random field isomorphism $\kappa : K \to K'$ to both P and C so as to obtain $P' = \kappa(P)$ and $C' = \kappa(C)$. We then compute $Q' = [d]P$ and then compute $Q = \kappa^{-1}(Q')$.

Goubin [6] defines a special point $P = (x, y) \in C$ to be one in which either $x = 0$ or $y = 0$. Goubin's attack works by feeding suitable multiples P', depending on ones guess for a given bit of d, of a special point into the smart card. Then when the smart cards computes $[d]P'$, the special point will occur within the computation assuming the guess is correct. The existence of the special point will be picked up with a DPA trace since the property of being a special point is preserved under the three randomizations above.

Elliptic curves in cryptography are usually chosen to have order

$$\#E(K) = h \cdot q$$

where q is a large prime and h is a small integer called the cofactor. In practice one usually has h chosen from the set $\{1, 2, 3, 4, 6\}$. The values of h correspond to the orders of the small subgroups of $E(K)$. We say that a special point has small order if it has order dividing h, otherwise we say it has large order.

In Table 1 we examine the various cases for different curves. A "?" in the Order column denotes that the point could be one of large order, whilst a "?" in the characteristic column means the curve can be used in any characteristic.

Table 1. Table of special points on Various Elliptic Curves

Curve Equation	Char	Special Point	Order
$y^2 + xy = x^3 + ax^2 + b$	2	$(0, \theta)$	2
$y^2 + xy = x^3 + ax^2 + b$	2	$(\theta, 0)$	?
$y^2 = x^3 + ax + b$	> 3	$(\theta, 0)$	2
$y^2 = x^3 + ax + b$	> 3	$(0, \theta)$	?
$x^3 + y^3 + 1 = Dxy$	?	$(\theta, 0)$	3
$x^3 + y^3 + 1 = Dxy$	?	$(0, \theta)$	3

3   "Special Points" of Small Order

Special points of small order can be dealt with by careful implementation of the protocols used in elliptic curves. Note, that since Goubin's attack is a DPA attack we need only consider protocols in which the same secret multiplier is used on multiple runs of the protocol. Hence, we only need to consider protocols such as ECIES, single-pass ECDH and single-pass ECMQV. To deal with small subgroup attacks various standards for these protocols make use of the co-factor h as a final postprocessing step before any point multiple is used in a protocol, see [1] or [2].

For example in the one-pass Diffie–Hellman protocol; if Alice has the long term key a and Bob sends her the ephemeral public key P, then Alice will compute $Q = [a]P$ followed by the (optional) postprocessing of $[h]Q$. If the cofactor is used then one calls the protocol cofactor-Diffie–Hellman. It is step of computing $Q = [a]P$ which is used by Goubin in his power analysis attack, by sending Alice a special point P of small order.

If we insist on implementors using the cofactor variant of Diffie–Hellman then we can avoid Goubin's attack by simply reversing the order of multiplication by a and h. In other words Alice first computes $Q = [h]P$ and then computes the shared secret via $[a]Q$, if and only if $Q \neq \mathcal{O}$. Goubin's attack then no longer applies since only genuine points in the subgroup of order q are passed into the point multiplication routine with the secret exponent a.

Similar arguments, involving insisting on cofactor variants of all protocols and reversing the order of multiplication by the cofactor and the secret key, can be applied to ECMQV and ECIES as defined in [1] and [2].

Recently Shoup has proposed a new variant of ECIES [18] for inclusion in a draft ISO standard. This new variant processes the cofactor in a completely different way to the old version of ECIES. A quick look at the new ECIES reveals that the new version processes the cofactor before the secret key multiplication as we recommend, hence the new version is already protected against Goubin's attack for special points of small order.

4   Recap on Isogenies

We recap, for use in the next section, some basics on isogenies between elliptic curves. All of what we require can be found in [3].

Let E_1 and E_2 denote elliptic curves over a finite field K of characteristic p. An isogeny

$$\psi : E_1 \longrightarrow E_2$$

is a non-constant rational map which respects the group structure of E_1 and E_2, i.e.

$$\psi(P + Q) = \psi(P) + \psi(Q).$$

Every isogeny has a finite kernel and the size of this kernel is called the degree of the isogeny. If E_1 and E_2 are isogenous then we have that $\#E_1(K) = \#E_2(K)$.

If j_1 and j_2 are the j-invariants of the two curves then an isogeny of prime degree l exists (over the base field) if and only if j_1 and j_2 are a solution of the modular equation of degree l, i.e.

$$\Phi_l(j_1, j_2) = 0.$$

The equation $\Phi_l(X, Y) = 0$ defines the modular curve $X_0(l)$ which parametrizes all elliptic curves for which there is a degree l isogeny between them.

These modular equations $\Phi_l(X, Y)$ grow large quite quickly as one increases l. This has led to the introduction of more suitable modular curves (and hence equations) for larger values of l (say $l > 41$). But, since in our application we are only interested in small values of l, we will not consider these generalized modular equations and so will restrict ourselves to the standard modular curves.

As a subprocedure of the Schoof-Elkies-Atkin algorithm [3][Chapter VII] one takes an elliptic curve E_1 and then determines whether there is an elliptic curve E_2 which is l-isogenous to E_1. This is done by solving the modular equation

$$\Phi_l(X, j_1) = 0$$

over the field K. One can then determine E_2 and then, via a rather involved procedure, determine the mapping ψ. See [3][Chapter VII] for the precise algorithm for determining E_2 and ψ.

The mapping ψ for a degree l isogeny is of the form

$$\psi : \begin{cases} E_1 & \longrightarrow & E_2 \\ (x, y) & \longmapsto & \left(\frac{f_1(x)}{g(x)^2}, \frac{y \cdot f_2(x)}{g(x)^3} \right) \end{cases}$$

where f_1, f_2 and g are polynomials over K of degree $2d + 1$, $3d + 1$ and d, for $d = (l - 1)/2$ respectively.

5   "Special Points" of Large Order

In characteristic two the special points $(\theta, 0)$ of large order can easily be defended against by use of the Montgomery ladder [13], since in that case the y-coordinate is not used. Hence, we will restrict our discussion to large prime characteristic and to curves in Weierstrass form, since these are more important for applications.

Special points of large order have been shown to exist by Goubin on a large number of the curves of large prime characteristic defined in standards. The existence of special points of large order is due to the equation

$$y^2 = x^3 + ax + b$$

being such that b is a square in $\mathbb{F}_p^*$. We propose to manage the problem of special points of large order by transferring the cryptographic protocol over to an isomorphic group (but not an isomorphic curve) via an isogeny

$$\psi : E_1 \longrightarrow E_2.$$

Note, the curve E_2 and the isogeny we will use are all defined over the base field $\mathbb{F}_p$. For each curve in the standards, which exhibits a special point, we then need to determine a fixed (low degree) isogeny to an elliptic curve which does not exhibit a special point.

In Table 2 we list all recommended/mandated curves over fields of large prime characteristic in the main standards. For each curve we list the minimal degree of an isogeny to a curve which does not exhibit a special point. If the original curve does not exhibit a special point then we specify this degree as one. We also list the degree of the minimal isogeny to a curve one would prefer, i.e. a curve which has a particularly efficient model for computational purposes, by this we mean in odd characteristic a model of the form

$$y^2 = x^3 - 3x + b.$$

If the curve has minimal isogeny degree one and if the original curve was not of the above special form then we do not give a figure for the preferred minimal degree.

The data in Table 2 was computed via the Magma computer algebra system. Given a curve and the minimal isogeny degree it is relatively straightforward, see [3][Chapter VII], to compute the equation of the isogenous curve and the isogeny itself using Vélu's formulae [21]. Indeed Magma will compute this isogeny for you if required.

Hence, all that the smart card need do to protect against special points of large order is to store along with the original curve from the standard, the equation of the isogenous curve and the equation of the isogeny and its inverse. Then input points can be mapped over to the isogenous curve for computation and then mapped back again to the original curve for further processing. Clearly all the standard defences such as randomized projective representation and randomized curve isomorphisms can then be applied to the computation on the isogenous curve.

Table 2. Minimal Isogeny Degree Needed to Remove a Special Point

Curve Name	Minimal Isogeny Degree	Preferred Minimal Degree
secp112r1	1	1
secp112r2	11	11
secp128r1	7	7
secp128r2	1	-
secp160k1	1	-
secp160r1	13	13
secp160r2	19	41
secp192k1	1	-
secp192r1	23	73
secp224k1	1	-
secp224r1	1	-
secp256k1	1	-
secp256r1	3	11
secp384r1	19	19
secp521r1	5	5

6   Relative Cost of the Isogeny Defence

To apply the isogeny defence it would be better to alter the standards so that the curves are replaced with isogenous ones. However, since this is unlikely to be an option the smart card needs to convert the input point to the isogenous curve. If we assume the isogeny is of degree l this means we need to evaluate three polynomials of degree $2d + 1$, $3d + 1$ and d, where $d = (l - 1)/2$. Using Horner's rule this implies a maximum number of field multiplications of

$$(2d + 1) + (3d + 1) + d = 6d + 2 \approx 3l.$$

Of course one problem is to actually store the coefficients of the polynomials defining the isogenies, which could be a problem in a device with limited memory.

In [5] Coron mentions other defences against DPA which could be used to thwart Goubin's attack. We discuss each of these in turn:

Randomization of the Private Exponent: Here one sets $d' = d + k \cdot \#E(\mathbb{F}_p)$ for some random 20-bit (say) value k. One then computes $Q = [d']P$. On average this will require an additional

$$20M_D + 10M_A$$

field multiplications, where M_D is the number of field multiplications required in an elliptic curve doubling operation and M_A is the number of field multiplications required in an elliptic curve addition operation. Typically, with projective coordinates, we have $M_D = 8$ and $M_A = 16$. Hence, one requires on average 320 extra field multiplications, which becomes more efficient than the isogeny method as soon as $l > 106$. Whilst this method appears to be slower than the isogeny method, it should be noted however that randomizing the private exponent is easier to implement than the isogeny method.

Point Blinding: Here one first computes $S = [d]R$, for some random value of R of large order. Then at each request for the calculation of $Q = [d]P$ one computes

$$R = (-1)^b 2R, \; S = (-1)^b 2S$$

and

$$Q = [d](P + R) - S.$$

Hence, at each iteration one needs to perform two extra point additions and two extra point doublings. This corresponds to a typical cost of 48 field multiplications, which is more efficient than the isogeny method when $l > 16$.

As we have already mentioned, Coron also proposed the use of randomized projective coordinates. This does protect against other forms of DPA, but not Goubin's attack. Hence, combining randomized projective coordinates and the isogeny method one could achieve an efficient defence against all known forms of DPA against an elliptic curve implementation.

7   Conclusion

We have shown why Goubin's refined power analysis attack can be discounted for many elliptic curve systems either by use of the cofactor variant of many protocols or the use of isogenies.

The author would like to thank Marc Joye for useful comments on an earlier draft of this paper. The observation that the Montgomery ladder removes the need to consider special points of large order in characteristic two is due to him.

References

1. ANSI X9.63, Public Key Cryptography for the Financial Services Industry: Elliptic Curve Key Agreement and Transport Protocols. *American National Standards Institute*, Draft 2001.
2. SECG SEC 1: Elliptic Curve Cryptography. *Standards for Efficient Cryptography Group*, 1999.
3. I.F. Blake, G. Seroussi and N.P. Smart. *Elliptic Curves in Cryptography*. Cambridge University Press, 1999.
4. E. Brier and M. Joye. Weierstrass elliptic curves and side-channel analysis. In *Public Key Cryptography – PKC 2002*, Springer-Verlag LNCS 2274, 335–345, 2002.
5. J.-S. Coron. Resistance against differential power analysis for elliptic curve cryptosystems. In *Cryptographic Hardware and Embedded Systems – CHES '99*, Springer-Verlag LNCS 1717, 292–302, 1999.

6. L. Goubin. A refined power analysis attack on elliptic curve cryptosystems. In *Public Key Cryptography – PKC 2003*, Springer-Verlag LNCS 2567, 199–211, 2003.

7. M. Joye and J.-J. Quisquater. Hessian elliptic curves and side-channel attacks. In *Cryptographic Hardware and Embedded Systems – CHES 2001*, Springer-Verlag LNCS 2162, 412–420, 2001.

8. M. Joye and C. Tymen. Protections against differential attacks for elliptic curve cryptography – An algebraic approach. In *Cryptographic Hardware and Embedded Systems – CHES 2001*, Springer-Verlag LNCS 2162, 377–390, 2001.

9. N. Koblitz. Elliptic curve cryptosystems. *Math. Comp.*, **48**, 203–209, 1987.

10. P. Kocher, J. Jaffe and B. Jun. Introduction to differential power analysis and related attacks. Technical Report, Cryptography Research Inc., 1998.

11. P. Kocher, J. Jaffe and B. Jun. Differential power analysis. In *Advances in Cryptology – CRYPTO '99*, Springer-Verlag LNCS 1666, 388–397, 1999.

12. P.-Y. Liardet and N.P. Smart. Preventing SPA/DPA in ECC systems using the Jacobi form. In *Cryptographic Hardware and Embedded Systems – CHES 2001*, Springer-Verlag LNCS 2162, 401–411, 2001.

13. J. López and R. Dahab. Fast Multiplication on Elliptic Curves over $GF(2^m)$ without Precomputation In *Cryptographic Hardware and Embedded Systems – CHES 1999*, Springer-Verlag LNCS 1717, 316–327, 1999.

14. V. Miller. Uses of elliptic curves in cryptography. In *Advances in Cryptology – CRYPTO '85*, Springer-Verlag LNCS 218, 417–426, 1986.

15. P.L. Montgomery. Speeding the Pollard and Elliptic Curve Methods for factorization. *Math. Comp.*, **48**, 243–264, 1987.

16. K. Okeya, H. Kurumatani and K. Sakurai. Elliptic curve with Montgomery form and their cryptographic applications. In *Public Key Cryptography – PKC 2000*, Springer-Verlag LNCS 1751, 238–257, 2000.

17. K. Okeya and K. Sakurai. Power analysis breaks elliptic cryptosystem even secure against timing attack. In *INDOCRYPT 2000*, Springer-Verlag LNCS 1977, 178–190, 2000.

18. V. Shoup. A proposal for an ISO standard for public key encryption, v2.1. Preprint, 2001.

19. N.P. Smart. The Hessian form of an elliptic curve. In *Cryptographic Hardware and Embedded Systems – CHES 2001*, Springer-Verlag LNCS 2162, 118–125, 2001.

20. N.P. Smart and E.J. Westwood. Point Multiplication on Ordinary Elliptic Curves over Fields of Characteristic Three *Applicable Algebra in Engineering, Communication and Computing*, **13**, 485–497, 2003.

21. J. Vélu. Isogégnies entre courbes elliptiques. *Comptes Rendus l'Acad. Sci. Paris, Ser A*, **273**, 238–241, 1971.

A New Type of Timing Attack: Application to GPS

Julien Cathalo, François Koeune, and Jean-Jacques Quisquater

Université catholique de Louvain
Place du Levant 3
1348 Louvain-la-Neuve, Belgium
{cathalo,fkoeune,jjq}@dice.ucl.ac.be

Abstract. We investigate side-channel attacks where the attacker only needs the Hamming weights of several secret exponents to guess a long-term secret. Such weights can often be recovered by SPA, EMA, or simply timing attack. We apply this principle to propose a timing attack on the GPS identification scheme. We consider implementations of GPS where the running time of the exponentiation (commitment phase) leaks the exponent's Hamming weight, which is typical of a square and multiply algorithm for example. We show that only 800 time measures allow the attacker to find the private key in a few seconds on a PC with a success probability of 80%. Besides its efficiency, two other interesting points in our attack are its resistance to some classical countermeasures against timing attacks, and the fact that it works whether the Chinese Remainder Technique is used or not.

Keywords: Side-Channel Attacks, Timing Attacks, GPS, Identification Schemes

1 Introduction

Timing attacks are certainly less powerful than power or electromagnetic analysis. On the other hand, the very limited equipment they require and the simplicity of the measurements make them much easier to deploy, even for a non-skillful adversary with very limited resources. Moreover, there are situations in which power consumption or electromagnetic radiations cannot be measured while the running time may be obtained, for example by measuring the delay between question and answer [3].

GPS is an identification scheme initially proposed by Girault [7]. It was recently selected in Nessie's portfolio of cryptographic primitives [4]. This protocol was designed for smart cards, allowing fast identification even on low-cost processors. The security of GPS was proven in [11]. The scheme is complete, sound (under the hypothesis that computing discrete logarithms with small exponents modulo $n = pq$ is hard), and statistically zero-knowledge. Careless implementations might nevertheless be subject to side-channel attacks.

Timing attacks against exponentiation schemes have been known for several years, and various countermeasures have been proposed against them. However,

C.D. Walter et al. (Eds.): CHES 2003, LNCS 2779, pp. 291–303, 2003.

since some of these countermeasures are precisely based on randomizing the exponent, one may feel tempted to believe that an algorithm where the exponent is random by nature, and where a different exponent is used for each execution may not be subject to a timing attack. We show in this paper that this is not true.

We propose a timing attack on GPS allowing recovering the prover's private key provided the exponentiation's running time is dependent (in our example, linear) in the exponent's Hamming weight. In our scenario, the attacker impersonates the verifier, and is able to measure precisely the computation time for the commitment step. Apart from this, the attacker has no knowledge of the implementation, such as multiplication algorithm, time needed for an individual multiplication, . . .

To our knowledge, none of the previously known timing attacks needed only the leakage of the exponent's Hamming weight ([5] and [12], for example, were based on modular reductions occurring in Montgomery multiplications). As a consequence, several of the countermeasures proposed against timing attacks turn out to be useless in our context.

Lastly, previous timing attacks required big amounts of data to work. Our attack only needs 600 timings to obtain the key with a probability of success of 72% after a few hours of treatment on a single PC, while 1000 timings allow 89% success after a few seconds of treatment. Collecting such a number of timings is easy (we recall that [8] states that the total time of the commitment followed by the answer takes less than 100 milliseconds with a crypto-processor card).

Notations. For any integer b, we denote by $|b|$ its size in bits, and by b_j the j-th bit of b, starting from the least significant bit, i.e. $b = b_{|b|-1} \ldots b_0$. For an interval I, $b \in_u I$ means that b is chosen at random in I. $l \oplus m$ denotes the addition $l + m$ modulo 2.

2   The GPS Identification Scheme

This section briefly reminds the principle of GPS (see [2] for more details) then focuses on the commitment step.

2.1   Short Description of GPS

An authority generates two strong primes p and q and computes $n = pq$. It also chooses an integer g. GPS authors [8] recommend the value $g = 2$, as it fits the security requirements while allowing good performance. Note that this choice has no effect on our attack.

Three parameters A, B and S are needed. We recall their minimum sizes as recommended in [8].

- S is the size of the (long-term) private key, $|S| \geq 160$
- B is the size of the challenge, $|B| = 16, 32$ or 64

– A is the size of the ephemeral keys, $|A| = |B|+|S|+64$ or $|A| = |B|+|S|+80$

We set $E = A + (B - 1)(S - 1)$.
Two cases may occur:

– Each user has his own modulus. In this case, $|n|$ should be at least 1024, and the factors of n, p and q, can be revealed to the prover, allowing her to use the Chinese Remainder Technique to speed-up the commitment phase.
– Several users share the same modulus n. In that case, $|n|$ should be at least 2048.

The prover's private key is a random element $x \in [0, S[$. Her public key is $X = g^{-x} \bmod n$. An identification with GPS proceeds as follows:

1. *Commitment:* The prover picks a random $y \in [0, A[$. She computes $Y = g^y \bmod n$ and sends Y to the verifier.
2. *Challenge:* The verifier sends a random integer $c \in [0, B[$ to the prover.
3. *Response:* The prover checks that $c \in [0, B[$ and computes $z = y + cx$. She sends z to the prover.
4. *Verification:* The verifier checks that $z \in [0, E[$ and $g^z X^c \equiv Y \pmod{n}$.

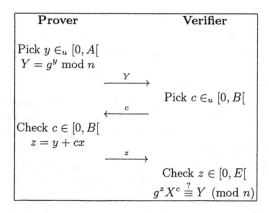

Fig. 1. A round of GPS

2.2   About the Commitment Step

Two methods are possible to manage the commitment issue. The first one consists in pre-computing commitments outside the card and then storing them in the card, using several tricks to save memory space (see [2] for details). This solution is interesting since it can be implemented on a card without crypto-processor, but it becomes a problem for applications that require a lot of identifications to be performed (such as pay-TV, for example): the card has to be reloaded with fresh commitments in a secure way periodically, or can only perform a limited number of identifications (an optimized version with 6000 precomputed commitments needs about 36 kb of memory space).

The second method consists in making the card compute the commitments itself. This solution allows it to perform enough identifications for any application. Two variants of this method are possible: the computation can be done online, i.e. during the identification, or offline. The online variant requires a crypto-processor. In that case, the commitment can be computed in less that 100 milliseconds [8]. In the offline variant, the card computes the commitment before the execution of the identifcation itself. Speed is not an issue anymore, thus no crypto-processor is needed; however, the card needs to be supplied with current – and therefore, for classical smart cards, to be connected to a reader – during this computation.

3   Context and Attack Overview

In this section, we introduce the context of our attack, then draw its main principles.

3.1   Scenario

We assume that the attacker is able to accurately measure the computation time of the commitment. We furthermore assume that this computation time is roughly linear in the Hamming weight of the exponent. This section briefly discusses the realism of these assumptions.

First of all, this will of course only be possible if the commitments are computed by the card, either online or offline. In the offline case, running times may be difficult to obtain by direct measurements, but indirect methods (e.g. based on power consumption) are still possible.

We also assume that the attacker is able to impersonate a honest verifier, which is a weak assumption: as resistance to dishonest verifier attacks is a natural requirement for an identification scheme, there is usually no need to bother about the verifier's identity. In its core version, GPS does not perform any verification step on the verifier before entering the commitment phase.

As far as accuracy of the measurement is concerned, we believe our assumption to be realistic, for example in the context of a smart card (a typical target for side-channel attacks) since smart cards are usually not equipped with an internal clock, but have their clock signal provided by the reader they are put in; in this context, accurate running time is easy to obtain with a rogue reader. To resist side-channel attacks, some modern smart cards get equipped with an internal clock. However, our attack seems to be robust against small imprecision in running time, and could therefore be carried out simply by measuring the elapsed time between startup and commitment.

Finally, the running time will clearly be a linear function of the Hamming weight of the exponent in the case of a classical square and multiply algorithm.

Other methods, such as sliding windows or Walter's division chains [15], are discussed in section 6.

Remark: For the sake of simplicity, we will assume in this description that A is a power of 2 (in fact, it is likely that many implementations choose this value, because it makes generation of random elements in $[0, A[$ easier). Taking this value allows us to use a very simple formula linking the Hamming weight of the ephemeral key y and the probability for one of its bits to be 0. However, a different A makes computations more fastidious, but has no effect on the attack's efficiency.

3.2   Test Platform

To validate our attack, we performed timing measurements using a smart card development kit [10,1]. Although not strictly practical, this scenario should be very close to the reality. Since the emulator is designed to allow implementors to optimize their code before "burning" actual smart cards, its predictions almost perfectly match the smart card's behaviour. We therefore believe physical attack of an actual smart card should not induce much more measurement errors than the ones we encountered here.

3.3   Attack Overview

The core principle of the attack goes as follows: suppose the attacker has obtained $w(y)$, the Hamming weight of some y generated by the prover. Impersonating the verifier and sending the challenge $c = 1$, he has also obtained the prover's response, $z = y + x$.

He starts by attacking the least significant bit of the secret, x_0. If the Hamming weight of y is smaller than half of y's bit length, he may assume that the least significant bit of y is more likely to be equal to 0 than to 1. More precisely, he assumes that the probability $P_0 = P(y_0 = 0)$ is equal to $1 - \frac{w(y)}{|A|}$.

Therefrom, and knowing the value of z, he can easily deduce a probability for the least significant bit of x to be zero.

Of course, this single guess has a non-negligible chance to be wrong. However, if the attacker is able to repeat this experiment over several identifications and compute the average of these probabilities, then this error risk shrinks.

Once he has obtained x_0, the attacker can deduce the actual value of the least significant bit $y_0^{(i)}$ of every message of his sample, which will allow him to attack the second bit x_1. Note that it is not necessary to collect new sets of samples.

The next section takes a deeper look at the attack.

4   How to Recover the Key: Full Details

The attack proceeds in three phases: time measurements, computation of a candidate for the key, key recovering.

4.1   Phase I: Time Measurements

In phase I, the attacker impersonates the honest verifier k times, always sending the same challenge[1] $c = 1$, and collects a list of couples $(t^{(i)}, z^{(i)}), i = 1..k$, where $t^{(i)}$ is the computation time for the commitment of the i-th interaction, and $z^{(i)}$ is the corresponding answer $(z^{(i)} = y^{(i)} + x)$.

Assuming the exponentiation's running time is roughly a linear function of the Hamming weight, i.e.

$$t^{(i)} = \alpha \times w(y^{(i)}) + \beta + \epsilon^{(i)},$$

with unknown parameters α and β ($\epsilon^{(i)}$ represents the error), the attacker estimates the Hamming weights of the set of ephemeral keys.

Figure 2 gives a clear idea of how parameters α and β can be estimated. The left graph shows the sorted running times of 80 exponentiations performed on our test platform (with the following parameters: $g = 2$, $|n|=1024$, $A = 2^{240}$ and without CRT) and the right graph shows the corresponding Hamming weights. Linear regression techniques on the left graph immediately provide good estimates for α and β.

These figures might deserve a bit more attention. As we can see, running times are grouped by "steps", and the attacker can safely assume that these "steps" correspond to several exponents having the same Hamming weight, whereas the average height between two "steps" is the time for a modular multiplication (i.e. α). The right graph shows that this estimate is pretty accurate: actual Hamming weight is rarely further than 2 from its estimate.

Remarks:

- We purposely took a small number of samples in order not to overload the figures. In practice, using a much larger set (say, 1000 samples) will of course reduce the error.
- We emphasize that this estimation is possible *whether the Chinese Remainder Technique is used or not*. Without CRT, the exponentiation time leaks information on $w(y)$. But with CRT, the card first computes $g^y \bmod p$ then $g^y \bmod q$ (since $y \ll p, q$, we have $y = y \bmod (p - 1) = y \bmod (q - 1)$), and the timing leaks information on $2 \times w(y)$.
- Smart card implementations of GPS probably require a constant time between the insertion of the card in the reader and the beginning of the computation of the commitment, meaning that the attacker gets a list $(t^{(i)} + R)$, where R is constant and unknown to the attacker, instead of a list $(t^{(i)})$. Thanks to the linear regression, this has no effect on the attack's efficiency.

---

[1] Always sending $c = 1$ improves both the precision and the simplicity of the attack, but this is not a necessary condition. For example, the principle of the attack also works when the verifier always sends powers of two as challenges: the pair $(w(y), z = y + 2^v x)$ will leak information on bits $x_{|S|-1} \ldots x_v$ of x.

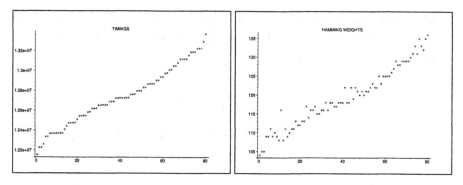

Fig. 2. Sorted timings for 80 samples and corresponding Hamming weights

4.2   Phase II: Computing a Candidate for the Key

In phase II, the attacker uses the collected data $(w^{(i)}, z^{(i)})$ to recover a candidate $\bar{x}$ for the private key.

Remark: Our procedure obviously works independently from the way the attacker obtained the Hamming weights. Although we put ourselves in the context of a timing attack, other methods, such as power or electromagnetic analysis, would work as well.

Estimation of $P(y_j^{(i)} = 0)$. At step j of the procedure, we must estimate $P(y_j^{(i)} = 0)$ for each $i = 1, \ldots, k$. A basic estimation would be $P(y_j^{(i)} = 0) = 1 - (w^{(i)}/|A|)$, but we use two tricks to refine it:

- since $y^{(i)}$ is at least 64 bits longer than x, the first bits of $y^{(i)}$ are very likely to be the same as the first bits of $z^{(i)}$. We have $z_{|A|-1}^{(i)} \cdots z_{|S|+10}^{(i)} = y_{|A|-1}^{(i)} \cdots y_{|S|+10}^{(i)}$ with probability $1 - \frac{1}{2^{10}}$. For simplicity, we will assume in the following that this relationship always holds.
- if we already know $x_{j-1} \ldots x_0$, then we can use $z^{(i)}$ to compute $y_{j-1}^{(i)} \ldots y_0^{(i)}$.

This leads to the following estimation of $P(y_j^{(i)} = 0)$:

$$P(y_j^{(i)} = 0) = 1 - \frac{w(y_{|S|+9}^{(i)} \cdots y_j^{(i)})}{|S| + 10 - j}$$

with

$$w(y_{|S|+9}^{(i)} \cdots y_j^{(i)}) = w^{(i)} - w(z_{|A|-1}^{(i)} \cdots z_{|S|+10}^{(i)}) - w(y_{j-1}^{(i)} \ldots y_0^{(i)})$$

Our procedure will only store k values $w^{(i)}$ corresponding to the weight of the part of $y^{(i)}$ that has not been guessed yet.

Guessing x_0. For each $i = 1, \ldots, k$, we have $z^{(i)} = y^{(i)} + x$. Thus $z_0^{(i)} = y_0^{(i)} \oplus x_0$. Therefore we have $P(x_0 = 0) = P(z_0^{(i)} = y_0^{(i)})$. For each i, we estimate $P(y_0^{(i)} = 0)$, then $P(z_0^{(i)} = y_0^{(i)})$:

$$P(z_0^{(i)} = y_0^{(i)}) = \begin{cases} P(y_0^{(i)} = 0) \text{ if } z_0^{(i)} = 0 \\ \\ 1 - P(y^{(i)} = 0) \text{ if } z_0^{(i)} = 1 \end{cases}$$

Using all the couples $(w^{(i)}, z^{(i)})$, we can use the following estimation:

$$P(x_0 = 0) = \frac{1}{k} \sum_{i=1}^{k} P(z_0^{(i)} = y_0^{(i)})$$

If we get $P(x_0 = 0) > 1/2$, we set $\overline{x}_0 = 0$; else we set $\overline{x}_0 = 1$.

Guessing x_j for $j > 0$. To try to guess x_j for $j > 0$, the situation is a little different, because we now have to deal with a carry: $z_j^{(i)} = y_j^{(i)} \oplus x_j \oplus carry_j^{(i)}$. We use our previous guesses $(\overline{x}_{j-1}, \ldots, \overline{x}_0)$ to guess the carry. As we already explained, we also use $(\overline{x}_{j-1}, \ldots, \overline{x}_0)$ to refine our estimation of $P(y_j^{(i)} = 0)$.

Each couple $(w^{(i)}, z^{(i)})$ leads to an estimation of $P(x_j = 0)$:

$$P(x_j = 0) \simeq P(y_j^{(i)} = z_j^{(i)} \oplus carry_j^{(i)})$$

We take the mean of these estimations for $i = 1, \ldots, k$.

$$P(x_j = 0) = \frac{1}{k} \sum_{i=1}^{k} P(y_j^{(i)} = z_j^{(i)} \oplus carry_j^{(i)})$$

Again, if we get $P(x_j = 0) > 1/2$, we set $\overline{x}_j = 0$; else we set $\overline{x}_j = 1$.

Our procedure *FindCandidate* stores the list $carry^{(i)}, i = 1, \ldots, k$, where $carry^{(i)}$ equals $carry_{j-1}^{(i)}$ at the beginning of procedure *PartialEstimate*(i, j) and $carry_j^{(i)}$ at the end of this procedure.

Remark: If the estimation is wrong at step j, meaning that $\overline{x}_j \neq x_j$, the main consequence is an error on the next carry [2]. It is easy to see that this error will essentially propagate like a carry error in a classical addition, meaning that only a small block of the output bits will be erroneous. Our experiments confirmed this fact.

---

[2] Actually, a wrong estimation on x_j also induces an error on the updated estimation $w^{(i)}$. This error is small and does not affect the success of our attack.

Algorithm 1 $PartialEstimate(i,j)$: computes the i-th estimation of $P(x_j = 0)$ and the carry $carry_j^{(i)}$

$\overline{y}_{j-1}^{(i)} := z_{j-1}^{(i)} \oplus carry_{j-1}^{(i)} \oplus \overline{x}_{j-1}^{(i)}$

$w^{(i)} := w^{(i)} - \overline{y}_{j-1}^{(i)}$

if $\overline{x}_{j-1} + \overline{y}_{j-1}^{(i)} + carry_{j-1}^{(i)} \geq 2$ **then**

   $carry_j^{(i)} := 1$

else

   $carry_j^{(i)} := 0$

end if

Clear $carry_{j-1}^{(i)}$

Store $carry_j^{(i)}$

$P_j^{(i)} := 1 - (w^{(i)}/(|S| + 10 - j))$

if $z_j^{(i)} \oplus carry_j^{(i)} = 0$ **then**

   Return $P_j^{(i)}$

else

   Return $1 - P_j^{(i)}$

end if

Algorithm 2 $FindCandidate$: computes $\overline{x}$

for i=1 to k **do**

   $w^{(i)} := w^{(i)} - w(z_{|A|-1}^{(i)} \cdots z_{|S|+10}^{(i)})$

end for

for $j = 0$ to $|S| - 1$ **do**

   $P := 0$

   for $i = 1$ to k **do**

      $P := P + PartialEstimate(i,j)$

   end for

   $P := P/k$

   if $P > 0.5$ **then**

      $\overline{x}_j := 0$

   else

      $\overline{x}_j := 1$

   end if

   Store $\overline{x}_j$

end for

Return $\overline{x}_{|S|-1} \ldots \overline{x}_0$

4.3   Phase III: Using the Candidate to Find the Key

At the end of Phase II, the attacker finds a candidate $\overline{x}$. If $g^{-\overline{x}} \equiv X \pmod{n}$, the attack is a success. Even if $\overline{x} \neq x$, the attacker can make an exhaustive search on values x' such that $d(x', \overline{x})$, the Hamming distance between x' and $\overline{x}$, is small, testing whether $g^{-x'} \equiv X \pmod{n}$. The maximum distance such that the attack can succeed obviously depends on the capacity of the attacker; practical values are given in the next section.

5   Practical Results

We chose the following values to perform our tests: $g = 2$, $|n| = 1024$, $A = 2^{240}$, then used the test platform described in section 3.2 to obtain running times. We did not use the Chinese Remainder Technique.

For Phase III, we assumed that the attacker can perform 400 tests per second (which roughly corresponds to computations on a 2 Ghz PC with the GMP library). This leads to the following classification for the candidates:

1. If $\overline{x} = x$, $\overline{x}$ is already the correct key
2. $d(\overline{x}, x) \leq 2$, the attacker will need less than 16 seconds on average to find the correct key; we say that $\overline{x}$ is a "seconds" candidate
3. $d(\overline{x}, x) \leq 4$, the attacker will need less than 9 hours on average to find the correct key; we say that $\overline{x}$ is a "hours" candidate
4. $d(\overline{x}, x) \leq 5$, the attacker will need less than 12 days on average to find the correct key; we say that $\overline{x}$ is a "days" candidate

Our attack's results are summarized in Table 1.

Table 1. Simulation Results

k (number of samples)	200	400	600	800	1000
Immediate keys	0%	2%	21%	52%	72%
"seconds" candidates	0%	3%	54%	80%	89%
"hours" candidates	0%	6%	72%	94%	97%
"days" candidates	0%	10%	77%	96%	98%
avg. distance $d(\overline{x}, x)$	45.97	16.11	3.88	1.43	0.7

Phase II itself completes in a few seconds. When the number of samples is very small, it may be worth noticing that even if the attack does not produce directly exploitable results, it does anyway reveal substantial information about the key.

6   Countermeasures

Several countermeasures against timing attacks are known today. However, some care must be taken, as they will not all be efficient against our attack.

First of all, most timing attacks known so far exploited the modular reduction occurring in a Montgomery (resp. Barrett, ...) multiplication. Therefore, several countermeasures [6,13,14] typically consisted in removing the time variation in this multiplication. Since our attack is not based on the same property, these countermeasures are pointless.

Another frequently suggested countermeasure consists in randomizing the exponent by adding a random multiple of $\varphi(n)$ to it, a modification that does not affect the final result. However, GPS, that was designed for efficiency, uses a short (typically, between 240 and 304 bits) exponent. Clearly, adding a multiple of $\varphi(n)$ (typically, 1024 or 2048 bits) will imply a very serious performance drawback.

Simple sliding window techniques are probably too basic to make the attack impossible (in short, there is still a correlation between running time and Hamming weight).

Nevertheless, some other countermeasures are possible.

Pre-computed Commitments A natural countermeasure is to use pre-computed commitments. This obviously makes the attack impossible; the advantages and the drawbacks of such a method have already been discussed in section 2.2.

Square and Multiply Always Using dummy multiplications during the exponentiation allows to hide the hamming weight of the ephemeral keys. This countermeasure increases the computation time by about 30%.

MIST algorithm: A much more efficient countermeasure could be the use of Walter's MIST exponentiation algorithm [15]. MIST is not strictly constant-time (the greater the exponent, the longer the division chain). However, the information it could leak is probably limited to the strong bits of the exponent [16], which are already known to any eavesdropper (as they correspond to the strong bits of the answer). Thus MIST, which was designed to resist power analysis for RSA-like systems, i.e. when a same exponent is used many times, seems to fit out needs with GPS, where a new exponent is used for each commitment.

Remark: We were also suggested to modify the pseudo-random number generator, as a possible countermeasure. The most straightforward way to do this is to take $A = 2^a$ and ensure that the outputs $y \in [0, 2^a[$ of the generator are such that $w(y) = a/2$. The corresponding information on individual bits of y is null. However, we believe this countermeasure must be considered with great caution: the security of identification schemes such as GPS is strongly related to the randomness of the commitments, and tampering with "randomness modifications" is always very risky.

7   Conclusion

We proposed a strategy for a side-channel attack on GPS, and showed that it is realistic and efficient in a timing attack context. We believe that several refinements of the attack are still possible to further improve its efficiency. As usual with side-channel attacks – that do not target cryptographic primitives themselves, but rather specific implementations – this paper does not question the security of the GPS primitive. Instead, we showed how straightforward implementations can easily be broken, and pointed out the question of efficient countermeasures.

Acknowledgements. The authors wish to thank Colin Walter, Gaël Hachez and Mathieu Ciet for fruitful discussions and Sylvie Baudine for her permanent English support.

References

1. Cascade (Chip Architecture for Smart CArds and portable intelligent DEvices). Project funded by the European Community, see http://www.dice.ucl.ac.be/crypto/cascade.
2. O. Baudron, F. Boudot, P. Bourel, E. Bresson, J. Corbel, L. Frisch, H. Gilbert, M. Girault, L. Goubin, J.-F. Misarsky, P. Nguyen, J. Patarin, D. Pointcheval, J. Stern, and J. Traoré. GPS, an asymetric identification scheme for *on the fly* authentification of low cost smart cards. Submitted to the European NESSIE Project, 2000. Available at https://www.cryptonessie.org/.
3. D. Boneh and D. Brumley. Remote timing attacks are practical. Submitted to Usenix Security. Available at http://crypto.stanford.edu/~dabo/papers/ssl-timing.pdf, 2003.
4. Nessie consortium. Portfolio of recommended cryptographic primitives. Available at https://www.cryptonessie.org/deliverables/decision-final.pdf, 2003.
5. J.-F. Dhem, F. Koeune, P.-A. Leroux, P. Mestré, J.-J. Quisquater, and J.-L. Willems. A practical implementation of the timing attack. In J.-J. Quisquater and B. Schneier, editors, *Proc. CARDIS 1998, Smart Card Research and Advanced Applications*, LNCS. Springer, 1998.
6. J.F. Dhem. *Design of an efficient public-key cryptographic library for RISC-based smart cards.* PhD thesis, Université catholique de Louvain – UCL Crypto Group – Laboratoire de microélectronique (DICE), May 1998.
7. M. Girault. Self-Certified Public Keys. In D.W. Davies, editor, *Advances in Cryptology - Proceedings of EUROCRYPT 1990*, volume 0547 of *Lecture Notes in Computer Science*, pages 490–497. Springer, 1991.
8. Marc Girault, Guillaume Poupard, and Jacques Stern. Some modes of use of the GPS identification scheme. In *Proceedings of the third Nessie workshop*, November 6–7 2002.
9. P. Kocher. Timing attacks on implementations of Diffie-Hellman, RSA, DSS, and other systems. In N. Koblitz, editor, *Advances in Cryptology - CRYPTO '96, Santa Barbara, California*, volume 1109 of *LNCS*, pages 104–113. Springer, 1996.
10. Advanced RISC Machines Ltd. *ARM Software Developpment Toolkit version 2.11: User guide.* Advanced RISC Machines Ltd, 1997. Document number: ARM DUI 0040C.
11. G. Poupard and J. Stern. Security Analysis of a Practical *on the fly* Authentification and Signature Generation. In K. Nyberg, editor, *Advances in Cryptology – Proceedings of EUROCRYPT 1998*, volume 1403 of *Lecture Notes in Computer Science*, pages 422–436. Springer, 1998.
12. W. Schindler. A timing attack against RSA with the Chinese remainder theorem. In Ç. Koç and C. Paar, editors, *Proc. of Cryptographic Hardware and Embedded Systems (CHES 2000)*, volume 1965 of *LNCS*, pages 109–124. Springer, 2000.

13. Colin D. Walter. Montgomery Exponentiation Needs no Final Subtractions. *Electronics Letters*, 35(21):1831–1832, October 1999.
14. Colin D. Walter. Montgomery's Multiplication Technique: How to Make It Smaller and Faster. In Çetin K. Koç and Christof Paar, editors, *Cryptographic Hardware and Embedded Systems - CHES '99*, volume 1717 of *Lectures Notes in Computer Science (LNCS)*, pages 80–93. Springer-Verlag, August 1999.
15. Colin D. Walter. MIST: An efficient, randomized exponentiation algorithm for resisting power analysis. In *Topics in Cryptology – CT-RSA 2002*, Lecture Notes in Computer Science. Springer, April 2002.
16. Colin D. Walter. Seeing through MIST given a small fraction of an RSA private key. In M. Joye, editor, *Topics in Cryptology – CT-RSA 2003*, volume 2612 of *Lectures Notes in Computer Science (LNCS)*. Springer, 2003.

Unified Hardware Architecture for 128-Bit Block Ciphers AES and Camellia

Akashi Satoh and Sumio Morioka

IBM Research, Tokyo Research Laboratory, IBM Japan Ltd., 1623-14,
Shimotsuruma, Yamato-shi, Kanagawa 242-8502, Japan
{akashi, e02716}@jp.ibm.com

Abstract. We proposed unified hardware architecture for the two 128-bit block ciphers AES and Camellia, and evaluated its performance using a 0.13-μm CMOS standard cell library. S-Boxes are the biggest hardware components in block ciphers, and some times they consume more than half of the design area. The S-Boxes in AES and Camellia are the combination of affine transformations and multiplicative inversions on a Galois fields. The size of the fields is same, but their structures are different. Therefore we converted the basis between the fields by using isomorphism transformations, and shared the inverter between AES and Camellia. The affine transformations were also merged by factoring common terms. In addition to the S-Box sharing, many other components such as permutation layers and key whiting are also merged. As a result, a compact hardware of 14.9K gates with throughputs of 469 Mbps for AES and of 661 Mbps for Camellia was achieved. The hardware synthesized with speed optimization obtained throughputs of 794 Mbps and 1.12 Gbps for each algorithm with 24.4K gates.

1 Introduction

The AES (Advanced Encryption Standard) project [1] for the new US federal standard block cipher algorithm replacing DES (Data Encryption Standard) [2] was started in 1997, and Rijndael [3] was standardized as FIPS PUB 197 [4] in 2001. After then, many block ciphers that have the AES compatible interface, supporting a 128-bit data block and 128/192/256-bit keys, have been proposed for other organizations [5-9]. Camellia [9, 10] was developed by NTT (Nippon Telegraph and Telephone Corp.) and Mitsubishi Electric is the one of them. In February 2003, the NESSIE (New European Schemes for Signatures, Integrity and Encryption) project [6] chose it as a recommended algorithm and decided to input it to ISO and IETF.

Camellia has good performance in both software and hardware implementations, and a promising alternative of AES. However, supporting multiple algorithms simply multiplies the hardware costs, while it is not a big issue in software implementation. The algorithm structures of AES and Camellia are completely different; the former uses SPN (Substitution Permutation Network) and the latter does a Feistel network.

C.D. Walter et al. (Eds.): CHES 2003, LNCS 2779, pp. 304–318, 2003.

However, they use very similar basic components, such as multiplicative inversion on Galois field $GF(2^8)$ and affine transformation on $GF(2)$. Therefore, it is possible to reduce hardware cost by reusing these common components between two algorithms.

In this paper, we first propose an unified S-Box architecture sharing a GF inverter and merging affine transformations between AES and Camellia, and factoring techniques for the permutation layers are also shown. Then entire data path architecture including a key scheduler is described. Finally ASIC hardware performance in size and speed of the proposed architecture is compared with the discrete implementations of the two algorithms.

2   S-Box Structures

2.1   AES S-Box

The AES S-Boxes are combinations of a multiplicative inversion on $GF(2^8)$ and affine transformations. The irreducible polynomial of Equation (1) is used to define the field.

$$m(x) = x^8 + x^4 + x^3 + x + 1 \tag{1}$$

Following the inversion, an affine transformation A defined by Equation (2) is executed in the S-Box for encryption. In the equation, an operator $\oplus$ means XOR (Exclusive-OR). In the decryption S-Box, the multiplicative inversion follows the inverse affine transformation A^{-1} defined by Equation (3) that is not shown in the AES specification [4].

$$A: \begin{pmatrix} b_0 \\ b_1 \\ b_2 \\ b_3 \\ b_4 \\ b_5 \\ b_6 \\ b_7 \end{pmatrix} = \begin{pmatrix} 1 & 0 & 0 & 0 & 1 & 1 & 1 & 1 \\ 1 & 1 & 0 & 0 & 0 & 1 & 1 & 1 \\ 1 & 1 & 1 & 0 & 0 & 0 & 1 & 1 \\ 1 & 1 & 1 & 1 & 0 & 0 & 0 & 1 \\ 1 & 1 & 1 & 1 & 1 & 0 & 0 & 0 \\ 0 & 1 & 1 & 1 & 1 & 1 & 0 & 0 \\ 0 & 0 & 1 & 1 & 1 & 1 & 1 & 0 \\ 0 & 0 & 0 & 1 & 1 & 1 & 1 & 1 \end{pmatrix} \begin{pmatrix} a_0 \\ a_1 \\ a_2 \\ a_3 \\ a_4 \\ a_5 \\ a_6 \\ a_7 \end{pmatrix} \oplus \begin{pmatrix} 1 \\ 1 \\ 0 \\ 0 \\ 0 \\ 1 \\ 1 \\ 0 \end{pmatrix} \tag{2}$$

$$A^{-1}: \begin{pmatrix} b_0 \\ b_1 \\ b_2 \\ b_3 \\ b_4 \\ b_5 \\ b_6 \\ b_7 \end{pmatrix} = \begin{pmatrix} 0 & 0 & 1 & 0 & 0 & 1 & 0 & 1 \\ 1 & 0 & 0 & 1 & 0 & 0 & 1 & 0 \\ 0 & 1 & 0 & 0 & 1 & 0 & 0 & 1 \\ 1 & 0 & 1 & 0 & 0 & 1 & 0 & 0 \\ 0 & 1 & 0 & 1 & 0 & 0 & 1 & 0 \\ 0 & 0 & 1 & 0 & 1 & 0 & 0 & 1 \\ 1 & 0 & 0 & 1 & 0 & 1 & 0 & 0 \\ 0 & 1 & 0 & 0 & 1 & 0 & 1 & 0 \end{pmatrix} \begin{pmatrix} a_0 \oplus 1 \\ a_1 \oplus 1 \\ a_2 \\ a_3 \\ a_4 \\ a_5 \oplus 1 \\ a_6 \oplus 1 \\ a_7 \end{pmatrix} \tag{3}$$

A 128-bit nonlinear function that contains 16 encryption S-Boxes is called SubBytes, and that contains 16 decryption S-Boxes is called InvSubBytes. Example hardware implementations of each S-Box is shown in Fig. 1. An XOR operation with '1' in Equations (2) and (3) equals to a NOT operation. XOR followed by NOT and XOR following NOT can be replaced by XNORs (Exclusive NORs). Circuit cost (transistor counts and operation delay) is basically same between XOR and XNOR gates, so hardware performance can be improved by using XNOR instead of XOR with NOT. However the circuits shown in Fig. 1 do not use XNOR and are straightforward implementations of Equations (2) and (3), because they are much suitable for examples.

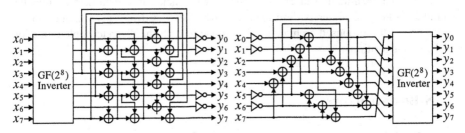

Fig. 1. Straightforward hardware Implementation of AES S-Boxes

2.2 Camellia S-Box

S-Boxes of Camellia use multiplicative inversion on a Galois field and affine transformations in similar fashion of AES. The Camellia description [9, 10] only shows a truth table of the inversion, but its field structure is not clearly described. So we investigated a lot of fields, and found that the field extended by using irreducible polynomials in Equation (4) satisfies the truth table.

$$\begin{cases} GF(2^4): & g_0(x) = x^4 + x + 1 \\ GF((2^4)^2): g_1(x) = x^2 + x + \omega \quad (\omega = \{1001\}) \end{cases} \tag{4}$$

Two sets of four S-Boxes (S1~S4) are used every iteration round. In the S-Box S1, affine transformations F and H defined by Equations (5) and (6) are executed before and after the multiplicative inversion respectively.

$$F: \begin{pmatrix} b_0 \\ b_1 \\ b_2 \\ b_3 \\ b_4 \\ b_5 \\ b_6 \\ b_7 \end{pmatrix} = \begin{pmatrix} 0 & 1 & 0 & 0 & 0 & 1 & 0 & 0 \\ 1 & 0 & 0 & 0 & 0 & 0 & 1 & 0 \\ 0 & 0 & 1 & 0 & 1 & 0 & 0 & 1 \\ 0 & 0 & 1 & 0 & 0 & 0 & 0 & 1 \\ 0 & 0 & 0 & 1 & 0 & 0 & 1 & 0 \\ 0 & 1 & 0 & 0 & 1 & 0 & 0 & 0 \\ 1 & 0 & 0 & 0 & 0 & 0 & 0 & 1 \\ 0 & 0 & 0 & 1 & 0 & 1 & 0 & 0 \end{pmatrix} \begin{pmatrix} a_0 \oplus 1 \\ a_1 \oplus 1 \\ a_2 \\ a_3 \\ a_4 \\ a_5 \oplus 1 \\ a_6 \\ a_7 \oplus 1 \end{pmatrix} \tag{5}$$

$$H: \begin{pmatrix} b_0 \\ b_1 \\ b_2 \\ b_3 \\ b_4 \\ b_5 \\ b_6 \\ b_7 \end{pmatrix} = \begin{pmatrix} 0 & 1 & 0 & 0 & 1 & 1 & 0 & 0 \\ 0 & 1 & 0 & 0 & 0 & 1 & 0 & 0 \\ 0 & 0 & 0 & 1 & 0 & 0 & 1 & 0 \\ 0 & 1 & 0 & 0 & 0 & 0 & 0 & 1 \\ 0 & 0 & 1 & 0 & 0 & 0 & 1 & 0 \\ 1 & 0 & 0 & 0 & 0 & 0 & 0 & 1 \\ 1 & 0 & 0 & 0 & 1 & 0 & 0 & 0 \\ 0 & 0 & 1 & 0 & 0 & 1 & 0 & 0 \end{pmatrix} \begin{pmatrix} a_0 \\ a_1 \\ a_2 \\ a_3 \\ a_4 \\ a_5 \\ a_6 \\ a_7 \end{pmatrix} \oplus \begin{pmatrix} 0 \\ 1 \\ 1 \\ 0 \\ 1 \\ 1 \\ 1 \\ 0 \end{pmatrix} \qquad (6)$$

The S-Boxes S2 and S3 are defined as S1 followed by 1-bit right rotation $(b_7, b_0, b_1, b_2, b_3, b_4, b_5, b_6)$ and 1-bit left rotation $(b_1, b_2, b_3, b_4, b_5, b_6, b_7, b_0)$ respectively. The input bits of S1 are rotated as $(a_0, a_1, a_2, a_3, a_4, a_5, a_6, a_7)$ for the S-Box S4. Fig. 2 shows an example circuit of the Camellia S-Boxes. Also here, it is possible to combine XOR with NOT into XNOR.

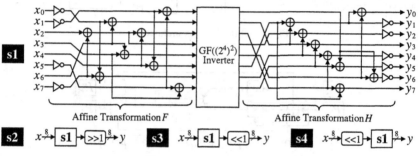

Fig. 2. Camellia S-Boxes

3   Unified S-Box

3.1   Construct Unified S-Box

In this section, we propose the shared S-Box architecture where a multiplicative inverter is reused and affine transformations are merged between SubBytes, InvSub-Bytes and S1~S4. Fig. 3 shows the process of S-Box sharing. The right arrows in the figure are all 8-bit data buses. In Fig. 3 (1), two S-Boxes SubBytes and InvSubBytes are independently implemented, and have the same $GF(2^8)$ inverter. In (2), the inverter is shared between the S-Boxes by switching affine transformations A and A^{-1} using 2:1 selectors. We also want to share the inverter with Camellia, but AES and Camellia use different Galois field for their S-Boxes. However, all fields who have same size are isomorphic, so we map all elements on the AES's field to the Camellia's composite field, and use the $GF((2^4)^2)$ inverter for all S-Boxes. It is possible to use the AES's $GF(2^8)$ inverter in the Camellia S-Boxes, but the $GF(2^8)$ inverter generated from a look-up table is much bigger than the $GF((2^4)^2)$ inverter where sub-field arithmetic for compact implementation can be applied. The structure of the $GF((2^4)^2)$ inverter is detailed in Fig. 4, where the width of all data buses is 4 bits. The box $[x^{-1}]$ shows a GF(2^4) inverter, and is designed as SOP (Sum of Products) logic in our implementations.

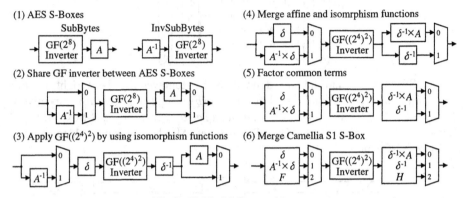

Fig. 3. Unified S-Box architecture

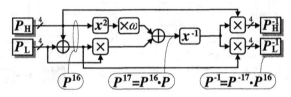

Fig. 4. GF($(2^4)^2$) inverter

δ and δ^{-1} in Equations (7) and (8) are isomorphism functions from GF(2^8) to GF($(2^4)^2$) and from GF($(2^4)^2$) to GF(2^8) respectively. The functions are defined as 8×8 XOR matrices as same as the affine transformations used in the S-Boxes. In Fig. 3(3), the isomorphism functions δ and δ^{-1} placed before and after the GF($(2^4)^2$) inverter expand the critical path. In order to shorten the path, the isomorphism functions δ and δ^{-1} are combined with affine transformations A and A^{-1} in Fig. 3(4). The combined functions $A^{-1}\times\delta$ and $\delta^{-1}\times A$ defined by Equations (9) and (10) require 43 2-input XOR gates, while 48 gates are used for A and A^{-1}. Therefore, circuit size is slightly reduced.

By comparing the matrices between Equations (7) and (9), and between (8) and (10), many common terms (1s in same columns) can be found. Therefore, hardware size can be reduced by sharing XOR gates corresponding to these common terms. However, the half of input bits of Equation (9), (a_1, a_2, a_6, a_7), are XORed with '1', and thus these values are reversed before the matrix operation. Therefore the common terms for these bits cannot be shared between Equations (7) and (9). In order to share these bits, replace these reverse operations on the input bits with those on the output bits as shown in Equation (11). This is possible because these matrix operations are all linear functions. The AES S-Box circuit after merging the matrices becomes Fig. 3(5).

$$\delta: \begin{pmatrix} b_0 \\ b_1 \\ b_2 \\ b_3 \\ b_4 \\ b_5 \\ b_6 \\ b_7 \end{pmatrix} = \begin{pmatrix} 1 & 0 & 1 & 0 & 0 & 0 & 0 & 0 \\ 1 & 0 & 1 & 0 & 1 & 1 & 0 & 0 \\ 1 & 1 & 0 & 1 & 0 & 0 & 1 & 0 \\ 0 & 1 & 1 & 1 & 0 & 0 & 0 & 0 \\ 1 & 1 & 0 & 0 & 0 & 1 & 1 & 0 \\ 0 & 1 & 0 & 1 & 0 & 0 & 1 & 0 \\ 0 & 0 & 0 & 0 & 1 & 0 & 1 & 0 \\ 1 & 1 & 0 & 1 & 1 & 1 & 0 & 1 \end{pmatrix} \begin{pmatrix} a_0 \\ a_1 \\ a_2 \\ a_3 \\ a_4 \\ a_5 \\ a_6 \\ a_7 \end{pmatrix} \tag{7}$$

$$\delta^{-1}: \begin{pmatrix} b_0 \\ b_1 \\ b_2 \\ b_3 \\ b_4 \\ b_5 \\ b_6 \\ b_7 \end{pmatrix} = \begin{pmatrix} 0 & 0 & 1 & 0 & 0 & 1 & 0 & 0 \\ 1 & 1 & 1 & 0 & 1 & 1 & 1 & 0 \\ 1 & 0 & 1 & 0 & 0 & 1 & 0 & 0 \\ 0 & 1 & 0 & 1 & 1 & 0 & 1 & 0 \\ 1 & 0 & 1 & 1 & 0 & 0 & 1 & 0 \\ 0 & 1 & 1 & 1 & 0 & 0 & 1 & 0 \\ 1 & 0 & 1 & 1 & 0 & 0 & 0 & 0 \\ 0 & 1 & 0 & 1 & 0 & 0 & 0 & 1 \end{pmatrix} \begin{pmatrix} a_0 \\ a_1 \\ a_2 \\ a_3 \\ a_4 \\ a_5 \\ a_6 \\ a_7 \end{pmatrix} \tag{8}$$

$$\begin{matrix} A^{-1} \\ \times \delta' \end{matrix}: \begin{pmatrix} b_0 \\ b_1 \\ b_2 \\ b_3 \\ b_4 \\ b_5 \\ b_6 \\ b_7 \end{pmatrix} = \begin{pmatrix} 1 & 1 & 0 & 0 & 0 & 1 & 1 & 0 \\ 0 & 1 & 1 & 1 & 0 & 0 & 0 & 1 \\ 0 & 1 & 1 & 1 & 1 & 0 & 0 & 0 \\ 1 & 1 & 1 & 1 & 0 & 1 & 1 & 1 \\ 1 & 0 & 1 & 0 & 0 & 0 & 0 & 0 \\ 0 & 0 & 1 & 0 & 1 & 0 & 1 & 0 \\ 0 & 1 & 1 & 0 & 1 & 1 & 0 & 0 \\ 0 & 0 & 1 & 0 & 0 & 0 & 1 & 0 \end{pmatrix} \begin{pmatrix} a_0 \\ a_1 \oplus 1 \\ a_2 \oplus 1 \\ a_3 \\ a_4 \\ a_5 \\ a_6 \oplus 1 \\ a_7 \oplus 1 \end{pmatrix} \tag{9}$$

$$\begin{matrix} \delta^{-1} \\ \times A \end{matrix}: \begin{pmatrix} b_0 \\ b_1 \\ b_2 \\ b_3 \\ b_4 \\ b_5 \\ b_6 \\ b_7 \end{pmatrix} = \begin{pmatrix} 1 & 0 & 0 & 0 & 0 & 1 & 1 & 0 \\ 1 & 1 & 0 & 1 & 0 & 0 & 0 & 0 \\ 1 & 0 & 0 & 0 & 1 & 1 & 1 & 0 \\ 0 & 1 & 1 & 1 & 1 & 0 & 1 & 1 \\ 0 & 0 & 0 & 0 & 0 & 1 & 0 & 1 \\ 0 & 1 & 0 & 1 & 1 & 0 & 0 & 1 \\ 1 & 0 & 0 & 0 & 1 & 1 & 1 & 1 \\ 0 & 1 & 1 & 0 & 0 & 1 & 0 & 1 \end{pmatrix} \begin{pmatrix} a_0 \\ a_1 \\ a_2 \\ a_3 \\ a_4 \\ a_5 \\ a_6 \\ a_7 \end{pmatrix} \oplus \begin{pmatrix} 0 \\ 1 \\ 1 \\ 0 \\ 0 \\ 0 \\ 1 \\ 1 \end{pmatrix} \tag{10}$$

$$\begin{matrix} A^{-1} \\ \times \delta' \end{matrix}: \begin{pmatrix} b_0 \\ b_1 \\ b_2 \\ b_3 \\ b_4 \\ b_5 \\ b_6 \\ b_7 \end{pmatrix} = \begin{pmatrix} 1 & 1 & 0 & 0 & 0 & 1 & 1 & 0 \\ 0 & 1 & 1 & 1 & 0 & 0 & 0 & 1 \\ 0 & 1 & 1 & 1 & 1 & 0 & 0 & 0 \\ 1 & 1 & 1 & 1 & 0 & 1 & 1 & 1 \\ 1 & 0 & 1 & 0 & 0 & 0 & 0 & 0 \\ 0 & 0 & 1 & 0 & 1 & 0 & 1 & 0 \\ 0 & 1 & 1 & 0 & 1 & 1 & 0 & 0 \\ 0 & 0 & 1 & 0 & 0 & 0 & 1 & 0 \end{pmatrix} \begin{pmatrix} a_0 \\ a_1 \\ a_2 \\ a_3 \\ a_4 \\ a_5 \\ a_6 \\ a_7 \end{pmatrix} \oplus \begin{pmatrix} 0 \\ 1 \\ 0 \\ 0 \\ 1 \\ 0 \\ 0 \\ 0 \end{pmatrix} \tag{11}$$

Finally, we also merged the Camellia affine transformations F and H according to the same manner, and obtained the shared S-Box shown in Fig. 3(6). Before merging F defined by Equation (5), here we also transform it to Equation (12). In the Camellia

S-Boxes S1~S3, only the order of the output bits is different. Therefore, the circuit shown in Fig. 3(6) can be used for all of them by only twisting the output wires. On the other hand, the input bits are twisted in the S4 S-Box. Therefore, we use the affine transformation F ' defined by Equation (13) instead of F, where the columns of the matrix is rotated to the right by one bit

$$F: \begin{pmatrix} b_0 \\ b_1 \\ b_2 \\ b_3 \\ b_4 \\ b_5 \\ b_6 \\ b_7 \end{pmatrix} = \begin{pmatrix} 0 & 1 & 0 & 0 & 0 & 1 & 0 & 0 \\ 1 & 0 & 0 & 0 & 0 & 0 & 1 & 0 \\ 0 & 0 & 1 & 0 & 1 & 0 & 0 & 1 \\ 0 & 0 & 1 & 0 & 0 & 0 & 0 & 1 \\ 0 & 0 & 0 & 1 & 0 & 0 & 1 & 0 \\ 0 & 1 & 0 & 0 & 1 & 0 & 0 & 0 \\ 1 & 0 & 0 & 0 & 0 & 0 & 0 & 1 \\ 0 & 0 & 0 & 1 & 0 & 1 & 0 & 0 \end{pmatrix} \begin{pmatrix} a_0 \\ a_1 \\ a_2 \\ a_3 \\ a_4 \\ a_5 \\ a_6 \\ a_7 \end{pmatrix} \oplus \begin{pmatrix} 0 \\ 1 \\ 1 \\ 1 \\ 0 \\ 1 \\ 0 \\ 1 \end{pmatrix} \quad (12)$$

$$F': \begin{pmatrix} b_0 \\ b_1 \\ b_2 \\ b_3 \\ b_4 \\ b_5 \\ b_6 \\ b_7 \end{pmatrix} = \begin{pmatrix} 0 & 0 & 1 & 0 & 0 & 0 & 1 & 0 \\ 0 & 1 & 0 & 0 & 0 & 0 & 0 & 1 \\ 1 & 0 & 0 & 1 & 0 & 1 & 0 & 0 \\ 1 & 0 & 0 & 1 & 0 & 0 & 0 & 0 \\ 0 & 0 & 0 & 0 & 1 & 0 & 0 & 1 \\ 0 & 0 & 1 & 0 & 0 & 1 & 0 & 0 \\ 1 & 1 & 0 & 0 & 0 & 0 & 0 & 0 \\ 0 & 0 & 0 & 0 & 1 & 0 & 1 & 0 \end{pmatrix} \begin{pmatrix} a_0 \\ a_1 \\ a_2 \\ a_3 \\ a_4 \\ a_5 \\ a_6 \\ a_7 \end{pmatrix} \oplus \begin{pmatrix} 0 \\ 1 \\ 1 \\ 1 \\ 0 \\ 1 \\ 0 \\ 1 \end{pmatrix} \quad (13)$$

3.2   Hardware Performance of Unified S-Boxes

In this section, hardware performance of the unified S-Box is compared with the AES and Camellia S-Boxes that are independently implemented.

Table 1 shows the number of XOR gates required for each matrix operation described in the previous section. Common terms are not shared between the matrices for the numbers in "original matrices", and are shared for the number in "sharing common terms." While the original matrices require 102 XOR gates in total, the number is reduced by more than 40% for the shared S-Boxes (60 XORs or 56 XORs).

The S-Box performances in size and speed are shown in Table 2, where a 0.13-μm CMOS standard cell library is used. One gate is equivalent to 2-input NAND gate, and the speed is estimated under the worst case conditions. The $GF((2^4)^2)$ inverter shown in the Fig. 4 is used in all S-Boxes. Two discrete S-Boxes (SubBytes and InvSubBytes) shown in Fig. 3(1), and one unified S-Box (SubBytes + InvSubBytes) in Fig 3(5) are implemented for AES. The performances between four Camellia S-Boxes are all same, and those between three shared S-Boxes (AES+S1~S3) are also same. When number of merged S-Boxes is increased (SubBytes with InvSubBytes, then with S1~S4), critical path becomes longer, because number of selectors is increased. The shared S-Box uses 411~414 gates that is almost half of 816 (=280+280+256) gates required for discrete implementation of two AES S-Boxes and one Camellia S-Box. In the actual use, a 3:1 selector is additionally needed for discrete implementation to switch three S-Boxes.

Table 1. Numbers of XOR gates required for each matrix operation

Original matrices							Sharing common terms	
δ	δ^{-1}	$A^{-1}{\times}\delta$	$\delta^{-1}{\times}A$	F	H	Total	AES+ S1~S3	AES+ S4
20	21	22	21	9	9	102	60	56

Table 2. ASIC Performance of each S-Box circuit

	S-Box type	Gate counts	Delay (ns)
AES	SubBytes	280	3.65
	InvSubBytes	280	3.56
	Merged	349	3.99
Camellia	S1~S4	256	3.45
AES+Canellia	AES + S1~S3	411	4.29
	AES + S4	414	4.65

(0.13-μm CMOS, 1gate = 2-input NAND, worst condition)

4   Unified Permutation Layer

AES uses permutation layers MixColumns and InvMixColumns in encryption and decryption respectively. MixColumns and InvMixColumns are inverse functions each other, and each function is defined as four 4-byte (4×4×8 bits = 128 bits) matrix operations. On the other hand, Camellia has only one permutation layer called P-function that is single 8-byte (8×8 bits = 64 bits) matrix operation. In order to merge these functions, compose two 8-byte matrices by gathering two 4-byte MixColumns and two 4-byte InvMixColimns respectively, and factorize them into a few 8-byte matrices as shown in Equations (14) and (15). Multiplications with the constant valued {8, 4, 3, 2, 1} in the matrices are defined over modulo $m(x)$ of Equation (1). By comparing Equations (14) for MixColumns and (15) for InvMixColumns, it is found that MixColumns is completely included in InvMixColumns [11]. Equation (16) is the matrix representation of P-function whose elements are '0' and '1', and thus modular arithmetic is not required. By breaking P-function into two matrices, many common terms with Mix-Columns are found. After the factorization and common term sharing, the permutation functions are represented as Equations (17)~(20). The basic structure of shared permutation circuit is shown in Fig. 5.

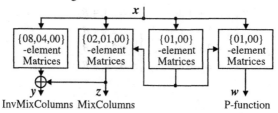

Fig. 5. Unified permutation circuit

$$
\begin{pmatrix} z_0 \\ z_1 \\ z_2 \\ z_3 \\ z_4 \\ z_5 \\ z_6 \\ z_7 \end{pmatrix} =
\begin{pmatrix}
2\,3\,1\,1 & & & & & & & \\
1\,2\,3\,1 & & & 0 & & & & \\
1\,1\,2\,3 & & & & & & & \\
3\,1\,1\,2 & & & & & & & \\
& & & & 2\,3\,1\,1 & & & \\
& 0 & & & 1\,2\,3\,1 & & & \\
& & & & 1\,1\,2\,3 & & & \\
& & & & 3\,1\,1\,2 & & &
\end{pmatrix}
\begin{pmatrix} x_0 \\ x_1 \\ x_2 \\ x_3 \\ x_4 \\ x_5 \\ x_6 \\ x_7 \end{pmatrix} =
\begin{pmatrix}
2\,2\,0\,0 & & & & & & & \\
0\,2\,2\,0 & & & 0 & & & & \\
0\,0\,2\,2 & & & & & & & \\
2\,0\,0\,2 & & & & & & & \\
& & & & 2\,2\,0\,0 & & & \\
& 0 & & & 0\,2\,2\,0 & & & \\
& & & & 0\,0\,2\,2 & & & \\
& & & & 2\,0\,0\,2 & & &
\end{pmatrix}
\begin{pmatrix} x_0 \\ x_1 \\ x_2 \\ x_3 \\ x_4 \\ x_5 \\ x_6 \\ x_7 \end{pmatrix} +
\begin{pmatrix}
0\,1\,1\,1 & & & & & & & \\
1\,0\,1\,1 & & & 0 & & & & \\
1\,1\,0\,1 & & & & & & & \\
1\,1\,1\,0 & & & & & & & \\
& & & & 0\,1\,1\,1 & & & \\
& 0 & & & 1\,0\,1\,1 & & & \\
& & & & 1\,1\,0\,1 & & & \\
& & & & 1\,1\,1\,0 & & &
\end{pmatrix}
\begin{pmatrix} x_0 \\ x_1 \\ x_2 \\ x_3 \\ x_4 \\ x_5 \\ x_6 \\ x_7 \end{pmatrix}
\tag{14}
$$

$$
\begin{pmatrix} y_0 \\ y_1 \\ y_2 \\ y_3 \\ y_4 \\ y_5 \\ y_6 \\ y_7 \end{pmatrix} =
\begin{pmatrix}
E\,B\,D\,9 & & & & & & & \\
9\,E\,B\,D & & & 0 & & & & \\
D\,9\,E\,B & & & & & & & \\
B\,D\,9\,E & & & & & & & \\
& & & & E\,B\,D\,9 & & & \\
& 0 & & & 9\,E\,B\,D & & & \\
& & & & D\,9\,E\,B & & & \\
& & & & B\,D\,9\,E & & &
\end{pmatrix}
\begin{pmatrix} x_0 \\ x_1 \\ x_2 \\ x_3 \\ x_4 \\ x_5 \\ x_6 \\ x_7 \end{pmatrix}
\tag{15}
$$

$$
=
\begin{pmatrix}
8\,8\,8\,8 & & & & & & & \\
8\,8\,8\,8 & & & 0 & & & & \\
8\,8\,8\,8 & & & & & & & \\
8\,8\,8\,8 & & & & & & & \\
& & & & 8\,8\,8\,8 & & & \\
& 0 & & & 8\,8\,8\,8 & & & \\
& & & & 8\,8\,8\,8 & & & \\
& & & & 8\,8\,8\,8 & & &
\end{pmatrix}
\begin{pmatrix} x_0 \\ x_1 \\ x_2 \\ x_3 \\ x_4 \\ x_5 \\ x_6 \\ x_7 \end{pmatrix} +
\begin{pmatrix}
4\,0\,4\,0 & & & & & & & \\
0\,4\,0\,4 & & & 0 & & & & \\
4\,0\,4\,0 & & & & & & & \\
0\,4\,0\,4 & & & & & & & \\
& & & & 4\,0\,4\,0 & & & \\
& 0 & & & 0\,4\,0\,4 & & & \\
& & & & 4\,0\,4\,0 & & & \\
& & & & 0\,4\,0\,4 & & &
\end{pmatrix}
\begin{pmatrix} x_0 \\ x_1 \\ x_2 \\ x_3 \\ x_4 \\ x_5 \\ x_6 \\ x_7 \end{pmatrix} +
\begin{pmatrix}
2\,3\,1\,1 & & & & & & & \\
1\,2\,3\,1 & & & 0 & & & & \\
1\,1\,2\,3 & & & & & & & \\
3\,1\,1\,2 & & & & & & & \\
& & & & 2\,3\,1\,1 & & & \\
& 0 & & & 1\,2\,3\,1 & & & \\
& & & & 1\,1\,2\,3 & & & \\
& & & & 3\,1\,1\,2 & & &
\end{pmatrix}
\begin{pmatrix} x_0 \\ x_1 \\ x_2 \\ x_3 \\ x_4 \\ x_5 \\ x_6 \\ x_7 \end{pmatrix}
$$

$$
\begin{pmatrix} w_0 \\ w_1 \\ w_2 \\ w_3 \\ w_4 \\ w_5 \\ w_6 \\ w_7 \end{pmatrix} =
\begin{pmatrix}
1\,0\,1\,1 & 0\,1\,1\,1 \\
1\,1\,0\,1 & 1\,0\,1\,1 \\
1\,1\,1\,0 & 1\,1\,0\,1 \\
0\,1\,1\,1 & 1\,1\,1\,0 \\
1\,1\,0\,0 & 0\,1\,1\,1 \\
0\,1\,1\,0 & 1\,0\,1\,1 \\
0\,0\,1\,1 & 1\,1\,0\,1 \\
1\,0\,0\,1 & 1\,1\,1\,0
\end{pmatrix}
\begin{pmatrix} x_0 \\ x_1 \\ x_2 \\ x_3 \\ x_4 \\ x_5 \\ x_6 \\ x_7 \end{pmatrix} =
\begin{pmatrix}
1\,0\,0\,0 & 0\,1\,1\,1 \\
0\,1\,0\,0 & 1\,0\,1\,1 \\
0\,0\,1\,0 & 1\,1\,0\,1 \\
0\,0\,0\,1 & 1\,1\,1\,0 \\
1\,1\,0\,0 & \\
0\,1\,1\,0 & 0 \\
0\,0\,1\,1 & \\
1\,0\,0\,1 &
\end{pmatrix}
\begin{pmatrix} x_0 \\ x_1 \\ x_2 \\ x_3 \\ x_4 \\ x_5 \\ x_6 \\ x_7 \end{pmatrix} +
\begin{pmatrix}
0\,0\,1\,1 & & & & \\
1\,0\,0\,1 & & & & \\
1\,1\,0\,0 & & 0 & \\
0\,1\,1\,0 & & & \\
& & & 0\,1\,1\,1 \\
& 0 & & 1\,0\,1\,1 \\
& & & 1\,1\,0\,1 \\
& & & 1\,1\,1\,0
\end{pmatrix}
\begin{pmatrix} x_0 \\ x_1 \\ x_2 \\ x_3 \\ x_4 \\ x_5 \\ x_6 \\ x_7 \end{pmatrix}
\tag{16}
$$

$$
\begin{cases}
y_0 = E_0 + z_0 \\
y_1 = E_1 + z_1 \\
y_2 = E_0 + z_2 \\
y_3 = E_1 + z_3
\end{cases}
\qquad
\begin{cases}
y_4 = E_2 + z_4 \\
y_5 = E_3 + z_5 \\
y_6 = E_2 + z_6 \\
y_7 = E_3 + z_7
\end{cases}
\tag{17}
$$

$$
\begin{cases}
z_0 = 2A_0 + A_2 + x_1 \\
z_1 = 2A_1 + A_3 + x_2 \\
z_2 = 2A_2 + A_0 + x_3 \\
z_3 = 2A_3 + A_1 + x_0
\end{cases}
\qquad
\begin{cases}
z_4 = 2A_4 + B_0 \\
z_5 = 2A_5 + B_1 \\
z_6 = 2A_6 + B_2 \\
z_7 = 2A_7 + B_3
\end{cases}
\tag{18}
$$

$$\begin{cases} w_0 = A_2 + B_0 + x_0 \\ w_1 = A_3 + B_1 + x_1 \\ w_2 = A_0 + B_2 + x_2 \\ w_3 = A_1 + B_3 + x_3 \end{cases} \begin{cases} w_4 = A_0 + B_0 \\ w_5 = A_1 + B_1 \\ w_6 = A_2 + B_2 \\ w_7 = A_3 + B_3 \end{cases} \tag{19}$$

$$\begin{cases} A_0 = x_0 + x_1 \\ A_1 = x_1 + x_2 \\ A_2 = x_2 + x_3 \\ A_3 = x_3 + x_4 \end{cases} \begin{cases} A_4 = x_4 + x_5 \\ A_5 = x_5 + x_6 \\ A_6 = x_6 + x_7 \\ A_7 = x_7 + x_4 \end{cases} \begin{cases} B_0 = A_6 + x_5 \\ B_1 = A_7 + x_6 \\ B_2 = A_4 + x_7 \\ B_3 = A_5 + x_4 \end{cases} \tag{20}$$

$$\begin{cases} C_0 = 4(x_0 + x_2) \\ C_1 = 4(x_1 + x_3) \\ C_2 = 4(x_4 + x_6) \\ C_3 = 4(x_5 + x_7) \end{cases} \begin{cases} D_0 = 2(C_0 + C_1) \\ D_1 = 2(C_2 + C_3) \end{cases} \begin{cases} E_0 = D_0 + C_0 \\ E_1 = D_0 + C_1 \\ E_2 = D_1 + C_2 \\ E_3 = D_1 + C_3 \end{cases}$$

Table 3 indicates the hardware size and number of stages of critical path in XOR gates for the permutation functions. The total size of our unified permutation circuit are only 476 XORs, while the original functions require 1,482 XORs. Therefore, the hardware cost is reduced down to less than 1/3 with only additional 2 XOR-gate delay.

Table 3. Performance of permutation functions

	Original matrices				Sharing common terms
	2 Mix-Columns	2 InvMix-Columns	P-func	Total	
XORs	304	880	288	1,482	476
Delay (gates)	3	5	3	5	7

5   Unified Data Path Architecture

Fig. 6 shows the unified data path architecture of the data randomization block. In addition to sharing S-Boxes and permutation, FL/FL^{-1} and key whitening functions are also merged. Only 128-bit key is supported in the current design, but 192- and 256-bit keys can be easily supported by modifying the key scheduler shown in Fig. 7. Many components are shared in the data randomization block, but only registers can be re-used in the key scheduler, because the key scheduling methods of two algorithms are much different.

128 bits are processed in one clock cycle by FL/FL^{-1} and key whitening in Camellia and the first AddRoundKey (using key whitening path) in AES. The other function blocks handle a 64 bits at one time. A straightforward implementation of AES that has 128-bit data path takes 11 cycles for the one block encryption or decryption. But the unified hardware of Fig. 6 processes 64 bits / cycle for the 2~11 rounds, and thus, it takes 20 cycles for these rounds. The key scheduler reusing the S-Boxes in the main data path requires additional 10 cycles. Therefore, our unified hardware takes

1+20+10=31 cycles for AES. If four 8-bit S-Boxes dedicated for the key scheduling are attached, the number of cycle is reduced to 21. On the other hand, Camellia takes 2 cycles for FL/FL^{-1}, 2 cycles for key whitening, and 18 cycles for the Feistel rounds (F function), therefore 22 cycles in total.

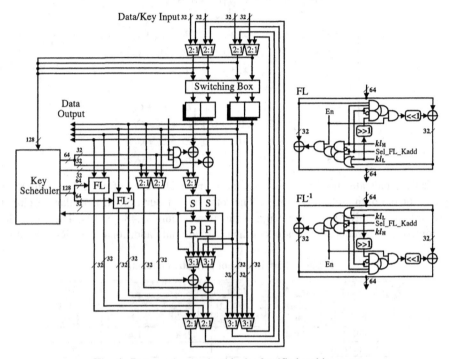

Fig. 6. Data randomization block of unified architecture

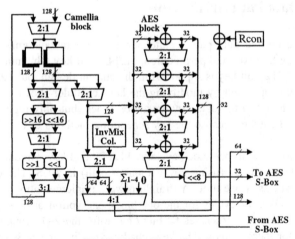

Fig. 7. Unified key scheduler

6   ASIC Performance Comparison

Table 4 shows the performance comparison between our unified hardware and independent implementations of the two algorithms [11, 12], where same 0.13-μm CMOS standard cell library is used for all. Two circuits were generated from each design source by indicating area and speed optimizations to a synthesis tool. The number of S-Boxes are eight (64 bits) in all designs.

In comparison with the AES hardware of the reference [11], the number of cycles of the unified hardware is one cycle fewer. As mentioned before, this is because a 128-bit block is processed at once in the first AddRoundKey while it is executed by 64 bits and takes two cycles in [11]. On the other hand, the Camellia operation takes 22 cycles in the unified hardware, while it is 18 cycles in [12]. Because the Camellia hardware in [12] executes the FL/FL^{-1} functions or the key whitening, and the F function in a same cycle. This approach is suitable for high-speed implementation, but requires additional hardware. Therefore we did not use it for the unified hardware whose priority is compactness. The maximum operation frequency of the unified hardware is lower than that of references [11, 12]. This is because the critical path became longer due to the additional hardware such as selectors to merge two data paths of different algorithms.

Discreet implementations require 21.6K (=8.0K+13.6K) gates for compact versions of two algorithms and 34.6K (=14.8K+19.8K) gates for high-speed versions. On the other hand, our unified hardware is 30% smaller, 14.9K gates and 24.4K gates respectively. The throughputs of the unified hardware are 9~14% lower for AES and 31~40% lower for Camellia. Therefore, the proposed architecture is much suitable for the application such as embedded use where hardware resource is more critical than speed.

Table 4. Hardware performance comparison

	Algorithms	Cycles	Gate counts	Max. frequency (MHz)	Throughput (Mbps)	Synthesis optimization
This work	AES	31	14,918	113.64	469.22	Area
	Camellia	22			661.18	
	AES	31	24,424	192.31	794.05	Speed
	Camellia	22			1,118.89	
Reference [11]	AES	32	7,998	137.17	548.68	Area
			14,777	218.82	875.28	Speed
Reference [12]	Camellia	18	13,557	153.85	1,094.04	Area
			19,783	227.27	1,616.14	Speed

(0.13-μm CMOS, 1gate = 2-input NAND, worst condition)

7 Conclusion

Unified hardware architecture for the 128-bit block ciphers AES and Camellia was proposed and its performance was evaluated in comparison with non-unified implementations. To merge the biggest hardware component S-Box between two algorithms, a multiplicative inverter on $GF((2^4)^2)$ was shared by using isomorphism transformation, and factoring technique was applied on affine transformations. The permutation layers were also merged by sharing common terms of the operator matrix. Our architecture was synthesized by using a 0.13-μm CMOS standard cell library, and compact implementations of 14.9K~24.4K gates were obtained with throughputs of 469M~794Mbps and 661M~1,119Mbps for AES and Camellia respectively. The gate counts were 30% smaller than the conventional implementations where two algorithms were discreetly designed.

References

1. National Institute of Standards and Technology (NIST), "AES Home Page," http://csrc.nist.gov/encryption/aes.
2. National Institute of Standards and Technology (NIST), "Data Encryption Standard (DES)," FIPS Publication 46-3, http://csrc.nist.gov/publications/fips/fips46-3/fips46-3.pdf, Oct. 1999.
3. "The Block Cipher Rijndael," http://www.esat.kuleuven.ac.be/~rijmen/rijndael.
4. National Institute of Standards and Technology (NIST), "Advanced Encryption Standard (AES) FIPS Publication 197," http://csrc.nist.gov/publications/fips/fips197/fips-197.pdf, Nov. 2001.
5. "NESSIE (New European Scheme for Signatures, Integrity and Encryption)", https://www.cosic.esat.kuleuven.ac.be/nessie.
6. ISO/IEC JTC 1/SC27, "Information technology – Security techniques," http://www.din.de/ni/sc27.
7. TV-Anytime Forum, "WG Rights Managements and Protection (RMP)," http://www.tv-anytime.org.
8. CRYPTREC, http://www.ipa.go.jp/security/enc/CRYPTREC.
9. S. Moriai, "Proposal of addition of new cipher suites to TLS to support Camellia, EPOC, and PSEC," Proc. the Forty-Eighth IETF, http://www.ietf.org/proceedings/00jul/SLIDES/tls-cep/index.html, 2000.
10. K. Aoki, T. Ichikawa, M. Kanda, M. Matsui, S.Moriai, J. Makajima, and T. Tokita, "Specification of Camellia – a 128-bit Block Cipher Version 2.0," http://info.isl.ntt.co.jp/camellia/CRYPTREC/2001/01espec.pdf, 2001.
11. A. Satoh, S. Morioka, K. Takano, and S. Munetoh, "A Compact Rijndael Hardware Architecture with S-box Optimization," ASIACRYPT 2001, LNCS 2248, pp.239–254, 2001.
12. A. Satoh and S. Morioka, "Compact Hardware Architecture for 128-bit Block Cipher Camellia," Proc. third NESSIE workshop, 2002.

Appendix 1   AES Algorithm

Fig. A1 shows an AES encryption process under a 128-bit secret key. 11 sets of round keys are generated from the secret key, and are fed to each round of the SPN block. The round operation is combination of four primitive functions, SubBytes (sixteen 8-bit S-Boxes), ShiftRows (byte boundary rotations), MixColumns (4-byte × 4-byte matrix operation), and AddRoundKeys (bit-wise XOR). In decryption, the inverse functions (AddRoundKey is identical) are executed in reverse order

The key scheduler uses four S-Boxes and 4-byte constant values Rcon[i] (i=1~10). The highest byte of Rcon[i] is the bit representation of the polynomial x^i mod $m(x)$, and the other three bytes are all zeros. In decryption, these sets of keys are used in reverse order.

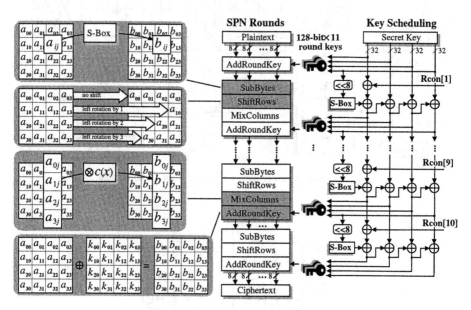

Fig. A1. Encryption process of AES algorithm

Appendix 2   Camellia Algorithm

Fig. A2 shows the encryption process of Camellia for a 128-bit secret key. At the initial and final stages, 128-bit data is XORed with 128-bit round keys. A 22-round data randomization part consists of three 6-round Feistel networks, and two FL/FL^{-1} functions placed between the networks. The 128-bit data input to the Feistel network is divided into two 64-bit data blocks, and the left half is fed into the F function with a 64-bit round key, and its output is XORed with the right half. The left and right half are swapped every round. 64-bit data input to the F function is XORed with the 64-bit round key. The result is divided into eight 8-bit blocks, and they are fed to eight S-Boxes (S1~S4) followed by the P-function. Same data path can be used in decryption by just changing order of round keys.

As shown in Fig. A3, a 128-bit intermediate key K_A is generated from the 128-bit secret key K_L by using the F function 4 times. The round keys are generated from K_A and K_L by bit rotations.

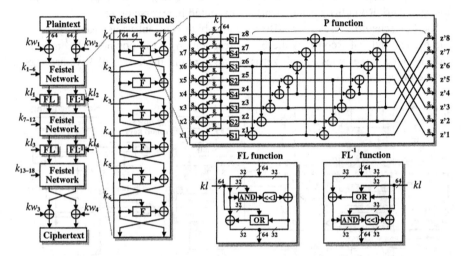

Fig. A2. Encryption process of Camellia for a 128-bit key

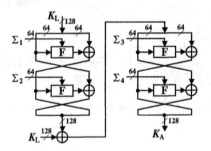

Fig. A3. Intermediate key K_A generating process for a 128-bit key

Very Compact FPGA Implementation of the AES Algorithm

Paweł Chodowiec and Kris Gaj

George Mason University, MS1G5, 4400 University Drive, Fairfax, VA 22030, USA
{pchodow1, kgaj}@gmu.edu
http://ece.gmu.edu/crypto-text.htm

Abstract. In this paper a compact FPGA architecture for the AES algorithm with 128-bit key targeted for low-cost embedded applications is presented. Encryption, decryption and key schedule are all implemented using small resources of only 222 Slices and 3 Block RAMs. This implementation easily fits in a low-cost Xilinx Spartan II XC2S30 FPGA. This implementation can encrypt and decrypt data streams of 150 Mbps, which satisfies the needs of most embedded applications, including wireless communication. Specific features of Spartan II FPGAs enabling compact logic implementation are explored, and a new way of implementing *MixColumns* and *InvMixColumns* transformations using shared logic resources is presented.

1 Introduction

The National Institute of Standards and Technology (NIST) selected the Rijndael algorithm as the new Advanced Encryption Standard (AES) [29] in 2001. Numerous FPGA [2] [15] [16] [17] [18] [19] [20] [24] [25] [26] [27] [28] and ASIC [4] [6] [7] [8] [10] [11] implementations of the AES were previously proposed and evaluated. To date, most implementations feature high speeds and high costs suitable for high-end applications only.

The need for secure electronic data exchange will become increasingly more important. Therefore, the AES must be extended to low-end customer products, such as PDAs, wireless devices, and many other embedded applications. In order to achieve this goal, the AES implementations must become very inexpensive.

Most of the low-end applications do not require high encryption speeds. Current wireless networks achieve speeds up to 60 Mbps. Implementing security protocols, even for those low network speeds, significantly increases the requirements for computational power. For example, the processing power requirements for AES encryption at the speed of 10 Mbps are at the level of 206.3 MIPS [12]. In contrast, a state-of-the-art handset processor is capable of delivering approximately 150 MIPS at 133 MHz, and 235 MIPS at 206 MHz.

This paper attempts to create a bridge between performance and cost requirements of the embedded applications. As a result, a low-cost AES implementation for FPGA devices, capable of supporting most of the embedded applications, was developed and evaluated.

C.D. Walter et al. (Eds.): CHES 2003, LNCS 2779, pp. 319–333, 2003.

2   Related Work

Early AES designs were mostly straightforward implementations of various loop
unrolled and pipelined architectures [24] [25] [26] [27] [28] with limited number
of architectural optimizations, which resulted in poor resource utilization. For
example, AES 8x8 S-boxes were implemented on LUTs as huge tables left for
synthesizers to optimize.

Later FPGA implementations demonstrate better utilization of FPGA resources. Several architectures using dedicated on-chip memories implementing
S-boxes and T-boxes were developed [15] [17] [18] [19] [20].

Recent research focused on fast pipelined implementations in both FPGA [2]
[3] [14] [18] [19] [20] and ASIC [4] [6] [7] [11] worlds. Unfortunately, most of those
implementations are too costly for practical applications.

The first significant step in compacting the AES implementation was made
when V. Rijmen proposed an AES S-box implementation based on composite
fields [31]. A similar solution was proposed by J. Wolkerstorfer [13]. Rijmen's idea
has already been implementated in FPGA [2], and in ASICs [4] [6] [8]. S. Morioka
et al. [10] went even farther and proposed a low-power compact S-box design
suited for ASIC designs.

3   Architecture of the Compact Implementation

We began the design of the compact architecture by analyzing the basic architecture, as introduced in [26]. The basic architecture unrolls only one full cipher
round, and iteratively loops data through this round until the entire encryption
or decryption transformation is completed. Only one block of data is processed
at a time making it equally suited for feedback and non-feedback modes of operation.

The structure of the AES round for encryption is shown in Fig. 1. The decryption round looks very similar, except it employs inverted operations in the following order: *InvShiftRows*, *InvSubBytes*, *AddRoundKey* and *InvMixColumns*.
The *SubBytes* and *ShiftRows* operations in Fig. 1 are reordered compared to the
cipher round depicted in the standard [29]. Their order is not significant because
SubBytes operates on single bytes, and *ShiftRows* reorders bytes without altering
them. This feature was used in our implementation.

The AES round shown in Fig. 1 reveals a great deal of parallelism. The
data bytes are ordered from the most significant (byte 0) to the least significant
(byte F) assuming big-endian representation. The round is composed of 16 8-bit
S-boxes computing *SubBytes*, and four 32-bit *MixColumns* operations, working
independent of each other. The only operation that spans throughout the entire
128-bit block is *ShiftRows*.

It is possible to implement only four *SubBytes* and one *MixColumns* in order
to compact the AES implementation. Ideally, the resources should be cut by
four, while execution of one round should take four clock cycles. This approach

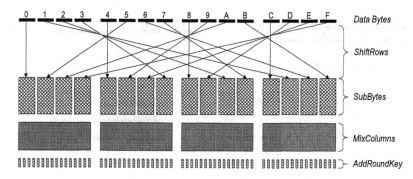

Fig. 1. Operations within AES encryption round

would result in approximately four times lower performance than for the basic architecture.

Cutting the resources by 75% may not appear easy. The *folded round*, as we call the modified round, still must transform 128 bits, and storage for all 128 bits of the data block must exist. Another complication is related to the implementation of the *ShiftRows* operation. The data bytes processed in the AES round cannot return to the same positions in the block register because it would not execute the *ShiftRows* operation. On the other hand, those same bytes cannot be placed into locations indicated by *ShiftRows* because those locations are occupied by other bytes that have not yet been processed. Therefore, additional bits of intermediate results must be stored, and more logic resources are needed.

One of the possible architectures for a folded implementation is shown in Fig. 2a. This architecture requires one 128-bit register, one 96-bit register and one 32-bit wide 4-to-1 multiplexer on top of the main cipher operations. The multiplexer becomes even bigger when both *ShiftRows* and *InvShiftRows* are implemented using same logic resources. The execution of one round takes four clock cycles. We believe that this, or very similar architecture, was implemented by A. Satoh et al. [23], but we cannot be sure since the authors do not provide enough detail. Their results show that the 4-cycle round takes 50% of the resources required by the 1-cycle round, and yields four times lower throughput.

Another possible architecture is shown in Fig. 2b. The 96-bit register is implemented as three 32-bit registers inserted into round operations creating a pipeline. In the case of FPGAs, those 32-bit registers will most likely be placed in the same Slices as logic operations yielding better resource utilization. The critical path is also shortened which permits the execution at a higher clock rate; however, the execution of the entire round requires seven, instead of four, clock cycles. We believe that this architecture was implemented by S. McMillan et al. [21], but again, we cannot be certain since the authors did not provide enough detail. S. McMillan et al. reported only slight difference of 48 Slices (16%), and large difference of 24 Block RAMs (75%), between 1-round unrolled and folded architecture.

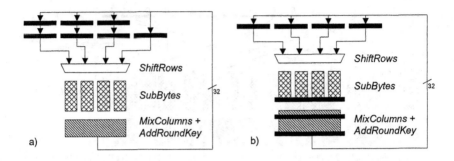

Fig. 2. Folded architecture. a) by A. Satoh et al. [23]; b) by S. McMillan et al. [21]

3.1 Implementation of a Folded Register

The two folded architectures described above are very straightforward and resulted in small logic savings. In order to create a folded architecture with better parameters, we decided to explore fine details of FPGA devices. We arranged data bytes into rows as shown in Fig. 3. This data arrangement is consistent with a *state* introduced in [30]. The following exercise can now be executed in steps:

1. Read input bytes: 0, 5, A, F; execute *SubBytes*, *MixColumns* and *AddRound-Key* on them; write results to the output at locations: 0, 1, 2, 3. This step is highlighted in the Fig. 3.
2. Repeat above operations for input bytes: 4, 9, E, 3; write results at output locations: 4, 5, 6, 7.
3. Repeat above operations for bytes: 8, D, 2, 7; write results at locations: 8, 9, A, B.
4. Repeat above operations for bytes: C, 1, 6, B; write results at locations: C, D, E, F. Output now becomes input for the next step.

In those four steps the entire AES round was executed including *ShiftRows* operation. At each step only one byte was read from each input row, and one byte was written to each output row. A similar exercise with identical conclusions can be executed for decryption transformation. Each row can be viewed as an addressable 8-bit wide memory. The correct execution of *ShiftRows* and *InvShiftRows* is now resolved to the proper addressing of each of the memories at the consecutive clock cycles. At the fourth clock cycle output memories become input memories and vice versa.

Dual-Port RAM Based Implementation. Each CLB Slice in Spartan II FPGA contains two look-up tables (LUT), which are the primary resources for logic implementation. Typically LUTs are configured as small 16x1 ROM tables

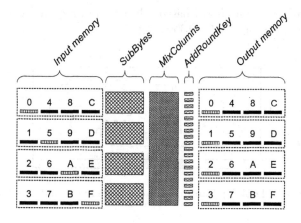

Fig. 3. Data arrangement in the folded architecture. Data bytes involved in the first step of calculation are highlighted

implementing logic functions of up to four inputs; however, other configurations are also possible. Two LUTs within the same Slice can implement a 16x1 Dual-Port RAM. An 8-bit wide Dual-Port RAM can be implemented using eight CLB Slices. This memory can be divided into two banks; each addressed by a different port. One port is used for reading data from the memory, while the other one for writes results back to the same memory. The switching between banks can be achieved by fliping one address bit in both ports every fourth clock cycle.

The Dual-Port RAM based solution has major advantages over solutions presented in Fig. 2:

- The logic resources required for storing intermediate results are far smaller.
- The multiplexer used before for *ShiftRows* and *InvShiftRows* is no longer needed.
- The complicated routing resulting from implementation of *ShiftRows* and *InvShiftRows* is avoided, yielding better performance.

Shift Register Based Implementation. A better solution may result from the following observation: all bytes from the output of *AddRoundKey* are written into consecutive locations in the output memory in consecutive clock cycles. Therefore, we could use a simple shift-register to shift computed data in without generating any addresses. Fortunately, LUTs can also be configured as 16-bit shift registers with variable taps, as shown in Fig. 4. Four Slices can implement an 8-bit wide, 16-bit long shift register. The input of the shift register is used for shifting results in while the output, selected dynamically by changing tap address, is used for reading data out. This solution encompasses all of the advantages of the Dual-Port RAM based solution, and requires less than a half of the logic resources than the Dual-Port RAM.

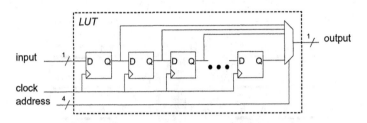

Fig. 4. Look-Up Table (LUT) configured as a shift register

3.2   Implementation of the *SubBytes* and *InvSubBytes*

Various area efficient implementations of AES S-boxes were proposed in [2] [4] [6] [8] [10] [13] [22] [23] [31]. All of those implementations are based on an idea of transforming the original $GF(2^8)$ field into a composite of smaller fields $GF((2^4)^2)$. It is a very attractive solution especially from the perspective of an ASIC because its implementation occupies a smaller area than a ROM. In the case of FPGAs, S-boxes can be mapped into dedicated Block RAMs treated as ROMs, or into LUTs. The latter approach could utilize the idea of composite fields. We decided to keep a good balance between utilization of LUTs and Block RAMs for the entire design, and implemented our S-boxes on dedicated Block RAMs.

Each Block RAM represents a dual-port memory of 4096 bits. Each port can be independently configured for different width and depth [34]. We selected a 512x8 configuration for each port, which provides access to the same memory space in the same way from both ports. A single *SubBytes* or *InvSubBytes* implementation requires a 256x8 ROM. A Block RAM has enough space to implement both *SubBytes* and *InvSubBytes*, as shown in Fig. 5. Each port has access to the entire memory space, and can perform a *SubBytes* or *InvSubBytes* transformation independently of each other. The folded architecture requires only 2 Block RAMs to implement four *SubBytes* and four *InvSubBytes* operations all together.

The Block RAM is a fully synchronous memory. Reading from it requires supplying the address one clock cycle before the data appears at the output. This feature can be viewed as a pipeline stage introducing a delay of one clock cycle. Execution of the entire round in such a circuit would take five clock cycles; however, a simple modification can be applied to maintain the execution rate at the level of four clock cycles per round. The trick is based on the fact that the folded register, described in section 3.1, does not transform data bytes in any

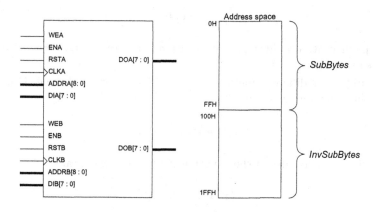

Fig. 5. Block RAM based implementation of *SubBytes* and *InvSubBytes*

other way than just reordering them. Therefore, this stage can be safely skipped if necessary. It apperars that forwarding of only one byte from the input to the folded register to the input of S-boxes is sufficient to maintain the execution rate of four clock cycles per round. Unfortunately, different bytes are forwarded in the case of encryption and in the case of decryption, as shown in Fig. 7.

3.3 Implementation of the *MixColumns* and *InvMixColumns*

The 32-bit input to the *MixColumns* transformation is represented as a polynomial of the form $a(x) = a_3x^3 + a_2x^2 + a_1x + a_0$, with coefficients in $GF(2^8)$. The coefficients of $a(x)$ are also polynomials of the form $b(x) = b_7x^7 + b_6x^6 + b_5x^5 + b_4x^4 + b_3x^3 + b_2x^2 + b_1x + b_0$, with their own coefficients in $GF(2)$.

The *MixColumns* multiplies the input polynomial by a constant polynomial

$$c(x) = \{03\}x^3 + \{01\}x^2 + \{01\}x + \{02\} \tag{1}$$

modulo $x^4 + 1$. The coefficients in $GF(2^8)$ are multiplied modulo $x^8 + x^4 + x^3 + x + 1$. The *InvMixColumns* multiplies the input polynomial by another constant polynomial:

$$d(x) = c^{-1}(x) = \{0b\}x^3 + \{0d\}x^2 + \{09\}x + \{0e\} \tag{2}$$

The implementation of the *MixColumns* is very simple because the coefficients of $c(x)$ are small. On the other hand, the *InvMixColumns* is far more complex and occupies larger area.

A. Satoh et al. [23] proposed an implementation based on the following idea:

$$d(x) = c(x) + e(x) + f(x) \tag{3}$$

where

$$e(x) = \{08\}x^3 + \{08\}x^2 + \{08\}x + \{08\} \tag{4}$$

$$f(x) = \{04\}x^2 + \{04\} \tag{5}$$

This implementation yields logic optimizations since *InvMixColumns* shares logic resources with *MixColumns*.

We propose a different method for exploring resource sharing. Our implementation is derived as follows:

$$c(x) \bullet d(x) = \{01\} \tag{6}$$

If we multiply both sides of the equation (6) by $d(x)$ we obtain:

$$c(x) \bullet d^2(x) = d(x) \tag{7}$$

where

$$d^2(x) = \{04\}x^2 + \{05\} \tag{8}$$

Note that two of the coefficients of the $d^2(x)$ are equal to $\{00\}$.

The *MixColumns* and *InvMixColumns* can be implemented using shared logic resources as shown in Fig. 6.

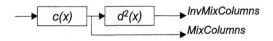

Fig. 6. Implementation of *MixColumns* and *InvMixColumns*

The multiplication by $\{04\}$ and $\{05\}$ lead to following equations:

$$b(x) \bullet \{04\} = b_5x^7 + b_4x^6 + (b_7 + b_3)x^5 + (b_7 + b_6 + b_2)x^4 +$$

$$+ (b_6 + b_1)x^3 + (b_7 + b_0)x^2 + (b_7 + b_6)x + b_6 \tag{9}$$

$$b(x) \bullet \{05\} = (b_7 + b_5)x^7 + (b_6 + b_4)x^6 + (b_7 + b_5 + b_3)x^5 + (b_7 + b_6 + b_4 + b_2)x^4 +$$

$$+ (b_6 + b_3 + b_1)x^3 + (b_7 + b_2 + b_0)x^2 + (b_7 + b_6 + b_1)x + (b_6 + b_0) \tag{10}$$

Their implementation appears area efficient since 4-input XOR gates are the widest gates involved in computations, and they get efficiently implemented in 4-input LUTs of the FPGA.

At the time this paper was written we learned that this technique was first discovered and proposed for software implementations by P. Barreto [5]. V. Fischer and F. Gramain were the first to apply it in hardware [1].

3.4   Encryption/Decryption Unit

Our circuit is capable of performing encryption and decryption. The AES encryption and decryption rounds substantially differ from the point of view of hardware implementations. One of the inconveniences arises from the fact that the *AddRoundKey* is executed after *MixColumns* in the case of encryption, and before *InvMixColumns* in the case of decryption. Therefore, a switching logic is required to select appropriate data paths, which affects the performance, as shown in Fig. 7.

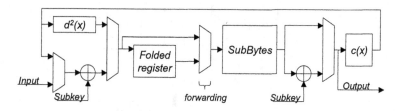

Fig. 7. Implementation of the encryption/decryption unit

It is possible to reorder the *InvMixColumns* and *AddRoundKey* and avoid some of the switching. In this case, the key schedule would need to perform additional *InvMixColumns* transformation on most of the subkeys. The *InvMixColumns* requires much more area than the switching logic. Our implementation delivers sufficient performance with the switching logic in place, therefore we implemented the architecture shown in Fig. 7.

3.5   Implementation of the Key Schedule

The key schedule is typically implemented using one of the two methods: computing keys on-the-fly for every block of encrypted data, or precomputing them in advance and storing. The computation of keys on-the-fly has an obvious advantage of changing keys fast with low or no delay. This performance comes for a price of increased power consumption as the key schedule computes over and over again for each data block.

In the case of the AES it is easy to perform key schedule transformations in the forward direction, and this is the order the round keys are applied in the case of encryption. In the case of decryption round keys are applied in the reversed order. The key schedule could compute round keys in the backward direction, but it is possible only by starting from the last key, not the main key. Unfortunately, the last key can be obtained from the main key only by computing the entire key schedule in the forward direction first. For this reason, the key schedule computing keys on-the-fly completely looses its advantage when decryption is performed.

Our AES implementation is designed to perform encryption and decryption. Since we did not see any advantage in computing round keys on-the-fly, we selected to implement the key schedule that precomputes all round keys. The implementation of the key schedule is shown in Fig. 8. It computes 32-bits of the key material per clock cycle, therefore, full key schedule execution takes 44 clock cycles. The computed round keys are stored in a single Block RAM.

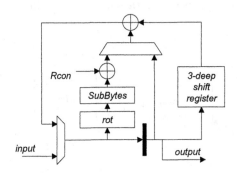

Fig. 8. Implementation of the key schedule

The key schedule uses *SubBytes* operation that is identical to the one used in the encryption circuit. Since key schedule does not work simultaneously with the encryption unit, it is possible to time share S-boxes between both circuits. This approach saves two Block RAMs at the expense of additional switching logic, and degraded performance. The performance is affected by the presence of the switching logic in the critical path, and by slightly more complicated floorplanning and routhing, as encryption/decryption and key schedule units are no longer separated. We implemented the switching logic using tri-state buffers in order to minimize its influence on the overall performance; however, this solution may not be the most desired for various reliability and testability related reasons. In the case when tri-state buffers are not allowed in the design, a multiplexer should be used for switching.

4   Targeted Device, Synthesis, and Implementation Results

The goal for this design was to create a low-cost implementation of AES in the FPGA targeted for real life applications. Much of the previous research targets state-of-the-art technologies forgetting that the individual cost of those devices ranges in hundreds of US dollars. We shifted our attention to older technologies and smaller devices. Xilinx Inc. produces two low-cost families of devices called Spartan II, and Spartan IIE. Pricing for Spartan II FPGAs starts from less than $10 per unit [35].

Spartan II FPGAs are manufactured in $0.22\mu m$ CMOS process. Their architecture is derived from a bigger family of Virtex devices. Spartan IIE are based on a newer VirtexE family, and are manufactured in $0.18\mu m$ CMOS process. The smallest device from the Spartan IIE family was too large for our needs. The device we selected for our implementation is Spartan II XC2S30; second smallest in its family.

The synthesis of our design was done using Synplify Pro 7.2 from Synplicity. We set the constraints for target clock frequency to 60MHz, fanout guide to 100, and enabled resourse sharing. We performed synthesis for speed grades -5 and -6.

The mapping, placing and routing was done using Xilinx ISE 5.2i package. Mapper optimized circuit for area, and router worked with effort level 5.

The results are given in the Table 1. The maximum frequencies come from static timing analysis only. The performance is nearly equally affected by logic and routing. The routing vs. logic delays ratio in the critical path is 54/46. Better results could be demonstrated with manual floorplanning.

Table 1. Implementation results

| Device | Area | | Max. clock | Throughput |
	CLB Slices	Block RAMs	frequency [MHz]	[Mbps]
XC2S30-5	222	3	50	139
XC2S30-6	222	3	60	166

5   Comparison with Other Designs

Despite our intensive search we encountered suprisingly few compact implementations of the AES algorithm in FPGAs. There exist commercial compact cores from Amphion [32] and Helion [33] companies. Both companies provide compact cores in encryption or decryption version only, and a 128-bit key schedule. We also encountered a JBits implementation by S. McMillan et al. [21]. Their implementation uses JBits to tailor the bitstream for particular key, and encryption or decryption operation. Therefore, encryption and decryption are never simultaneously present in the circuit, and the key schedule is not implemented in the hardware.

We also collected information about other existing architectures capable of encrypting or decrypting data in feedback modes of operation [15] [16] [17] [24] [26]. We did not take into account any implementations based on *T-boxes* as they give greater throughput at the expense of much larger area. The basic features of all the implementations are collected in Table 2, and their performance characteristics in Table 3.

Table 2. Basic features of compared architectures

	Device	Encryption	Decryption	Key Schedule 128	192	256
0.22μm						
Our	**Spartan II-6**	•	•	•		
P. Chodowiec et al. [15]	Virtex-6	•	•	•	•	•
A. Dandalis et al. [24]	Virtex-6	•	•	•		
A.J. Elbirt et al. [16]	Virtex-6	•				
V. Fischer et al. [17]	FLEX 10KE-1	•	•			
	ACEX 1K-1	•	•			
K. Gaj et al. [26]	Virtex-6	•	•			
S. McMillan et al. [21]	Virtex	•				
0.18μm						
Amphion CS5220XV [32]	VirtexE-8	•		•		
CS5230XV	VirtexE-8	•		•	•	•
Helion compact [33]	Spartan IIE-6	•		•		
fast	VirtexE-8	•	•	•		
V. Fischer et al. [17]	APEX 20KE-1	•	•			
0.15μm						
Amphion CS5220XV[32]	Virtex2-5	•		•		
CS5230XV	Virtex2-5	•		•	•	•
Helion fast [33]	Virtex2-5	•	•	•		

Among compact architectures, our design is one of the smallest and offers richer functionality than cores from Amphion and Helion because it supports both encryption and decryption. Both commercial cores are faster than ours; however, they are implemented in a better, thus more expensive technology. The implementation by S. McMillan et al. is also very compact and fast; however, it benefits from the JBits application which is not likely to work in an embedded environment.

We notice large differences among results for basic architecture. The implementation by P. Chodowiec et al. offers the most complete functionality and has nearly identical size with the fast Helion core. The implementation by V. Fischer et al. on FLEX and APEX also have similar parameters, but do not include key schedule. The Amphion core CS5230XV is the smallest implementation in the basic architecture, but does not support decryption.

Relating the results for our compact implementation to the implementations in the basic architecture, we can see that the goal of reducing the required logic resources by 75% was achieved. Moreover, the throughput of our design is higher than the 25% of the best throughput reported for the basic architecture in the same technology.

Table 3. Performance of all compared cores

	Area		Throughput	clock cycles
	CLB Slices	Block RAMs	[Mbps]	per round
0.22μm				
Our	**222**	**3**	**166**	**4**
P. Chodowiec et al.	~1230	18	577	1
A. Dandalis et al.	5673	0	353	1
A.J. Elbirt et al.	3528	0	294.2	1
V. Fischer et al. FLEX	2530 LE[1]	24 EAB[2]	451	1
ACEX	2923 LE[1]	12 EAB[2]	212	1
K. Gaj et al.	2902	0	331.5	1
S. McMillan et al.	240	8	250	7
0.18μm				
Amphion CS5220XV	421	4	294	4
CS5230XV	573	10	1061	1
V. Fischer et al. APEX	2493 LE[1]	50 ESB[2]	612	1
Helion compact	392 LUT[1]	3	223	4
fast	2259 LUT[1]	18	1001	1
0.15μm				
Amphion CS5220XV	403	4	350	4
CS5230XV	573	10	1323	1
Helion fast	2259 LUT[1]	18	1408	1

[1] 2 LE $\approx$ 2 LUT $\approx$ 1 Slice   [2] 1 EAB = 2 ESB = 1 BRAM

We intentionally did not provide Throughput/Area ratios for any of the compared designs as this measure can be very misleading when dedicated memories are present in the design.

6   Conclusions

In this paper the feasibility of creating a very compact, low-cost FPGA implementation of the AES was examined. The proposed folded architecture achieves good performance and occupies less area than previously reported designs. This compact design was developed by thorough examination of each of the components of the AES algorithm and matching them into the architecture of the FPGA.

The demonstrated implementation fits in a very inexpensive, off-the-shelf Xilinx Spartan II XC2S30 FPGA, which cost starts below $10 per unit. Only 50% of the logic resources available in this device were utilized, leaving enough area for additional glue logic. This implementation can encrypt and decrypt data streams up to 166 Mbps. The encryption speed, functionality, and cost make this solution perfectly practical in the world of embedded systems and wireless communication.

References

1. Fischer V. and Gramain F.: *Resource sharing in a Rijndael implementation based on a new MixColumn and InvMixColymn relation*, submitted to Electronic Letters, reference number: ELL 39 395, April 14, 2003
2. Järvinen K.U., Tommiska M.T., Skyttä J.O.: *A fully pipelined memoryless 17.8 Gbps AES-128 encryptor*, International Symposium on Field-Programmable Gate Arrays (FPGA 2003), Monterey, CA, 2003
3. Standaert F.X., Rouvroy G., Quisquater J.J., Legat J.D., *A methodology to implement block ciphers in reconfigurable hardware and its application to fast and compact AES RIJNDAEL*, International Symposium on Field-Programmable Gate Arrays (FPGA 2003), Monterey, CA, 2003
4. Verbauwhede I., Schaumont P., Kuo H.: *Design and performance testing of a 2.29-GB/s rijndael processor*, IEEE Journal of Solid-State Circuits, Volume: 38 Issue: 3, March 2003
5. Daemen J. and Rijmen V.: *The design of Rijndael: AES – The Advanced Encryption Standard*, Springer-Verlag, ISBN 3-540-42580-2, 2002
6. Lin T.F., Su C.P., Huang C.T., Wu C.W.: *A high-throughput low-cost AES cipher chip*, IEEE Asia-Pacific Conference on ASIC, 2002
7. Lutz A.K., Treichler J., Gürkaynak F.K., Kaeslin H., Basler G., Erni A., Reichmuth S., Rommens P., Oetiker S., Fichtner W., *2Gbit/s Hardware Realizations of RIJNDAEL and SERPENT: A Comparative Analysis*, Cryptographic Hardware and Embedded Systems (CHES 2002), San Francisco Bay, CA, 2002
8. Mayer U., Oelsner C., Kohler T.: *Evaluation of different rijndael implementations for high end servers*, IEEE International Symposium on Circuits and Systems (ISCAS 2002), 2002
9. Mitsuyama Y., Andales Z., Onoye T., Shirakawa I.: *Burst mode: a new acceleration mode for 128-bit block ciphers*, IEEE Custom Integrated Circuits Conference, 2002
10. Morioka S. and Satoh A., *An Optimized S-Box Circuit Architecture for Low Power AES Design*, Cryptographic Hardware and Embedded Systems (CHES 2002), San Francisco Bay, CA, 2002
11. Morioka S. and Satoh A.: *A 10 Gbps full-AES crypto design with a twisted-BDD S-Box architecture*, IEEE International Conference on Computer Design: VLSI in Computers and Processors, 2002
12. Ravi S., Raghunathan A., Potlapally N.: *Securing Wireless Data: System Architecture Challenges*, Symposium on System Synthesis, 2002
13. Wolkerstorfer J., Oswald E., Lamberger M.: *An ASIC Implementation of the AES SBoxes*, The Cryptographer's Track at the RSA Conference, San Jose, CA, 2002
14. Chodowiec P., Khuon P., Gaj K.: *Fast implementations of secret-key block ciphers using mixed inner- and outer-round pipelining*, International Symposium on Field-Programmable Gate Arrays (FPGA 2001), Monterey, CA, 2001
15. Chodowiec P., Gaj K., Bellows P., Schott B.: *Experimental Testing of the Gigabit IPSec-Compliant Implementations of Rijndael and Triple DES Using SLAAC-1V FPGA Accelerator Board*, Information Security Conference (ISC 2001), Malaga, Spain, 2001
16. Elbirt A.J., Yip W., Chetwynd B., Paar C.: *An FPGA-based performance evaluation of the AES block cipher candidate algorithm finalists*, IEEE Transactions on Very Large Scale Integration (VLSI) Systems, Volume: 9 Issue: 4, August 2001
17. Fischer V. and Drutarovský M.: *Two Methods of Rijndael Implementation in Reconfigurable Hardware*, Cryptographic Hardware and Embedded Systems (CHES 2001), Paris, France, 2001

18. McLoone M. and McCanny J.V.: *High Performance Single-Chip FPGA Rijndael Algorithm Implementations*, Cryptographic Hardware and Embedded Systems (CHES 2001), Paris, France, 2001
19. McLoone M. and McCanny J.V.: *Single-Chip FPGA Implementation of the Advanced Encryption Standard Algorithm*, Field-Programmable Logic and Applications (FPL 2001), Belfast, Northern Ireland, UK, 2001
20. McLoone W., McCanny J.V.: *Rijndael FPGA implementation utilizing look-up tables*, IEEE Workshop on Signal Processing Systems, 2001
21. McMillan S. and Patterson C.: *JBits Implementations of the Advanced Encryption Standard (Rijndael)*, Field-Programmable Logic and Applications (FPL 2001), Belfast, Northern Ireland, UK, 2001
22. Rudra A., Dubey P.K., Jutla C.S., Kumar V., Rao J.R., Rohatgi P.: *Efficient Rijndael Encryption Implementation with Composite Field Arithmetic*, Cryptographic Hardware and Embedded Systems (CHES 2001), Paris, France, 2001
23. Satoh A., Morioka S., Takano K., Munetoh S.: *A Compact Rijndael Hardware Architecture with S-Box Optimization*, Theory and Application of Cryptology and Information Security (ASIACRYPT 2001), Gold Coast, Australia, 2001
24. Dandalis A., Prasanna V.K., Rolim J.D.: *A Comparative Study of Performance of AES Final Candidates Using FPGAs*, Cryptographic Hardware and Embedded Systems Workshop (CHES 2000), Worcester, Massachusetts, 2000
25. Elbirt A.J., Yip W., Chetwynd B., Paar C.: *An FPGA Implementation and Performance Evaluation of the AES Block Cipher Candidate Algorithm Finalists*, Third Advanced Encryption Standard (AES3) Candidate Conference, New York, 2000
26. Gaj K. and Chodowiec P.: *Comparison of the hardware performance of the AES candidates using reconfigurable hardware*, Third Advanced Encryption Standard (AES3) Candidate Conference, New York, 2000
27. Gaj K. and Chodowiec P.: *Hardware performance of the AES finalists-survey and analysis results*, Technical Report, George Mason University, 2000, available at http://ece.gmu.edu/crypto/AES_survey.pdf
28. Ichikawa T. and Matsui T.: *Hardware Evaluation of the AES Finalists*, Third Advanced Encryption Standard (AES3) Candidate Conference, New York, 2000
29. National Institute of Standards and Technology: *FIPS 197: Advanced Encryption Standard*, November 2001
30. Daemen J. and Rijmen V.: *AES Proposal: Rijndael*, http://csrc.nist.gov/encryption/aes/rijndael/Rijndael.pdf
31. Rijmen V.: *Efficient Implementation of the Rijndael S-box*, available at: http://www.esat.kuleuven.ac.be/~rijmen/rijndael/sbox.pdf
32. Amphion: http://www.amphion.com/
33. Helion: http://www.heliontech.com/
34. Xilinx, Inc.: *Spartan II Data Sheet*, available at: http://www.xilinx.com
35. Xilinx, Inc.: *The New Spartan-II FPGA Family: Kiss Your ASIC Good-bye*, XCELL Journal, Q1, 2000, available at: http://www.xilinx.com/xcell/xl35/xl35_5.pdf

Efficient Implementation of Rijndael Encryption in Reconfigurable Hardware: Improvements and Design Tradeoffs

Francois-Xavier Standaert, Gael Rouvroy, Jean-Jacques Quisquater, and
Jean-Didier Legat

UCL Crypto Group
Laboratoire de Microelectronique
Universite Catholique de Louvain
Place du Levant, 3, B-1348 Louvain-La-Neuve, Belgium
standaert,rouvroy,quisquater,legat@dice.ucl.ac.be

Abstract. Performance evaluation of the Advanced Encryption Standard candidates has led to intensive study of both hardware and software implementations. However, although plentiful papers present various implementation results, it seems that efficiency could still be greatly improved by applying good design rules adapted to devices and algorithms. This paper addresses various approaches for efficient FPGA implementations of the Advanced Encryption Standard algorithm. As different applications of the AES algorithm may require different speed/area tradeoffs, we propose a rigorous study of the possible implementation schemes, but also discuss design methodology and algorithmic optimization in order to improve previously reported results. We propose heuristics to evaluate hardware efficiency at different steps of the design process. We also define an optimal pipeline that takes the place and route constraints into account. Resulting circuits significantly improve previously reported results: throughput is up to 18.5 Gbits/sec and area requirements can be limited to 542 slices and 10 RAM blocks with a ratio throughput/area improved by at least 25% of the best-known designs in the Xilinx Virtex-E technology.

1 Introduction

In October 2000, NIST (National Institute of Standards and Technology) selected Rijndael [2] as the new Advanced Encryption Standard. The selection process included performance evaluation on both software and hardware platforms. Many hardware architectures were proposed [3 − 16], but most of them were simple implementations according to the Rijndael specification. More recently, design strategies and implementation approaches were proposed for the implementation of block ciphers in reconfigurable hardware [17,18] while other papers focused on some interesting algorithmic optimizations, specially for the highly expensive substitution box of Rijndael [19,20,21]. This paper addresses various approaches for FPGA implementations of the Advanced Encryption Standard

C.D. Walter et al. (Eds.): CHES 2003, LNCS 2779, pp. 334–350, 2003.

algorithm and combines recent observations about Rijndael in efficient designs. As different applications of the AES algorithm may require different speed/area tradeoffs, we propose a rigorous study of the possible implementation schemes, but also discuss design methodology and algorithmic optimization in order to improve previously reported results. We first discuss the implementation of the substitution box and linear diffusion layer at the algorithmic level. Then we examine different possible architectures and optimizations. Finally, we present heuristics allowing to evaluate the efficiency of our architectures at different steps of the design process. Synthesis and implementation constraints of FPGAs are taken into account in order to define maximum and optimal pipeline. We apply these notions to loop and unrolled architecture in order to improve circuits performances and compare our results to the best designs reported in literature. The main contribution of this paper has to be found in the improvement of hardware efficiency that we define as the ratio throuhput/area: efficiency of best-known unrolled architectures is improved by 35% while efficiency of best-known loop architectures is improved by at least 25% in the Xilinx Virtex-E technology.

This paper is structured as follows. The description of the hardware, synthesis tool and implementation tool is in section 2. Section 3 gives a short mathematical description of Rijndael and we propose an efficient representation of the key schedule by means of a key round. The main contribution of this paper lies in section 4 where we discuss the possible implementation tradeoffs. Section 4.1 deals with design methodology and defines hardware efficiency and maximum pipeline for FPGAs. Section 4.2 presents possible algorithmic optimization of Rijndael. Different schemes for the substitution box are proposed and the diffusion layer is combined with the key addition. Section 4.3 proposes different architectures for various speed/area tradeoffs: loop architectures and unrolled architectures are studied and implemented. Finally, section 4.4 defines optimal pipeline for FPGAs as well as a heuristic rule to reach it. Practical results and comparisons with best known published designs are in section 5 and conclusions are in section 6.

2   Hardware Description

All our implementations were carried out on a XILINX VIRTEX3200ECG1156-8 FPGA. We chose this technology in order to allow relevant comparisons with the best-known FPGA implementations of Rijndael. In this section, we briefly describe the structure of a VIRTEX FPGA as well as the synthesis and implementation tools that were used to obtain our results.

Configurable Logic Blocks (CLB's): The basic building block of the VIRTEX logic block is the logic cell (LC). An LC includes a 4-input function generator, carry logic and a storage element. The output from the function generator in each LC drives both the CLB output and the D input of the flip-flop. Each VIRTEX CLB contains four LC's, organized in two similar slices. Figure 1, shows a detailed view of a single slice. Virtex function generators are implemented

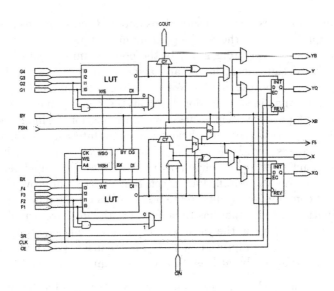

Fig. 1. The VIRTEX slice.

as 4-input look-up tables (LUTs). In addition to operate as a function generator, each LUT can provide a 16×1-bit synchronous RAM. Furthermore, the two LUTs within a slice can be combined to create a 16×2-bit or 32×1-bit synchronous RAM or a 16×1-bit dual port synchronous RAM. The VIRTEX LUT can also provide a 16-bit shift register.

The storage elements in the VIRTEX slice can be configured either as edge-triggered D-type flip-flops or as level-sensitive latches. The D inputs can be driven either by the function generators within the slice or directly from slice inputs, bypassing function generators.

The F5 multiplexer in each slice combines the function generator outputs. This combination provides either a function generator that can implement any 5-input function, a 4:1 multiplexer, or selected functions of up to nine bits. Similarly, the F6 multiplexer combines the outputs of all four function generators in the CLB by selecting one of the F5-multiplexer outputs. This permits the implementation of any 6-input function, an 8:1 multiplexer, or selected functions up to 19 bits. The arithmetic logic also includes a XOR gate that allows a 1-bit full adder to be implemented within an LC. In addition, a dedicated AND gate improves the efficiency of multiplier implementations.

Finally, VIRTEX FPGAs incorporate several large RAM blocks. These complement the distributed LUT implementations of RAM's. Every block is a fully synchronous dual-ported 4096-bit RAM with independent control signals for each port. The data widths of the two ports can be configured independently.

Our hardware: A VIRTEX3200ECG1156-8 FPGA contains 32448 slices and 208 RAM blocks, which means 64896 LUTs and 64896 flip-flops. In the next sections, we compare the number of LUTs, registers and slices. We also evaluate the delays and frequencies thanks to our synthesis tool. The synthesis was performed with FPGA Compiler 2 3.7.1 (SYNOPSYS) and the implementation with XILINX ISE-5. Finally, our circuit models were described using VHDL.

3   Block Cipher Description

Rijndael is an iterated block cipher that operates on a 128-bit cipher state and uses a 128-bit key[1]. It consists of a serie of 10 applications of a key-dependent round transformation to the cipher state. In the following, we will individually define the component mappings and constants that build up Rijndael, then specify the complete cipher in terms of these components.

Representation: The state and key are represented as a square array of 16 bytes. This array has 4 rows and 4 columns. It can also be seen as a vector in $GF(2^8)^{16}$. Let s be a cipher state or a key $\in GF(2^8)^{16}$, then s_i is the i-th byte of the state s and $s_i(j)$ is the j-th bit of this byte.

SubBytes, the non-linear layer γ: The SubBytes transformation is a non-linear byte substitution, operating on each byte independently. The substitution table (or s-box) is invertible and is constructed by the composition of two operations:

1. The multiplicative inverse in $GF(2^8)$.
2. An affine transform over $GF(2)$.

Every byte is therefore considered as a polynomial with coefficients in $GF(2)$:

$$b(x) = b_7 x^7 + b_6 x^6 + b_5 x^5 + b_4 x^4 + b_3 x^3 + b_2 x^2 + b_1 x^1 + b_0 x^0$$
$$b_7 b_6 b_5 b_4 b_3 b_2 b_1 b_0 \rightarrow b(x) \qquad (1)$$

Then SubBytes consists of the parallel application of this s-box S:

$$\gamma(a) = b \Leftrightarrow b_i = S[a_i], 0 \leq i \leq 15 \qquad (2)$$

The ShiftRows transformation δ: In ShiftRows, the rows of the state are cyclically shifted over different offsets. Row 0 is not shifted, row 1 is shifted over 1 byte, row 2 over 2 bytes and row 3 over 3 bytes.
The MixColumns transformation θ: In MixColumns, the columns of the state are considered as polynomials over $GF(2^8)$ and multiplied modulo $x^4 + 1$ with a fixed polynomial $c(x)$, given by:

$$c(x) =' 03' x^3 +' 01' x^2 +' 01' x +' 02' \qquad (3)$$

The polynomial is coprime to $x^4 + 1$ and therefore is invertible. This can be written as a matrix multiplication:

---

[1] Actually, there exist several versions of Rijndael with different block and key lengths, but we focus on this one.

$$\begin{bmatrix} b_0 \\ b_1 \\ b_2 \\ b_3 \end{bmatrix} = \begin{bmatrix} 02\ 03\ 01\ 01 \\ 01\ 02\ 03\ 01 \\ 01\ 01\ 02\ 03 \\ 03\ 01\ 01\ 02 \end{bmatrix} \times \begin{bmatrix} a_0 \\ a_1 \\ a_2 \\ a_3 \end{bmatrix}$$

Where (b_3, b_2, b_1, b_0) is a four-byte column of the state. An output byte of Mix-Columns (for example b_0) can be expressed as:

$$b_0 =' 02' \times a_0 \oplus' 03' \times a_1 \oplus' 01' \times a_2 \oplus' 01' \times a_3$$

We also define a function X, corresponding to the multiplication with '02' modulo the irreducible polynomial $m(x) = x^8 + x^4 + x^3 + x + 1$: $X : GF(2^8) \rightarrow GF(2^8) : X(a) = b \Leftrightarrow$

$$b(7) = a(6)$$
$$b(6) = a(5)$$
$$b(5) = a(4)$$
$$b(4) = a(3) \oplus a(7)$$
$$b(3) = a(2) \oplus a(7)$$
$$b(2) = a(1) \oplus a(7)$$
$$b(1) = a(0)$$
$$b(0) = 0 \oplus a(7)$$

The round key addition $\sigma[K]$: In this operation, a round key is applied to the state by a simple bitwise EXOR. The round key is derived from the cipher key by means of the key schedule. The round key length is equal to the block length.

$$\sigma[k](a) = b \Leftrightarrow b_i = a_i \oplus k_i, 0 \le i \le 15 \tag{4}$$

The round transformation $\rho[K]$: The round transformation can be written as a composition of the four previous transformations:

$$\rho[K] = \sigma[K] \circ \theta \circ \delta \circ \gamma = \sigma[K]\big(\theta(\delta(\gamma))\big) \tag{5}$$

The key schedule: The round keys are derived from the cipher key by means of the key schedule. This consists of two transformations: the key expansion and the round key selection. In our description, SubWord (SW) is a function that takes a 4-byte word in which each byte is the result of applying the Rijndael s-box. The function RotWord (RW) returns a word in which the bytes are a cyclic permutation of those in its inputs such that the input word (a, b, c, d) produces the output word (b, c, d, a). Finally, $RC(i)$ is an 8-bit round constant for the round i.

The key schedule can be easily described by the use of a key round β that takes four 4-byte input words, corresponding to a 128-bit key, and produces four 4-byte output words. The first round key K_0 is the cipher key, then, we have:

$$K_{i+1} = \beta(K_i), i = 0, ..., 10 \tag{6}$$

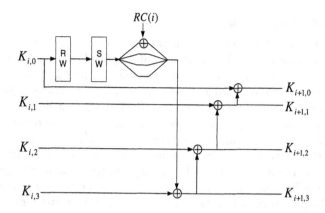

Fig. 2. The key round β.

Figure 2 illustrates the key round of Rijndael.

The complete cipher: Rijndael is defined for the cipher key K as the transformation Rijndael$[K]= \alpha[K_0, K_1, ..., K_{10}]$ applied to the plaintext where:

$$\alpha[K_0, K_1, ..., K_{10}] = \sigma[K_{10}] \circ \delta \circ \gamma \circ (\bigcirc_{r=1}^{9}\rho[K_r]) \circ \sigma[K_0] \qquad (7)$$

Our implementations are based on this description of AES Rijndael.

4   Implementation Tradeoffs

The optimization methods and the resulting implementation tradeoffs for the implementation of AES Rijndael can be divided into two classes: architectural and algorithmic optimization. Algorithmic optimization exploits algorithmic strength inside each round unit. Architectural optimization exploits design techniques such as pipelining, loop unrolling and sub-pipelining.

This paper first considers **loop architectures**, where only a small number m (typically $m = 1$) of rounds are independently implemented in hardware. Loop architectures enables small area circuits but have low throughput. Then we improve the throughput at the cost of increased area by the combination of loop unrolling and pipelining. **Unrolled architectures** have a large number m of rounds (typically all) that are independently implemented in hardware. **Pipelining** increases the encryption speed by processing multiple blocks of data simultaneously. It is achieved by inserting rows of registers among combinatorial logic. Parts of logic between two consecutive registers form pipeline stages. In case of block ciphers, each round constitutes a pipeline stage. Finally, **sub-pipelining** is similar to pipelining but also inserts registers inside the round functions.

Concerning algorithmic optimizations, we focused on the critical parts of Rijndael. Different schemes for the substitution box are proposed and compared. We also underline interesting combinations of the MixColumns θ with the key addition $\sigma[K]$. In this section, we propose a rigorous study of the possible tradeoffs for implementing Rijndael.

4.1   Design Methodology

In [18], a methodology to implement block ciphers in reconfigurable hardware is presented, based on simple digital design rules applied to iterated block ciphers. Looking at the round functions of iterated block ciphers, it is observed that they are mainly built on simple algebraic or logic operations. Therefore, the sub-pipelining of round functions is mandatory if efficient designs are wanted. Practically, the designer can easily keep his critical path inside one CLB slice. Moreover, looking at the CLB strcture, it can be seen that FPGAs involve specific constraints that have to be taken into account if an optimal design is wanted. As the slice of Figure 1 is divided into logic elements and storage elements, an efficient implementation will be the result of a better compromise between combinatorial logic used, sequential logic used and resulting performances. These observations lead to different definitions of implementation efficiency:

1. In terms of performances, let the efficiency of a block cipher be the ratio *Throughput (Mbits/s)/Area (slices)*.
2. In terms of resources, the efficiency is easily tested by computing the ratio *Nbr of LUTs/Nbr of registers*: it should be close to one.

Our implementations of Rijndael were designed in order to maximize these notions of hardware efficiency. It practically results in the sub-pipelining of every component of the round functions. The next section studies algorithmic optimizations combined with good sub-pipelining. For this purpose, we define the **maximum pipeline** as the pipeline of which number of stages implies that the ratio *Nbr of LUTs/Nbr of registers* is the closest to one (and lower than one).

4.2   Algorithmic Optimizations: A First Tradeoff

A. Implementing the substitution box: The Rijndael S-box is a non-linear byte substitution used 200 times in Rijndael with 128-bit block length and key length. It is invertible and is constructed by the composition of two transformations:

1. The mapping $x \rightarrow x^{-1}$, where x^{-1} represents the multiplicative inverse in the field $GF(2^8)$.
2. An affine transformation over $GF(2)$: $x \rightarrow Ax + b$, where A and b are constants.

In terms of hardware resources, the substitution box is the most expensive part of Rijndael. As a consequence, its implementation is a critical part in the design of an efficient encryption core. This transform can be implemented following different schemes. We propose to observe three possibilities and the resulting constraints.

A1. The multiplexor model: A first and obvious solution is to consider SubBytes as a large multiplexor and take advantage of special FPGA configurations to implement these ones. Figure 3 illustrates the implementation of an output bit of the Rijndael s-box. We pipelined γ by inserting two register levels so that the critical path corresponds to one 4-input LUT, one multiplexor F5 and one multiplexor F6. Table 1 summarizes the synthesis results for the non-linear transform γ where the s-box is repeated 16 times.

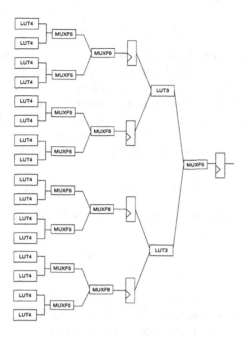

Fig. 3. The substitution box γ.

Table 1. Synthesis of the non-linear layer γ.

Component	Nbr of LUT	Nbr of registers
γ	$144 \times 16 = 2304$	$42 \times 16 = 672$

A2. RAM-based implementation: Another possibility is to use the RAM blocks available inside the VIRTEX to implement substitution boxes. The resulting SubBytes transform uses 8 RAM blocks and is performed in one clock cycle.

A3. Composite field solution: In AES Rijndael, every byte represents an element in the finite field $GF(2^8)$. It can also be represented as a polynomial of degree 8 in the field $GF(2)$: $b_7.x^7 + b_6.x^6 + b_5.x^5 + b_4.x^4 + b_3.x^3 + b_2.x^2 + b_1.x^1 + b_0$. Addition and substraction of polynomials are given by the sum modulo 2 of the coefficients of both terms (bitwise XOR). Multiplication in $GF(2^8)$ corresponds to multiplication of polynomials modulo an irreducible binary polynomial of degree 8. Rijndael uses $m(x) = x^8 + x^4 + x^3 + x + 1$. As the irreducible polynomial is used to construct the field and there are different irreducible polynomials of degree 8, several finite fields can be considered and generate different representations of Rijndael. These fields are isomorphic which means that there is a one-to-one mapping from one representation of Rijndael to another. Finally, the multiplicative inverse of a polynomial $b(x)$ is defined such that:

$$b(x).b^{-1}(x) = 1.mod.m(x) \tag{8}$$

In [19], subfield arithmetics are used to propose efficient implementations of Galois Field arithmetic, especially in the context of the Rijndael block cipher. Computations in the field $GF(2^8)$ are replaced by computations in the composite field $GF(2^4)^2$ in order to reduce the size of the tables needed for the inversion. Basically, the idea is to consider our polynomial of degree 8 in the field $GF(2)$ as a polynomial of degree 2 in the field $GF(2^4)$, say $a_1x + a_0$, where $a_0, a_1 \in GF(2^4)$. The multiplicative inverse of $a_1x + a_0$ is computed in the field $GF(2^4)^2$ as a polynomial $b_1x + b_0$ such that:

$$(a_1x + a_0) \times (b_1x + b_0) = 1.mod.P(x) \tag{9}$$

Where $P(x)$ is an irreductible polynomial and coefficients b_0, b_1 can be expressed as follows:

$$b_1 = a_1.(a_0^2 + a_1a_0 + \Delta a_1^2)^{-1}$$
$$b_0 = (a_0 + a_1).(a_0^2 + a_1a_0 + \Delta a_1^2)^{-1} \tag{10}$$

[19] gives details about parameter Δ and polynomial $P(x)$ as well as an affine transform that maps elements of $GF(2^8)$ to elements of $GF(2^4)^2$. We implemented the resulting composite field s-box as represented in Figure 4. We inserted seven pipeline levels in order to get the ratio $Nbr\ of\ LUTs/Nbr\ of\ registers$ close to one. Remark that this representation of the substitution box allows to keep the whole design unchanged as the Galois Field transform is used twice in order to be compatible with other transforms. Table 2 summarizes the synthesis results for the composite non-linear transform γ where the s-box is repeated 16 times. Compared to the multiplexor model, we have traded LUTs for registers, and obtained a better efficiency.

Table 2. Synthesis of the composite non-linear layer γ.

Component	Nbr of LUT	Nbr of registers
γ	$84 \times 16 = 1344$	$76 \times 16 = 1216$

B. Implementing the Other Components: The Mixadd Combination

B1. The ShiftRows transform δ: This is just routing information and takes no place in the design.

B2. The MixColumns transform θ: Mixcolum operates on a 4-byte column and corresponds to multiplications and additions in $GF(2^8)$. For example, for the output byte b_0, we have:

$$b_0 =' 02'a_0 +' 03'a_1 +' 01'a_2 +' 01'a_3 \tag{11}$$

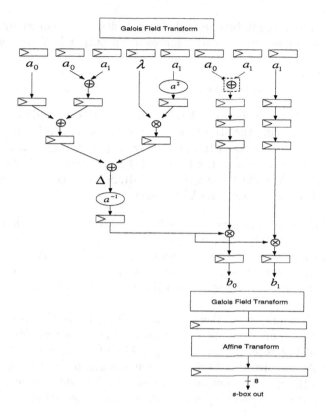

Fig. 4. The composite substitution box.

We implemented multiplications with a function X that corresponds with the multiplication with '02', modulo the irreducible polynomial $m(x) = x^8 + x^4 + x^3 + x + 1$. Figure 5(a) illustrates the function X. Note that output bits 0,2,5,6,7

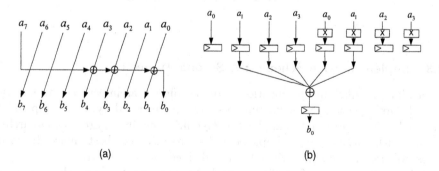

(a)   (b)

Fig. 5. (a) The function X. (b) Output byte b_0 of MixColumns.

just correspond to input bits shifted. Only 3 bits are modified by an EXOR operation. From this, we can easily represent an output byte of θ as shown in Figure 5(b):

$$b_0 = X(a_0) \oplus X(a_1) \oplus a_1 \oplus a_2 \oplus a_3 \tag{12}$$

Interesting combinations between MixColumns and the key addition can be performed when observing the structure of the Virtex slice (see Figure 1). Indeed, we observe that a slice offers the possibility to perform an EXOR between 5 bits: four bits are managed by the LUT and the last one by an EXOR gate next to the LUT. Our Mixadd transform takes advantage of this configuration and keeps the critical path inside one Virtex slice.

B3. The Mixadd transform ϵ: In Figure 5(b), we observe that an output byte of θ is obtained by a bitwise EXOR between 5 bytes: 3 are input bytes and the remaining ones are output bytes of function X. However, looking at the bit level, we know that 5 output bits of X are just shifted input bits. For these ones, only one register is needed to pipeline the diffusion layer.

For the 3 remaining bits, there is an additional EXOR inside the function X. Therefore, for these bits, we compute the bitwise EXOR between the 3 left bytes of Figure 5(b) and the output bits of X independently. Then we insert a register. A bitwise EXOR operation remains to be carried out and we combine it with the key addition. The resulting Mixadd transformation only needs two register levels to keep a critical path inside one slice.

Figure 6(a) illustrates the combination of MixColumns and Addroundkey at the bit level. Finally, Table 3 summarizes the synthesis results for the Mixadd transformation.

Table 3. Synthesis of the Mixadd transform ϵ.

Component	Nbr of LUT	Nbr of registers
ϵ	304	304

4.3   Implementation Schemes: A Second Tradeoff

Depending on different optimization criteria, different architectures can be employed. Optimization for maximum speed can be achieved by a fully pipelined unrolled architecture. In the applications requiring minimum area, a loop architecture with only one round implemented seems to be the best choice. In both cases, we tried to maximize the efficiency defined in section 4.1.
Our implementations of AES Rijndael directly results from the previous component descriptions. For high throughput constraints, we implemented a

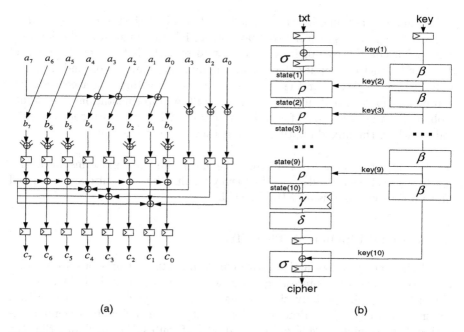

Fig. 6. (a) Mixadd transform at the bit level. (b) AES Rijndael: unrolled architecture.

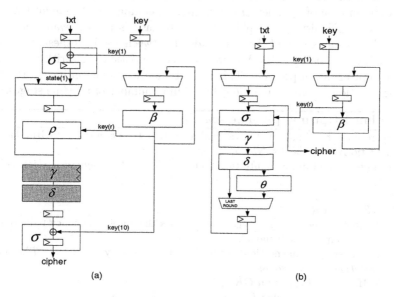

Fig. 7. AES Rijndael: loop architecture (a) and (b).

pipeline version that unrolls the 10 AES rounds illustrated in Figure 6(b). For low area constraints, we propose sequential implementations with only one unrolled round. Figure 7(a) uses the optimized combination of MixColumns and addkey and its grey functions are actually included into the round ρ. Figure 7(b) modifies the round structure so that the initial and final key additions are managed inside the round. It is important to remark that the modification of the round structure implies the loss of our mixadd combination and needs an additional multiplexor for the last round of the algorithm. As a consequence it presents no practical advantage in our FPGA implementations but would probably be the best choice for ASICs where the CLB structure does not exist and for which the mixadd optimization is therefore not relevant.

For all our proposals, we evaluated the hardware cost in terms of LUTs, registers and slices as well as the frequency results. These results are estimated after implementation, using XILINX ISE5.

4.4   Optimal Pipeline: A Third Tradeoff

The design methodology and algorithmic optimization of previous sections allowed us to reach very interesting frequencies after synthesis. However, the implementation (and specially the routing task) of large designs was a critical constraint in our designs. Practically, our most pipelined circuits presented surprising delays including 20% of logic and 80% of routes. We concluded that the real bottleneck of such large ciphers is the difficulty of having an efficient place and route: in case of complex circuits, high pipelining is not mandatory. Moreover, as the difficulty of the place and route task is hardly evaluated, a new practical problem is to find the best tradeoff between good synthesis results and good implementation results. We propose the heuristic of Algorithm 1 to solve this last optimization problem. This heuristic led us to the optimized results of the next section where we mention the optimal number of pipeline stages.

Algorithm 1 Optimal pipeline search

---

1. Start from the maximal pipeline defined in section 4.1, i.e. implement Rijndael with the best ratio *Nbr of LUTs/Nbr of registers*;
2. After implementation, compute the efficiency $E_{cur} = Throughput\ (Mbits/s)/Area$ (*slices*);
3. $OK = 0$;
While $OK = 0$ **do** {
    1. Remove the pipeline stage that involves the lowest frequency reduction and re-implement Rijndael;
    2. After implementation, compute the efficiency $E_{nxt} = Throughput$ (*Mbits/s*)/*Area* (*slices*);
    3. **If** $E_{cur} \geq E_{nxt}$ **then** $OK = 1$;
                else $E_{cur} = E_{nxt}$;
    }
4. The final efficiency E_{cur} specifies the optimal pipeline;

---

5 Practical Results and Comparisons

In order to take every possible tradeoff into account in this section, we list our results for different architectures and different substitution boxes. The tables presented are based on the optimal pipeline defined in previous section. Loop architectures (Figure 7(a)) are in Table 4. Unrolled architectures (Figure 6(b)) are in Table 5.

Table 4. Rijndael encryption: loop architectures on VIRTEX3200E.

Type	Nbr of LUT	Nbr of reg.	Nbr of slices	RAM blocks	Latency (cycles)	Output every (cycles)	Freq. after Impl. (Mhz)	Throughput (Mbits/sec)
LUT-based γ	3846	2517	2257	0	52	5/52	169	2008
RAM-based γ	877	668	542	10	21	2/21	119	1450
Composite γ	2524	2185	1767	0	82	8/82	167	2085

Table 5. Rijndael encryption: unrolled architectures on VIRTEX3200E.

Type	Nbr of LUT	Nbr of reg.	Nbr of slices	RAM blocks	Latency (cycles)	Output every (cycles)	Freq. after Impl. (Mhz)	Throughput (Mbits/sec)
LUT-based γ	33712	14592	19072	0	42	1	86	11008
RAM-based γ	3516	3840	2784	100	21	1	92	11776
Composite γ	19752	13479	15112	0	72	1	145	18560

Table 6. Comparisons with other implementations on VIRTEX-E technology.

Type	Nbr of LUTs	Nbr of slices	RAM blocks	Throughput (Mbits/s)	Throughput/Area ($\frac{Mbits/s}{slices,LUTs}$)
McLoone et al. [9]	/	2222	100	6956	3.1
Our design	**3516**	**2784**	**100**	**11776**	**4.2**
Helion tech. [10]	899	/	10	1187	1.32
Our design	**877**	**542**	**10**	**1450**	**1.65**
Satoh et al. composite [15]	/	1880	0	589	0.31
Our design	**2524**	**1767**	**0**	**2085**	**1.17**
Satoh et al. mux [15]	/	2529	0	833	0.33
Our design	**3846**	**2257**	**0**	**2008**	**0.88**

Finally, in Table 6, we compare our results with the best implementations of Rijndael encryption on VIRTEX-E technology found in literature. For RAM based substitution boxes, McLoone and McCanny had the best unrolled implementation in CHES 2001 [9] while Helion Technologies [10] had the best loop architecture. For LUT-based substitution boxes, we have no knowledge of any unrolled architecture but Satoh and Morioka presented in ASIACRYPT 2001 and in the Third NESSIE workshop the best results for loop implementations [15, 20]. They studied mux-modeled s-boxes as well as composite ones. Finally, concerning older technologies[2], we report in Table 7 an old result of our LUT-based loop architecture and compare it with results of the last AES conference [3,4,5, 6]. It is obvious that the methodology applied allowed us to significantly improve previously reported performances of Rijndael implemented in FPGAs.

6   Conclusion

When implementing block ciphers, several strategies can produce effective designs. Based on recently published works and observations about Rijndael, we studied different possible implementation tradeoffs. Inherent constrainst of FPGAs were taken into account in order to define an efficient methodology. We defined notions of hardware efficiency and optimal pipeline and our circuits were designed in order to optimize different possible architectures: loop and unrolled. Inside these architectures, we proposed algorithmic optimizations for the substitution box but also efficient combinations between the diffusion layer and the key addition.

Upon comparison, our circuits offer better performance than previously reported in literature. Compact and high speed architectures are proposed and implemented on VIRTEX-E technology. Throughput is up to 18.5 Gbits/sec and area requirements can be limited to 542 slices and 10 RAM blocks with an improved ratio throughput/area. Optimized efficiency was obtained by applying heuristic rules in order to deal with place and route constraints.

Table 7. Comparisons with the last AES conference.

Type	Nbr of slices	Device	Throughput (Mbits/s)	Throughput/Area ($\frac{Mbits/s}{slices}$)
Gaj et al.	2900	VIRTEX1000	331.5	0.11
Dandalis et al.	5673	VIRTEX1000	353	0.06
Elbirt et al.	9004	VIRTEX1000	1940	0.22
Our design	**2257**	**VIRTEX1000**	**1563**	**0.69**

References

1. Xilinx: *Virtex 2.5V Field Programmable Gate Arrays Data Sheet,*
   http://www.xilinx.com.
2. J.Daemen and V.Rijmen, *AES Proposal: Rijndael,* NIST's AES home page,
   http://www.nist.gov/aes.

---

[2] Most of the AES performance evaluation was done on VIRTEX1000 FPGAs.

3. A.J. Elbirt et Al, *An FPGA Implementation and Performance Evaluation of the AES Block Cipher Candidate Algorithm Finalists*, The Third Advanced Encryption Standard (AES3) Candidate Conference, April 13–14 2000, New York, USA.
4. K. Gaj and P. Chodowiec, *Comparison of the Hardware Performance of the AES Candidates using Reconfigurable Hardware*, The Third Advanced Encryption Standard (AES3) Candidate Conference, April 13–14 2000, New York, USA.
5. P. Chodowiec et al, *Experimental Testing of the Gigabit IPSec-Compliant Implementations of Rijndael and Triple-DES Using SLAAC-1V FPGA Accelerator Board*, in the proceedings of ISC 2001: Information Security Workshop, LNCS 2200, pp. 220-234, Springer-Verlag.
6. A. Dandalis et al, *A Comparative Study of Performance of AES Candidates Using FPGAs*, The Third Advanced Encryption Standard (AES3) Candidate Conference, April 13–14 2000, New York, USA.
7. T. Ichikawa et al, *Hardware Evaluation of the AES Finalists*, The Third Advanced Encryption Standard (AES3) Candidate Conference, April 13–14 2000, New York, USA.
8. O. Kwon et al, *Implementation of AES and Triple-DES Cryptography using a PCI-based FPGA Board*, in the proceedings of ITC-CSCC 2002: The International Technical Conference On Circuits/Systems, Computers and Communications.
9. M. McLoone and J.V. McCanny, *High Performance Single Ship FPGA Rijndael Algorithm Implementations*, in the proceedings of CHES 2001: The Third International CHES Workshop, Lecture Notes In Computer Science, LNCS 2162, pp 65–76, Springer-Verlag.
10. Helion Technology, *High Performance AES (Rijndael) Cores for XILINX FPGA*, http://www.heliontech.com.
11. V. Fischer and M. Drutarovsky, *Two Methods of Rijndael Implementation in Reconfigurable Hardware*, in the proceedings of CHES 2001: The Third International CHES Workshop, Lecture Notes In Computer Science, LNCS 2162, pp 65–76, Springer-Verlag.
12. CAST, *AES Encryption Cores*, http://www.cast-inc.com.
13. Amphion Semiconductor, *CS5210-40: High Performance AES Encryption Cores*, 2001. http://www.amphion.com/cs5210.html
14. N. Sklavos, O. Koufopavlou, *Architecutre and VLSI Implementations of the AES-Proposal Rijndael*, in IEEE Transactions on Computers, Volume 51, Number 12, pp 1454–1459, December 2002.
15. A. Satoh et al, *Compact Hardware Architecture for 128-bit Block Cipher Camellia*, in the Proceedings of the Third NESSIE Workshop, november 6–7, 2002, Munich, Germany.
16. N. Weaver and J. Wawrzynek *High Performance Compact AES Implementations in Xilinx FPGAs*, http://www.cs.berkeley.edu/ nweaver/Rijndael.
17. Xinmiao Zhang, Parhi K.K., *Implementation approaches for the advanced encryption standard algorithm*, in IEEE Circuits and Systems Magazine, pp 24–46, Fourth Quarter 2002.
18. FX. Standaert, G. Rouvroy, JD. Legat, JJ. Quisquater, *A Methodology to Implement Block Ciphers in Reconfigurable Hardware and its Application to Fast and Compact AES Rijndael*, in the proceedings of FPGA 2003: the Field Programmable Logic Array Conference, February 23–25 2003, Monterey, California.
19. A. Rudra et al, *Efficient Rijndael Encryption Implementation with Composite Field Arithmetic*, in the proceedings of CHES 2001: The Third International CHES Workshop, Lecture Notes In Computer Science, LNCS2162, pp 65–76, Springer-Verlag.

20. A. Satoh et al, *A Compact Rijndael Hardware Architecture with S-Box Optimization*, Advances in Cryptology – ASIACRYPT 2001, LNCS 2248, pp 239–254, Springer-Verlag.
21. J. Wolkerstorfer, E. Oswald, M. Lamberger, *An ASIC Implementation of the AES SBoxes*, in the proceedings of CT-RSA 2002, LNCS 2271, pp 67–78, Springer-Verlag.

Hyperelliptic Curve Cryptosystems: Closing the Performance Gap to Elliptic Curves

Jan Pelzl, Thomas Wollinger, Jorge Guajardo, and Christof Paar

Department of Electrical Engineering and Information Sciences
Communication Security Group (COSY)
Ruhr-Universitaet Bochum, Germany
{pelzl, wollinger, guajardo, cpaar}@crypto.rub.de

Abstract. For most of the time since they were proposed, it was widely believed that hyperelliptic curve cryptosystems (HECC) carry a substantial performance penalty compared to elliptic curve cryptosystems (ECC) and are, thus, not too attractive for practical applications. Only quite recently improvements have been made, mainly restricted to curves of genus 2. The work at hand advances the state-of-the-art considerably in several aspects. First, we generalize and improve the closed formulae for the group operation of genus 3 for HEC defined over fields of characteristic two. For certain curves we achieve over 50% complexity improvement compared to the best previously published results. Second, we introduce a new complexity metric for ECC and HECC defined over characteristic two fields which allow performance comparisons of practical relevance. It can be shown that the HECC performance is in the range of the performance of an ECC; for specific parameters HECC can even possess a lower complexity than an ECC at the same security level. Third, we describe the first implementation of a HEC cryptosystem on an embedded (ARM7) processor. Since HEC are particularly attractive for constrained environments, such a case study should be of relevance.

Keywords: Hyperelliptic curves, explicit formulae, comparison HECC vs. ECC, efficient implementation

1 Introduction

In 1976 Diffie and Hellman [DH76] revolutionized the field of cryptography by introducing the concept of public-key cryptography. Their key exchange protocol is based on the difficulty of solving the discrete logarithm (DL) problem over a finite field. Years later, [Mil86,Kob87] introduced a variant of the Diffie-Hellman key exchange, based on the difficulty of the DL problem in the group of points of an elliptic curve (EC) over a finite field. Since their introduction, elliptic curve cryptosystems (ECC) have been extensively studied not only by the research community but also in industry. In particular, there are several standards involving EC, such as the IEEE P1363 [P1399] standardization effort and the bank industry standards [ANS99]. It is important to point out that ECC

C.D. Walter et al. (Eds.): CHES 2003, LNCS 2779, pp. 351–365, 2003.
© Springer-Verlag Berlin Heidelberg 2003

benefit from shorter operand sizes when compared to RSA or DL based systems. This fact makes ECC particularly well suited for small processors and memory constrained environments.

In 1988 Koblitz suggested for the first time the generalization of EC to curves of higher genus, namely hyperelliptic curves (HEC) [Kob88]. In contrast to the EC case, it has only been until recently that Koblitz's idea to use HEC for cryptographic applications, has been analyzed and implemented both in software [Kri97,SS98,SSI98,Eng99,SS00] and in more hardware-oriented platforms such as FPGAs [Wol01,BCLW02]. In 1999, [Sma99] concluded that there seems to be little practical benefit in using HEC, because of the difficulty of finding hyperelliptic curves and their relatively poor performance when compared to EC. However, quite recently the efficiency of the HEC group operation has been improved [Har00,MDM$^+$02,Tak02,Lan02a]. It is well known that the best algorithm to compute the discrete logarithm in generic groups such as the Jacobian of a HEC is Pollard's rho method or one of its parallel variants [Pol78,vOW99]. For curves of genus higher than four, [Gau00] showed that there exists an algorithm with complexity $O(q^2)$ where F_q is the field over which the HEC is defined. Thus, in this work, we only consider HEC of genus less than four, as curves of higher genus are potentially insecure from a cryptographic point of view.

It is widely accepted that for most cryptographic applications based on EC or HEC one needs a group order of size at least $\approx 2^{160}$. Thus, for HECC over $\mathbb{F}_q$ we will need at least $g \cdot \log_2 q \approx 2^{160}$, where g is the genus of the curve. In particular, for a curve of genus two, we will need a field $\mathbb{F}_q$ with $|\mathbb{F}_q| \approx 2^{80}$, i.e., 80-bit long operands. Similarly, for curves of genus three, our discussion above implies 54-bit long operands. These field sizes make HEC specially promising for use in embedded environments where memory and speed are constrained, and where the above operand sizes seem well suited to their *small* processor architectures.

Our Main Contributions

Genus 3 group operations: The work at hand presents for the first time generalized explicit formulae for genus-3 curves including fields of characteristic 2. For certain curves our group doubling formula saves more than 66% of the field multiplications compared to [KGM$^+$02].

New complexity metric for HECC and ECC: We introduce a new metric for HECC and ECC over characteristic two fields which is based on an atomic operation count rather than on the (theoretical) bit complexity or specific timings. The most interesting results: (a) under certain conditions genus-3 hyperelliptic curves are faster than ECC at the same level of security and (b) these HEC are faster than genus-2 curves.

HECC implementation on an embedded platform: We support our theoretical findings with a HECC implementation on an ARM7TDMI, which is one of the most popular embedded processors. Our implementation uses the best explicit formulae for genus-2 and genus-3 curves.

The remainder of the paper is organized as follows. Section 2 summarizes contributions dealing with previous implementations and comparisons of HECC and ECC. Section 3 gives a brief overview of the mathematical background related to HECC. Section 4 and 5 present our new explicit formulae for genus-3 curves and a theoretical comparison between ECC and HECC. Section 6 introduces the implementation of HECC on embedded processors. Finally, we end this contribution with a discussion of our results and some conclusions.

2   Previous Work

In this section, we summarize previous improvements of the group operation of genus-2 and genus-3 curves, earlier theoretical comparisons between ECC and HECC, and other HECC implementations.

Improvements to HECC Group Operations

Table 1 summarizes the efforts made to date to speed up genus-2 curves. I refers to inversion, M to multiplication, S to squaring, and M/S to multiplications or squarings, since squarings are assumed to be of the same complexity as a multiplication in these publications. For more details on previous improvements made to the explicit formulae the interested reader is referred to [PWGP03]

For genus-3 hyperelliptic curves of odd characteristic the only improvement over Cantor's algorithm was presented in [KGM+02]. The authors adopted the methods from [MDM+02,Har00] to obtain the speed-up. The operation complexity for genus-3 curves is summarized in Table 3.

Theoretical Comparisons

In [SSI98], the authors clarified practical advantages of hyperelliptic cryptosystems when compared to ECC and to RSA. To our knowledge this is the first and only contribution that investigates in detail the theoretical complexity of ECC and HECC. They estimated the cost of different cryptosystems based on the number of bit operations. In their work they used Cantor's formula and the cost of one multiplication in $\mathbb{F}_{2^n}$ was assumed to take n^2 bit operations. One of the estimated theoretical results shows that genus-3 curves needed three times as many bit operations as elliptic curves. We want to point out that this publication used supersingular curves[1] and curves of genus higher than 4 which today are believed to be insecure due to the attacks presented in [FR94,Gau00,Gal01].

In the following years further analyses of the complexity of HECC were published. A theoretical analysis of the computational efficiency of the arithmetic on hyperelliptic curves is derived in [Eng99]. In [SS00], the authors implemented hyperelliptic curve cryptosystems and analyzed the complexity of the group law on Jacobians $\mathbb{J}_C(\mathbb{F}_p)$ and $\mathbb{J}_C(\mathbb{F}_{2^n})$. Moreover, they verified their theoretical complexity estimates with a HECC implementation and with the theoretical analysis

---

[1] [Gal01] gives some arguments against using supersingular hyperelliptic curves in cryptographic applications.

done by Enge in [Eng99]. More recent papers present timings for HECC using explicit formulae and compared HECC to ECC [Lan02a]. However, these comparisons were based on the implementation timings.

Table 1. Speeding up group operations on hyperelliptic curves of genus two.

	field charac.	curve properties	cost addition	doubling
Cantor [Nag00]	general		$3I + 70M/S$	$3I + 76M/S$
Nagao [Nag00]	odd	$h(x) = 0,\ f_i \in \mathbb{F}_2$	$1I + 55M/S$	$1I + 55M/S$
Harley [Har00]	odd	$h(x) = 0$	$2I + 27M/S$	$2I + 30M/S$
Matsuo et al. [MCT01]	odd	$h(x) = 0$	$2I + 25M/S$	$2I + 27M/S$
Miyamoto et al. [MDM$^+$02]	odd	$h(x) = 0,\ f_4 = 0$	$I + 26M/S$	$I + 27M/S$
Takahashi [Tak02]	odd	$h(x) = 0$	$I + 25M/S$	$I + 29M/S$
Lange [Lan02a]	general	$h_i \in \mathbb{F}_2,\ f_4 = 0$	$I + 22M + 3S$	$I + 22M + 5S$
	two	$h_i \in \mathbb{F}_2,\ f_4 = 0$	$I + 22M + 2S$	$I + 20M + 4S$
Lange [Lan02b]	general	$h_i \in \mathbb{F}_2,\ f_4 = 0$	$47M + 4S(40M + 3S)^2$	$40M + 6S$
	two	$h_i \in \mathbb{F}_2,\ f_4 = 0$	$46M + 2S$	$33M + 6S$
Lange [Lan02c]	odd	$h_i \in \mathbb{F}_2,\ f_4 = 0$	$47M + 7S(36M + 5S)^2$	$34M + 7S$
	even	$h_2 \neq 0,\ h_i \in \mathbb{F}_2,\ f_4 = 0$	$46M + 4S(35M + 5S)^2$	$35M + 6S$
	even	$h_2 = 0,\ h_i \in \mathbb{F}_2,\ f_4 = 0$	$44M + 6S(34M + 6S)^2$	$29M + 6S$

To our knowledge there is no theoretical complexity comparison between ECC and HECC published that uses the explicit formulae for HECC and compares HECC and ECC in terms of processor instructions, such as shift and XOR operations. Hence, this comparison is processor independent and can be adapted to any platform.

HECC Implementations

Since HEC cryptosystems were proposed, there have been several software implementations on general purpose machines and, only recently, publications dealing with hardware implementations of HECC. To our knowledge there has not been any work dealing with the implementation of HEC on embedded systems. The results of previous HECC software implementations are summarized in Table 2. Detailed information about previously made HECC implementations can be found in [PWGP03].

The first HECC hardware architectures were proposed in [Wol01]. The performance of a hardware-based genus two hyperelliptic curve coprocessor over $\mathbb{F}_{2^{113}}$ was presented in [BCLW02]. The FPGA was clocked at 45 MHz and required 4750 clock cycles for a group addition and 4050 clock cycles for a group doubling operation.

---

[2] mixed addition

Table 2. Execution times of recent HEC implementations in software.

reference	processor	genus	field	$t_{scalarmult.}$ in ms
[Kri97]	Pentium@100MHz	2	$\mathbb{F}_{2^{64}}$	520
		3	$\mathbb{F}_{2^{42}}$	1200
		4	$\mathbb{F}_{2^{31}}$	1100
[SS98]	Alpha@467MHz	3	$\mathbb{F}_{2^{59}}$	83.3
		3	$\mathbb{F}_{2^{89}}$	25700
		3	$\mathbb{F}_{2^{113}}$	37900
		4	$\mathbb{F}_{2^{41}}$	96.6
	Pentium-II@300MHz	3	$\mathbb{F}_{2^{59}}$	11700
		4	$\mathbb{F}_{2^{41}}$	10900
[SS00]	Alpha21164A@600MHz	3	$\mathbb{F}_p(\log_2 p = 60)$	98
		3	$\mathbb{F}_{2^{59}}$	40
		4	$\mathbb{F}_{2^{41}}$	43
[MCT01]	PentiumIII@866MHz	2	186-bit OEF	1.98
[MDM$^+$02]	PentiumIII@866MHz	2	186-bit OEF	1.69
[KGM$^+$02]	Alpha21264@667MHz	3	$\mathbb{F}_{2^{61}-1}$	0.932
[Lan02a]	Pentium-IV@1.5GHz	2	$\mathbb{F}_{2^{160}}$	18.875
		2	$\mathbb{F}_{2^{180}}$	25.215
		2	$\mathbb{F}_p(\log_2 p = 160)$	5.663
		2	$\mathbb{F}_p(\log_2 p = 180)$	8.162

3   Mathematical Background

In this section we present an elementary introduction to some of the theory of hyperelliptic curves over finite fields of arbitrary characteristic, restricting attention to material that is relevant for this work. For more details the reader is referred to [Kob89,Kob98].

3.1   HECC and the Jacobian

Let $\mathbb{F}$ be a finite field, and let $\overline{\mathbb{F}}$ be the algebraic closure of $\mathbb{F}$. A hyperelliptic curve C of genus $g \geq 1$ over $\mathbb{F}$ is the set of solutions $(u, v) \in \mathbb{F} \times \mathbb{F}$ to the equation

$$C : v^2 + h(u)v = f(u)$$

Such a curve is said to be non-singular if there are no pairs $(u, v) \in \overline{\mathbb{F}} \times \overline{\mathbb{F}}$ which simultaneously satisfy the equation of the curve C and the partial differential equations $2v + h(u) = 0$ and $h'(u)v - f'(u) = 0$. The polynomial $h(u) \in \mathbb{F}[u]$ is of degree at most g and $f(u) \in \mathbb{F}[u]$ is a monic polynomial of degree $2g + 1$. For odd characteristic it suffices to let $h(u) = 0$ and to have $f(u)$ square free.

A divisor $D = \sum m_i P_i$, $m_i \in \mathbb{Z}$, is a finite formal sum of $\overline{\mathbb{F}}$-points. Its degree is the sum of the coefficients $\sum m_i$. The set of all divisors form an Abelian group denoted by $\mathbb{D}(C)$. The set of divisors of degree zero will be denoted by $\mathbb{D}^0 \subset \mathbb{D}(C)$.

Every rational function on the curve gives rise to a divisor of degree zero, consisting of the formal sum of the poles and zeros of the function. Such divisors are called principal and the set of all principal divisors is denoted by $\mathbb{P}$. If $D_1, D_2 \in \mathbb{D}^0$ then we write $D_1 \sim D_2$ if $D_1 - D_2 \in \mathbb{P}$; D_1 and D_2 are said to be equivalent divisors. Now, we can define the Jacobian of C as the quotient group $\mathbb{D}^0/\mathbb{P}$. If we want to define the Jacobian over $\mathbb{F}$, denoted by $\mathbb{J}_C(\mathbb{F})$, we say that a divisor $D = \sum m_i P_i$ is defined over $\mathbb{F}$ (sometimes also called a $\mathbb{F}$-divisor or rational divisor) if $D^\sigma = \sum m_i P_i^\sigma$ is equal to D for all automorphisms σ of $\overline{\mathbb{F}}$ over $\mathbb{F}$. Notice that this does not mean that each P_i^σ is equal to P_i, σ may permute the points.

In [Can87], Cantor shows that each element of the Jacobian can be represented in the form $D = \sum_{i=1}^r P_i - r \cdot \infty$ such that for all $i \neq j$, P_i and P_j are not symmetric points. Such a divisor is called a semi-reduced divisor. Cantor concludes that from the Riemann-Roch Theorem follows that each element of the Jacobian can be represented uniquely by such a divisor, subject to the additional constraint $r \leq g$. Such divisors are referred to as reduced divisors. Finally, [Can87] shows that the divisors of the Jacobian can be represented as a pair of polynomials $a(u)$ and $b(u)$ with $\deg b(u) < \deg a(u) \leq g$, with $a(u)$ dividing $v^2 + h(u)v - f(u)$ and where the coefficients of $a(u)$ and $b(u)$ are elements of $\mathbb{F}$ [Mum84] (notice that in our particular application $\mathbb{F}$ is a finite field). In the remainder of this paper, a divisor D represented by polynomials will be denoted by $div(a, b)$.

3.2   Group Operations on a Jacobian

This section gives a brief description of the algorithms used for adding and doubling divisors on $\mathbb{J}_C(\mathbb{F})$. These group operations will be performed in two steps. First we have to find a semi-reduced divisor $D' = div(a', b')$, such that $D' \sim D_1 + D_2 = div(a_1, b_1) + div(a_2, b_2)$ in the group $\mathbb{J}_C(\mathbb{F})$. In the second step we have to reduce the semi-reduced divisor $D' = div\,(a', b')$ to an equivalent divisor $D = (a, b)$. Algorithm 1 describes the group addition.

Algorithm 1 Group addition

Require: $D_1 = div(a_1, b_1)$, $D_2 = div(a_2, b_2)$
Ensure: $D = div(a, b) = D_1 + D_2$
1: $d = \gcd(a_1, a_2, b_1 + b_2 + h) = s_1 a_1 + s_2 a_2 + s_3(b_1 + b_2 + h)$
2: $a'_0 = a_1 a_2 / d^2$
3: $b'_0 = [s_1 a_1 b_2 + s_2 a_2 b_1 + s_3(b_1 b_2 + f)]d^{-1}(\bmod a'_0)$
4: **while** $\deg a'_k > g$ **do**
5:     $a'_k = \dfrac{f - b'_{k-1}h - (b'_{k-1})^2}{a'_{k-1}}$
6:     $b'_k = (-h - b'_{k-1}) \bmod a'_k$
7: **end while**
8: Output $(a = a'_k, b = b'_k)$

Doubling a divisor is easier than general addition and therefore, Steps 1,2, and 3 of Algorithm 1 can be simplified. The formulae given for the group operation of HECC can be written explicitly as previously mentioned. In Section 4 we develop explicit formulae of Cantor's Algorithm for genus-3 curves. For security considerations of the HEC used see [PWGP03].

4   Speed-up for Genus-3 Curves

In this section we briefly outline the ideas of [GH00] and [KGM+02] which are the starting point for our improvements. In [GH00], the authors noticed that one can reduce the number of operations required to add/double divisors by distinguishing between possible cases according to the properties of the input divisors. This technique is combined with the use of the Karatsuba multiplication algorithm [KO63] and the Chinese remainder theorem to further reduce the complexity of the overall group operations. The work of [GH00] was generalized by [KGM+02] to genus-3 curves defined over odd characteristic fields. In particular, they notice that for genus-3 curves there are 6 possible choices for the degree of the input polynomials to Algorithm 1 and that further classification according to the common factors of the polynomials would lead to about 70 subcases. However, they only consider the most frequent cases[3] which occur with overwhelming probability of $1 - O(1/q) \approx 1 - 2^{-60}$ for genus-3 curves over $\mathbb{F}_{2^{60}}$. For the remaining cases, they use Cantor's algorithm.

In this work, we further optimize the formulae of [KGM+02] and generalize them to arbitrary characteristic. Table 7 presents the explicit formulae for a group addition and Table 8 those for a group doubling. The formulae shown in the tables are based on the assumption that $h_i \in \{0, 1\}$, where $i = 0, 1, 2, 3$, and that f_6 is equal to zero. The latter can be achieved by substituting $x' = x + \frac{f_6}{7}$. The coefficient is still included in the algorithm for completeness.

Our improvements are based on the following techniques [PWGP03]:

1. Montgomery's trick of simultaneous inversions [Coh93, Algorithm 10.3.4]
2. Reordering of normalization step [Tak02]
3. Karatsuba multiplication
4. Calculation of the resultant using Bezout's matrix
5. Choice of HEC

Table 3. Comparing the complexity of the group operations on HEC of genus three.

	field characteristic	curve properties	cost addition	doubling
Cantor [Nag00]	general	$h(x) = 0,\ f_i \in \mathbb{F}_2$	$4I + 200M/S$	$4I + 207M/S$
Nagao [Nag00]	odd		$2I + 154M/S$	$2I + 146M/S$
Kuroki et al. [KGM+02]	odd	$h(x) = 0,\ f_6 = 0$	$I + 81M/S$	$I + 74M/S$
This work (Tables 7, 8)	general	$h_i \in \mathbb{F}_2,\ f_6 = 0$	$I + 70M + 6S$	$I + 61M + 10S$
	two	$h_i \in \mathbb{F}_2,\ f_6 = 0$	$I + 65M + 6S$	$I + 53M + 10S$
	two	$h(x) = 1,\ f_6 = 0$	$I + 65M + 6S$	$I + 14M + 11S$

---

[3] For addition the inputs are two co-prime polynomials of degree 3, for doubling the input is a square free polynomial of degree 3

As a summary we include the computational cost of all the published results for genus-3 curves in Table 3. Compared to [KGM$^+$02], we save 5 multiplications in the addition algorithm and 3 multiplications in the doubling algorithm even though our formulae are more general.

5   Comparing ECC and HECC

In the past, providing complexity measures and, thus, comparisons between ECC and HECC was a difficult undertaking. The operations involved in both systems were very different (different field orders, field operations vs. operations with polynomials, etc.). Furthermore, measures such as the bit complexity often provide very little information about the *de facto* complexity in actual implementations. The underlying motivation for the work described in the following was the development of a more accurate metric for practical purposes. All operations which are computationally expensive will be expressed in terms of *atomic operations* (AOPS), such as processor word-SHIFTs and XORs. In particular, we will decompose field multiplications into AOPS. This provides a metric which allows a comparison of fields of different sizes which is crucial for comparing ECC and HECC with equal level of security. The approach possesses the advantage that it accurately counts the actual elementary processor operations (as opposed to the more theoretical bit complexity), while at the same time avoiding processor and implementation-dependent "tricks" which can skew comparisons that are merely based on timings. In summary, we believe we developed a method which allows accurate predictions of the performance on a given processor without the laborious task of actually implementing the cryptosystem. The accuracy of the new metric is demonstrated by a mere 12% difference between our theoretical and practical results.

The number of atomic operations is denoted as AOPS. In our comparison we make the following assumptions:

1. We only consider fields of characteristic two and thus neglect the cost of squaring.
2. We perform field multiplications with Algorithm 5 published[4] in [LD00]. This algorithm requires $3 + 2(w/4 - 1)$ word-SHIFTs and $s(11 + n/4) + 8(2s - 1)$ word-XORs, where w is the word size of processor and $s = \lceil \frac{n}{w} \rceil$ is the number of words needed to represent an element of the underlying field $\mathbb{F}_{2^n}$.
3. We express the cost of one field inversion as m field multiplications and denote the ratio of multiplications to inversions as *MI*-ratio.

Based on the assumptions stated above, the complexity of the group operations of HEC and EC are summarized. Referring to Tables 7 and 8, a divisor addition for a genus-3 curve requires $1I + 65M$ and doubling needs $1I + 53M$ (using a curve with $h = 1$, doubling needs only $1I + 14M$). Assuming that the cost

---

[4] To our knowledge this is the fastest published multiplication algorithm for finite fields of characteristic two.

of one field inversion is equivalent to m field multiplications, leads to $(65+m)M$ and $(53+m)M$ for addition and doubling, respectively. Due to the higher extension of the underlying field used for genus-2 curves, a different MI-ratio l is used. This leads to $(22+l)M$ for a divisor addition and $(20+l)M$ for a divisor doubling. The number of inversions and multiplications for a group operation on EC heavily depends on the chosen coordinate system. For completeness we summarize the number of required operations in Table 4.

Table 4. Field operations required in each coordinate system [HHM00]

Coordinate system	EC Addition		EC Doubling
	general	mixed coord.	
Affine coordinates	$1I+2M$		$1I+2M$
Standard projective coordinates [CC87,CMO98]	$13M$	$12M$	$7M$
Jacobian projective coordinates [CC87,CMO98]	$15M$		$5M$
New projective coordinates [LD99]	$14M$	$9M$	$4M$

Figure 1 illustrates the number of operations for a scalar multiplication on a 32-bit processor depending on the MI-ratios. The scalar multiplication with an n-bit scalar is realized by the sliding window method with an approximated cost of $n \cdot \text{doublings} + 0.2 \cdot n \cdot \text{additions}$ for a 4-bit window size [BSS99]. Figure 1 allows to estimate the efficiency of an ECC or a HECC built on top of a given field library by comparing the different MI-ratios.

In general we can draw the following conclusions from this comparison:

1. ECC with projective coordinates is in almost all cases the most efficient cryptosystem.
2. Scalar multiplication of genus-3 HEC with $h(x) = 1$ always outperforms genus-2 HEC.
3. Genus-3 HECC scalar multiplication is in most cases faster than ECC using affine coordinates.
4. For field libraries with very high MI-ratio, ECC using Jacobian projective is more efficient than genus-3 HEC. However, for low MI-ratios the HECC scalar multiplication becomes less expensive.

6   HECC on Embedded Systems

With the predicted advent of ubiquitous computing, embedded processors will play an increasingly important role in providing security functions. Due to their relatively short operand lengths, HECC are particularly well suited for embedded processors which are typically computationally constrained. We chose a representative of the popular ARM processor family for our implementation. The

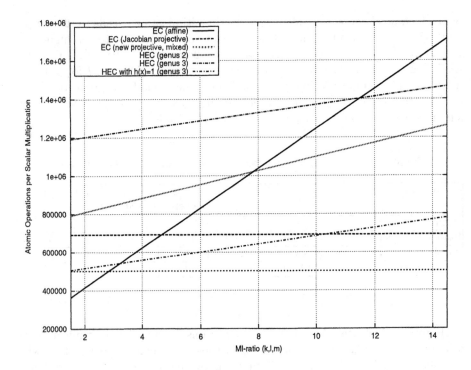

Fig. 1. Cost of a scalar multiplication for different MI-ratios and cryptosystems in AOPS (32-bit μP, group order $\approx 2^{190}$)

purpose was twofold. First, we wanted to provide actual timings of a highly optimized HEC implementation. Secondly, we wanted to validate our complexity metric.

The ARM7TDMI@80MHz[5] processor environment was chosen to implement both elliptic and hyperelliptic curve cryptosystems. For the elliptic curve case we used curves over $\mathbb{F}_{2^{191}}$ and Jacobian projective coordinates. The most efficient explicit formulae were implemented in the case of HECC. For genus-3 curves the polynomial $h(x)$ equals one. The group orders range from 2^{162} to 2^{190}. Table 5 presents timings for divisor addition, divisor doubling and scalar multiplication on the ARMulator. To our knowledge these are the first published timings for HECC on an embedded processor.

To theoretically determine the most efficient cryptosystem based on the timings given in Table 6, one can either use Figure 1 or calculate the necessary number of AOPS. Considering a finite field $\mathbb{F}_{2^{63}}$ for a genus-3 HEC, $619,402$ AOPS are needed to calculate one scalar multiplication. HECC of genus 2 with the underlying field $\mathbb{F}_{2^{95}}$ will take $1,049,028$ AOPS, and ECC over $\mathbb{F}_{2^{191}}$ using Jacobian projective coordinates requires $699,060$ AOPS. Thus, we expect

---

[5] Depending on the features of processor board, the performance numbers can differ.

HECC of genus-2 to be a factor of 1.5 slower and genus-3 HECC a factor of 1.1 faster than ECC. Genus-2 HECC is expected to be 1.5-times slower than genus-3 HECC.

Table 5. Timings of group operations with ARMulator ARM7TDMI@80MHz (explicit formulae)

Genus	Field	Group order	Group addition in μs	Group doubling in μs	Scalar. mult. in ms^6
	$\mathbb{F}_{2^{54}}$	2^{162}	914	317	90
	$\mathbb{F}_{2^{55}}$	2^{165}	917	319	91
3	$\mathbb{F}_{2^{59}}$	2^{177}	1180	415	126
	$\mathbb{F}_{2^{60}}$	2^{180}	921	324	100
	$\mathbb{F}_{2^{61}}$	2^{183}	1183	417	130
	$\mathbb{F}_{2^{63}}$	2^{189}	925	329	106
	$\mathbb{F}_{2^{81}}$	2^{162}	618	628	128
	$\mathbb{F}_{2^{83}}$	2^{166}	732	756	157
2	$\mathbb{F}_{2^{88}}$	2^{176}	749	774	170
	$\mathbb{F}_{2^{91}}$	2^{182}	754	778	177
	$\mathbb{F}_{2^{95}}$	2^{190}	641	650	155
1	$\mathbb{F}_{2^{191}}$	2^{191}	598	358	100

The timings for a scalar multiplication of genus-3 curves over $\mathbb{F}_{2^{63}}$ and of genus-2 curves over $\mathbb{F}_{2^{95}}$ are compared with the performance of the ECC scalar multiplication over $\mathbb{F}_{2^{191}}$. HECC of genus 3 is a factor of 1.1 and HECC of genus 2 is a factor of 1.5 slower than ECC. Furthermore, a divisor scalar multiplication on a HEC of genus 2 performs a factor of 1.5 worse than a genus-3 HECC. The deviation of our implementation and the theoretical findings is at most 12%. Thus, we can conclude that our theoretical estimates were quite accurate.

Table 6. Timings of the field library and corresponding MI-ratios. All timings in μs assuming a 80MHz clock rate.

Field	Multiplication	Inversion	MI-ratio
$\mathbb{F}_{2^{63}}$	11.5	73.7	6.4
$\mathbb{F}_{2^{95}}$	19.3	157.2	8.2
$\mathbb{F}_{2^{191}}$	50.7	469.9	9.3

---

[6] A further speed-up can be achieved by the use of special reduction routines targeting a fixed irreducible polynomial.

7   Conclusions

In this contribution, we were able to close the gap between the performance of HECC and ECC. In particular, an improvement of the explicit formulae for arbitrary characteristic for the case of genus-3 hyperelliptic curves was presented. For certain curves over fields of characteristic 2, the efficiency of the doubling algorithm could be enhanced drastically. This increased the performance of a scalar multiplication by over 50% compared to [KGM+02].

A theoretical comparison of ECC to HECC with coefficients in $\mathbb{F}_{2^m}$ assuming the currently fastest algorithms for field operations was also presented. An important finding is that HECC can reach about the same throughput than ECC and that genus-3 HECC with $h(x) = 1$ are always faster than genus-2 HECC. However, the properties of the field libraries are the key to determine the overall performance of ECC and HECC.

The theoretical results are confirmed by the first implementation of genus-2 and genus-3 curves on an embedded processor.

References

[ANS99]   ANSI X9.62-1999. The Elliptic Curve Digital Signature Algorithm. Technical report, ANSI, 1999.

[BCLW02]   N. Boston, T. Clancy, Y. Liow, and J. Webster. Genus Two Hyperelliptic Curve Coprocessor. In *Workshop on Cryptographic Hardware and Embedded Systems — CHES 2002*, New York, 2002. Springer Verlag. LNCS 2523.

[BSS99]   I.F. Blake, G. Seroussi, and N.P. Smart. *Elliptic Curves in Cryptography*. London Mathematical Society Lecture Notes Series 265. Cambridge University Press, Reading, Massachusetts, 1999.

[Can87]   D.G. Cantor. Computing in Jacobian of a Hyperelliptic Curve. In *Mathematics of Computation*, volume 48(177), pages 95–101, January 1987.

[CC87]   D.V. Chudnovsky and G.V. Chudnovsky. Sequences of numbers generated by addition in formal groups and new primality and factorization tests. In *Advances in Applied Mathematics*, volume 7, pages 385–434, 1987.

[CMO98]   H. Cohen, A. Miyaji, and T. Ono. Efficient elliptic curve exponentiation using mixed coordinates. In K. Ohta and D. Pei, editors, *Advances in Cryptology — ASIACRYPT 98*, pages 51–65. Springer Verlag, 1998. LNCS 1514.

[Coh93]   H. Cohen. *A course in computational number theory*. Graduate Texts in Math. 138. Springer-Verlag, Berlin, 1993. Third corrected printing 1996.

[DH76]   W. Diffie and M. E. Hellman. New directions in cryptography. *IEEE Transactions on Information Theory*, IT-22:644–654, 1976.

[Eng99]   A. Enge. The extended Euclidean algorithm on polynomials, and the computational efficiency of hyperelliptic cryptosystems, November 1999. Preprint.

[FR94]   G. Frey and H.-G. Rück. A remark concerning m-divisibility and the discrete logarithm in the divisor class group of curves. *Mathematics of Computation*, 62(206):865–874, April 1994.

[Gal01]   S.D. Galbraith. Supersingular curves in cryptography. In *Advances in Cryptology — ASIACRYPT 2001*, pages 495–517, 2001. LNCS 2248.

[Gau00]   P. Gaudry. An algorithm for solving the discrete log problem on hyperelliptic curves. In Bart Preneel, editor, *Advances in Cryptology — EUROCRYPT 2000*, volume LNCS 1807, pages 19–34, Berlin, Germany, 2000. Springer-Verlag.

[GH00]   P. Gaudry and R. Harley. Counting Points on Hyperelliptic Curves over Finite Fields. In W. Bosma, editor, *The 4th Algorithmic Number Theory Symposium — ANTS IV*, pages 297–312, Berlin, 2000. Springer Verlag. LNCS 1838.

[Har00]   R. Harley. Fast Arithmetic on Genus Two Curves. Available at http://cristal.inria.fr/~harley/hyper/, 2000.

[HHM00]   D. Hankerson, J. López Hernandez, and A. Menezes. Software Implementation of Elliptic Curve Cryptography Over Binary Fields. In Çetin K. Koç and Christof Paar, editors, *Workshop on Cryptographic Hardware and Embedded Systems — CHES 2000*, pages 1–24. Springer Verlag, August 2000. LNCS 1717.

[KGM$^+$02]   J. Kuroki, M. Gonda, K. Matsuo, Jinhui Chao, and Shigeo Tsujii. Fast Genus Three Hyperelliptic Curve Cryptosystems. In *The 2002 Symposium on Cryptography and Information Security, Japan - SCIS 2002*, Jan.29-Feb.1 2002.

[KO63]   A. Karatsuba and Y. Ofman. Multiplication of multidigit numbers on automata. *Sov. Phys. Dokl. (English translation)*, 7(7):595–596, 1963.

[Kob87]   N. Koblitz. Elliptic curve cryptosystems. *Mathematics of Computation*, 48:203–209, 1987.

[Kob88]   N. Koblitz. A Family of Jacobians Suitable for Discrete Log Cryptosystems. In Shafi Goldwasser, editor, *Advances in Cryptology — Crypto '88*, pages 94–99, Berlin, 1988. Springer-Verlag. LNCS 403.

[Kob89]   N. Koblitz. Hyperelliptic Cryptosystems. In Ernest F. Brickell, editor, *Journal of Cryptology*, pages 139–150, 1989.

[Kob98]   N. Koblitz. *Algebraic Aspects of Cryptography*. Algorithms and Computation in Mathematics. Springer-Verlag, 1998.

[Kri97]   Uwe Krieger. signature.c, February 1997. Diplomarbeit, Universität Essen, Fachbereich 6 (Mathematik und Informatik).

[Lan02a]   T. Lange. Efficient Arithmetic on Genus 2 Hyperelliptic Curves over Finite Fields via Explicit Formulae. Cryptology ePrint Archive, Report 2002/121, 2002. http://eprint.iacr.org/.

[Lan02b]   T. Lange. Inversion-Free Arithmetic on Genus 2 Hyperelliptic Curves. Cryptology ePrint Archive, Report 2002/147, 2002. http://eprint.iacr.org/.

[Lan02c]   T. Lange. Weighted Coordinates on Genus 2 Hyperelliptic Curves. Cryptology ePrint Archive, Report 2002/153, 2002. http://eprint.iacr.org/.

[LD99]   J. Lopéz and R. Dahab. Improved algorithms for elliptic curve arithmetic in $GF(2^n)$. In *Selected Areas in Cryptography - SAC '98*, pages 201–212, 1999. LNCS 1556.

[LD00]   J. Lopez and R. Dahab. High-speed software multiplication in $\mathbb{F}_{2^m}$. In *INDOCRYPT*, pages 203–212, 2000.

[MCT01]   K. Matsuo, J. Chao, and S. Tsujii. Fast Genus Two Hyperelliptic Curve Cryptosystems. In *ISEC2001-31, IEICE*, 2001.

[MDM⁺02]   Y. Miyamoto, H. Doi, K. Matsuo, J. Chao, and S. Tsuji. A Fast Addition
Algorithm of Genus Two Hyperelliptic Curve. In *SCIS, IEICE Japan*,
pages 497–502, 2002. in Japanese.

[Mil86]   V. Miller. Uses of elliptic curves in cryptography. In H. C. WIlliams,
editor, *Advances in Cryptology — CRYPTO '85*, volume LNCS 218, pages
417–426, Berlin, Germany, 1986. Springer-Verlag.

[Mum84]   D. Mumford. Tata lectures on theta II. In *Prog. Math.*, volume 43.
Birkhäuser, 1984.

[Nag00]   K. Nagao. Improving group law algorithms for Jacobians of hyperelliptic
curves. In W. Bosma, editor, *ANTS IV*, volume 1838 of *Lecture Notes in
Computer Science*, pages 439–448, Berlin, 2000. Springer Verlag.

[P1399]   *IEEE P1363 Standard Specifications for Public Key Cryptography*,
November 1999. Last Preliminary Draft.

[Pol78]   J. M. Pollard. Monte carlo methods for index computation mod p. *Mathematics of Computation*, 32(143):918–924, July 1978.

[PWGP03]   Jan Pelzl, Thomas Wollinger, Jorge Guajardo, and Christof Paar. Hyperelliptic Curve Cryptosystems: Closing the Performance Gap to Elliptic
Curves (Update). Cryptology ePrint Archive, Report 2003/026, 2003.
http://eprint.iacr.org/.

[Sma99]   N.P. Smart. On the Performance of Hyperelliptic Cryptosystems. In *Advances in Cryptology – EUROCRYPT '99*, volume 1592 of *Lecture Notes
in Computer Science*, pages 165–175, Berlin, 1999. Springer-Verlag.

[SS98]   Y. Sakai and K. Sakurai. Design of Hyperelliptic Cryptosystems in small
Characteristic and a Software Implementation over $\mathbb{F}_{2^n}$. In *Advances
in Cryptology — ASIACRYPT '98*, pages 80–94, Berlin, 1998. Springer
Verlag. LNCS 1514.

[SS00]   Y. Sakai and K. Sakurai. On the Practical Performance of Hyperelliptic
Curve Cryptosystems in Software Implementation. In *IEICE Transactions on Fundamentals of Electronics, Communications and Computer
Sciences*, volume E83-A NO.4, pages 692–703, April 2000. IEICE Trans.

[SSI98]   Y. Sakai, K. Sakurai, and H. Ishizuka. Secure Hyperelliptic Cryptosystems and their Performance. In *Public Key Cryptography*, pages 164–181,
Berlin, 1998. Springer-Verlag. LNCS 1431.

[Tak02]   M. Takahashi. Improving Harley Algorithms for Jacobians of Genus 2
Hyperelliptic Curves. In *SCIS, IEICE Japan*, 2002. in Japanese.

[vOW99]   P. C. van Oorschot and M. J. Wiener. Parallel collision search with cryptanalytic applications. *Journal of Cryptology*, 12(1):1–28, Winter 1999.

[Wol01]   T. Wollinger. Computer Architectures for Cryptosystems Based on Hyperelliptic Curves, 2001. Master Thesis, Worcester Polytechnic Institute.

A   Explicit Formulae for Genus Three HEC

The explicit formulae for the group operations on HEC of genus 3 and arbitrary characteristic as well as the most efficient formulae for doubling on a HEC with $h(x) = 1$ for characteristic two is presented in Tables 7 and 8.

Table 7. Explicit formulae for adding on a HEC of genus three

Input	Weight three reduced divisors $D_1 = (u_1, v_1)$ and $D_2 = (u_2, v_2)$	
	$h = x^3 + h_2 x^2 + h_1 x + h_0$, where $h_i \in \mathbb{F}_2$;	
	$f = x^7 + f_5 x^5 + f_4 x^4 + f_3 x^3 + f_2 x^2 + f_1 x + f_0$;	
Output	A weight three reduced divisor $D_3 = (u_3, v_3) = D_1 + D_2$	
Step	**Procedure**	**Cost**
1	Resultant r of u_1 and u_2 (Bezout)	$12M + 2S$
2	Almost inverse $inv = r/u_1 \bmod u_2$	$4M$
3	$s' = rs \equiv (v_2 - v_1)inv \bmod u_2$ (Karatsuba)	$11M$
4	$s = (s'/r)$ and make s monic	$I + 6M + 2S$
5	$z = su_1$	$6M$
6	$u' = [s(z + w_4(h + 2v_1)) - w_5((f - v_1 h - v_1^2)/u_1)]/u_2$	$15M$
7	$v' = -(w_3 z + h + v_1) \bmod u'$	$8M$
8	u', i.e. $u_3 = (f - v'h - v'^2)/u'$	$5M + 2S$
9	$v_3 = -(v' + h) \bmod u_3$	$3M$
Total	in fields of arbitrary characteristic	$I + 70M + 6S$
	in fields of characteristic 2	$I + 65M + 6S$

Table 8. Explicit formulae for doubling on HEC of genus three

Input	A weight three reduced divisors $D_1 = (u_1, v_1)$	
	$h = x^3 + h_2 x^2 + h_1 x + h_0$, where $h_i \in \mathbb{F}_2$;	
	$f = x^7 + f_5 x^5 + f_4 x^4 + f_3 x^3 + f_2 x^2 + f_1 x + f_0$;	
Output	A weight three reduced divisor $D_2 = (u_2, v_2) = [2]D_1$	
Step	**Procedure**	**Cost**
1	Resultant r of u_1 and $h + 2v_1$ (Bezout)	$6M + 2S$
2	Almost inverse $inv = r/(h + 2v_1) \bmod u_1$	$4M$
3	$z = ((f - hv_1 - v_1^2)/u_1) \bmod u_1$	$7M + 2S$
4	$s' = zinv \bmod u_1$ (Karatsuba)	$11M$
5	$s = (s'/r)$ and make s monic	$I + 6M + 2S$
6	$G = su_1$	$6M$
7	$u' = u_1^{-2}[(G + w_4 v_1)^2 + w_4 hG + w_5(hv_1 - f)]$	$5M + 2S$
8	$v' = -(Gw_3 + h + v_1) \bmod u'$	$8M$
9	u', i.e. $u_2 = (f - v'h - v'^2)/u'$	$5M + 2S$
10	$v_2 = -(v' + h) \bmod u_2$	$3M$
Total	in fields of arbitrary characteristic	$I + 61M + 10S$
	in fields of characteristic 2	$I + 53M + 10S$
I	$d = gcd(u_1, 1) = 1 = s_1 a + s_3 h (s_3 = 1, s_1 = 0)$	—
II	$u' = u_1^2$	$3S$
III	$v' = v_1^2 + f \bmod u'$	$3S$
IV	$u'' = ((f - hv' - v'^2)/u')$	$3M + 3S$
V	$u_2 = u''$ made monic	$1I + 2M$
VI	$v_2 = -(v' + h) \bmod u_2$ (Karatsuba)	$5M$
VII	$u_3 := (f - v_2 * h - v_2^2)/u_2$	$1M + 2S$
VIII	$v_3 := -(v_2 + h) \bmod u_3$	$3M$
Total	in fields of characteristic 2 and with $h(x) = 1$	$I + 14M + 11S$

Countermeasures against Differential Power Analysis for Hyperelliptic Curve Cryptosystems

Roberto M. Avanzi[*]

Institut für Experimentelle Mathematik
University of Duisburg-Essen (Essen site)
Ellernstraße 29 – 45326 Essen, Germany
mocenigo@exp-math.uni-essen.de

Abstract. In this paper we describe some countermeasures against differential side-channel attacks on hyperelliptic curve cryptosystems. The techniques are modelled on the corresponding ones for elliptic curves. The first method consists in picking a random group isomorphic to the one where we are supposed to compute, transferring the computation to the random group and then pulling the result back. The second method consists in altering the internal representation of the divisors on the curve in a random way. The impact of the recent attack of L. Goubin is assessed and ways to avoid it are proposed.

Keywords. Public-key cryptography, Side-channel attacks, Differential power analysis (DPA), Timing attacks, Hyperelliptic curves, Smart cards.

1 Introduction

The use of Jacobian varieties of hyperelliptic curves in discrete logarithm cryptosystems was proposed by N. Koblitz as early as 1988 [17,18] as an alternative to elliptic curves. Hyperelliptic curves are a generalisation of elliptic curves: the latter are just the hyperelliptic curves of genus one.

Until very recently, however, elliptic curve cryptosystems (short: ecc) have been perceived as faster than hyperelliptic systems (short: hecc, but some other authors prefer abbreviations like hec or hcc) of genus at least two and offering comparable security. An important milestone in the road to change this perception happened in September 2002: at the ECC 2002 Workshop in Essen, K. Nguyen of Philips Research reported on his implementation on a hardware simulator of T. Lange's projective formulae for genus 2 [25]. This showed for the first time that the performance of hecc can be competitive, even for smart card applications. Shortly afterwards J. Pelzl, T. Wollinger, J. Guajardo and C. Paar [38] obtained efficient formulae for genus 3 hyperelliptic Jacobians in all characteristics improving on the work of [23].

---

[*] The work described in this paper has been supported by the Commission of the European Communities through the IST Programme under Contract IST-2001-32613 (see http://www.arehcc.com or http://www.arehcc.org).

C.D. Walter et al. (Eds.): CHES 2003, LNCS 2779, pp. 366–381, 2003.

This raises immediately the issue of the security of hecc against side-channel attacks, first introduced in the form of timing attacks in [20] and then simple and differential power analysis (SPA and DPA) [21,22]. These attacks measure some leaked information of a cryptographic device (e.g. timing, power consumption, electromagnetic radiation) while it processes its inputs. For historical reasons we just write DPA also when exploiting leaked information other than power consumption. If a single input is used, the process is referred to as a *Simple Power Analysis* (SPA), and if several different inputs are used together with statistical tools, it is called *Differential Power Analysis* (DPA). We are concerned here with the second type of analysis.

SPA attempts to recover the secret scalar from one observation of the sequence of operations: For example, in a simple double-and-add algorithm the number of consecutive group doublings minus one is the amount of zeros between two ones in the binary representation of the scalar. For ecc there exist two anti-SPA strategies.

The first strategy aims at making the sequence of group operations seemingly independent from the scalar. In the "double-and-add-always" [7] scalar multiplication method an addition is performed after each doubling, even if the corresponding digit of the scalar is zero: This can be done of course in any group, including the Jacobians of hyperelliptic curves. For curves in odd characteristic admitting a particular form, the "Montgomery" method [33,37] allows a very fast computation where the y-coordinate is not used. Analogues of this idea exists for binary curves [1,30] and for all elliptic curves over prime fields [3,8].

The second strategy relies on indistinguishable addition and doubling formulae. They exist for many classes of curves, such as those in Hessian [15,40] or in Jacobi Form [28]. E. Brier and M. Joye found such formulae for elliptic curves over all fields [3]. Another way of pursuing this strategy is to insert dummy operations: for an even characteristic example see [2].

At the moment of this writing little has been done to protect specifically a hecc against SPA. The only currently known methods are the generic ones such as: (i) the insertion of dummy group additions in the scalar multiplication algorithm (as in the "double-and-add-always" method) or (ii) the insertion of dummy field operations in the addition and doubling formulae. T. Lange [27] remarked that the latter can be realized easily and efficiently with the genus 2 affine formulae: this is particularly important for the applications, since the formulae are simpler than in the genus 3 and 4 cases, and the security of genus 2 curves is better understood.

Henceforth we shall always assume that the scalar multiplication algorithm has been made immune from SPA by at least one of these two techniques.

In a DPA the side-channel information collected upon processing of several different inputs is correlated with the value of a boolean function χ of the internal representation of the operands in the cryptographic hardware. The attacker, which is assumed to know the algorithm, *guesses* that the hardware will perform a specific operation at a given point – for example which operand from a table is reused, or which branch is taken – depending on some part of the secret information to be elicited. The inputs are then sorted in two sets according to the values of χ on the corresponding guessed outputs. If the statistical correlation with the leaked information is good, the guess was correct. This leads to attacks which require time linear in the length of the cryptographic operation. We refer the reader to [20,21,22] for more details. Short descriptions can also be found in [7, §3] and [16, §§ 3.2 and 3.3].

The present work is a first attempt to harden hecc *against DPA. In the next section we develop hyperelliptic curve analogues of Coron's third countermeasure [7] (point randomisation) and of the curve randomisation method of M. Joye and C. Tymen [16]. The impact of the recent results of L. Goubin [13] is discussed. We also discuss the applicability of such techniques in light of: (i) the state of the art of explicit formulae for divisor addition and (ii) security results for specific classes of varieties. An appendix contains an example of explicit transformations for the curve randomisations in genus 2.*

2   The Techniques

2.1   Curve Randomisation

An excellent, low brow introduction to the subject of hyperelliptic curves, with a detailed derivation of the facts used below, is given in [31]: Our notation is slightly different, but conforms to that of [24,25,26,27,38].

The idea behind curve randomisation techniques is to "scramble" all the bits of the computation in a (hopefully) unpredictable way. It consists in picking a random group isomorphic to the one on which the cryptosystem is based, transferring the computation to it and then pulling the result back.

More formally, let $\mathcal{C}$ and $\mathcal{C}'$ be two hyperelliptic curves of genus $g \geqslant 1$ over a finite field $\mathbb{F}_q$. Suppose that $\phi : \mathcal{C} \to \tilde{\mathcal{C}}$ is an $\mathbb{F}_q$-isomorphism which is easily extended to an $\mathbb{F}_q$-isomorphism of the Jacobians $\phi : \mathcal{J}(\mathcal{C}) \to \mathcal{J}(\tilde{\mathcal{C}})$. Let us further assume that ϕ, together with its inverse, is computable in a reasonable amount of time, *i.e.* small with respect to the time of a scalar multiplication. We do not require *a priori* the computation time of ϕ to be negligible with respect to a single group operation. Then instead of computing $Q = nD$ in $\mathcal{J}(\mathcal{C})(\mathbb{F}_q)$, where n is an integer and $D \in \mathcal{J}(\mathcal{C})(\mathbb{F}_q)$, we perform:

$$Q = \phi^{-1}\big(n\,\phi(D)\big) \tag{1}$$

so that the bulk of the computation is done in $\mathcal{J}(\tilde{\mathcal{C}})(\mathbb{F}_q)$, or, since a picture is worth a thousand words, we note that the following diagram commutes

$$
\begin{array}{ccc}
\mathcal{J}(\mathcal{C})(\mathbb{F}_q) & \xrightarrow{\text{multiplication by } n} & \mathcal{J}(\mathcal{C})(\mathbb{F}_q) \\
\phi \downarrow & & \uparrow \phi^{-1} \\
\mathcal{J}(\tilde{\mathcal{C}})(\mathbb{F}_q) & \xrightarrow{\text{multiplication by } n} & \mathcal{J}(\tilde{\mathcal{C}})(\mathbb{F}_q)
\end{array}
$$

and we follow it along the longer path.

The countermeasure is effective if the representations of the images under ϕ of the curve coefficients and of the elements of $\mathcal{J}(\mathcal{C})(\mathbb{F}_q)$ are unpredictably different from those of their sources. This can be achieved by multiplying the quantities involved in a computation with randomly chosen numbers (but: see Subsection 2.3). We are going to show that, in the case of hyperelliptic curves, this can be done by a small number of elementary field operations.

We do not discuss the use of random field isomorphisms according to [16, § 4.2]. The treatment carries over with little or no changes, and the method is computationally heavy, considerably slowing down all field operations. It is not clear whether it can even be done on a smart card in the ecc case. In hecc, the ground field being smaller, it is possible that this countermeasure could be implemented. As there is a potential performance/security trade-off in even characteristic with curve randomisations (see § 2.1), especially in the genus 2 case, one might be tempted to reconsider the use of field isomorphisms: However, divisor randomisation (see § 2.2) makes them superfluous.

General curve isomorphisms. We now put in practice the idea just sketched. Let $g \geqslant 1$ be an integer, and $\mathbb{F}_q$ be a finite field. Let $\mathcal{C}, \tilde{\mathcal{C}}$ be two hyperelliptic curves of genus g defined by *Weierstrass equations*

$$\mathcal{C} \ : \ y^2 + h(x)y - f(x) = 0 \tag{2}$$

$$\tilde{\mathcal{C}} \ : \ y^2 + \tilde{h}(x)y - \tilde{f}(x) = 0 \tag{3}$$

over $\mathbb{F}_q$, where $f, \tilde{f}$ are monic polynomials of degree $2g + 1$ in x and $h(x), \tilde{h}(x)$ are polynomials in x of degree at most g. $\mathcal{C}$ (and $\tilde{\mathcal{C}}$) has no singular affine points, *i.e.* there are no solutions $(x, y) \in \mathbb{F}_q \times \mathbb{F}_q$ which simultaneously satisfy the equation $y^2 + h(x)y - f(x) = 0$ and the partial derivative equations $2y + h(x) = 0$ and $h'(x)y - f'(x) = 0$. This is equivalent to saying that the discriminant of $4f + h^2$ does not vanish [29, Theorem 1.7]. Analogous conditions holds for $\tilde{\mathcal{C}}$. Denote by ∞ the non affine point in the projective completions of $\mathcal{C}$ and $\tilde{\mathcal{C}}$. All $\mathbb{F}_q$-isomorphisms of curves $\phi : \mathcal{C} \to \tilde{\mathcal{C}}$ are, by [29, Proposition 1.2], of the type

$$\phi \ : \ (x, y) \mapsto \left(s^{-2}x + b, s^{-(2g+1)}y + A(x)\right) \tag{4}$$

for some $s \in \mathbb{F}_q^\times$, $b \in \mathbb{F}_q$ and a polynomial $A(x) \in \mathbb{F}_q[x]$ of degree at most g. Upon substituting $s^{-2}x + b$ and $s^{-(2g+1)}y + A(x)$ in place of x and y in equation (3) and comparing with (2) we obtain

$$\begin{cases} h(x) = s^{2g+1}\left(\tilde{h}(s^{-2}x + b) + 2\,A(x)\right) \\ f(x) = s^{2(2g+1)}\left(\tilde{f}(s^{-2}x + b) - A(x)^2 - \tilde{h}(s^{-2}x + b)\,A(x)\right) \end{cases} \tag{5}$$

whose inversion is

$$\begin{cases} \tilde{h}(x) = s^{-(2g+1)}h(\hat{x}) - 2A(\hat{x}) \\ \tilde{f}(x) = s^{-2(2g+1)}f(\hat{x}) + s^{-(2g+1)}h(\hat{x})A(\hat{x}) - A(\hat{x})^2 \\ \qquad \text{where} \quad \hat{x} = s^2(x - b) \ . \end{cases} \tag{6}$$

Now ϕ is an isomorphism of $\mathcal{C}$ onto $\tilde{\mathcal{C}}$ and it induces an isomorphism (which we also call ϕ) of their Jacobians, which is also a group isomorphism. It is a well known fact that the Jacobian of a curve $\mathcal{C}$ is isomorphic to the ideal class group $\mathrm{Cl}^0(\mathcal{C})$, which is more suitable for direct computations, and for this reason we want to see see how ϕ operates on the elements of $\mathrm{Cl}^0(\mathcal{C})$.

D. Mumford [35] has introduced a representation of the elements of the latter group as polynomial pairs, for which D. Cantor [4] provided an explicit arithmetic algorithm. Any divisor can be written as $D = \sum_{P \in \mathcal{S}} m_P P - (\sum_{P \in \mathcal{S}} m_P)\infty$ for a finite subset of points $\mathcal{S}$ of $\mathcal{C}(\overline{\mathbb{F}_q})$ called the *support* of D, the m_i being positive integers, and the *degree* of D is the integer $\deg(D) = \sum_{P \in \mathcal{S}} m_P$. Let D be the unique principal divisor of degree at most g in a given divisor class on $\mathcal{C}$. Then (the ideal class associated to) D is represented by a unique pair of polynomials $U(t), V(t) \in \mathbb{F}_q[t]$ with $g \geqslant \deg_t U > \deg_t V$, U monic and such that:

$$\begin{cases} U(t) = \prod_{P \in \mathcal{S}} (t - x_P)^{m_P} \\ V(x_P) = y_P \quad \text{for all } P \in \mathcal{S} \\ U(t) \text{ divides } V(t)^2 + V(t)h(t) - f(t) \ . \end{cases} \tag{7}$$

We say that the pair $[U(t), V(t)]$ represents the *reduced divisor* D. It is $\deg(D) = \deg(U)$.

It is clear that we want to find a pair of polynomials $\tilde{U}(t), \tilde{V}(t) \in \mathbb{F}_q[t]$ which satisfy similar conditions, but for the divisor $\phi(D) = \sum_{P \in \mathcal{S}} m_P \phi(P) - (\sum_{P \in \mathcal{S}} m_P)\infty$ in place of D. In other words, we must have:

$$D = \sum_{P \in \mathcal{S}} m_P P - \left(\sum_{P \in \mathcal{S}} m_P \right)\infty \xrightarrow{\quad \phi \quad} \sum_{P \in \mathcal{S}} m_P \phi(P) - \left(\sum_{P \in \mathcal{S}} m_P \right)\infty = \phi(D)$$

$$\|$$

$$[U(t), V(t)] \qquad \xrightarrow{\quad \phi \quad} \qquad [\tilde{U}(t), \tilde{V}(t)]$$

This is very straightforward to obtain. Clearly

$$\tilde{U}(t) = \prod_{P \in \mathcal{S}} \left(t - x_{\phi(P)} \right)^{m_P} = \prod_{P \in \mathcal{S}} \left(t - s^{-2} x_P - b \right)^{m_P}$$

$$= s^{-2 \sum_{P \in \mathcal{S}} m_P} U\left(s^2(t - b) \right) = s^{-2 \deg_t U} U\left(s^2(t - b) \right) \ . \tag{8}$$

Then, $\tilde{V}$ must satisfy $\tilde{V}(x_{\phi(P)}) = y_{\phi(P)}$ for all $P \in \mathcal{S}$, in other words

$$\tilde{V}(s^{-2} x_P + b) = s^{-(2g+1)} y_P + A(x_P) = s^{-(2g+1)} V(x_P) + A(x_P)$$

i.e.

$$\tilde{V}(t) = s^{-(2g+1)} V\left(s^2(t - b) \right) + A\left(s^2(t - b) \right) \ . \tag{9}$$

Equations (8) and (9) give the correct $\tilde{U}(t)$ and $\tilde{V}(t)$. This follows from the uniqueness of the representation of reduced divisors: In fact $\tilde{U}(t)$ and $\tilde{V}(t)$ are defined over $\mathbb{F}_q$, $\deg \tilde{V} = \deg V < \deg U = \deg \tilde{U}$, and it is straightforward to verify that $\tilde{U}(t)$ divides $\tilde{V}(t)^2 + \tilde{V}(t)\tilde{h}(t) - \tilde{f}(t)$.

Odd characteristic. Here we consider the case where $\mathbb{F}_q$ is a finite field of odd characteristic. We assume that $h(x) = \tilde{h}(x) = 0$, since we can transform the equations by the variable change $y \mapsto y - h(x)/2$ and $y \mapsto y - \tilde{h}(x)/2$. The advantage in doing so is that Cantor's algorithm will run faster, and for the same reason explicit formulae for odd characteristic have only been developed under this assumption. Then the equations of $\mathcal{C}, \tilde{\mathcal{C}}$ are of the form

$$\mathcal{C} \ : \ y^2 - f(x) = 0 \tag{10}$$
$$\tilde{\mathcal{C}} \ : \ y^2 - \tilde{f}(x) = 0 \tag{11}$$

which imply, by (6), that $A(x) = 0$.

If, furthermore, $\operatorname{char} \mathbb{F}_q \nmid 2g + 1$, we can assume that the second most significant coefficient of $f(x)$ (and of $\tilde{f}(x)$), i.e. the coefficient f_{2g} of x^{2g}, vanishes too, since we can perform the variable change $x \mapsto x - f_{2g}/(2g+1)$. In this case, moreover, by (6) it must be $b = 0$, so the isomorphism ϕ takes the simple form

$$\phi \ : \ (x, y) \mapsto \left(s^{-2}x, s^{-(2g+1)}y\right) \tag{12}$$

where $s \in \mathbb{F}_q^\times$. (For simplicity, we shall consider only isomorphisms of this kind, even if $\operatorname{char} \mathbb{F}_q \mid 2g + 1$.) The formula for $\tilde{f}$ is

$$\tilde{f}(x) = s^{-2(2g+1)} f\left(s^2 x\right) \ .$$

This randomisation modifies all non-zero coefficients of the Weierstrass equation (that is, all those who are used in the computation) and of the two polynomials representing a reduced divisor (except for the leading coefficient of U, which must remain equal to 1), namely

$$\tilde{U}(t) = s^{-2 \deg_t U} U\left(s^2 t\right), \quad \tilde{V}(t) = s^{-(2g+1)} V\left(s^2 t\right) \ .$$

Explicit description, an implementation trick. The method is very fast. First, we pick a random $s \in \mathbb{F}_q^\times$ and compute its multiplicative inverse. They are both needed: s^{-1} for ϕ and s for ϕ^{-1}. We make the computation of ϕ explicit. If

$$f(x) = x^{2g+1} + \sum_{i=0}^{2g-1} f_i x^i$$

then

$$\tilde{f}(x) = x^{2g+1} + \sum_{i=0}^{2g-1} s^{2i-2(2g+1)} f_i x^i \ .$$

For $U(t)$ and $V(t)$ in the general case it is

$$U(t) = t^g + \sum_{i=0}^{g-1} U_i t^i \quad \text{and} \quad V(t) = \sum_{i=0}^{g-1} V_i t^i$$

so that

$$\tilde{U}(t) = t^g + \sum_{i=0}^{g-1} s^{2i-2g} U_i t^i \quad \text{and} \quad \tilde{V}(t) = \sum_{i=0}^{g-1} s^{2i-(2g+1)} V_i t^i \ .$$

To apply ϕ to the equation of the curve and to the basis divisor $[U(t), V(t)]$ we proceed as follows: Assume we have already s and s^{-1}. We compute s^{-k} for $k = 2, 3, \ldots 2g+1$ and $k = 2(g+1), 2(g+2), \ldots, 2(2g+1)$. This requires $3g+1$ multiplications (some can be replaced with squarings). For even k we compute $\tilde{f}_{2g+1-k/2} = s^{-k} f_{2g+1-k/2}$ (if $k \neq 2$) and $\tilde{U}_{g-k/2} = s^{-k} U_{g-k/2}$ (with $k \leqslant 2g$). If k is odd and $\leqslant 2g+1$ we multiply $V_{g-(k-1)/2}$ by s^{-k} to obtain $\tilde{V}_{g-(k-1)/2}$. Computing $\tilde{f}, \tilde{U}$ and $\tilde{V}$ requires $4g$ multiplications, hence the total amount of operations required to apply ϕ is $7g+1$ multiplications. Computing ϕ^{-1} requires only $4g$ multiplications in $\mathbb{F}_q$, bringing the total to $11g+1$.

In the cases $g = 2$, resp. 3 this randomisation needs 23, resp. 34 field multiplications (and possibly one inversion), which compares favorably to the costs of one group addition: in the genus 2 case, according to T. Lange [24] one group addition requires 25 multiplications and 1 inversion, and in the genus 3 case J. Pelzl et al. [38] need 76 multiplications and 1 inversion.

We mention an implementation trick to save an inversion each time the device is used at the price of a multiplication. During the initialisation of the device, a set $(\kappa_i, \kappa_i^{-1})$ of randomly chosen elements of $\mathbb{F}_q^\times$ together with their inverses is stored in the E^{2}PROM. Before each cryptographic operation, two random indices $i \neq j$ are picked, and the i-th pair is replaced by $(\kappa_i \cdot \kappa_j, \kappa_i^{-1} \cdot \kappa_j^{-1})$. The result is used as the (s, s^{-1}) for the curve randomisation in the current session.

Partial conclusions. Curve randomisation in odd characteristic is a fast countermeasure. The total amount of operations required to apply this technique is either comparable with that of a single group operation or much smaller.

Even characteristic. The discussion in § 2.1 applies in particular to the case of even characteristic. Let $d = [\mathbb{F}_q : \mathbb{F}_2]$. Since in this case one must have $h(x)\tilde{h}(x) \neq 0$ in equations (2) and (3), it is clear that applying the isomorphisms in general will not be as efficient as in the odd characteristic case.

In place of the fully general isomorphisms (4) we assume $b = 0$ and $A(x) = 0$, and proceed as at the end of § 2.1. The isomorphisms of the form

$$\phi : (x, y) \mapsto \left(s^{-2}x, s^{-(2g+1)}y \right) \tag{12'}$$

for generic $s \in \mathbb{F}_{2^d} \setminus \mathbb{F}_2$ randomise the coefficients of the equation as follows

$$\begin{cases} \tilde{h}(x) = s^{-(2g+1)} h(s^2 x) \\ \tilde{f}(x) = s^{-2(2g+1)} f(s^2 x) \ . \end{cases} \tag{13}$$

As in §2.1 we make this explicit: if

$$f(x) = x^{2g+1} + \sum_{i=0}^{2g-1} f_i x^i \quad \text{and} \quad h(x) = \sum_{i=0}^{g} h_i x^i$$

then

$$\tilde{f}(x) = x^{2g+1} + \sum_{i=0}^{2g-1} s^{2i-2(2g+1)} f_i x^i \quad \text{and} \quad \tilde{h}(x) = \sum_{i=0}^{g} s^{2i-(2g+1)} h_i x^i$$

and the formulae for $\tilde{U}, \tilde{V}$ are the same as in §2.1. All the coefficients of the equation and of the divisor are then multiplied by random constants. In even characteristic we must compute also the coefficients of $\tilde{h}(x)$ from those of $h(x)$. Hence, at most $g+1$ field operations more are required than in the odd characteristic case, bringing the cost of the computation of ϕ to at most $8g+2$ multiplications, after s has been randomly chosen and s^{-1} computed. The computation of ϕ^{-1} still requires $4g$ multiplications. The total cost of this randomisation is thus $12g+2$ field multiplications and one inversion: The implementation trick described in §2.1 not necessary in even characteristic, inversion being much faster in this case.

Restricting h: h constant. In even characteristic often the coefficients of $h(x)$ are restricted for performance reasons. In this paragraph we consider the case where $h(x)$ is a non-zero constant. Equation (6) implies that $\tilde{h}(x)$ will also be a non-zero constant.

It is an established fact in algebraic geometry that curves of equation $y^2 + cy = f(x)$ with $\deg f = 5$ and $c \neq 0$ are supersingular [11, Theorem 9] and so are not suitable for the cryptographic applications we have in mind.

On the other hand there are no hyperelliptic supersingular curves of genus 3 in characteristic 2 [39], so curves of the form $y^2 + cy = f(x)$ where $\deg f = 7$ and $c \neq 0$ do not appear to be weak provided that parameters as extension degree and group order are suitably chosen. Now, even though in [38] a very fast doubling formula is given for the doubling in the case $h(x) = 1$, J. Pelzl has privately communicated to us that in the generic case where $h(x)$ is a non-zero constant $h(x) = c \in \mathbb{F}_{2^d}$ doubling speed can still be improved dramatically. Trivially, $\tilde{h}(x) = s^{-(2g+1)} c = s^{-7} c$. This makes the genus 3 case important.

Restricting h: h non-constant but defined over $\mathbb{F}_2$. Another technique for gaining performance is to choose $h(x)$ non-constant but defined over $\mathbb{F}_2$ (see for example [25] and [26]). By (6) this leads to the question: *if $h(x) \in \mathbb{F}_2[x]$, for which elements $b \in \mathbb{F}_q$ and $s \in \mathbb{F}_q^{\times}$ is it $\tilde{h}(x) = s^{-(2g+1)} h(s^2(x-b)) \in \mathbb{F}_2[x]$?*

The leading coefficient of $\tilde{h}(x)$ equals s^{-r} where $r = (2g+1) - 2 \deg h$, and since it cannot vanish, it is 1, *i.e.* $s^r = 1$. Now r is an odd positive integer bounded by $2g - 1$. The cryptosystem must withstand P. Gaudry's low genus algorithm for computing discrete logarithms in hyperelliptic Jacobians [12]. Hence g must be small, in fact $g \leqslant 4$, so $r \leqslant 7$. This implies that s can take only very few possible values, making superfluous the effort of randomising it.

Remark: *In order to make Weil Descent attacks [9,10] infeasible, the extension degree d is usually taken to be a large prime number $p \gtrsim 160/g$ or twice a prime $p \gtrsim 80/g$. Recall also that $g \leqslant 4$. The possible values of s are limited to the roots of irreducible factors of $X^r - 1$ of degree dividing d. If $d = p \gtrsim 160/g \geqslant 40$, which is also the preferred case, then $s = 1$. If $d = 2p$ with $p \gtrsim 80/g \geqslant 20$, s can only be a root of a factor of $X^r - 1$ of degree at most 2 and irreducible over $\mathbb{F}_2$. A quick verification of such factors (recall that r is odd and $\leqslant 7$) implies that either $s = 1$ or $r = 3$ and $s^2 + s + 1 = 0$. If two coefficients of $h(x)$ are equal to 1, forcing the corresponding coefficients of $\tilde{h}(x)$ to be also equal to 1 implies always $s = 1$.*

Let σ be the Frobenius automorphism of $\mathbb{F}_q/\mathbb{F}_2$, i.e. $\alpha \mapsto \alpha^2$. Now $\tilde{h}(x) = h(x-b) \in \mathbb{F}_2[x]$, hence $h(-b^{\sigma^j}) = h(-b)^{\sigma^j} = h(-b) \in \mathbb{F}_2$ for all j. In other words all distinct conjugates of $-b$ are roots of $h(x) - h(-b) = 0$, and if $b \notin \mathbb{F}_2$ there are at least $p \gtrsim 80/g \geqslant 20$ of such conjugates, including $-b$. But the degree of h, as we already know, is at most $g \leqslant 4$, and this forces $b \in \mathbb{F}_2$. There are only two choices for b, making useless to consider its randomisation.

We see that the isomorphisms we can use are of the form

$$\phi \; : \; (x, y) \mapsto (x, y + A(x))$$

where the polynomial $A(x) \in \mathbb{F}_q[x]$ has degree at most g. The situation is similar to that for elliptic curves as described in the already cited paper of M. Joye and C. Tymen: we can efficiently randomise only one of the two polynomials (V, whereas U will be left untouched), or, in other words, only a half of the coordinates. In fact, by (6) not all coefficients of f are randomised in $\tilde{f}$, increasing the likelihood of successful bit-correlations if this countermeasure is used alone.

Partial conclusions. We conclude that for genus 2 hyperelliptic curves in characteristic 2, curve randomisation is not adequate if one wants to force the coefficients of $\tilde{h}$ to lie outside $\mathbb{F}_2$.

In the genus 3 case curves of equation $y^2 + cy = f(x)$ can be randomised obtaining good performance and security.

In all other cases, we recommend other techniques, such as divisor randomisation, which also works in odd characteristic. We sketch it in the next section in the case of genus 2.

2.2   Divisor Randomisation in Genus 2

Divisor randomisation works by randomising the bits of the representation of a reduced divisor, which can be either the base group element of the cryptosystem or any intermediate result of the computation of a scalar multiplication. This technique does not scramble the bits of the internal representations of the coefficients of the curve. It can be used whenever a group element can be represented in several different ways. Notable examples are the projective coordinates on elliptic curves: two triples (X, Y, Z) and (X', Y', Z') represent the same point if and only if there exists a non zero element s in the base field such that $X = sX'$, $Y = sY'$ and $Z = sZ'$. With Jacobian coordinates [5], two triples (X, Y, Z) and (X', Y', X') represent the same point if and only if $X = s^2 X'$, $Y = s^3 Y'$ and $Z = sZ'$.

Recently, alternative coordinate systems for genus 2 hyperelliptic curves have been proposed: An inversion-free system by Miyamoto et al. [32] which operates on the hyperelliptic analogue of projective coordinates, later extended and improved by Lange [25], who also developed an analogue of Jacobian coordinates, called the *new (or weighted) coordinates* [26]. We are not aware of similar coordinate systems for genus 3 curves. Furthermore, as the genus of the considered curve increases, the size of the base field decreases, and the cost of a field inversion relative to a field multiplication also decreases quickly. This makes inversion-free formulae in genus at least 3, not so desirable from the point of view of raw performance, because they trade a single inversion for a lot more multiplications than the affine formulae.

In projective coordinates a divisor class D with associated reduced polynomial pair $[U(t), V(t)]$ is represented as a quintuple $[U_1, U_0, V_1, V_0, Z]$ where

$$U(t) = t^2 + \frac{U_1}{Z} t + \frac{U_0}{Z} \quad \text{and} \quad V(t) = \frac{V_1}{Z} t + \frac{V_0}{Z} .$$

The randomisation in this case consists in picking a random $s \in \mathbb{F}_q^\times$ and by performing the following replacement

$$[U_1, U_0, V_1, V_0, Z] \mapsto [sU_1, sU_0, sV_1, sV_0, sZ] .$$

In new (weighted) coordinates a divisor class is represented by means of six coordinates $[U_1, U_0, V_1, V_0, Z_1, Z_2]$ where

$$U(t) = t^2 + \frac{U_1}{Z_1^2} t + \frac{U_0}{Z_1^2} \quad \text{and} \quad V(t) = \frac{V_1}{Z_1^3 Z_2} t + \frac{V_0}{Z_1^3 Z_2} .$$

To blind the base point or an intermediate computation, two elements s_1, s_2 are picked in $\mathbb{F}_q^\times$ at random and the following substitution is performed:

$$[U_1, U_0, V_1, V_0, Z_1, Z_2] \mapsto [s_1^2 U_1, s_1^2 U_0, s_1^3 s_2 V_1, s_1^3 s_2 V_0, s_1 Z_1, s_2 Z_2] .$$

If (some or all of) the additional coordinates z_1, z_2, z_3 and z_4 are used – which satisfy $z_1 = Z_1^2$, $z_2 = Z_2^2$, $z_3 = Z_1 Z_2$ and $z_4 = z_1 z_2$ – then they must also be updated: the fastest way is to recompute them from Z_1 and Z_2 by two squarings and two multiplications.

2.3 Goubin's Attack May Force Further Blinding

Recently L. Goubin [13] has pointed out a potential weakness of some ecc randomisation procedures, including Coron's third and Joye-Tymen's, when implemented on systems where the secret scalar is fixed and the base of the scalar
multiplication (the message) can be chosen. Since our techniques generalise the
above ones, it is natural to investigate how Goubin's ideas might affect our work.

His attack is based on the randomisation of 0 by multiplication by a constant
or by field isomorphism being still 0. It relies also on the fact that the scalar
multiplication algorithm has a fixed sequence of group operations for a given
scalar – even after removing any dummy operations. (It should work also if the
number of possible operations sequences for a given scalar is small enough.)

Suppose that the most significant bits $n_r, n_{r-1}, \dots, n_{j+1}$ of the secret scalar
n are known and that we want to discover the next bit n_j. Assume also that a
chosen message attack can be set up to obtain in a specific step of the scalar
multiplication – namely the one corresponding to the processing of n_j – a point
or a divisor with one or more coordinates equal to zero, provided that n_j has
been guessed correctly (that divisor can be tD where D is the "message" and $t =
(n_r, n_{r-1}, \dots, n_{j+1}, n_j)_2$). The side-channel trace correlation may reveal if the
guess was correct even in presence of multiplicative randomisation procedures,
because some multiplications by zero will occur in any case. In particular, this
can affect the random isomorphisms of the form $\phi : (x, y) \mapsto \left(s^{-2}x, s^{-(2g+1)}y \right)$
and the divisor randomisation techniques of Subsection 2.2.

An approach to thwart Goubin's attack could use the more general isomorphisms (4) with b, $A(x) \neq 0$ to randomise also the vanishing coefficients of the
divisors: this has the disadvantage of requiring curve equations in general form
and thus slower formulae for the operations.

There is a development of Goubin's ideas which might be even more serious.
We first fix some notation: Λ is the large prime order subgroup of $\mathrm{Cl}^0(\mathcal{C})(\mathbb{F}_q)$
used in the cryptosystem and ℓ its order.

A variant of Goubin's attack may exploit the fact that the basic explicit
formulae for small genus hyperelliptic curves only deal with the most common
cases (cfr. [24,27,38]). They do not hold if the divisors given as input to a group
operation satisfy exceptional conditions, such as:

(i) If the reduced divisor D_i, for $i = 1, 2$, is represented by the polynomial pair
     $[U_i, V_i]$, then the greatest common divisor of U_1 and U_2 is non-constant or,
     equivalently, their resultant is vanishing.
     In this case we say that D_1 and D_2 *collide*. This happens if the supports
     of D_1 and of either D_2 or $-D_2$ have at least one point in common.
(ii) $\deg(D_1) < g$ or, equivalently, $\deg(U_1) < g$ (this applies to additions as well
      as to doublings).

Such situations occur in practice with very small probability ($O(q^{-1})$ for curves
over $\mathbb{F}_q$), hence no separate formulae for these cases are implemented and either

Cantor's algorithm or formulae with quite different characteristics are used[1]. Since the characterising properties of these divisors are of geometric nature, they are preserved under curve isomorphisms. Their occurrence may thus be induced at prescribed points to verify the guesses of the bits of the scalar. At least in theory, the attacker guesses that at some point the scalar multiplication algorithm adds D to tD (resp. doubles tD) and therefore chooses D to collide with tD (resp. $\deg(tD) < g$). We do not know, except for very simple cases (i.e. $|t| \leqslant g$), how to produce for a given t (with $\ell \nmid t$) a divisor D colliding with tD (reduced) – we suspect that it is in general a hard problem. On the other hand it is very easy to find D such that $\deg(tD) < g$: just pick any $D' \in \Lambda$ with $\deg(D') < g$, find an integer s with $st \equiv 1 \pmod{\ell}$ and put $D = sD'$. Then $tD = D'$ and, if the doubling formula for the exceptional case is distinguishable from the generic one, the attack can be launched.

This represents an obvious danger with affine coordinates: if one of the above exceptional conditions occurs, the most common case formulae cannot be used, to avoid a division by zero. With inversion-free coordinate systems the situation is only apparently different: one can just use only the most common case formulae and check at the end of the scalar multiplication if the divisor belongs to the curve – but also in that case anomalous behaviour of the device at the end of the scalar multiplication could be detected.

We therefore need additional scalar and message blinding methods.

We briefly discuss scalar blinding methods. Their purpose is to render unpredictable the addition chain used in the scalar multiplication, thus preventing the attacker to guess for which integers t group operations of the type $D + tD$ or $2(tD)$ are actually performed.

The first is Coron's first countermeasure [7], *i.e.* the replacement of the scalar n with $n + k\ell$ in nD for a random integer k. This technique can be traced back to [20], and was shown [36] to leave a bias in the least significant bits of the scalar. B. Möller [34] combines it (only in the ecc case) with an idea of C. Clavier and M. Joye, and suggests the computation of $nD = (n + k_1 + k_2\ell)D - k_1D$, where k_1 and k_2 are two suitably sized random numbers: k_1 and k_2 should be large enough to make L. Goubin's attack not palatable, yet not too big, to leave the overhead tolerable (for example $k_1, k_2 \approx 2^{32}$ are good choices if $\ell \approx 2^{160}$).

For a completely different technique see [41].

For message blinding, a hecc analogue of Coron's second method [7] consists in replacing the computation of nD with that of $n(D + R) - S$, where $R \in \Lambda$ is a secret divisor for which $S = nR$ is known. A set of secret divisor pairs $(R_i, S_i) \in \Lambda \times \Lambda$ with $S_i = nR_i$ can be stored in the smart card at initialisation time, and at each run both elements of a randomly chosen pair are multiplied by the same small signed scalar and added to the respective elements of another pair. The result is then used to randomise the scalar multiplication.

---

[1] To provide explicit *and* indistinguishable formulae for all cases would be a formidable feat – and would probably slow down considerably the cryptosystem.

Suppose that a computation involving tD has to be done during the scalar multiplication, either $D + tD$ or $2(tD)$, and that either D collides with tD (this is relevant only to the addition $D + tD$) or $\deg(tD) < g$ – as wished by the attacker. If D has been replaced at the beginning by $D + R$ for a randomly chosen point R, then $D + R$ and $t(D + R)$ will collide with probability $O(q^{-1})$ (this is actually a conjecture which has been extensively confirmed experimentally on small curves), resp. $\deg(t(D + R)) = g$ also with probability $O(q^{-1})$: The last statement holds because $t(D + R)$ is in practice a random point, which implies also that, even if tD had some zero coordinates, a fixed coordinate of $t(D + R)$ would be zero with probability q^{-1}.

We infer that this type of message blinding (which, if used alone, might arouse suspicion) thwarts Goubin's attack. Due to a similar underlying philosophy, additional message blinding should be effective also against some hecc analogue of the "exceptional procedure attack" [14].

To prevent a variant of Goubin's attack, we recommend to use at least an additional scalar or message blinding method besides our randomisation procedures. The hyperelliptic analogue of Coron's second countermeasure, being less expensive than scalar blinding, looks particularly attractive. The isomorphism ϕ need not be of the most general type described in 2.1, but the conclusions of 2.1 and the caveats of 2.1 still apply.

3   Conclusions

We proposed two methods to blind the base divisor class for hyperelliptic curve cryptosystems, in order to provide resistance against DPA.

The first method consists in transferring the critical computation to the Jacobian of a different randomly chosen isomorphic curve. It can be applied to curves of all genera.

The second method is a hyperelliptic analogue of Coron's third countermeasure. It applies only to families of curves for which we know explicit formulae for hyperelliptic analogues of elliptic curve projective and Jacobian coordinates. Explicit examples in the case of genus 2 have been worked out in detail.

These techniques are easy to implement and do not impact the performance significantly. In fact their cost is at most that of a single group addition.

In conjunction with suitable additional scalar and message blinding techniques, they can be made resistant against Goubin's recent chosen message attack, as well as against a possibly more serious variant of the latter based on the structure of the divisors on hyperelliptic curves.

Acknowledgement. The author wishes to express his gratitude to Tanja Lange, Kim Nguyen, Bertrand Byramjee and the anonymous referees for their valuable remarks and suggestions. The author thanks also Alessandro Baldaccini for partially reviewing his English.

References

1. G.B. AGNEW, R.C. MULLIN AND S.A. VANSTONE, *An Implementation of Elliptic Curve Cryptosystems over $F_{2^{155}}$*. IEEE Journal on Selected Areas in Communications, vol. 11, n. 5, pp. 804–813, 1993.
2. A. BELLEZZA, *Countermeasures against Side-Channel Attacks for Elliptic Curve Cryptosystems*. Cryptology ePrint Archive, Report 2001/103. Available from: http://eprint.iacr.org/
3. E. BRIER AND M. JOYE, *Weierstrass Elliptic Curves and Side-Channel Attacks*. In: *Proceedings of PKC'2002*, LNCS 2274, pp. 335–345. Springer-Verlag, 2002.
4. D. CANTOR, *Computing in the jacobian of a hyperelliptic curve*. Mathematics of Computation, **48** (1987), pp. 95–101.
5. D.V. CHUDNOVSKY AND G.V. CHUDNOVSKY, *Sequences of numbers generated by addition in formal groups and new primality and factoring tests*, Advances in Applied Mathematics, **7** (1987), pp. 385–434.
6. C. CLAVIER AND M. JOYE, *Universal exponentiation algorithm - a first step towards provable SPA-resistance*. In: *Proceedings of CHES 2001*, LNCS 2162, pp. 300–308. Springer-Verlag, 2000.
7. J.-S. CORON, *Resistance against Differential Power Analysis for Elliptic Curve Cryptosystems*. In: *Proceedings of CHES '99*, LNCS 1717, pp. 292–302. Springer-Verlag, 1999.
8. W. FISCHER, C. GIRAUD, E.W. KNUDSEN AND J.-P. SEIFERT, *Parallel Scalar Multiplication on General Elliptic Curves over F_p hedged against Differential Side Channel Attacks*. Cryptology ePrint Archive, Report 2002/007, 2002. Available from: http://eprint.iacr.org/
9. G. FREY, *How to disguise an elliptic curve (Weil descent)*. Talk at ECC '98, Waterloo, 1998. Slides available from: http://www.cacr.math.uwaterloo.ca/conferences/1998/ecc98/slides.html
10. G. FREY, *Applications of arithmetical geometry to cryptographic constructions*. In: *Finite fields and applications (Augsburg, 1999)*, pp. 128–161. Springer-Verlag, 2001.
11. S.D. GALBRAITH, *Supersingular curves in cryptography*. In: *Proceedings of ASIACRYPT 2001*, LNCS 2248, pp. 495–513. Springer-Verlag 2001.
12. P. GAUDRY, *An algorithm for solving the discrete log problem on hyperelliptic curves*. In: *Advances in Cryptology – Eurocrypt 2000*, LNCS 1807, pp. 19–34. Springer-Verlag, 2000.
13. L. GOUBIN, *A Refined Power-Analysis Attack on Elliptic Curve Cryptosystems*. In: *Proceedings of PKC 2003*, LNCS 2567, pp. 199–211. Springer-Verlag, 2003.
14. T. IZU AND T. TAKAGI, *Exceptional Procedure Attack on Elliptic Curve Cryptosystems*. In: *Proceedings of PKC 2003*, LNCS 2567, pp. 224–239. Springer-Verlag, 2003.
15. M. JOYE, J.-J. QUISQUATER, *Hessian Elliptic Curves and Side-Channel Attacks*. In: *Proceedings of CHES'2001*, LNCS 2162, pp. 412–420, Springer–Verlag, 2001
16. M. JOYE AND C. TYMEN, *Protections against Differential Analysis for Elliptic Curve Cryptography – An Algebraic Approach*. In: *Proceedings of CHES 2001*, LNCS 2162, pp. 377–390. Springer-Verlag, 2001.
17. N. KOBLITZ, *A family of Jacobians suitable for discrete log cryptosystems*. In: *Advances in Cryptology – Proceedings of CRYPTO '88*, LNCS 403, pp. 94-99. Springer-Verlag, 1990.

18. N. KOBLITZ, *Hyperelliptic Cryptosystems*, Journal of Cryptology **1** (1989), pp. 139–150.

19. N. KOBLITZ, *Algebraic aspects of cryptography*. Springer, 1998.

20. P. KOCHER, *Timing attacks on implementations of Diffie-Hellman, RSA, DSS and other systems*. In: *Advances in Cryptology, Proceedings of Crypto'96*, LNCS 1109, pp. 104–113, Springer-Verlag, 1996.

21. P. KOCHER, J. JAFFE AND B. JUN, *Introduction to Differential Power Analysis and Related Attacks*, 1998.
    Available from: http://www.cryptography.com/dpa/technical

22. P. KOCHER, J. JAFFE AND B. JUN, *Differential Power Analysis*. In: *Proceedings of CRYPTO'99*, LNCS 1666, pp. 388–397. Springer-Verlag, 1999.

23. J. KUROKI, M. GONDA, K. MATSUO, J. CHAO AND S. TSUJII, *Fast Genus Three Hyperelliptic Curve Cryptosystems*. In: *The 2002 Symposium on Cryptography and Information Security, Japan - SCIS 2002*, Jan. 29–Feb. 1 2002.

24. T. LANGE, *Efficient Arithmetic on Genus 2 Hyperelliptic Curves over Finite Fields via Explicit Formulae*. Cryptology ePrint Archive, Report 2002/121. Available from: http://eprint.iacr.org/ – See also [27].

25. T. LANGE, *Inversion-Free Arithmetic on Genus 2 Hyperelliptic Curves*. Cryptology ePrint Archive, Report 2002/147. Available from: http://eprint.iacr.org/ – See also [27].

26. T. LANGE, *Weighted Coordinates on Genus 2 Hyperelliptic Curves*. Cryptology ePrint Archive, Report 2002/153. Available from: http://eprint.iacr.org/ – See also [27].

27. T. LANGE, *Formulae for Arithmetic on Genus 2 Hyperelliptic Curves*. Preprint. Available from: http://www.ruhr-uni-bochum.de/itsc/tanja/
    It partially contains and extends the material of the previous three papers [24, 25,26].

28. P.-Y. LIARDET, N.P. SMART, *Preventing SPA/DPA in ECC system using the Jacobi Form*. In: *Proceedings of CHES'2001*, LNCS 2162, pp. 401–411. Springer-Verlag, 2001.

29. P. LOCKHART, *On the discriminant of a hyperelliptic curve*. Trans. Amer. Math. Soc. **342** (1994), no. 2, pp. 729–752.

30. J. LÓPEZ AND R. DAHAB, *Fast Multiplication on Elliptic Curves over $GF(2^m)$ without Precomputation*. In: *Proceedings of CHES'99*, LNCS 1717, pp. 316–327. Springer-Verlag, 1999.

31. A. MENEZES, Y.-H. WU AND R. ZUCCHERATO, *An Elementary Introduction to Hyperelliptic Curves*. In [19].

32. Y. MIYAMOTO, H. DOI, K. MATSUO, J. CHAO AND S. TSUJI, *A fast addition algorithm of genus two hyperelliptic curve*. In: *Proceedings of SCIS 2002*, IEICE Japan, pp. 497–502, 2002. In Japanese.

33. P.L. MONTGOMERY, *Speeding the Pollard and Elliptic Curve Methods for Factorizations*. Mathematics of Computation, vol. 48, pp. 243–264, 1987.

34. B. MÖLLER, *Securing Elliptic Curve Point Multiplication against Side-Channel Attacks*. In: *Proceedings of ISC '2001*, LNCS 2200, pp. 324–334. Springer-Verlag, 2001.

35. D. MUMFORD, *Tata Lectures on Theta II*. Birkhäuser 1984.

36. K. OKEYA AND K. SAKURAI, *Power analysis breaks elliptic curve cryptosystems even secure against the timing attack*. In: *Progress in Cryptology - INDOCRYPT 2000*, LNCS 1977, pp. 178–190. Springer-Verlag, 2000.

37. K. Okeya, H. Kurumatani and K. Sakurai, *Elliptic curves with the Montgomery–form and their cryptographic applications.* In: *Public Key Cryptography PKC 2000*, LNCS 1751, pp. 238–257. Springer-Verlag, 2000.
38. J. Pelzl, T. Wollinger, J. Guajardo and C. Paar, *Hyperelliptic Curve Cryptosystems: Closing the Performance Gap to Elliptic Curves.* This Volume.
39. J. Scholten and H.J. Zhu, *Hyperelliptic curves in characteristic 2.* Inter. Math. Research Notices, **17** (2002), pp. 905–917.
40. N.P. Smart, *The Hessian Form of an Elliptic Curve.* In: *Proceedings of CHES '2001*, LNCS 2162, pp. 118–125, Springer-Verlag, 2001.
41. C.D. Walter, *MIST: An Efficient, Randomized Exponentiation Algorithm for Resisting Power Analysis.* In: *Topics in Cryptology – CT-RSA 2002, The Cryptographer's Track at the RSA Conference, 2002, San Jose, CA, USA, February 18-22, 2002, Proceedings*, LNCS 2271, pp. 53–66. Springer-Verlag, 2002.

Appendix: Explicit Transformations for the Curve Randomisation for Genus 2

As an example, in this appendix we write down the transformations for the curve randomisation method explained in Subsection 2.1, for $g = 2$. In view of the results of §2.1, we consider here only the curve isomorphism of type (12) where the equations of C and $\tilde{C}$ are given by

$$C \ : \ y^2 + h(x)y - f(x) = 0 \quad \text{and} \quad \tilde{C} \ : \ y^2 + \tilde{h}(x)y - \tilde{f}(x) = 0 \ .$$

The polynomials $f(x)$ and $h(x)$ are of the form

$$f(x) = x^5 + f_4x^4 + f_3x^3 + f_2x^2 + f_1x + f_0 \quad \text{and}$$
$$h(x) = h_2x^2 + h_1x + h_0 \ .$$

Their images are

$$\tilde{f}(x) = x^5 + s^{-2}f_4x^4 + s^{-4}f_3x^3 + s^{-6}f_2x^2 + s^{-8}f_1x + s^{-10}f_0 \quad \text{and}$$
$$\tilde{h}(x) = s^{-1}h_2x^2 + s^{-3}h_1x + s^{-5}h_0 \ .$$

If the base divisor is given by $D = [U(t), V(t)]$ with $\deg(U) = 2$,

$$U(t) = t^2 + U_1t + U_0 \quad \text{and} \quad V(t) = V_1t + V_0$$

then its image $[\tilde{U}(t), \tilde{V}(t)]$ is

$$\tilde{U}(t) = t^2 + s^{-2}U_1t + s^{-4}U_0 \quad \text{and} \quad \tilde{V}(t) = s^{-3}V_1t + s^{-5}V_0 \ .$$

If $\deg(U) = 1$, *i.e.* $U(t) = t + U_0$, then its image is $\tilde{U}(t) = t + s^{-2}U_0$, whereas the image of $V(t)$ is independent of the degree: in this case $V(t) = V_0$ and thus $\tilde{V}(t) = s^{-5}V_0$. The inverse transformation from $\phi(D)$ to D is obvious.

The total number of field operations is at most 26 multiplications in the even characteristic case, 23 multiplications in odd characteristic (because $h = 0$), and one inversion (but: see end of 2.1).

A Practical Countermeasure against Address-Bit Differential Power Analysis

Kouichi Itoh[1], Tetsuya Izu[2], and Masahiko Takenaka[1]

[1] FUJITSU LABORATORIES Ltd.,
64, Nishiwaki, Okubo-cho, Akashi, 674-8555, Japan
[2] FUJITSU LABORATORIES Ltd.,
4-1-1, Kamikodanaka, Nakahara-ku, Kawasaki, 211-8588, Japan
{kito,izu,takenaka}@labs.fujitsu.com

Abstract. The differential power analysis (DPA) enables an adversary to reveal the secret key hidden in a smart card by observing power consumption. The address-bit DPA is a typical example of DPA which analyzes a correlation between addresses of registers and power consumption. In this paper, we propose a practical countermeasure, the *randomized addressing countermeasure*, against the address-bit DPA which can be applied to the exponentiation part in RSA or ECC with and without pre-computed table. Our countermeasure has almost no overhead for the protection, namely the processing speed is no slower than that without the countermeasure. We also report experimental results of the countermeasure in order to show its effect. Finally, a complete comparison of countermeasures from various view points including the processing speed and the security level is given.

Keywords. Differential Power Analysis (DPA), address-bit DPA, countermeasure, exponentiation, RSA, ECC

1 Introduction

Smart cards are becoming a new infrastructure of the coming IT society for their plenty of attractive applications such as identification cards, telephone cards and electronic tickets. However, the side channel attacks are real threats for them [19,20]. In the attack, an adversary observes side channel information such as computing time and power consumption. The adversary can obtain the secret information if there is a tight relation between the side channel information and the secret information (secret key) hidden in the smart card. Especially, if there is a irregular procedure (short-cuts) in the computation, the adversary can easily detect it. Thus the adversary could reveal the secret key without tampering the device physically. The simple power analysis (SPA) and the differential power analysis (DPA) are typical examples of the side channel attacks. Implementers of cryptographic schemes should take countermeasures against these attacks.

In 1999, Messerges et al. proposed a new powerful attack against the secret key cryptosystems, the *address-bit DPA* (ADPA) (from now on, we call the previous DPA as the *data-bit DPA* (DDPA)), which analyzes a correlation between

C.D. Walter et al. (Eds.): CHES 2003, LNCS 2779, pp. 382–396, 2003.
© Springer-Verlag Berlin Heidelberg 2003

the secret information and addresses of registers [26]. Then, in 2002, Itoh et al. extended the attack to Elliptic Curve based Cryptosystems (ECC) [11]. These results suggest that implementers should consider the correlation of the secret information to not only data of registers but also addresses of registers. Itoh et al. also gave several countermeasures against the attack, but those countermeasures require at least twice computing time than without them.

In this paper, we propose a practical countermeasure against the address-bit DPA by randomizing addresses of registers for ECC and RSA. Our countermeasure does not change the scalar to be multiplied; an overhead is very small and the processing speed is as fast as before, namely, a scheme resistant against the data-bit DPA can be easily converted to that against the address-bit DPA with almost no penalty. The conversion can be applied not only binary methods but window-based methods. Moreover, we show the concrete security evaluation result of our countermeasure by theoretically and experimentally.

An approach of our countermeasure is similar to Random Register Renaming (RRR), a DPA countermeasure proposed by May et al. [27]. RRR is supposed to be implemented on a processor called NDISC, which can execute instructions in parallel, while our countermeasure does not require special hardware because it can be implemented by only software with very simple program code.

Side channel attack are so powerful that numerous countermeasures have been proposed (a brief overview is in [34]). However, some of them are proved or shown insecure by newer attacks. Finding a good countermeasure, which satisfies a certain security level and requires a compromisable processing speed, is becoming a hard task for implementers. In this paper, we give a complete comparison of countermeasures from several view points including the security level and the processing speed. As a result, our proposed method (combined with other countermeasures) can provide a good practical solution for resisting the side channel attacks.

In the following, we basically deal with scalar exponentiations in ECC, however, most of algorithms, especially proposed countermeasures, can be applied to other exponentiation based cryptosystems such as RSA. The rest of this paper is organized as follows: In section 2, we give a brief overview of side channel attacks and countermeasures. Then section 3 describes our proposed countermeasure and experimental results. A comparison of countermeasures are in section 4.

2   Preliminaries

In this section, we give a brief introduction of Elliptic Curve based Cryptosystems (ECC) and side channel attacks (SCA) against them (however, most attacks can applied to other exponentiation based cryptosystems such as RSA).

2.1   Elliptic Curve Based Cryptosystems (ECC)

Elliptic curve based cryptosystems (ECC) are one of the standard technology in the area of cryptography [10,28,37]. The most advantage of ECC is the key length; it is currently chosen much shorter than those of existing other cryptosystems (RSA and ElGamal). This feature is quite suitable for implementing on smart cards.

Let K be a finite field with elements a power of a prime. An elliptic curve over K can be represented by the Weierstrass equation

$$E(K) := \{(x, y) \in K \times K \mid y^2 + a_1 xy + a_3 y = x^3 + a_2 x^2 + a_4 x + a_6\} \cup \mathcal{O}, \tag{1}$$

where $a_i \in K$. A special point $\mathcal{O}$ is called the point of infinity. An elliptic curve $E(K)$ has an additive group structure, in which an neutral element is the point of infinity. We call $P_1 + P_2$ ($P_1 \neq P_2$) the elliptic curve addition (ECADD) and $P_1 + P_2$ ($P_1 = P_2$), that is $2P_1$, the elliptic curve doubling (ECDBL). Let d be an integer and P be a point on the elliptic curve $E(K)$. A *scalar exponentiation* is to compute the point $dP = P + \cdots + P$ ($d-1$ additions). A dominant computation of all encryption/decryption and signature generation/verification algorithms of ECC is the computation of dP, where d is a secret integer and P is a base point. Numerous researches have been dedicated to improve the computing time of this part (see [8] for a survey) by omitting unnecessary computations or by finding various short-cuts.

Let $d = d_{n-1} 2^{n-1} + \cdots + d_1 2^1 + d_0$ be a binary expression of d with $d_{n-1} = 1$. Then the binary method (from the most significant bit, MSB) for a scalar exponentiation is in Alg. 1. Similar method from the least significant bit (LSB) is easily constructed, but we omit this case in this paper for simplicity.

INPUT: d[], P
OUTPUT: dP
1: T[0] = P
2: for i=n-2 downto 0 {
3:     T[0] = ECDBL(T[0])
4:     if(d[i]==1){
5:       T[0] = ECADD(T[0],P)
6:     }
7: }
8: return T[0]

Alg. 1. Binary method (from MSB)

INPUT: d[], P
OUTPUT: dP
1: T[0] = P
2: for i=n-2 downto 0 {
3:     T[0] = ECDBL(T[0])
4:     T[1] = ECADD(T[0],P)
5:     T[0] = T[d[i]]
6: }
7: return T[0]

Alg. 2. Add-and-double-always method (from MSB)

2.2   Side Channel Attack

The *side channel attacks* (SCA) are powerful attacks against implementations of cryptographic schemes on smart cards. An adversary observes side channel information such as computing time, or power consumption of the device. Then he/she tries to reveal the secret information by analyzing side channel information. The *simple power analysis* (SPA) [19] and the differential power analysis (DPA) [20] are typical examples of SCA. SPA only uses a single observed information, while DPA uses a lot of observed information together with statistic tools.

Simple Power Analysis. The binary method of Alg. 1 computes ECADD only when the bit of the secret key d_i is 1. Therefore an adversary easily detects this irregular procedure by observing side channel information and obtain the bit information of d_i. This is a basic idea of the *simple power analysis* (SPA) [19].

There are three approaches to resist SPA. The first one uses the special addition formula, in which ECDBL and ECADD are computed by same order of operations (the indistinguishable addition formula [6]). Brier-Joye proposed the formula for the Weierstrass form [3], but the security hole was pointed out in [14]. The second one uses so called the *add-and-double-always* method [5], (Alg. 2 for example), in which both ECDBL and ECADD are computed in every bit (in step 3 and step 4), and a pattern of the side channel information is fixed. Thus the adversary cannot obtain the bit information of d by SPA.

```
INPUT: d[], P
OUTPUT: dP
1: T[0] = P, T[1] = ECDBL(P)
2: for i=n-2 downto 0 {
3:     T[2] = ECDBL(T[d[i]])
4:     T[1] = ECADD(T[0],T[1])
5:     T[0] = T[2-d[i]]
6:     T[1] = T[1+d[i]]
7: }
8: return T[0]
```
Alg. 3. Montgomery ladder

```
INPUT: d[], P
OUTPUT: dP
1: T[0] = RPC(P)
2: for i=n-2 downto 0 {
3:     T[0] = ECDBL(T[0])
4:     T[1] = ECADD(T[0],P)
5:     T[0] = T[d[i]]
6: }
7: return invRPC(T[0])
```
Alg. 4. Add-and-double-always method (from MSB) and RPC

The third approach to resist SPA is the Montgomery ladder [24] which essentially computes ECDBL and ECADD repeatedly [3,7,12,13,18,31,32]. This method can be viewed as a variant of the add-and-double-always method, but provides a good processing speed [12,13]. A sample algorithm of the Montgomery ladder is in Alg. 3.

Differential Power Analysis. Even if a scheme is SPA-resistant, it is not always secure, because the *differential power analysis* (DPA) might reveal the secret key by analyzing observed information statistically [20,25]. In DPA, an adversary makes an assumption on d_i ($d_i = 0$, for example) and simulates the computation repeatedly. Then he/she divides the side channel information into two groups depending on the assumption, in order to make the bias of the hamming weights of the internal information between these groups. If the assumption is correct, then a difference of the information of two groups (a *spike*) can be observed in the trace. Countermeasures against DPA aim to make the simulation impossible by using random numbers [5]. By the randomization we are able to enhance an SPA-resistant scheme to be DPA-resistant easily. Earlier DPA (the *data-bit DPA* (DDPA)) only considered a correlation of data of registers to the secret information [20,25], while newer DPA (the *address-bit DPA* (ADPA)) also considers a correlation of addresses of registers [11,26].

Data-bit DPA: DDPA analyzes a relation between the secret key and data of registers. For example, after finishing step 5 in Alg. 2, data of T[0] is same as that of T[0] if $d_i = 0$; otherwise it is same as that of T[1] if $d_i = 1$. Coron proposed the randomized projective coordinate (RPC) countermeasure [5] in order to resist DDPA. Let $P = (X : Y : Z)$ be a base point represented by the projective coordinate. Then $(X : Y : Z)$ equals to $(rX : rY : rZ)$ for all $r \in K^*$ mathematically; but they all are different data as bit sequences. The side channel information of a scalar exponentiation is randomized if $(X : Y : Z)$ is randomized to $(rX : rY : rZ)$. An example of RPC for the add-and-double-always method (from MSB) is given in Alg. 4, where a function RPC outputs the randomized point and a function invRPC denotes its inverse map.

Joye-Tymen proposed another countermeasure against DDPA [16], the randomized curve (RC) countermeasure, which uses an isomorphism of elliptic curves with which a curve equation and a base point are transformed with holding the group structure. Two isomorphic curves are same mathematically; but different as bit sequences. Thus the side channel information will be randomized if the curve and the base point is randomized. A sample algorithm is easily obtained similarly to Alg 4 by changing functions RPC, invRPC to RC, invRC.

Instead of changing the expression of a base point, Coron also proposed the randomized exponent and the randomized base point countermeasures, in which the scalar is randomized. However, Okeya-Sakurai showed the bias of these countermeasures [32]. Messerges et al. proposed the randomized start point countermeasure, in which a scalar is divided to two parts and computed by different methods [25]. Oswald-Aigner proposed another approach to randomize the scalar [30], but security problems are pointed out [33,35,39]. Walter proposed the MIST algorithm [38], which randomizes the intermediate data T (and U) by repeating $T = (d_i \bmod r_i)U + T$, $U = r_iU$ and $d_{i+1} = \lfloor d_i/r_i \rfloor$, where r_i are random numbers and initial values of T, U, d_i are $\mathcal{O}$, P, d respectively. Hasan proposed an approach by randomizing a scalar representation in the Koblitz curve [9].

Address-bit DPA: DDPA analyzes a relation between the secret key and data of registers [20,25]. Messerges et al. proposed the *address-bit DPA* (ADPA) for symmetric-key cryptosystems, which analyzes a relation between the key and addresses of registers [26]. Recently, Itoh et al. extended the analysis to public-key cryptosystems [11]. For example, in step 5 in Alg. 4, a correlation of data are given by $T[0] \leftarrow T[0]$ if $d_i = 0$, and $T[0] \leftarrow T[1]$ if $d_i = 1$; as these operations are same, substituted data are loaded from different registers and ADPA detects this relation. They concluded that only the exponent splitting countermeasure [6], in which the scalar is divided into $d = (d - r) + r$ for a random number r, and the randomized window method [15] are resistant against the attack [11]. But as a drawback, required computing time become at least twice than that of without countermeasures. When a special hardware is available, RRR [27], proposed by May et al., resists ADPA.

2.3   Window-Based Method

In a scalar exponentiation, a pre-computed table sometimes deduces the computing time if extra registers are available (the *window-based method*). The most

INPUT: d[], P
OUTPUT: dP

```
 1: W[0] = 0, W[1] = P
 2: W[2] = ECDBL(W[1])
 3: for i=3 upto 15 {
 4:    W[i] = ECADD(W[i-1],W[1])
 5: }
 6: T = W[d[n-1,n-4]]
 7: for i=n-5 downto 0 step -4 {
 8:    T = ECDBL(T), T = ECDBL(T)
 9:    T = ECDBL(T), T = ECDBL(T)
10:    T = ECADD(T,W[d[i,i-3]])
11: }
12: return T
```

Alg. 5. 4-bit window method

INPUT: d[], P
OUTPUT: dP

```
 1: W[0] = 0, W[1] = RPC(P)
 2: W[2] = ECDBL(W[1])
 3: for i=3 upto 15 {
 4:    W[i] = ECADD(W[i-1],W[1])
 5: }
 6: T = W[d[n-1,n-4]]
 7: for i=n-5 downto 0 step -4 {
 8:    T = ECDBL(T), T = ECDBL(T)
 9:    T = ECDBL(T), T = ECDBL(T)
10:    T = ECADD(T,W[d[i,i-3]])
11: }
12: return invRPC(T)
```

Alg. 6. 4-bit window method and RPC

simplest example is in Alg. 5 with window size 4 just for simplicity (where n is supposed to be a multiple of 4).

Similar to the binary method (Alg. 1), the window method is also vulnerable to SPA, because ECADD in step 10 is not computed if $W[i,i-3] = \mathcal{O}$. Möller proposed a method to construct an addition chain in which $W[]$ is never equal to $\mathcal{O}$ [22,23], which assures the SPA-resistance. One can combine RPC or RC countermeasures with Möller's method (Alg. 6 for RPC case) in order to resist DDPA. However, Okeya-Sakurai showed the insecureness against the second-order data-bit DPA [34]. Other approach to resist both DDPA and ADPA is proposed by Itoh et al. [15], which randomize the window to be added in step 10 in Alg. 6.

3   Proposed Countermeasure

In this section, we propose a practical countermeasure, the *randomized addressing method*, against the address-bit DPA by randomizing addresses of registers used in a scalar exponentiation. An overhead of the countermeasure is small and the effectiveness will be shown by experimental results described in this section.

An approach of our countermeasure is similar to that of Randomized Register Renaming (RRR), a hardware-based DPA countermeasure proposed by May et al. [27]. However, implementation method is quite different. That is, our countermeasure can be implemented on various processors because it requires no special hardware, and can be implemented by only software with very simple program codes. These specifications make our countermeasure very practical. On the other hand, when a program code is implemented on the processor with RRR, physical registers are randomly chosen by hardware as far as the computation result is unchanged. Because execution timing, instruction order and physical registers are randomly changed, RRR is secure against DPA. However they did not show concrete results of security evaluation.

3.1 Outline

The address-bit DPA is based on the dependency of addresses of registers on the secret key. In other word, addresses of registers are determined by the secret key uniquely, because in Alg. 4, for example, $T[0] \leftarrow T[0]$ if $d_i = 0$ and $T[0] \leftarrow T[1]$ if $d_i = 1$ so that if d_i changes, then registers will change, too.

In order to resist ADPA, previous countermeasures randomize the scalar value [11]. However, the weakness lies on the direct correlation between the secret key and addresses of registers. What we should hide is this relation rather than the scalar value. So we randomize addresses of registers by a one-time random number $r_{n-1}2^{n-1} + \cdots + r_1 2 + r_0$ ($r_i \in \{0, 1\}$). We change all parameters d_i to $d_i \oplus r_i$, where $\oplus$ denotes the XOR operation. Then all addresses of registers are randomized so that the side channel information will be randomized for each scalar exponentiation. This is a basic idea of our proposing countermeasure, the *randomized addressing method* (RA), against ADPA.

The most advantage of our method is the small overhead; the random number is easily generated and required additional operations are just XORs. However, our countermeasure has no DDPA-resistance. We have to combine other countermeasures to resist all side channel attacks. A DDPA-resistant scheme can be converted to an ADPA-scheme with almost no cost. Moreover, our countermeasure can be easily applied to window methods.

3.2 Description of Algorithms

Example algorithms of our countermeasure combined with the add-and-double-always method, the Montgomery ladder, and a window method (with 4-bit window) are in Alg. 7-9. All sample algorithms are resistant against SCA, namely SPA, DDPA, and ADPA. Here $r_{n-1}2^{n-1} + \cdots + r_1 2 + r_0$ ($r_i \in \{0, 1\}$) is a random number and r in Alg. 9 is a 4-bit random number. The functions R, invR denote RPC, invRPC or RC, invR described in section 2.2, respectively.

```
INPUT: d[], P
OUTPUT: dP
1: T[2] = R(P)
2: T[r[n-1]] = T[2]
3: for i=n-2 downto 0 {
4:    T[r[i+1]] = ECDBL(T[r[i+1]])
5:    T[1-r[i+1]] =
            ECADD(T[r[i+1]],T[2])
6:    T[r[i]] = T[d[i]⊕r[i+1]]
7: }
8: return invR(T[r[0]])
```

Alg. 7. Proposed countermeasure (add-and-double-always method from MSB)

```
INPUT: d[], P
OUTPUT: dP
1: T[r[n-1]] = R(P)
2: T[1-r[n-1]] = ECDBL(T[r[n-1]])
3: for i=n-2 downto 0 {
4:    T[2] = ECDBL(T[d[i]⊕r[i+1]])
5:    T[1] = ECADD(T[T[0]],T[1])
6:    T[0] = T[2-(d[i]⊕r[i])]
7:    T[1] = T[1+(d[i]⊕r[i])]
8: }
9: return invR(T[r[0]])
```

Alg. 8. Proposed countermeasure (Montgomery ladder)

Note 1. In the above algorithms, a random number r can be computed on the fly for efficiency rather than generated and stored in memory at the beginning.

```
INPUT: d[], P
OUTPUT: dP
 1: W[0] = 0, W[1⊕r] = P
 2: W[2⊕r] = ECDBL(W[1⊕r])
 3: for i=3 upto 15 {
 4:    W[i⊕r] = ECADD(W[(i-1)⊕r],W[1⊕r])
 5: }
 6: T = W[d[n-1,n-4]⊕r]
 7: for i=n-5 downto 0 step -4 {
 8:    T = ECDBL(T), T = ECDBL(T)
 9:    T = ECDBL(T), T = ECDBL(T)
10:    T = ECADD(T,W[d[i,i-3]⊕r])
11: }
12: return invR(T)
```

Alg. 9. Proposed countermeasure (4-bit window method)

Note 2. A similar countermeasure by randomizing registers is also proposed by May et al. [27]. Their approach is to construct a specialized hardware, while ours is in the software level.

3.3 Security Analysis

Let us discuss the security of our countermeasure. Basically our scheme is designed to combine other countermeasures to totally resist SCA; we only consider the security against ADPA. In Alg. 7-9, addresses of registers are determined by $d_i \oplus r_i$. ADPA can distinguish whether two addresses corresponding to d_i and d_j are same or not. If these addresses are same, an adversary can know $d_i \oplus r_i = d_j \oplus r_j$. But he/she cannot determine $d_i = d_j$ or not, because r_i, r_j are chosen randomly. Conversely, even if $d_i = d_j$, addresses are not always same by r_i and r_j. Thus addresses of registers are randomized and our countermeasure is secure against ADPA.

3.4 Experimental Results

We performed an experiment for verifying the effect on the security by using our countermeasure. We used the address-bit DPA attack against an implementation of Montgomery ladder using RPC with and without the register randomization. In the experiment, the target processor was run at 10 MHz, the sampling ratio was set to 100 MHz, and we made a differential power trace as

$$(means\ of\ 10000\ traces\ when\ loading\ Q[addr[d_a]])$$
$$- (means\ of\ 10000\ traces\ when\ loading\ Q[addr[d_b]])$$

for $d_a \neq d_b$ where $addr[d_x](x = a\ or\ b)$ represents the address value determined from d_x in each implementation. Fig.1 shows a differential power trace without register randomization, and Fig.2 shows that with register randomization. In Fig.1, some spikes showing the evidence for $d_a \neq d_b$ are observed, but they are not found in Fig.2. Hence we confirmed the effect of our countermeasure for protecting against address-bit DPA attack.

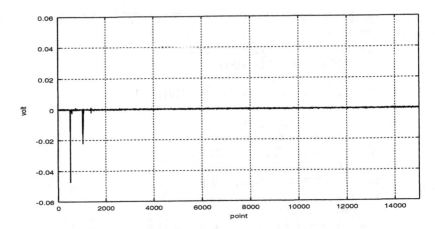

Fig. 1. Differential power traces without register randomization

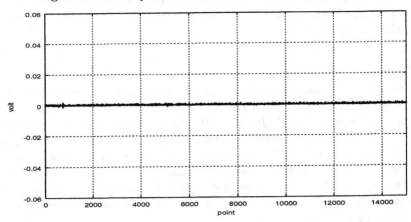

Fig. 2. Differential power traces with register randomization (proposed method)

4   Comparison

In order to resist the side channel attacks, numerous numbers of countermeasures have been proposed. However, finding a good countermeasure which has a certain security level and compromisable processing speed is a hard task for implementers, for there were no standard rule to compare these countermeasures. In this section, we provide a complete comparison of these countermeasures from viewpoints of security level, processing speed, and amount of required registers. Moreover combined countermeasures can be evaluated by the comparison. As a result, our proposed countermeasure can be combined to establish a practical solution for resisting side channel attacks in real world.

4.1 Evaluation Technique

We examine an exponentiation in 160-bit ECC from the security level, processing speed, and amount of required registers. In the followings, the binary methods and the Montgomery Ladder are called the *base algorithms* to which countermeasures are applied. We did not discuss the window based cases and the methods unavailable in the generic elliptic curve for space limitation.

Security Level. The security of a countermeasure against each of SPA, DDPA, and ADPA is measured by the *Attenuation Ratio* (AR)[1] proposed by Itoh et al. [15]. AR is given by a ratio of heights of spikes with and without the countermeasure. We denote the ratio against SPA, DDPA and ADPA by AR_S, AR_d and AR_a respectively. All AR ratios are between 0 and 1 and AR is desired to be smaller. If AR = 0, an adversary cannot observe spikes by any cost and the method is secure; if AR = 1, the adversary can always observe spikes and the method is totally insecure. AR_S takes 0 or 1, while AR_d and AR_a take arbitrary value from the interval [0, 1].

Processing Speed. Processing speed of a scalar exponentiation dP is measured by the numbers of ECADDs denoted N_A and ECDBLs denoted N_D. For the base algorithms, N_D and N_A are given by integers. If a countermeasure requires ECADDs a times than before, we denote it by "$\times a$". We ignore the cost for randomizations or transformations required in the countermeasure because they are relatively small compared to N_A and N_D.

Register. Amount of required registers are measured by the number of points R_P and that of scalars R_s in a scalar exponentiation. We assume all points are represented by the projective coordinate just for simplicity. For the base algorithms, R_P and R_s are given by integers. For countermeasures, R_P and R_s are evaluated as extra points and scalars required in the exponentiation. If a countermeasure requires b extra points than before, R_P is denoted by "$+b$". We ignore the temporary registers for ECADD and ECDBL.

4.2 Countermeasures

In this section, we evaluate these values for each countermeasure. We only deal with countermeasures whose recommended parameters are explicitly given in the original papers, so we exclude RRR and some other countermeasures. We did not deal with the randomized addition-subtraction chain [30] because it is shown to be insecure [33,35,39].

SPA Countermeasure. The add-and-double-always method (Alg. 1) computes ECDBL and ECADD for every bit and is SPA-resistant ($AR_S = 0$). But data computed in step 5 is predictable ($AR_d = 1$) and a correlation of addresses

---

[1] AR is originally introduced to evaluate the security against DDPA, however it is easily to be applied to other attacks.

in step 5 is related to d ($\mathrm{AR}_a = 1$). Required ECDBLs are same ($\mathrm{N}_D = \times 1$) but ECADDs are doubled in average ($\mathrm{N}_A = \times 2$). Extra points and scalars are not required ($\mathrm{R}_P = +0$, $\mathrm{R}_s = +0$). Same discussion can be applied to the LSB case.

The Montgomery Ladder (Alg. 3) computes ECDBL and ECADD for every bit and is SPA-resistant ($\mathrm{AR}_S = 0$). But data computed in step 5 and 6 are predictable ($\mathrm{AR}_d = 1$) and a correlation of addresses of registers in step 3,5,6 are related to d ($\mathrm{AR}_a = 1$). Required ECDBLs and ECADDs are 160 ($\mathrm{N}_D = \mathrm{N}_A = 160$). Required registers are $\mathrm{R}_P = 3$, $\mathrm{R}_s = 1$.

DPA Countermeasure by Data Randomization. In the Randomized Projective Coordinate (RPC) [5], a point (X, Y, Z) is randomized to (rX, rY, rZ) by a 160-bit random number r and it resists DDPA ($\mathrm{AR}_d = 2^{-160}$). As it is vulnerable to SPA ($\mathrm{AR}_S = 1$), it must be used with the add-and-double-always method or the Montgomery Ladder. It is also vulnerable to ADPA ($\mathrm{AR}_a = 1$). Processing speed is unchanged ($\mathrm{N}_A = \mathrm{N}_D = \times 1$). An extra point for the randomized point and an extra scalar for random number are required ($\mathrm{R}_P = +1$, $\mathrm{R}_s = +1$).

In the Randomized Curve (RC) [16], a point (X, Y, Z) is randomized to $(r^2 X, r^3 Y, Z)$ on an isomorphic curve by a 160-bit random number r and it resists DDPA ($\mathrm{AR}_d = 2^{-160}$). But it is vulnerable to SPA and ADPA ($\mathrm{AR}_S = \mathrm{AR}_a = 1$). Processing speed is unchanged ($\mathrm{N}_A = \mathrm{N}_D = \times 1$). Extra scalars for r and coefficients of isomorphic curves are required ($\mathrm{R}_P = +0$, $\mathrm{R}_s = +3$).

In the Randomized Base Point [5], a scalar exponentiation dP is computed by $d(P+R) - dR$ for a random point R. As R is chosen for each exponentiation, the method is DDPA-resistant ($\mathrm{AR}_d = 2^{-160}$). But it is vulnerable to SPA and ADPA ($\mathrm{AR}_S = \mathrm{AR}_a = 1$). The countermeasure requires two exponentiations and an extra ECADD ($\mathrm{N}_D = \times 2$, $\mathrm{N}_A = \times 2 + 1$). Extra registers for R and $P+R$ are required ($\mathrm{R}_P = +2$, $\mathrm{R}_s = +0$).

DPA Countermeasure by Scalar Randomization. In the Randomized Exponent [5], a scalar d is randomized to $d + r\phi$ for a random number r and the order ϕ of the base point P. In the original paper, the length of r is 20-bit and, thus, the countermeasure is DDPA-resistant ($\mathrm{AR}_d = 2^{-20}$) (There is an analysis which claims the 20-bit randomization is not sufficient [32]. Of course, we can relax the condition to 160-bit, however, the processing speed becomes much slower). It is also ADPA-resistant ($\mathrm{AR}_a = 0$), however it is vulnerable to SPA ($\mathrm{AR}_S = 1$). Processing speed becomes slower for the scalar is 20-bit longer ($\mathrm{N}_D = \mathrm{N}_A = \times 180/160 = \times 1.13$). An extra register for r is required ($\mathrm{R}_P = +0$, $\mathrm{R}_s = +1$).

In the Randomized Start Point [25], a start bit is chosen from a 160-bit scalar and an exponentiation is computed from the chosen bit by MSB for upper bits and by LSB for lower bits. However, the effect is rather small ($\mathrm{AR}_d = 1/160 = 2^{-7.3}$). It is also ADPA-resistant ($\mathrm{AR}_a = 0$), however it is vulnerable to SPA ($\mathrm{AR}_S = 1$). There requires no extra process ($\mathrm{N}_D = \mathrm{N}_A = \times 1$) and register ($\mathrm{R}_P = \mathrm{R}_s = +0$).

In the Exponent Splitting [6], a scalar d is divided into r and $d - r$ for a 160-bit random number r. As the secret information d is randomized, it is resistant

Table 1. Comparison of countermeasures (non-window methods)

No.	Method	AR_S	AR_d	AR_a	N_D	N_A	R_P	R_s
1	*binary method (form MSB)*	1	1	1	160	80	1	0
2	*binary method (from LSB)*	1	1	1	160	80	2	0
3	add-and-double-always method	0	1	1	×1	×2	+1	+0
4	*Montgomery ladder*	0	1	1	160	160	3	0
5	Randomized Projective Coordinate (RPC)	1	2^{-160}	1	×1	×1	+1	+1
6	Randomized Curve (RC)	1	2^{-160}	1	×1	×1	+0	+3
7	Randomized Base Point	1	2^{-160}	1	×2	×2	+2	+0
8	Randomized Exponent ($|r| = 20$)	1	2^{-20}	0	×1.13	×1.13	+0	+1
9	Randomized Start Point	1	$2^{-7.3}$	0	×1	×1	+0	+0
10	Exponent Splitting	1	2^{-160}	0	×2	×2 + 1	+1	+2
11	Randomized Addressing (RA)	1	1	0	×1	×1	+0	+2
1+3+5+11	Best Combination	0	2^{-160}	0	160	160	3	3
1+3+6+11	Best Combination	0	2^{-160}	0	160	160	2	5
4+5+11	Other Solution	0	2^{-160}	0	160	160	4	3
4+6+11	Other Solution	0	2^{-160}	0	160	160	3	5

against DDPA and ADPA ($AR_a = 0$, $AR_d = 2^{-160}$). However, it is vulnerable to SPA ($AR_S = 1$). The countermeasure requires two exponentiations and an extra ECADD ($N_D = \times 2$, $N_A = \times 2 + 1$). Extra registers for $(d - r)P$, r, d are required ($R_P = +1$, $R_s = +2$).

In the Randomized Addressing (RA) (Alg. 7), proposed in this paper, registers in step 4,5,6 are determined by a random bit r_i. It is ADPA-resistant ($AR_a = 0$), however, it is vulnerable to SPA and DDPA ($AR_S = AR_d = 1$). There requires no extra process ($N_D = N_A = \times 1$). Extra registers for r and $r \oplus d$ are required ($R_P = +0$, $R_s = +2$).

4.3 Combining Countermeasures

As we saw in the previous section, some all countermeasures only resist specific attacks. Implementers should combine them to resist all side channel attacks. In our setting, we can easily evaluate and compare each combined countermeasures. The security level AR_S, AR_d, AR_a can be evaluated by their product. Similarly, the processing speed is evaluated by their product and amount of registers are evaluated by their sum. For example, if we use the Montgomery Ladder combined with RPC and RA, which is denoted as 4+5+11 in the following Table 1, the security levels, processing speed and amount of registers are given by $AR_S = 0$, $AR_d = 2^{-160}$, $AR_a = 0$, $N_D = 160$, $N_A = 160$, $R_P = 4$, $R_s = 3$.

4.4 Comparison

This section provides a complete comparison of countermeasures (Table 1). The base algorithms are described in *italic*. As in the previous section, the security level can be evaluated for combined countermeasures. By this table, we can easily choose the most suitable countermeasure(s) from the security level, processing speed and amount of required registers.

From Table 1, we can conclude that combinations 1+3+5+11 and 1+3+6+11 provide the best combination from security level and processing speed. Other possible solutions are 4+5+11 and 4+6+11. In these combinations, more effective addition formula for ECADD and ECDBL are applicable and faster processing speed and fewer amount of registers can be expected [13]. In these combinations, our proposed countermeasure are used.

Note 3. Similar discussion for window based methods can be established. We do not compare them here for a space limitation.

5   Concluding Remarks

In this paper, a practical countermeasure, the randomized addressing method, against the address-bit DPA, was proposed. As an overhead is quite small, the method provides no slower scalar exponentiation algorithm with improving the security. Although an approach of our proposal is similar to the previous work by May et al. [27], implementational methodology is quite different. Our approach requires no special hardware and can be implemented on various processors with very simple program codes. Moreover, we showed a concrete security evaluation results by theoretically and experimentally.

In order to resist the side channel attacks, considering countermeasures against each attack is an important factor for implementers. However, when they are to establish a total security, combinations of some countermeasures are more important. For this purpose, our comparison table (Table 1) will be a great help.

Acknowledgment. The authors are grateful to Naoya Torii for his supports and many invaluable suggestions. The authors also thank to Colin Walter and anonymous referees for their helpful comments and references.

References

1. M. Akkar, P. Dischamp, and D. Moyart, "Power Analysis, What is Now Possible...", *Asiacrypt 2000*, LNCS 1976, pp. 489–502, Springer-Verlag, 2000.
2. O. Billet, and M. Joye, "The Jacobi Model of an Elliptic Curve and Side-Channel Analysis", Cryptology ePrint Archive, 2002/125, 2002. Available from http://eprint.iacr.org/2002/125/
3. E. Brier, and M. Joye, "Weierstraß Elliptic Curves and Side-Channel Attacks", *PKC 2002*, LNCS 2274, pp. 335–345, Springer-Verlag, 2002.
4. I. Blake, G. Seroussi, and N. Smart, *Elliptic Curves in Cryptography*, Cambridge University Press, 1999.
5. J. Coron, "Resistance against differential power analysis for elliptic curve cryptosystem", *CHES'99*, LNCS 1717, pp. 292–302, Springer-Verlag, 1999.
6. C. Clavier, and M. Joye, "Universal exponentiation algorithm – A first step towards provable SPA-resistance –", *CHES 2001*, LNCS 2162, pp. 300–308, Springer-Verlag, 2001.

7. W. Fischer, C. Giraud, E. Knudsen, and J. Seifert, "Parallel Scalar Multiplication on General Elliptic Curves over F_p Hedged Against Non-Differential Side-Channel Attacks", Cryptology ePrint Archive, 2002/007, 2002. Available from http://eprint.iacr.org/2002/007/

8. D. Gordon, "A Survey of fast exponentiation methods", *J. Algorithms*, vol.27, pp. 129–146, 1998.

9. M. Hasan, "Power Analysis Attacks and Algorithmic Approaches to Their Countermeasures for Koblitz Curve Crypto-systems", *IEEE Trans. Computers*, pp. 1071–1083, October 2001.

10. IEEE P1363, Standard Specifications for Public-Key Cryptography, 2000.

11. K. Itoh, T. Izu, and M. Takenaka, "Address-bit Differential Power Analysis of Cryptographic Schemes OK-ECDH and OK-ECDSA", *CHES 2002*, LNCS 2523, pp. 129–143, Springer-Verlag, 2003.

12. T. Izu, B. Möller, and T. Takagi, "Improved Elliptic Curve Multiplication Methods Resistant against Side Channel Attacks", *Indocrypt 2002*, LNCS 2551, pp. 296–313, Springer-Verlag, 2002.

13. T. Izu, and T. Takagi, "A Fast Parallel Elliptic Curve Multiplication Resistant against Side Channel Attacks", *PKC 2002*, LNCS 2274, pp. 280–296, Springer-Verlag, 2002.

14. T. Izu, and T. Takagi, "Exceptional Procedure Attack on Elliptic Curve Cryptosystems", *PKC 2003*, LNCS 2567, pp. 224–239, Springer-Verlag, 2003.

15. K. Itoh, J. Yajima, M. Takenaka, and N. Torii, "DPA Countermeasures by Improving the Window Method", *CHES 2002*, LNCS 2523, pp. 303–317, Springer-Verlag, 2003.

16. M. Joye, C. Tymen, "Protections against Differential Analysis for Elliptic Curve Cryptography", *CHES 2001*, LNCS 2162, pp. 377–390, Springer-Verlag, 2001.

17. M. Joye, J. Quisquater, "Hessian Elliptic Curves and Side-Channel Attacks", *CHES 2001*, LNCS 2162, pp. 402–410, Springer-Verlag, 2001.

18. M. Joye, and S-M. Yen, "The Montgomery Powering Ladder", *CHES 2002*, LNCS 2523, pp. 291–302, Springer-Verlag, 2003.

19. C. Kocher, "Timing attacks on Implementations of Diffie-Hellman, RSA, DSS, and other systems", *Crypto'96*, LNCS 1109, pp. 104–113, Springer-Verlag, 1996.

20. C. Kocher, J. Jaffe, and B. Jun, "Differential power analysis", *Crypto'99*, LNCS 1666, pp. 388–397, Springer-Verlag, 1999.

21. P. Liardet, N. Smart, "Preventing SPA/DPA in ECC Systems Using the Jacobi From", *CHES 2001*, LNCS 2162, pp. 391–401, Springer-Verlag, 2001.

22. B. Möller, "Securing Elliptic Curve Point Multiplication against Side-Channel Attacks", *ISC 2001*, LNCS 2200, pp. 324–334, Springer-Verlag, 2001.

23. B. Möller, "Parallelizable Elliptic Curve Point Multiplication Method with Resistance against Side-Channel Attacks", *ISC 2002*, LNCS 2433, pp. 402–413, Springer-Verlag, 2002.

24. P. Montgomery, "Speeding the Pollard and elliptic curve methods for factorizations", *Math. of Comp*, vol.48, pp. 243–264, 1987.

25. T. Messerges, E. Dabbish, and R. Sloan, "Power Analysis Attacks of Modular Exponentiation in Smartcards", *CHES'99*, LNCS 1717, pp. 144–157, Springer-Verlag, 1999.

26. T. Messerges, E. Dabbish, and R. Sloan, "Investigations of Power Analysis Attacks on Smartcards", preprint, USENIX Workshop on Smartcard Technology, 1999.

27. D. May, H.L. Muller, and N.P. Smart, "Random Register Renaming to Foil DPA", *CHES 2001*, LNCS 2162, pp. 28–38, Springer-Verlag, 2001.

28. National Institute of Standards and Technology, Recommended Elliptic Curves for Federal Government Use, in the appendix of FIPS 186-2.

29. E. Oswald, "Enhancing Simple Power-Analysis Attacks on Elliptic Curve Cryptosystems", *CHES 2002*, LNCS 2523, pp. 82–97, Springer-Verlag, 2003.

30. E. Oswald, and M. Aigner, "Randomized Addition-Subtraction Chains as a Countermeasure against Power Attacks", *CHES 2001*, LNCS 2162, pp. 39–50, Springer-Verlag, 2001.

31. K. Okeya, H. Kurumatani, and K. Sakurai, "Elliptic curves with the Montgomery form and their cryptographic applications", *PKC 2000*, LNCS 1751, pp. 446–465, Springer-Verlag, 2000.

32. K. Okeya, and K. Sakurai, "Power analysis breaks elliptic curve cryptosystem even secure against the timing attack", *Indocrypt 2000*, LNCS 1977, pp. 178–190, Springer-Verlag, 2000.

33. K. Okeya, and K. Sakurai, "On Insecurity of the Side Channel Attack Countermeasure Using Addition-Subtraction Chains under Distinguishability between Addition and Doubling", *ACISP 2002*, LNCS 2384, pp. 420–435, Springer-Verlag, 2002.

34. K. Okeya, and K. Sakurai, "A Second-Order DPA Attack Breaks a Window-method based Countermeasure against Side Channel Attacks", *ISC 2002*, LNCS 2443, pp. 389–401, Springer-Verlag, 2002.

35. K. Okeya, and K. Sakurai, "A Multiple Power Analysis Breaks the Advanced Version of the Randomized Addition-Subtraction Chains Countermeasure against Side Channel Attacks", to appear in the proceedings of 2003 IEEE Information Theory Workshop.

36. N. Smart, "The Hessian Form of an Elliptic Curve", *CHES 2001*, LNCS 2162, pp. 118–125, Springer-Verlag, 2001.

37. Standards for Efficient Cryptography Group (SECG), Specification of Standards for Efficient Cryptography.

38. C. Walter, "MIST: An Efficient, Randomized Exponentiation Algorithm for Resisting Power Analysis", *CT-RSA 2002*, LNCS 2271, pp. 53–66, Springer-Verlag, 2002.

39. C. Walter, "Security Constraints on the Oswald-Aigner Exponentiation Algorithm", Cryptology ePrint Archive, Report 2003/013, 2003. Available from http://eprint.iacr.org/2003/013/

A More Flexible Countermeasure against Side Channel Attacks Using Window Method

Katsuyuki Okeya[1] and Tsuyoshi Takagi[2]

[1] Hitachi, Ltd., Systems Development Laboratory,
292, Yoshida-cho, Totsuka-ku, Yokohama, 244-0817, Japan
ka-okeya@sdl.hitachi.co.jp
[2] Technische Universität Darmstadt, Fachbereich Informatik,
Alexanderstr.10, D-64283 Darmstadt, Germany
ttakagi@cdc.informatik.tu-darmstadt.de

Abstract. Elliptic curve cryptosystem (ECC) is well-suited for the implementation on memory constraint environments due to its small key size. However, side channel attacks (SCA) can break the secret key of ECC on such devices, if the implementation method is not carefully considered. The scalar multiplication of ECC is particularly vulnerable to the SCA. In this paper we propose an SCA-resistant scalar multiplication method that is allowed to take any number of pre-computed points. The proposed scheme essentially intends to resist the simple power analysis (SPA), not the differential power analysis (DPA). Therefore it is different from the other schemes designed for resisting the DPA. The previous SPA-countermeasures based on window methods utilize the fixed pattern windows, so that they only take discrete table size. The optimal size is 2^{w-1} for $w = 2, 3, ...$, which was proposed by Okeya and Takagi. We play a different approach from them. The key idea is randomly (but with fixed probability) to generate two different patterns based on pre-computed points. The two distributions are indistinguishable from the view point of the SPA. The proposed probabilistic scheme provides us more flexibility for generating the pre-computed points — the designer of smart cards can freely choose the table size without restraint.

Keywords: Elliptic Curve Cryptosystem, Side Channel Attacks, Width-w NAF, Fractional window, Pre-computation Table, Smart Card, Memory Constraint

1 Introduction

We are standing to the beginning of the ubiquitous computing era. It is expected that we can accomplish lucrative applications by effectively synthesizing the ubiquitous computer with cryptography. The ubiquitous computer only has scarce computational environments, so that we have to make an effort to optimize the memory and efficiency of the cryptosystem. Elliptic curve cryptosystem (ECC) is suitable for the purpose because of its short key size [Kob87,Mil86]. However, several experimental tests show that side channel attacks (SCA) can

C.D. Walter et al. (Eds.): CHES 2003, LNCS 2779, pp. 397–410, 2003.
© Springer-Verlag Berlin Heidelberg 2003

break the ECC if the implementation on the devices is not carefully considered [Cor99,HaM02,IIT02].

The SCA tries to find a correlation between the side channel information and the operation related to the secret key. In this paper we discuss the SCA on ECC using the power analysis [KJJ99], which consists of the simple power analysis (SPA) and the differential power analysis (DPA). The SPA simply observes several power consumptions of the device, and the DPA is additionally allowed to use a statistical tool in order to guess the secret information. An SPA-resistant scheme can be converted to be a DPA-resistant one by randomizing the parameters of the underlying system (See for example [Cor99,JT01]). There are three different types of SPA-resistant scheme: (1) indistinguishable addition formula that uses one formula for both of elliptic addition and doubling [LS01,JQ01, BJ02]. (2) addition chain that always computes elliptic addition and doubling for each bit [Cor99,OS00,FGKS02,IT02,BJ02]. (3) window based addition chain with fixed pattern [Möl01a,Möl01b,Möl02a,OT03]. In this paper we deal with the third category. The optimal one in (3) is the scheme proposed by Okeya and Takagi [OT03].

We intend to propose an SPA-resistant scalar multiplication that allows us to choose any number of the pre-computed points. We try to reduce the table size of the Okeya-Takagi scheme using the fractional window method proposed by Möller [Möl02b]. The fractional window method reduces a part of pre-computed points to smaller window size. Therefore, the table length is not fixed anymore, and the corresponding addition chain has no fixed pattern. It is not obvious to construct an SPA-resistant scheme using the fractional window method. In order to overcome this bias we propose a novel approach. We generate the points with smaller window size as the probabilistic process, which are indistinguishable from the view point of the SPA. Indeed, all points in the table are classified: (*lower*) points (i.e. $uP, u < 2^{w-1}$) and (*upper*) points (i.e. $uP, u > 2^{w-1}$), where w and P is the underlying width and the base point, respectively. We control the reduction probability of (*lower*) based on that of (*upper*), namely the distribution of both (*lower*) and (*upper*) are indistinguishable against SPA. The pre-computed points for (*upper*) are randomly chosen for every scalar multiplication, and the points in class (*lower*) are randomly reduced with the above reduction probability. Thus the SPA cannot detect which point is used in each class (*upper*) and (*lower*).

In order to implement highly functional applications on memory constraint device such as smartcards, the cryptographic functions are usually required to be efficient and to use small memory. In addition, some applications are often appended to (deleted from) the smartcards, thus the memory space allowed to use for the cryptographic functions depends on such individual situations. Hence the cryptographic schemes should be optimized on such individual situations. The proposed scheme attains the SPA-resistant scheme with any size of the pre-computed table. The designer of ECC can flexibly choose the table size suitable for the smartcards.

This paper is organized as follows: In Section 2 we review the scalar multiplication of elliptic curves. The width-w NAF and the fractional window method are reviewed. In Section 3 the side channel attacks are discussed. The fast and memory efficient countermeasures are presented. In Section 4 we show the proposed scheme. The security and efficiency are discussed. In Section 5 we conclude the result of our paper.

2 Scalar Multiplication of ECC

In this section we review the scalar multiplication of elliptic curve cryptosystem (ECC). The width-w non-adjacent form (NAF) and the fractional window method are discussed

The scalar multiplication computes dP for a point P on the elliptic curve and a scalar d. A lot of algorithms of computing the scalar multiplication have been proposed. Because the inverse of a point P can be computed with little additional cost, the signed representation of d is usually deployed. The fastest method with less memory is the width-w non-adjacent form (NAF). The width-w NAF represents an n-bit integer $d = \sum_{i=0}^{n} d_w[i]2^i$, where $d_w[i]$ are odd integers with $|d_w[i]| < 2^{w-1}$ and there are at most one non-negative digit among w-consecutive digits. Therefore, we pre-compute the table with points $P, 3P, .., (2^{w-1}-1)P$, which has 2^{w-2} points including base point P. The points with the opposite sign are generated on the fly during the scalar multiplication.

Generating_Width-w_NAF	Scalar_Multiplication_with_Width-w_NAF
INPUT An n-bit d, a width w	INPUT $d_w[i]$, P, $(|d_w[i]|)P$
OUTPUT $d_w[n], d_w[n-1], ..., d_w[0]$	OUTPUT dP
1. $i \leftarrow 0$	1. $Q \leftarrow d_w[c]P$
2. While $d > 0$ do the following	for the largest c with $d_w[c] \neq 0$
2.1. if d is odd then do following	2. For $i = c-1$ to 0
2.1.1. $d_w[i] \leftarrow d$ mods 2^w	2.1. $Q \leftarrow \mathrm{ECDBL}(Q)$
2.1.2. $d \leftarrow d - d_w[i]$	2.2. if $d_w[i] \neq 0$
2.2. else $d_w[i] \leftarrow 0$	then $Q \leftarrow \mathrm{ECADD}(Q, d_w[i]P)$
2.3. $d \leftarrow d/2$, $i \leftarrow i+1$	3. Return Q
3: Return $d_w[n], d_w[n-1], ..., d_w[0]$	

Several methods for generating the width-w NAF have been proposed [KT92], [MOC97], [BSS99], [Sol00]. Generating_Width-w_NAF is an algorithm that generates the width-w NAF proposed by Solinas [Sol00]. Notation "mods 2^w" at Step 2.1.1 stands for the signed residue modulo 2^w, namely $\pm 1, \pm 3, .., \pm(2^{w-1}-1)$. Note that the next $(w-1)$ consecutive bits of non-zero bits in the width-w NAF are always zero. It is known that the density of the non-zero bits of the width-w NAF is asymptotically equal to $1/(1+w)$.

Scalar_Multiplication_with_Width-w_NAF is an algorithm of computing the scalar multiplication using the width-w NAF. It is calculated from the most significant bit — elliptic curve doubling (ECDBL) at Step 2.1 is executed for each bit and elliptic curve addition (ECADD) at Step 2.2 is executed if and only if $d_w[i]$ is non-zero. Therefore we have to compute $(c+1)$-time ECDBLs and

$(c+1)/(1+w)$-time ECADDs, where c is the largest integer with $d_w[c] \neq 0$. If we choose larger width w, then the scalar multiplication becomes faster, but with more memory.

2.1   Fractional Width-w NAF

The width-w NAF uses the table $P, 3P, .., (2^{w-1}-1)P$. The size of the table takes discrete values $1, 2, 4, 8, ...$ for $w = 2, 3, 4, ...$ The density of non-zero bits of the width-w NAF also takes the discrete values $1/(w+1)$. In order to interpolate their intermediate values, Möller discussed how to construct the NAF with fractional widths [Möl02b]. His idea is to utilize the degenerated width-w NAF — some table values of the width-w NAF are not pre-computed[1], where $w > 3$. We call it the fractional width-w NAF in this paper.

The fractional width-w NAF can be easily generated by modifying Generating_Width-w_NAF. Indeed we insert the following step between Step 2.1.1 and Step 2.1.2:

$$\text{if } |d_w[i]| > 2^{w-2} + B \text{ then } d_w[i] \leftarrow d_w[i] \text{ mods } 2^{w-1},$$

where B is an integer $0 \leq B \leq 2^{w-2}$ that determines the table size and the efficiency between width-w and width-$(w-1)$ NAF. If we choose $B = 0$ or $B = 2^{w-2}$, then it becomes the width-$(w-1)$ or width-w NAF, respectively.

We define the width-w suitable for our paper in the following. Let $w = (w_0 - 1) + w_1$, where $w_0 - 1$ and w_1 are the integral and fractional parts[2] of w, respectively; $w_0 = \lceil w \rceil, w_1 = w - (w_0 - 1)$. The fractional part w_1 takes one of $1/2^{w_0-2}, 2/2^{w_0-2}, ..., (2^{w_0-2}-1)/2^{w_0-2}, 2^{w_0-2}/2^{w_0-2}$. Here the pre-computed points are $P, 3P, .., (2^{w_0-1}-1)P, (2^{w_0-1}+1)P, ..., (2^{w_0-1}+w_1 2^{w_0-1}-1)P$. There are $(1 + w_1)2^{w_0-2}$ points. The non-zero density of the fractional width-w NAF is $1/(1 + w)$. The scalar multiplication using the fractional width-w NAF is computed as same for Scalar_Multiplication_with_Width-w_NAF.

3   Side Channel Attacks and Their Countermeasures

In this section we review side channel attacks and their countermeasures.

Side channel attacks (SCA) are allowed to access the additional information linked to the operations using the secret key, e.g., timings, power consumptions,

---

[1] Strictly speaking, Möller's idea is as follows: Some values of the width-$(w+1)$ NAF are appended to the table. This enhances the speed but additional memory is required. On the contrary, the degenerated width-w NAF provides efficient memory but reduces the speed. In other words, speed and memory have a trade-off relation. The expression in this paper is different from that in [Möl02b], however, they are equivalent in this point. We use the former for the sake of the description of the proposed scheme in the following sections.

[2] We may define $w = w_0 + w_1$, $w_0 = \lfloor w \rfloor, w_1 = w - w_0$. For the sake of simplicity and the easiness of the comparison between original and proposed schemes in the following sections, we use the notations of the former.

etc. The attack aims at guessing the secret key (or some related information). Scalar_Multiplication_with_Width-w_NAF can be broken by the SCA. It calculates the ECADD if and only if the i-th bit is not zero. The standard implementation of ECADD is different from that of ECDBL, and thus the ECADD in the scalar multiplication can be detected using SCA.

If the attacker is allowed to observe the side channel information only a few times, it is called the simple power analysis (SPA). If the attacker can analyze several side channel information using a statistical tool, it is called the differential power analysis (DPA). The standard DPA utilizes the correlation function that can distinguish whether a specific bit is related to the observed calculation. In order to resist DPA, we need to randomize the parameters of elliptic curves.

There are three standard randomizations [Cor99,JT01]: (1)the base point is masked by a random point, (2)the secret scalar is randomized with multiplier of the order of the curve. (3)the base point is randomized in the projective coordinate (or Jacobian coordinate). Some attacks or weak classes against each countermeasure have been proposed [Gou03,OS00]. However, if these randomization methods are simultaneously used, no attack is known to break the combined scheme. In other words, SPA-resistant schemes can be easily converted to be DPA-resistant ones using these randomizations.

On the contrary, there are some schemes which try to achieve the SPA- and DPA-resistance simultaneously without using the combinations, e.g. randomized window methods [Wal02a,IYTT02,LS01,OA01], etc. The security of these schemes causes many controversies — some of them have been broken [OS02a, Wal02b,Wal03a] or less secure than expected [Wal03b]. Therefore we are interested in the SPA-resistant schemes.

3.1   SPA-Resistant Methods

We review the SPA-resistant schemes of computing the scalar multiplication.

There are three different approaches to resist the SPA. We explain these schemes in the following. (1)We construct the indistinguishable addition formula [LS01,JQ01,BJ02]. (2)We use the addition formula that always computes ECADD and ECDBL for each bit [Cor99,OS00,FGKS02,IT02,BJ02]. (3)We generate the addition chain with fixed pattern [Möl01a,Möl01b,Möl02a,OT03].

(1)Whereas the indistinguishable addition formula conceals addition and doubling, the attacker can detect the number of additions and doublings in the computation. In other words, the indistinguishable addition formula pulls the SPA back to the timing attack. Hence, this type is imperfect. (2)Since the second type does not compute the pre-computed points, the memory consumption is small. In addition, if we are allowed to use a special form such as Montgomery-form, this type is the fastest. However, some international standards [ANSI,IEEE,NIST,SEC] do not support such a form. Without using the special form, this type is not so fast because it requires many ECADD operations. (3)The third type utilizes pre-computed points for speeding up the computation, since the pre-computed points reduce the number of ECADD operations. Whereas the large number of the pre-computed points achieves a fast computation, it requires large memory for storing the points. Okeya and Takagi proposed an SPA-resistant addition chain with small memory, which is based on the width-w NAF [OT03]. The algorithm is as follows (We modify it suitable for our proposed scheme):

SPA-resistant_Width-w_NAF_with_Odd_Scalar
___
INPUT An odd n-bit d
OUTPUT $d_w[n], d_w[n-1], ..., d_w[0]$
___
1. $r \leftarrow 0$, $i \leftarrow 0$, $r_0 \leftarrow w$
2. While $d > 1$ do the following
   2.1. $u[i] \leftarrow (d \bmod 2^{w+1}) - 2^w$
   2.2. $d \leftarrow (d - u[i])/2^{r_i}$
   2.3. $d_w[r + r_i - 1] \leftarrow 0, d_w[r + r_i - 2] \leftarrow 0, ..., d_w[r + 1] \leftarrow 0, d_w[r] \leftarrow u[i]$
   2.4. $r \leftarrow r + r_i$, $i \leftarrow i + 1$, $r_i \leftarrow w$
3. $d_w[n] \leftarrow 0, ..., d_w[r + 1] \leftarrow 0, d_w[r] \leftarrow 1$
4. Return $d_w[n], d_w[n-1], ..., d_w[0]$
___

The algorithm generates the SPA-resistant chain only for odd scalar, and the treatment for even scalar was discussed in [OT03]. We assume that the scalar d is odd in the following. At Step 2.1, the integer $u[i]$ is assigned as $(d \bmod 2^{w+1}) - 2^w$. The computation assures that $u[i]$ is odd whenever d is odd. Since $d - u[i] = d - (d \bmod 2^{w+1}) + 2^w = 2^w \bmod 2^{w+1}$, the resultant $(d - u[i])/2^w$ is odd. Thus, each integer $u[i]$ is odd. Note that d terminates with $d = 1$. Hence we can achieve the SPA-resistant chain, e.g., the fixed pattern

$$| \underbrace{0..0}_{w-1} x | \underbrace{0..0}_{w-1} x | ... | \underbrace{0..0}_{w-1} x | \text{ with odd integers } |x| < 2^w.$$

The number of the pre-computed points is 2^{w-1}, and the density of the non-zero bit is $1/w$. The scalar multiplication using this chain is computed as same for Scalar_Multiplication_with_Width-w_NAF.

Note that this scheme is optimal in respect of the memory, and the table size takes $1, 2, 4, 8, ...$ for $w = 1, 2, 3, 4, ...$. If the designer of smart cards allows to use the table sizes $1, 2, 4, 8, ...$, this scheme is one of the best solutions. However, if he allows to use just the sizes $3, 5, 6, ...$ not $1, 2, 4, 8, ...$, it compromises the memory and/or the speed. This situation often occurs because some restrictions about resources such as memory and cost are determined by the applications of the smart cards, not the specifications of the cryptographic schemes. Such restrictions impose the flexibility on the cryptographic schemes. Hence, we need to construct such a scheme.

4   Proposed Scheme

In this section we propose a new SPA-resistant scheme with *any* table size. After describing its main idea, we present the details of our algorithm. We then discuss the security, the efficiency, and the memory requirement of our proposed scheme.

4.1   Main Idea

We describe the main idea of our proposed algorithm. The proposed scheme is converted from SPA-resistant_Width-w_NAF_with_Odd_Scalar using the idea of the fractional window method.

First, we discuss the security of the straight-forwardly combined scheme between Okeya-Takagi scheme [OT03] and the fractional window method [Möl02b], and find that this combined scheme is *not* secure against SPA. For the sake of simplicity we explain it with $w = 4$. SPA-resistant_Width-w_NAF_with_Odd_Scalar with $w = 4$ pre-computes the signed odd integer modulo 2^w, i.e., $U_w = \{\pm 1, \pm 3, \pm 5, \pm 7, \pm 9, \pm 11, \pm 13, \pm 15\}$. The fractional width-$w$ NAF can reduce it to smaller one, for instance, $F = \{\pm 1, \pm 3, \pm 5, \pm 7, \pm 9\}$. Note that it still contains the representations of the smaller modulus 2^{w-1}, i.e., $U_{w-1} = \{\pm 1, \pm 3, \pm 5, \pm 7\}$. The fractional window using class F is constructed by inserting the following step between Step 2.1 and 2.2:

$$\text{if } |u[i]| > 2^{w-1} + B, \quad \text{then } u[i] \leftarrow (u[i] \bmod 2^w) - 2^{w-1}, \; r_i \leftarrow w - 1,$$

where B is an integer $0 < B < 2^{w-1}$ (in the case of F we choose $B = 1$). However, sequence $d_w[n], d_w[n-1], ..., d_w[0]$ generated by this fractional window method has no fixed pattern, so that it is not secure against the SPA. Indeed, we know $|u[i]| > 2^{w-1} + B$ if and only if $(w-2)$-consecutive zeros (i.e. $r_i = w - 1$) appear. In order to overcome this bias, we propose two novel ideas in the following.

The first one is to control the choice of two moduli 2^w and 2^{w-1} as the probabilistic process. We reduce $u[i]$ with the uniform probability from the view point of the SPA. Since $u[i]$ with $|u[i]| < 2^{w-1}$ is possible to utilize both moduli 2^w and 2^{w-1}, the use of the following trick achieves our aim:

If $|u[i]| < 2^{w-1}$,
then $u[i] \leftarrow (u[i] \bmod 2^w) - 2^{w-1}$, $r_i \leftarrow w - 1$ with probability $1 - P_w$,

where P_w is the probability that $u[i]$ within U_{w-1} remains the representation of mod 2^w, and we should select $P_w = \frac{\#F - \#U_{w-1}}{\#U_w - \#U_{w-1}}$. This means that we reduce $u[i]$ to the representation of mod 2^{w-1} with the same probability for both $|u[i]| < 2^{w-1}$ and $|u[i]| > 2^{w-1}$. Thus the SPA cannot distinguish the two distributions.

The second idea is to use a different representation of residue class modulo 2^w. The use of the different representation conceals the information that a specific $u[i]$ belongs to the class F. Instead of the "integer" B, we use the "subset" B of $U_w \setminus U_{w-1}$. Then the class F is chosen as $F = U_{w-1} \cup B$, since F contains any odd signed residue modulo 2^{w-1}. Because of $\#F = 10$ we should choose $\#B = 2$. Thus B is randomly chosen from one of $\pm 9, \pm 11, \pm 13$, or ± 15. The attacker cannot guess the value of B because of this random choice.

4.2   Proposed Algorithm

We present the algorithm of our proposed algorithm. The algorithm generates an SPA-resistant fractional width-w NAF for given n-bit odd scalar d and width w. The algorithm is as follows:

SPA-resistant_Fractional_Width-w_NAF_with_Odd_Scalar

INPUT An odd n-bit d, and a width w

OUTPUT $d_w[n + w_0 - 1], ..., d_w[n - 1], ..., d_w[0]$, and $B = \{\pm b_1, ..., \pm b_{w_1 2^{w_0 - 2}}\}$

1. $w_0 \leftarrow \lceil w \rceil$, $w_1 \leftarrow w - (w_0 - 1)$
2. Randomly choose distinct integers
   $b_1, ..., b_{w_1 2^{w_0 - 2}} \in_R U_{w_0}^+ \setminus U_{w_0 - 1}^+ = \{2^{w_0 - 1} + 1, 2^{w_0 - 1} + 3, ..., 2^{w_0} - 1\}$,
   and put $B = \{\pm b_1, ..., \pm b_{w_1 2^{w_0 - 2}}\}$, $P_w \leftarrow w_1$,
   where $U_v^+ = \{1, 3, 5, ..., 2^v - 1\}$ for positive integer v.
3. $r \leftarrow 0$, $i \leftarrow 0$
4. While $d > 1$ do the following
   4.1. $x[i] \leftarrow (d \bmod 2^{w_0 + 1}) - 2^{w_0}$, $y[i] \leftarrow (d \bmod 2^{w_0}) - 2^{w_0 - 1}$
   4.2. if $|x[i]| < 2^{w_0 - 1}$ then
        $r_i \leftarrow w_0$, $u[i] \leftarrow x[i]$ with P_w; $r_i \leftarrow w_0 - 1$, $u[i] \leftarrow y[i]$ with $1 - P_w$
        else if $x[i] \in B$ then
        $r_i \leftarrow w_0$, $u[i] \leftarrow x[i]$ else $r_i \leftarrow w_0 - 1$, $u[i] \leftarrow y[i]$
   4.3. $d \leftarrow (d - u[i])/2^{r_i}$
   4.4. $d_w[r + r_i - 1] \leftarrow 0, d_w[r + r_i - 2] \leftarrow 0, ..., d_w[r + 1] \leftarrow 0, d_w[r] \leftarrow u[i]$
   4.5. $r \leftarrow r + r_i$, $i \leftarrow i + 1$
5. $d_w[n + w_0 - 1] \leftarrow 0, ..., d_w[r + 1] \leftarrow 0, d_w[r] \leftarrow 1$
6. Return $d_w[n + w_0 - 1], ..., d_w[n - 1], ..., d_w[0]$, and $B = \{\pm b_1, ..., \pm b_{w_1 2^{w_0 - 2}}\}$

At Step 1 we assign the integral part w_0 and the fractional part w_1 of the width w. At Step 2 the pre-computed index $b_1, ..., b_{w_1 2^{w_0 - 2}}$ that belongs to upper set $U_{w_0}^+ \setminus U_{w_0 - 1}^+$ are randomly chosen. The random signed index B is returned as the part of output. The reduction probability P_w is assigned. At Step 3 integers r, i are initialized. Step 4 is the main loop of the proposed algorithm. At Step 4.1 we generate two different residue values $x[i] \bmod 2^{w_0}$ and $y[i] \bmod 2^{w_0 - 1}$. At Step 4.2 one of $x[i]$ and $y[i]$ is assigned for $u[i]$ based on both the size $|x[i]|$ and the probability P_w. At Step 4.3 we eliminate least r_i bits of d. At Step 4.4 bit information $d_w[i]$ is assigned. The $(r_i - 1)$ consecutive bits after the lowest bit $d_w[r]$ are zero. At Step 4.5 integers r, i are updated. Finally we return all bits of the proposed addition chain. The total bit could be at most w_0 bits larger than the original n bits.

The pre-computed points are calculated using not only a base point P and a width w but also the randomized index $B = \{b_1, ..., b_{w_1 2^{w_0 - 2}}\}$. For index B the pre-computed points are $P, 3P, ..., (2^{w_0 - 1} - 1)P$ and $b_1 P, ..., b_{w_1 2^{w_0 - 2}} P$. The scalar multiplication using the proposed chain is computed as same for Scalar_Multiplication_with_Width-w_NAF.

At Step 4.2 $u[i] = x[i]$ is assigned with probability $P_w = a/2^{w_0 - 2}$ for $a = 1, 2, .., 2^{w_0 - 2}$. We can easily generate the "probability" using a 1-bit random number generator as follows: First we obtain a random $(w_0 - 2)$-bit number $rand$ by executing the 1-bit random number generator $w_0 - 2$ times. Then we assign $u[i] = x[i]$ if and only if $rand \leq a$ holds. The probability of $rand \leq a$ is exactly $a/2^{w_0 - 2}$ due to the uniform distribution of $rand$ in $\{0, 1, ..., 2^{w_0 - 1} - 1\}$. A 1-bit random number generator is usually equipped on smart cards. We can generate the probability P_w with a small additional cost.

4.3   Security against SPA

We discuss the security of the proposed scheme against the SPA. We prove that the sequence $d_w[n], d_w[n-1], ..., d_w[0]$ arisen from the proposed algorithm has no correlation to the secret bit information in the sense of SPA.

Theorem 1. *The proposed scheme is secure against the SPA.*

Proof. $u[i]$ is a non-zero odd integer from the construction of the proposed algorithm. Thus, any subsequence of the consecutive zero bits in the sequence $d_w[n], d_w[n-1], ..., d_w[0]$ has the length $w_0 - 1$ or $w_0 - 2$;

$$..0u[i+1]\Big|\underbrace{0..0}_{r_i-1}u[i]\Big|00.., \quad r_i = w_0 \text{ or } w_0 - 1.$$

The corresponding AD sequence is

$$..DDA\Big|\underbrace{D..DD}_{r_i}A\Big|DD.., \quad r_i = w_0 \text{ or } w_0 - 1,$$

where A and D indicate ECADD and ECDBL, respectively. Hence, all the information that the SPA can obtain from the AD sequence is the length of the consecutive zero, namely r_i.

In the following we prove that r_i provides no information about the secret scalar d. Indeed we show that two AD sequences $\underbrace{D..DD}_{w_0}A$ and $\underbrace{D..DD}_{w_0-1}A$ are independently distributed from the secret scalar d. Here we can assume that d is randomly and uniformly distributed in all n-bit odd integers, because d is the secret key. Since $x[i]$ and $y[i]$ are assigned dependently on only the lower $(w_0 + 1)$ bits of d, they are random w_0-bit and $(w_0 - 1)$-bit signed odd integers, respectively, due to the uniform distribution of d. Thus, we consider the lower $(w_0 + 1)$ bits of the binary representations of d. Note that the lowest bit is always 1, and is converted by the preceding d. Thus, we do not need to consider the effect of the lowest bit.

We estimate the probability that $x[i]$ or $y[i]$ is assigned at Step 4.2 of the proposed algorithm. We have the following 4 cases: $\text{LSB}_{w_0+1}(d) = 00 * ... * 1, 01 * ... * 1, 10 * ... * 1$, and $11 * ... * 1$, where $\text{LSB}_{w_0+1}(d)$ denotes the lower $(w_0 + 1)$ bits of d. First we discuss the case that $\text{LSB}_{w_0+1}(d) = 00 * ... * 1$. In this case we have $-2^{w_0} < x[i] < -2^{w_0-1}$. That is, $|x[i]| \geq 2^{w_0-1}$. At Step 4.2, the lower half instructions are executed. Since the probability of $x[i] \in B$ is $\#B/(\#U_{w_0} - \#U_{w_0-1}) = P_w$, we have $r_i = w_0$ with the probability P_w, and $r_i = w_0 - 1$ with the probability $1 - P_w$. Next, we discuss the case of $\text{LSB}_{w_0+1}(d) = 01 * ... * 1$. At Step 4.2, the upper half instructions are executed, since $-2^{w_0-1} < x[i] < 0$. Thus we have $r_i = w_0$ with the probability P_w, and $r_i = w_0 - 1$ with the probability $1 - P_w$. In the case of $\text{LSB}_{w_0+1}(d) = 10 * ... * 1$, the upper half instructions at Step 4.2 are executed, since $0 < x[i] < 2^{w_0-1}$. Thus we have $r_i = w_0$ with the probability P_w, and $r_i = w_0 - 1$ with the probability

$1 - P_w$. Finally, in the case of $\text{LSB}_{w_0+1}(d) = 11*...*1$, the lower half instructions at Step 4.2 are executed, since $2^{w_0-1} < x[i] < 2^{w_0}$. Because the probability that $x[i] \in B$ is P_w, we have $r_i = w_0$ with the probability P_w, and $r_i = w_0 - 1$ with the probability $1 - P_w$. Therefore, the proposed scheme produces $r_i = w_0$ with probability P_w and $r_i = w_0 - 1$ with probability $1 - P_w$, which is independent from d.   □

We point out that each bit $d_w[i]$ is not randomly distributed in class $U_{w_0-1} \cup B$. The auxiliary variables $x[i]$ and $y[i]$ of the proposed algorithm are randomly distributed in U_{w_0} and U_{w_0-1}, respectively. The resulting $d_w[i]$ is assigned as $x[i]$ with probability P_w and $y[i]$ with probability $1 - P_w$, respectively. Thus some points in class $U_{w_0-1} \cup B$ appear with higher probability. However, the proposed scheme conceals such points, because B is randomly chosen and points in U_{w_0-1} are also randomly chosen. That is, the attacker might reveal the distribution, however he/she cannot detect the correspondence between the point with higher probability and the value of $d_w[i]$.

On the contrary, if we simply choose predetermined numbers like the fractional window method, then the scheme is not secure against SPA. For example, we consider the case of $w = 3 + 1/8$; $B = \{\pm 9\}$. If the length of the consecutive zero bit is 3, then the conditional probabilities that the next non-zero $d_w[i] = \pm 9$ are $1/4$ each, while the probabilities that $d_w[i] = \pm 1, \pm 3, \pm 5, \pm 7$ are $1/16$ each. Thus, the probabilities are not uniform. Since the attacker knows the predetermined numbers that belong to B, he/she has an advantage to guess $d_w[i]$. For example, the use of the attack proposed by Oswald [Osw02] reduces the cost of the exhaustive search for the candidates of the secret key which are not uniformly distributed.

4.4   Memory Consumption and Computation Cost

We discuss the memory and efficiency of the proposed scheme.

The efficiency of ECC is strongly depending on the representation of the base fields, the coordinate systems, and the definition equations. The proposed scheme aims at developing a secure encoding of the addition chain, and it can freely choose these parameters. We attach importance to the flexibility of cryptographic schemes, so that we estimate no computation cost of individual optimizations. We intend to estimate the trade-off between memory consumption (the size of pre-computed table) and the computation cost (the density of non-zero bits) for the proposed scheme. We have the following theorem.

Theorem 2. *The size of the pre-computed table is $(1 + w_1)2^{w_0-2}$. The density of the non-zero bits is asymptotically $\frac{w_0-P_w}{(w_0-1)w_0}$.*

Proof. In Step 2 we pre-compute set B whose size is exactly equal to $w_1 2^{w_0-2}$. In addition to the set B, the proposed scheme prepares the pre-computed points $P, 3P, ..., (2^{w_0-1} - 1)P$ for the base point P, the number of points is 2^{w_0-2}. Thus the number of all the pre-computed points is $(1 + w_1)2^{w_0-2}$.

In Step 4.2 we assign the length of the consecutive zero bit is always $w_0 - 1$ (i.e. $\underbrace{0..0}_{w_0-1} u[i]$) with probability P_w or $w_0 - 2$ (i.e. $\underbrace{0..0}_{w_0-2} u[i]$) with probability $1 - P_w$. Therefore the density of non-zero bits is asymptotically $\frac{w_0 - P_w}{(w_0-1)w_0}$. □

In Table 1 we summarize these results. The table size includes the base point P itself. If w is integral; $w = w_0$, then the size of the table and the density of non-zero bit of the proposed scheme are same as those of the scheme proposed by Okeya-Takagi, respectively. The proposed scheme interpolates the gap of the discrete table sizes 2^{w-1} for $w = 2, 3, 4, ...$, namely all possible numbers of pre-computed table can be used. Thus the designers of elliptic curve cryptosystems can flexibly choose the table size suitable for the smart card.

Table 1. Memory and Efficiency of the Proposed Scheme

Width	2	2.5	3	3.25	3.5	3.75	4	4.125	⋯
Table Size	2	3	4	5	6	7	8	9	⋯
Non-Zero Density	0.5	0.42	0.33	0.313	0.291	0.271	0.25	0.244	⋯

4.5 Other Security Properties

We discuss other security properties of our proposed scheme, namely a possible attack using the DPA and a security comparison with the randomized window methods.

The proposed scheme aims at resisting the SPA, but we discuss the security against the DPA. In Section 3, we mentioned that the SPA-resistant schemes can be easily converted to be DPA-resistant ones using randomization tricks [Cor99, JT01]. Thus, the proposed scheme can be converted to be DPA-resistant one. On the other hand, the window methods using the fixed secret scalar are vulnerable to the sophisticated DPA, e.g., the second order DPA [OS02b,OT03] and the address-bit DPA [IIT02]. These DPA can detect which pre-computed points are called for the ECADD, and the associated bits of the secret scalar can be revealed. Since also countermeasures against such sophisticated DPA attacks were proposed in the papers [OS02b,OT03,IIT02], the combined scheme is secure against the sophisticated DPA.

The addition chain of the proposed scheme is generated by randomly choosing two window lengths 2^{w_0} and 2^{w_0-1}. There are several window methods that intend to protect the DPA by randomizing the addition chain [Wal02a,IYTT02, LS01,OA01], etc. The goal of these schemes is different from ours, but we compare the security in the sense of the SPA. These schemes produce several AD sequences depending on the secret scalar d and random numbers. However, the distribution of the AD sequences are not uniform, but depends on the *secret* scalar. Indeed, some of them were broken [OS02a,Wal02b,Wal03a,OS03,HCJ+03] because of such bias. On the other hand, the randomization of the proposed scheme is independent from the *secret* scalar. There are two different AD sequences for the proposed scheme, and the probability of appearing the two AD sequences only depends on the width-w, which is a *public* parameter.

5   Conclusion

We proposed an SPA-resistant scalar multiplication for elliptic curve cryptosystem, which allows us to choose any size of the pre-computation points with efficient running time.

It is expected that smartcards are able to equip highly functional applications. In addition, in order to accomplish the aims of users, some applications are often appended to (deleted from) the smartcards. The memory space of cryptographic functions depends on these applications. In other words, the cryptographic schemes are imposed on the flexibility of the memory consumption and efficiency. Indeed, with our proposed scheme, (1)the designer of smart cards can flexibly choose the table size suitable for the individual situations, (2)the private information in the smart cards are protected against the side channel attacks.

References

[ANSI]   ANSI X9.62, Public Key Cryptography for the Financial Services Industry, *The Elliptic Curve Digital Signature Algorithm (ECDSA)*, (1998).

[BJ02]   Brier, É., Joye, M., *Weierstrass Elliptic Curves and Side-Channel Attacks*, Public Key Cryptography (PKC2002), LNCS 2274, (2002), 335–345.

[BSS99]   I. Blake, G. Seroussi, and N. Smart, *Elliptic Curves in Cryptography*, Cambridge University Press, 1999.

[Cor99]   Coron, J.S., *Resistance against Differential Power Analysis for Elliptic Curve Cryptosystems*, Cryptographic Hardware and Embedded Systems (CHES'99), LNCS 1717, (1999), 292–302.

[FGKS02]   Fischer, W., Giraud, C., Knudsen, E.W., Seifert, J.P., *Parallel scalar multiplication on general elliptic curves over $\mathbf{F}_p$ hedged against Non-Differential Side-Channel Attacks*, International Association for Cryptologic Research (IACR), Cryptology ePrint Archive 2002/007, (2002).
            http://eprint.iacr.org/2002/007/

[Gou03]   Goubin, L., *A Refined Power-Analysis Attack on Elliptic Curve Cryptosystems*, Public Key Cryptography, (PKC 2003), LNCS 2567, (2003), 199–211.

[HaM02]   Ha, J., and Moon, S., *Randomized Signed-Scalar Multiplication of ECC to Resist Power Attacks*, Workshop on Cryptographic Hardware and Embedded Systems 2002 (CHES 2002), LNCS 2523, (2002), 551–563.

[HCJ+03]   Dong-Guk Han, Nam Su Chang, Seok Won Jung, Young-Ho Park, Chang Han Kim, Heuisu Ryu, *Cryptanalysis of the Full version Randomized Addition-Subtraction Chains*, to appear in ACISP 2003.

[IIT02]   Itoh, K., Izu, T., and Takenaka, M., *Address-bit Differential Power Analysis on Cryptographic Schemes OK-ECDH and OK-ECDSA*, Workshop on Cryptographic Hardware and Embedded Systems 2002 (CHES 2002), LNCS 2523, (2002), 129–143.

[IYTT02]   Itoh, K., Yajima, J., Takenaka, M., and Torii, N., *DPA Countermeasures by improving the Window Method*, Workshop on Cryptographic Hardware and Embedded Systems 2002 (CHES 2002), LNCS 2523, (2002), 318–332.

[IEEE]   IEEE P1363, Standard Specifications for Public-Key Cryptography. http://groupe.ieee.org/groups/1363/

[IT02]   Izu, T., Takagi, T., *A Fast Parallel Elliptic Curve Multiplication Resistant against Side Channel Attacks*, Public Key Cryptography (PKC2002), LNCS 2274, (2002), 280–296.

[JQ01]   Joye, M., Quisquater, J.J., *Hessian elliptic curves and side-channel attacks*, Cryptographic Hardware and Embedded Systems (CHES'01), LNCS 2162, (2001), 402–410.

[JT01]   Joye, M., Tymen, C., *Protections against differential analysis for elliptic curve cryptography: An algebraic approach*, Cryptographic Hardware and Embedded Systems (CHES'01), LNCS2162, (2001), 377–390.

[Kob87]   Koblitz, N., *Elliptic curve cryptosystems*, Math. Comp. 48, (1987), 203–209.

[KJJ99]   Kocher, C., Jaffe, J., Jun, B., *Differential Power Analysis*, Advances in Cryptology – CRYPTO '99, LNCS 1666, (1999), 388–397.

[KT92]   K. Koyama and Y. Tsuruoka, *Speeding Up Elliptic Curve Cryptosystems using a Signed Binary Windows Method*, Advances in Cryptology – CRYPTO '92, LNCS 740, (1992), 345–357.

[LS01]   Liardet, P.Y., Smart, N.P., *Preventing SPA/DPA in ECC systems using the Jacobi form*, Cryptographic Hardware and Embedded System (CHES'01), LNCS2162, (2001), 391–401.

[Mil86]   Miller, V.S., *Use of elliptic curves in cryptography*, Advances in Cryptology – CRYPTO '85, LNCS218, (1986), 417–426.

[MOC97]   Atsuko Miyaji, Takatoshi Ono, Henri Cohen, *Efficient elliptic curve exponentiation*, Information and Communication Security (ICICS 1997), (1997), 282–291.

[Möl01a]   Möller, B., *Securing Elliptic Curve Point Multiplication against Side-Channel Attacks*, Information Security (ISC2001), LNCS2200, (2001), 324–334.

[Möl01b]   Möller, B., *Securing elliptic curve point multiplication against side-channel attacks*, addendum: Efficiency improvement. http://www.informatik.tu-darmstadt.de/TI/Mitarbeiter/moeller/ecc-scaisc01.pdf, (2001).

[Möl02a]   Möller, B., *Parallelizable Elliptic Curve Point Multiplication Method with Resistance against Side-Channel Attacks*, Information Security Conference (ISC 2002), LNCS 2433, (2002), 402–413.

[Möl02b]   Möller, B., *Improved Techniques for Fast Exponentiation*, The 5th International Conference on Information Security and Cryptology (ICISC 2002), LNCS 2587, (2003), 298–312.

[NIST]   National Institute of Standards and Technology, FIPS 186-2, http://csrc.nist.gov/publication/fips/fips186-2/fips186-2.pdf

[OA01]   Oswald, E., Aigner, M., *Randomized Addition-Subtraction Chains as a Countermeasure against Power Attacks*, Cryptographic Hardware and Embedded Systems (CHES'01), LNCS2162, (2001), 39–50.

[OS00]   Okeya, K., Sakurai, K., *Power Analysis Breaks Elliptic Curve Cryptosystems even Secure against the Timing Attack*, Progress in Cryptology – INDOCRYPT 2000, LNCS1977, (2000), 178–190.

[OS02a]   Okeya, K., Sakurai, K., *On Insecurity of the Side Channel Attack Countermeasure using Addition-Subtraction Chains under Distinguishability between Addition and Doubling*, The 7th Australasian Conference in Information Security and Privacy, (ACISP 2002), LNCS2384, (2002), 420–435.

[OS02b]   Okeya, K., Sakurai, K., *A Second-Order DPA Attack Breaks a Window-method based Countermeasure against Side Channel Attacks*, Information Security Conference (ISC 2002), LNCS 2433, (2002), 389–401.

[OS03]   Okeya, K., Sakurai, K., *A Multiple Power Analysis Breaks the Advanced Version of the Randomized Addition-Subtraction Chains Countermeasure against Side Channel Attacks*, in 2003 IEEE Information Theory Workshop (ITW 2003) (these proceedings), (2003).

[Osw02]   Oswald, E., *Enhancing Simple Power-Analysis Attacks on Elliptic Curve Cryptosystems*, Workshop on Cryptographic Hardware and Embedded Systems 2002 (CHES 2002), LNCS 2523, (2002), 82–97.

[OT03]   Okeya, K., Takagi, T., *The Width-w NAF Method Provides Small Memory and Fast Elliptic Scalar Multiplications Secure against Side Channel Attacks*, Topics in Cryptology, The Cryptographers' Track at the RSA Conference 2003 (CT-RSA 2003), LNCS2612, (2003), 328–342.

[SEC]   Standards for Efficient Cryptography Group (SECG), http://www.secg.org

[Sol00]   Solinas, J.A., *Efficient Arithmetic on Koblitz Curves*, Design, Codes and Cryptography, 19, (2000), 195–249.

[Wal02a]   Walter, C.D., *Some Security Aspects of the Mist Randomized Exponentiation Algorithm*, Workshop on Cryptographic Hardware and Embedded Systems 2002 (CHES 2002), LNCS 2523, (2002), 564–578.

[Wal02b]   Walter, C.D., *Breaking the Liardet-Smart Randomized Exponentiation Algorithm*, Proceedings of CARDIS'02, USENIX Assoc, (2002), 59–68.

[Wal03a]   Walter, C.D., *Security Constraints on the Oswald-Aigner Exponentiation Algorithm*, International Association for Cryptologic Research (IACR), Cryptology ePrint Archive 2003/013, (2003). http://eprint.iacr.org/2003/013/

[Wal03b]   Walter, C.D., *Seeing through Mist Given a Small Fraction of an RSA Private Key*, Topics in Cryptology, Topics in Cryptology, The Cryptographers' Track at the RSA Conference 2003 (CT-RSA 2003), LNCS2612, (2003), 391–402.

On the Security of PKCS #11

Jolyon Clulow

University of Natal, Department of Mathematical and Statistical Sciences, Durban,
South Africa clulow@icon.co.za

Abstract. Public Key Cryptography Standards (PKCS) #11 has
gained wide acceptance within the cryptographic security device community and has become the interface of choice for many applications.
The high esteem in which PKCS #11 is held is evidenced by the fact
that it has been selected by a large number of companies as the API for
their own devices. In this paper we analyse the security of the PKCS
#11 standard as an interface (e.g. an application-programming interface
(API)) for a security device. We show that PKCS #11 is vulnerable to a
number of known and new API attacks and exhibits a number of design
weaknesses that raise questions as to its suitability for this role. Finally
we present some design solutions.

1   An Introduction to PKCS #11

The Public Key Cryptography Standards (PKCS) were developed by RSA Security Inc. "in cooperation with representatives of industry, academia and government to provide a standard to allow interoperability and compatibility between vendor devices and implementations." [1] A significant factor in the success
of these standards can be attributed to this co-operative approach. The standards cover a variety of aspects of Public Key cryptography including PKCS #1:
RSA Encryption Standard, PKCS #11: Cryptographic Token Interface Standard
[18] and PKCS #8: Private-Key Information Syntax Standard. Many significant
APIs and protocols have been built upon PKCS #11 (e.g. SSL). Notable products with PKCS #11 support include Mozilla (the open source browser upon
which the Netscape browser is based) and SSL hardware accelerators from companies such as nCipher, IBM, Thales, Rainbow and AEP amongst others. Indeed,
this research was prompted by the question of the suitability of the PKCS #11
API as an interface to a hardware security module (or crypto coprocessor).

The designers of PKCS #11 described the design goals as follows: to "provide
a standard interface between applications and (portable) cryptographic devices"
and at the same time to "allow resource sharing" (a many-to-many relationship
between applications and devices). It was not intended to be a general interface
to cryptographic operations or security services. Rather it could be used to build
such services, operations or suitable APIs.

---

[1] Unless indicated otherwise, all quotations and figures are reproduced with permission
from [18].

C.D. Walter et al. (Eds.): CHES 2003, LNCS 2779, pp. 411–425, 2003.
© Springer-Verlag Berlin Heidelberg 2003

In PKCS #11 terminology, a token is a device that stores objects (e.g. Keys, Data and Certificates) and performs cryptographic operations. This is a logical rather than a physical characterization; where one device may have several, distinct logical tokens (e.g. akin to the concept of distinct domains). When intending to make use of a token (or to communicate with it), one must first establish a session with the token, which requires the user to 'login' and to be authenticated to the device. Thereafter, the user may make use of the functionality provided by the token by making calls through the interface or API. Objects are characterized as either token objects or session objects. Token objects are non-volatile in nature and exist (i.e., are stored) on a token. In addition, they possess the property that they are visible to all applications connected to the token. In contrast, session objects are volatile, existing only for the duration of the session between an application and a token. They only have scope within that session (i.e., are only visible to the application which created them).

Each object has a set of properties that describes the object and controls its use. For example, every key possesses the Key Type property which identifies it either as a public, private or secret key. Private and secret keys are recognised by the standard for the requirement to protect the secrecy thereof, and possess the properties sensitive, extractable, always sensitive and never extractable. "Sensitive keys cannot be revealed in plaintext off the token, and unextractable keys cannot be revealed off the token even when encrypted (though they can still be used as keys)."

PKCS #11 describes two types of users: security officers (SO) and normal users (users). The security officer is responsible for administering the users and for performing such operations as initially setting and changing passwords. Unlike normal users they cannot perform cryptographic operations. All users must 'login' (i.e., be authenticated to the token) before they can access the objects or capabilities of a token. This is achieved through the use of a personal identification number (PIN), which acts essentially as a password. The standard allows for this mechanism to be augmented with or replaced by an alternative, custom mechanisms in any given implementation (e.g. PIN entry via PINpad or the use of smarts cards). This does not, however, prevent access to other users' token objects although this could be made another implementation feature.

The Security of PKCS #11

The standard has the following stated security targets.

1. "Access to private objects on the token, ..., requires a PIN. Thus, possessing the cryptographic device that implements the token may not be sufficient to use it; the PIN may also be needed."
2. "Additional protection can be given to private keys and secret keys by marking them as 'sensitive' or 'unextractable'. Sensitive keys cannot be revealed in plaintext off the token, and unextractable keys cannot be revealed off the token even when encrypted (though they can still be used as keys)."

Implied within these statements is the intention that by marking objects as 'sensitive' and 'unextractable', another user is prevented from recovering the secret values thereof. It does not appear to be the intention to prevent one user from using another user's private objects.

The designers discuss several areas of concern including operating system security, the actions of rogue applications and the threat posed by Trojan linked libraries or device drivers that may subvert security, perhaps by stealing the password. Similar concerns related to the 'sniffing' of communication lines to the cryptographic device exist(eavesdropping). This leads to several possible compromises such as PIN recovery, unauthorized access to a session (and the ability to insert, modify or delete commands) and the impersonation of a token or device. However, the standard claims that "...none of the attacks just described can compromise keys marked 'sensitive,' since a key that is sensitive will always remain sensitive. Similarly, a key that is 'unextractable' cannot be modified to be extractable." Thus, in addition to examining the API for vulnerabilities, we are particularly interested in this claimed property.

A cryptographic device that supports a PKCS #11 faces the following potential threat models:

- a malicious security officer who abuses the authority of his position and his access to the device and user management functions,
- a cheating or malicious user who exploits his authorized access to the token, and
- a malicious third party who gains access to the token through some means.

Essentially, these threats resolve into either gaining access to a session, or gaining access to a device during a session (e.g. by injecting messages into communications lines) or having knowledge of a password.

There exist some obvious, well-known attacks that are, generally speaking, implementation dependant as opposed to weaknesses in the API itself. We briefly describe them for completeness. The C_Login function is potentially vulnerable to an exhaustive PIN (password) search since a user can try all possible passwords. One typical defence is to keep a count of the number of failed login attempts and 'lock' the card after a certain threshold of fails has been reached. Ideally, the counter should be incremented prior to testing the PIN and decremented thereafter only if successful. This can lead to a denial of service attack where a malicious party tries to prevent a valid user from being able to use the token. The attacker repeatedly and intentionally masquerades as the user and attempts to login with an incorrect PIN. An alternative approach is to make use of time delays during start up and between login attempts.

```
CK_DEFINE_FUNCTION(CK_RV, C_Login)
(
    CK_SESSION_HANDLE hSession,
    CK_USER_TYPE userType,
    CK_CHAR_PTR pPin,
    CK_ULONG ulPinLen
);
```

A malicious security officer could use the C_InitPIN function to change a given user's PIN to a known value, hence gaining security access to the token. Since all users have access to all objects on the token, another less detectable approach would be to make a new user with a known PIN. This new user would be able to gain access to the token objects. While the power inherently held by a security officer in a given system is understood, PKCS #11 fails to specify directly the use of dual control mechanisms, which would defeat a single malicious security officer, although not a conspiracy of security officers.

```
CK_DEFINE_FUNCTION(CK_RV, C_InitPIN)
(
    CK_SESSION_HANDLE hSession,
    CK_CHAR_PTR pPin,
    CK_ULONG ulPinLen
);
```

Key Management Functions

PKCS #11 provides a typical set of key management functionality including:

- C_GenerateKey that generates a secret key,
- C_GenerateKeyPair that generates a public/private key pair,
- C_WrapKey that wraps (i.e., encrypts) a private or secret key,
- C_UnwrapKey that unwraps (i.e. decrypts) a wrapped key, and
- C_DeriveKey that derives a key from a base key.

Let us consider the C_WrapKey function further. It has the following prototype:

```
CK_DEFINE_FUNCTION(CK_RV, C_WrapKey)
(
    CK_SESSION_HANDLE hSession,
    CK_MECHANISM_PTR pMechanism,
    CK_OBJECT_HANDLE hWrappingKey,
    CK_OBJECT_HANDLE hKey,
    CK_BYTE_PTR pWrappedKey,
    CK_ULONG_PTR pulWrappedKeyLen
);
```

hSession is the session's handle; pMechanism points to the wrapping mechanism; hWrappingKey is the handle of the wrapping key; hKey is the handle of the key to be wrapped; pWrappedKey points to the location that receives the wrapped key; and pulWrappedKeyLen points to the location that receives the length of the wrapped key.

C_WrapKey can be used in the following situations:

- To wrap any secret key with an RSA public key.
- To wrap any secret key with any other secret key.
- To wrap an RSA, Diffie-Hellman, or DSA private key with any secret key.

2 Symmetric Key API Attacks

A wrapped key or external encrypted key is commonly referred to as an encrypted key token (T). Keys are typically wrapped (or encrypted) under a key encrypting key (KEK) for exchange or under a master key (MK) for storage external to the device. Initially we shall consider the wrapping of a secret key with another secret key. The mechanism describes the method of the wrapping operation and follows a naming convention of the form CKM_<NAME>_<MODE>. For example, CKM_DES_ECB, CKM_DES_CBC, CKM_DES_CBC_PAD, CKM_DES3_ECB, CKM_DES3_CBC and CKM_DES3_CBC_PAD are the mechanisms that make use of either single DES or triple DES. Other ciphers are possible including RC2, RC4, RC5, CAST, IDEA, etc.

2.1 Key Conjuring

Key conjuring is any technique that leads to the unauthorized generation of keys in the device. It is so named owing to the fact that the keys are 'conjured' (magically created or appearing seemingly out of nowhere). Bond in [6] first identified key conjuring as a security risk. This is for two reasons. First, it defeats any access control that was placed on the official key generation function by providing an alternative and unauthorized mechanism to perform effectively the same operation. Secondly, a key conjuring mechanism can be exploited to build a large set of keys, which can then be attacked by a parallel search, as described in Section 2.6.

Bond observed that crypto coprocessor designs, which stored keys outside the tamper-proof device, were vulnerable to unauthorized key generation. For instance, a random 8 bytes submitted as an external encrypted DES key will be decrypted and used as key. For example, using random data (R), a user creates a token $T_{random} = R$ which is then supplied to the C_UnWrapKey function call to the device. The device decrypts T_{random} as $d_{MK}(T_{random})$, yielding a new key $k_{random} = d_{MK}(T_{random})$. If parity checking is enforced, then there is a 1 in 2^8 chance that this new 'key' will have the correct parity. By repeating this process on average 2^8 times, an attacker can expect to conjure successfully a new key into the system in this manner. In fact this method is available in some older devices as a key generation function. Instead of merely testing for parity, the function will correctly set the parity in the process.

Key conjuring can be defeated through the associated use of a MAC or hash. This has the property of authenticating the clear value of the key as valid.

2.2 Key Binding (Integrity)

We observe that the choice of mode for the C_WrapKey is left to the caller (the user). In addition, there is no enforced use of a MAC or other technique to ensure data authenticity. There is also no restriction on the use of keys with repeated halves. As a result of the lack of cryptographic binding, one can attack each half of a key independently in the following way:

1. Export the target double length key (under any key encrypting key and in any mode). We denote the double length key as the ordered pair $K = \langle K_1, K_2 \rangle$ and note that each half is encrypted independently to form the encrypted key token (T);

$$T = e_{KEK}(\langle K_1, K_2 \rangle)$$
$$= \langle e_{KEK}(K_1), e_{KEK}(K_2) \rangle$$
$$= \langle T_1, T_2 \rangle.$$

2. Re-import the first half of the exported key as a single length key encrypted in ECB mode (using the same key encrypting key); $d_{KEK}(T_1) = d_{KEK}(e_{KEK}(K_1)) = K_1$.
3. Re-import the second half as a single length key encrypted in ECB mode (using the same key encrypting key); $d_{KEK}(T_2) = d_{KEK}(e_{KEK}(K_2)) = K_2$.
4. Perform a key search attack against each single length key (K_1, K_2) individually.

Algorithm 1: Typical Key Binding Attack

The key binding issue for double (and triple) length DES keys is well known, having been documented in [6] and exploited by [7], [9] and [11]. Indeed, this flaw has prompted a warning from the ANSI X9 Financial Services Committee [3] and is the subject of several revised proposals [1] and [2].

The API should not allow an exported key to be modified (especially the 'cut and paste' action on key components). Ideally, it should prevent the importation of such a modified or 'Trojan' key by employing some technique to verify that it is a genuine and authentic key. A typical solution is the use of a MAC on the exported key.

2.3 Key Separation

The secret key objects of PKCS #11 do allow for the specification of the use of the key for the operations of encrypting, decrypting, signing (MAC generation), verifying (MAC verification), key wrapping and key unwrapping. This is done through the use of the following attributes:

Attribute	Value	Meaning
CKA_ENCRYPT	CK_BBOOL	TRUE if key supports encryption
CKA_DECRYPT	CK_BBOOL	TRUE if key supports decryption
CKA_SIGN	CK_BBOOL	TRUE if key supports signatures (i.e.,authentication codes)
CKA_VERIFY	CK_BBOOL	TRUE if key supports verification (i.e., of authentication codes)
CKA_WRAP	CK_BBOOL	TRUE if key supports wrapping
CKA_UNWRAP	CK_BBOOL	TRUE if key supports unwrapping

Unfortunately, the API allows the specification of conflicting properties in that these attributes can be independently specified. This leads to a typical separation attack:

1. Start with the key (K) having the ability to wrap keys (i.e., act as a key encrypting key) and decrypt data.

2. Export the target key (K_{target}) under any key encrypting key (K) using the C_WrapKey function yielding the token $T = e_K(K_{target})$.
3. Decrypt the resultant token using the C_Decrypt function with K (the key wrapping key) as a data decryption key. This returns $d_K(T) = d_K(e_K(K_{target})) = K_{target}$ (i.e., the clear value of the target key).

Algorithm 2: Typical Key Separation Attack

Since the values of the attributes may be modified using the C_SetAttributeValue call or in the process of copying an object using the C_CopyObject function, it is possible for an adversary to manipulate existing keys. The PKCS #11 documentation does note that a particular implementation or token may choose to " ...permit modification of the attribute, or may not permit modification of the attribute during the course of a C_CopyObject call".

The problem is exacerbated in the key export/import process since an exported (or wrapped) key contains no such separation information bound to the token. As a result, any given exported key could be imported twice with different attributes. For example, the key could be imported as a key wrapping key the first time, and then as a data decrypting key the second time, thus facilitating the attack.

Clearly, greater consideration must be paid to key separation issues in the API. Ideally, the choice of attribute combination must be restrictive in order to prevent such attacks. Furthermore, such information must be cryptographically bound to the wrapped key as in [1].

2.4   Weaker Key/Algorithm

The PKCS #11 specification allows for the wrapping of a key by a second key of shorter length. Thus one need only attack the weaker key in order to recover the original key.

1. Export the target double length DES key $(K_{target} = \langle K_1, K_2 \rangle)$ under a single length key (KEK) as

$$
\begin{aligned}
T &= e_{KEK}(K_{target}) \\
&= e_{KEK}(\langle K_1, K_2 \rangle) \\
&= \langle e_{KEK}(K_1), e_{KEK}(K_2) \rangle \\
&= \langle T_1, T_2 \rangle .
\end{aligned}
$$

2. Export the single length key (KEK) under itself yielding $T_{KEK} = e_{KEK}(KEK)$.
3. Attack the single length key by performing an exhaustive search.
4. Once the single length key has been recovered, one can trivially recover the original double length key.

Algorithm 3: Example Weaker Key Attack

PKCS #11 supports keys with particularly small key sizes (e.g. RC2), making the search feasible. It should not be possible to downgrade the security, by protecting a longer key with a shorter key. Similarly, it should not be possible to use a weaker algorithm when exporting keys.

We note that the previous attacks do not contradict the security claim that 'sensitive' and 'unextractable' keys cannot be compromised, since they require that the target key be exportable. What about other attacks? We focus our attention on the C_DeriveKey function, which has the following prototype:

```
CK_DEFINE_FUNCTION(CK_RV, C_DeriveKey)
(
    CK_SESSION_HANDLE hSession,
    CK_MECHANISM_PTR pMechanism,
    CK_OBJECT_HANDLE hBaseKey,
    CK_ATTRIBUTE_PTR pTemplate,
    CK_ULONG ulAttributeCount,
    CK_OBJECT_HANDLE_PTR phKey
);
```

The C_DeriveKey supports the following mechanisms:

- CKM_CONCATENATE_BASE_AND_KEY, which derives a secret key from the concatenation of two existing secret keys,
- CKM_CONCATENATE_BASE_AND_DATA, which derives a secret key by concatenating data onto the end of a specified secret key,
- CKM_CONCATENATE_DATA_AND_BASE, which derives a secret key by prepending data to the start of a specified secret key,
- CKM_XOR_BASE_AND_DATA, which is a mechanism that provides the capability for deriving a secret key by performing the exclusive-oring of a key pointed to by a base key handle and some data, and finally
- CKM_EXTRACT_KEY_FROM_KEY that provides the capability of creating one secret key from the bits of another secret key.

2.5   Reduced Key Space

Using the CKM_EXTRACT_KEY_FROM_KEY mechanism, one can extract a subset of the bits from a given key to create a shorter key. The can be used to reduce the key space required to be searched. For example, one could extract 40 bits from a DES key to create a 40-bit RC2 key, which can then be searched by exhaustive means. The actual key space may be smaller owing to the existence of parity bits in the DES key. The remaining 24 bits (less 3 parity bits) of the original DES key can then be searched for independently. This potentially dangerous mechanism relies on the 'unextractable' flag in the key token to prevent misuse. It does not prevent an attacker from using this method to obtain a known key in the system or from compromising extractable keys.

2.6 Parallel Search

The CKM_XOR_BASE_AND_DATA provides an easy method with which to exclusive-or known patterns onto a key. This can be used to reduce the key space required to be searched by generating a large number of (known) related keys as per the method suggested by [12] and [19] and exploited by [9].

1. Generate a set of 2^{16} known related keys of original target key $\{K_i | K_i = K_{target} \bigoplus \Delta_i, i = 1, ..., 2^{16}\}$ where $\Delta_i \neq \Delta_j$ for $i, j \leq 2^{16}$, $i \neq j$ and Δ_i is a non-zero known value.
2. Using each key, encrypt a known pattern (P) and store the result in searchable database $\{C_i | C_i = e_{K_i}(P), i = 1, ..., 2^{16}\}$.
3. Search for a key by iteratively performing trial encryptions of the known pattern (P) and compare result to entries in database.
4. After 2^{39} trial encryptions on average, we expect to find a match (i.e., we find a key K_i which produces an encrypted output in the database).
5. Recover the original target key K_{target} as $K_{target} = K_i \bigoplus \Delta_i$.

Algorithm 4 : Parallel Key Search Using Related Keys

Since we know how this key is related to all the others, we known all the 2^{16} keys including the original one. This clearly demonstrates the danger of being able to modify a key as well as the true threat posed by the seemingly benign key conjuring vulnerability. Knowledge of the modification makes the attack easier but is not a requisite for the attack.

2.7 Related Key Attack

Using the CKM_XOR_BASE_AND_DATA mechanism, one can create a set of related keys with which to perform a related key attack [5], [14], [15]. This can be used to reduce 3-key 3DES to only slightly stronger than single DES (reducing the key space search to 2^{56} operations to isolate a key component). The attack is elegantly simple and easily explained. Using the related key pair $K1 =< k1, k2, k3 >$, $K2 =< k1 \bigoplus \Delta, k2, k3 >$, encrypt a plaintext P with $K1$, and then decrypt the ciphertext with $K2$ yielding P'. Then $C = e_{K1}(P)$, $P' = d_{K2}(C)$, and hence $P' = d_{K2}(e_{K1}(P))$. Using 3DES in EDE mode (the mode itself doesn't matter):

$$P' = d_{k1 \bigoplus \Delta}(e_{k2}(d_{k3}(e_{k3}(d_{k2}(e_{k1}(P)))$$
$$= d_{k1 \bigoplus \Delta}(e_{k1}(P)) .$$

Thus, $k1$ has been successfully isolated and can be recovered independently of $k2$ and $k3$, typically by exhaustive key search. The work required on average to effect the search is 2^{56} single DES operations. Hence the cipher in triple mode has been reduced to only slightly greater than the strength of the cipher in single mode. This attack can be further enhanced by combining it with parallel key search techniques. For example, using a set of related key pairs $\{(< k_1 \oplus \Delta_i, k_2, k_3 >, < k_1 \oplus \Delta_i \Delta, k_2, k_3 > | i = 1, \ldots, 2_{16}\}$ would reduce the average search effort to 2^{40} DES operations.

The 2-key 3DES version of the attack described in [11] is not practically feasible. However, there exists a more efficient attack by first 'converting' a double length DES key into a triple length DES key using the CKM_CONCATENATE_BASE_AND_DATA mechanism. Following this, the three-key related key attack can be used as is.

Analysis and Implications

We return to the security claim made by the designers. Both the parallel search attack and the related key attack *contradict* the claims of the API designers. This has several implications for individual users who are reliant on the security of a PKCS #11 token. Any user with read and write access to the token has the ability to recover all token key objects. In addition, an adversary with the ability to gain access to a session (perhaps by injecting raw messages into the physical communications lines) likewise has the ability to recover keys from the token. To thwart the attack, one must prevent all unauthorized access to token objects. This intensifies the security concerns already listed by the designers and previously referred to.

We now consider a means to expand the scope of the attack to include sessions with read only access to token objects. The C_CopyObject provides a method to copy a read only token object and to produce as output a session object. However, since all session objects have read/write access to that session, the attacker successfully obtains a duplicate of the key object with write access. He can thus attack the session object using the methods previously described, despite only having read access to the original target object. Therefore, it is advisable to reconsider the functionality of the C_CopyObject call particularly with respect to the preservation of properties such as write access.

Finally, it is worth noting the work done in [9] as it directly reflects on the feasibility and speed of performing these attacks in practice. Bond and Clayton devised a parallel exhaustive key search machine using an 'off the shelf' FPGA evaluation card costing approximately $1000, which was capable of performing a 2^{39} search in 22 hours.

3   Public Key API Attacks

We now extend our focus to consider attacks involving the use of (or against) Public Key API functionality. We start by revisiting the C_WrapKey function and consider first the wrapping of private RSA keys by symmetric keys.

Wrapping/Unwrapping of Private Keys Using Symmetric Keys

In PKCS #11, a private key can only be exported (and imported) if it contains not only the private exponent and modulus, but also the public exponent and

CRT info. This information is BER-encoded according to PKCS #1's RSAPrivateKey ASN.1 type. The resulting string of bytes is encrypted with a secret key in CBC mode and with PKCS padding.

Attribute	Data Type	Meaning
CKA_MODULUS	Big integer	Modulus n
CKA_PUBLIC_EXPONENT	Big integer	Public exponent e
CKA_PRIVATE_EXPONENT	Big integer	Private exponent d
CKA_PRIME_1	Big integer	Prime p
CKA_PRIME_2	Big integer	Prime q
CKA_EXPONENT_1	Big integer	Private exponent d modulo $p-1$
CKA_EXPONENT_2	Big integer	Private exponent d modulo $q-1$
CKA_COEFFICIENT	Big integer	CRT coefficient $q-1 \bmod p$

The CBC-encrypted ciphertext is decrypted, and the PKCS padding is removed. The data thereby obtained are parsed as a PrivateKeyInfo type, and the wrapped key is produced. An error will result if the original wrapped key does not decrypt properly, or if the decrypted unpadded data does not parse properly, or its type does not match the key type specified in the template for the new key. The unwrapping mechanism contributes only those attributes specified in the PrivateKeyInfo type to the newly-unwrapped key; other attributes must be specified in the template, or will take their default values.

3.1 Weaker Key/Algorithm

Following this description we are immediately concerned with the choice of symmetric key algorithm (and key length) used to protect the RSA private key leading to equivalent attacks described in Section 2.4.

3.2 Private Key Modification

Consider the effect of replacing one block of the ciphertext (i.e., the wrapped key) with a different value. When the key is unwrapped, this will cause the corresponding block of plaintext as well as the following block to have different values. The rest of the key remains intact. The length of the BER encoded big number data types depends upon the size of the big numbers (typically 512, 1024 or 2048 bit numbers). In any event, they consist of at least a number of blocks. Thus an attacker can modify one of the big numbers independently of the other data in the wrapped private key (including the padding at the end). If the various key components (e.g. n, p, q, e, d, $d \bmod p-1$, $d \bmod q-1$ and $q-1 \bmod p$) are not explicitly tested for consistency, the attacker gains access to a modified 'Trojan' key in the system. This can be used to effect the Fault Analysis attacks of [8], [4] and [13]. A similar attack against PGP private keys is described in [16] and, more generally, against public key APIs in [17] and [10].

A possible solution is that encrypted private keys have a strong cryptographic method to ensure integrity of the key (e.g. MAC, hash or signature). In addition, the integrity of the key must be confirmed using simple arithmetic checks (for example, is $d_p \equiv d \mod p$ and $n = p \cdot q$).

Wrapping/Unwrapping of Symmetric Keys Using Public Keys Techniques

PKCS #11 supports two mechanisms for wrapping symmetric keys using Public Key techniques, namely:

- CKM_RSA_PKCS (PKCS #1 RSA), and
- CKM_RSA_X_509 (X.509 Raw RSA).

The CKM_RSA_X_509 mechanism performs no padding or manipulation of data prior to encryption. It merely "...encrypts a byte string by converting it to an integer, most-significant byte first, applying 'raw' RSA exponentiation, and converting the result to a byte string, most significant byte first." The encrypted token is $T = k^e \mod n$ where e is public exponent, k the key being exported and n the modulus. This simple method results in exported keys being vulnerable when encrypted under small public exponents.

3.3   Small Public Exponent with No Padding

The clear key is right justified in the field provided, and the field padded to the left with zeroes up to the size of the RSA encryption block (e.g. for 128-bit key $k = k_1 k_2 \ldots k_{128}$ is prepended with zero bits $0_1 0_2 \ldots 0_{l-128} k_1 k_2 \ldots k_{128}$, where l is the length of the modulus). The resultant field is encrypted yielding $T = k^e \mod n$. If $k^e < n$ (i.e., $e < \frac{log_2(n)}{log_2(k)} \leq \frac{log_2(n)}{128}$), then $T = k^e$. Thus k can be recovered as $k = T^{\frac{1}{e}}$.

Due to the speed advantages of having a small exponent with low Hamming weight, it is common for public keys to have exponents of 3 and $2^{16} + 1$. It is not uncommon to be able to specify this as an option in many APIs when generating a public key. It is thus possible that a suitable public key will exist in the system. In any event, the public keys in PKCS #11 are clear tokens and thus one can easily 'conjure' or create a public key with an exponent of 3. This weakness exists in a number of APIs [10].

3.4   Trojan Public Key

As previously mentioned, the public keys in the PKCS #11 API are clear tokens with no additional authentication checks. Thus it is possible to use any clear public key as input to the C_WrapKey function. This allows an attacker to use a 'Trojan' public key for which he knows the corresponding private key (typically the attacker will probably generate the key pair himself). He requests the PKCS #11 token exports the target key k under his supplied public key obtaining the response $T = k^e \mod n$. Since the attacker knows the corresponding private exponent d, he can easily recover the key as $T^d \mod n = (k^e)^d = k$. This simple method can be used to recover all exportable keys regardless of whether they are symmetric or private keys. It is thus clear that a public key needs to be authenticated before use to verify that it indeed has the authority to export a given key.

3.5 Trojan Wrapped Key

Similarly to the unauthenticated use of public keys, there is no method to verify that a wrapped key token is indeed authentic. Thus given a PKCS #11 device containing a private key $(< d, n >)$, and knowledge of the value of the public key $(< e, n >)$, the attacker proceeds as follows. He chooses an arbitrary key k, which he then 'wraps' under the known public key obtaining $T = k^e \bmod n$. He then calls the C_UnWrapKey function supplying this 'Trojan' wrapped key T and referencing the handle of the private key inside the device. The PKCS #11 token calculates $T^d \bmod n = (k^e)^d = k$ and imports the known k as a new key into the system. The attacker can then use k to export other keys from the device, which he can then decrypt and recover. Thus there exists a requirement to provide a means to verify the authenticity and origin of the wrapped key.

3.6 Key Separation

A symmetric key wrapped by a public key contains no separation information and can be exploited as described previously in Section 2.3.

4 Solutions

Some of these security issues can be easily addressed in the implementation of a PKCS #11 API. The more concerning issues unfortunately require a design change to the PKCS #11 standard. With the latter come the dual concerns of backwards compatibility and interoperability with other systems. A lack of backwards compatibility may be the price for a previously flawed design and a commitment to security.

The Key Conjuring and Key Binding attacks are perhaps best addressed through a change in the external key token format, particularly for wrapped keys. There exist proposals such as [1] and [2] and one can expect a decision and guidance from such influential bodies as ANSI Financial Services Committee, which will largely address the interoperability issues. Key Separation can be partially addressed by a given implementation that does not permit the conflicting use of key attributes (e.g. CKA_WRAP and CKA_DECRYPT). However, the fact that the wrapped key contains no separation information is a fundamental design flaw and like the Key Conjuring and Key Binding attacks must be addressed through a new external key token format. The Weaker Key/Algorithm attack can be prevented by a given implementation by understanding and obeying the principle that a key should not be protected by a weaker key or algorithm. The 'unextractable' and 'never extractable' flags do offer protection against the

Reduced Key Search attack. Regardless, the author is not convinced that the CKM_EXTRACT_KEY_FROM_KEY mechanism deserves consideration in the API. Similarly, the CKM_XOR_BASE_AND_DATA mechanism creates the opportunity for both the Parallel Search and Related Key attacks. Again one may question the need for such a function, particularly in its present form.

Prevention of the Private Key Modification attack requires either the use of a consistency check to confirm the integrity of the key components, which could be implementation specific, or else a revision of the encrypted RSA key token that ensures integrity through some cryptographic means, such as an encrypted hash or MAC over the token. The Small Public Exponent with No Padding attack highlights the dangers of providing raw RSA functionality. The most sensible solution is to enforce the use of a recognised padding scheme. The only concern here would be backwards compatibility. Interoperability should not be an issue since any device that uses this method to export a key is obviously vulnerable to the attack. The Trojan Public Key and Trojan Wrapped Key attacks exploit a lack of authentication of public keys used for export and wrapped keys being imported. This requires a significant change to the standard to achieve these goals.

5   Conclusions

This paper has shown the susceptibility of PKCS #11 used as an API to a number of attacks. The attacks are efficient, computationally trivial and easy to implement. Some possible solutions are presented to defend against the attacks.

Acknowledgements. I would like to express deep gratitude to my supervisor Henda Swart whose guidance and support were crucial to the successful completion of this project. I am also indebted to Alex Dent and Christine Swart for their rigorous reading of an earlier version of this paper and for their comments and suggestions, which shaped this work. I would also like to thank the current PKCS #11 editor, Simon McMahon, as well as Burt Kaliski and Magnus Nystrom for the positive and proactive manner with which they considered this research and for their helpful comments and continued efforts relating to the security of PKCS. The author is supported by the Cecil Renaud Scholarship.

References

1. ACI Worldwide, HP Atalla, Diebold, Thales e-Security, and VeriFone Inc. Global interoperable secure key exchange key block specification, 2002.
2. ACI Worldwide, HP Atalla, Diebold, Thales e-Security, and VeriFone Inc. Newly-formed payment consortium moves ahead with endorsement of secure 3DES implementation specification: Industry leaders align on new proposed key management standard, 2002.
3. American National Standards Institute (ANSI) Accredited Standards Committee (ASC) X9 - Financial Services (X9-F). Notice regarding TDES key wrapping techniques, 2002.

4. Feng Bao, Robert H. Deng, Yongfei Han, Albert B. Jeng, A. Desai Narasimhalu, and Teow-Hin Ngair. Breaking public key cryptosystems on tamper resistant devices in the presence of transient faults. In *Security Protocols Workshop*, volume 1361, pages 115–124, 1997.
5. Eli Biham. New types of cryptanalytic attacks using related keys. In *Advances in Cryptology EUROCRYPT '93*, volume 675, pages 398–409, 1994.
6. Mike Bond. Attacks on cryptoprocessor transaction sets. In *Cryptographic Hardware and Embedded System – CHES 2001 Third International Workshop*, volume 2162, pages 220–234, 2001.
7. Mike Bond and Ross J. Anderson. API-level attacks on embedded systems. *Computer*, 34(10):67–75, 2001.
8. Dan Boneh, Richard A. DeMillo, and Richard J. Lipton. On the importance of checking cryptographic protocols for faults. In *Advances in Cryptology EURO-CRYPT '97*, volume 1233, pages 37–51, 1997.
9. Richard Clayton and Mike Bond. Experience using a low-cost FPGA design to crack DES keys. In *Cryptographic Hardware and Embedded System - CHES 2002*, volume 2523, pages 579–592, 2003.
10. Jolyon Clulow. The design and security of public key crypto APIs, 2001.
11. Jolyon Clulow. The design and analysis of cryptographic application programming interfaces for devices. Master's thesis, University of Natal, Durban, 2003.
12. Electronic Frontier Foundation. *Cracking DES: Secrets of Encryption Research, Wiretap Politics, and Chip Design*. O'Reilly, Sebastopol, 1998.
13. M. Joye, A. K. Lenstra, and J.-J. Quisquater. Chinese remaindering based cryptosystems in the presence of faults. *Journal of Cryptology*, 12(4):241–246, 1999.
14. John Kelsey, Bruce Schneier, and David Wagner. Key-schedule cryptanalysis of IDEA, G-DES, GOST, SAFER, and Triple-DES. In *Advances in Cryptology – CRYPTO '96*, volume 1109, pages 237–251, 1996.
15. John Kelsey, Bruce Schneier, and David Wagner. Related-key cryptanalysis of 3-WAY, Biham-DES, CAST, DES-X NewDES, RC2, and TEA. In *1997 International Conference on Information and Communications Security, Beijing*, volume 1334, pages 233–246, 1997.
16. Vlastimil Klíma and Tomas Rosa. Attack on private signature keys of the OpenPGP format, PGPTM programs and other applications compatible with OpenPGP. *Cryptology ePrint Archive*, 2002.
17. Vlastimil Klíma and Tomas Rosa. Further results and considerations on side channel attacks on rsa. In *Cryptographic Hardware and Embedded System – CHES 2002*, volume 2523, pages 244–259, 2003.
18. RSA Security Inc. PKCS #11: Cryptographic Token Interface Standard. An RSA Laboratories Technical Note, Version 2.01, December 22, 1997.
19. F. Hoornaert Y. Desmedt and J. J. Quisquater. Several exhaustive key search machines and DES. In *EUROCRYPT '86*, pages 17–19, 1986.

Attacking RSA-Based Sessions in SSL/TLS

Vlastimil Klíma[1], Ondřej Pokorný, and Tomáš Rosa[1,2]

[1] ICZ, Prague, Czech Republic
[2] Dept. of Computer Science and Eng., FEE, Czech Technical University in Prague
vlastimil.klima@i.cz, ondrej.pokorny@i.cz, tomas.rosa@i.cz

Abstract. In this paper we present a practically feasible attack on RSA-based sessions in SSL/TLS protocols. We show that incorporating a version number check over PKCS#1 plaintext used in the SSL/TLS creates a side channel that allows an attacker to invert the RSA encryption. The attacker can then either recover the *premaster-secret* or sign a message on behalf of the server. Practical tests showed that two thirds of randomly chosen Internet SSL/TLS servers were vulnerable. The attack is an extension of Bleichenbacher's attack on PKCS#1 (v. 1.5). We introduce the concept of a *bad-version oracle* (BVO) that covers the side channel leakage, and present several methods that speed up the original algorithm. Our attack was successfully tested in practice and the results of complexity measurements are presented in the paper.

1 Introduction

In contemporary cryptography, it is widely agreed that one of the most important issues of all asymmetric schemes is the way in which the scheme encodes the data to be processed. In the case of RSA [14], the most widely used encoding methods are described in PKCS#1 [9]. This standard also underlies RSA-based sessions in the family of SSL/TLS protocols. These protocols became de facto the standard platform for secure communication in the Internet environment. In this paper we assume a certain familiarity with their architecture (c.f. §5). Since its complete description is far beyond the scope of this article, we refer interested readers to the excellent book [10] for further details. In 1998 Bleichenbacher showed that the concrete encoding method called EME-PKCS1-v1_5, which is also employed in the SSL/TLS protocols, is highly vulnerable to chosen ciphertext attacks [1]. The attack assumes that information about the course of the decoding process is leaking to an attacker. We refer to such attacks as *side channel attacks*, since they rely on *side information* that unintentionally leaks out from a cryptographic module during its common activity.

Bleichenbacher showed that it is highly probable that side information exists allowing the attacker to break the particular realization of the RSA scheme in many systems based on EME-PKCS1-v1_5. He has also shown how to use such information to decrypt any captured ciphertext or to sign any arbitrary message by using a common interaction with the attacked cryptographic module. As a countermeasure to his attack it was recommended to either use the EME-OAEP method (also defined in PKCS#1) or to steer attackers away from knowing details about the course of the decoding process. In the case of the SSL/TLS protocols it seemed to be possible to

C.D. Walter et al. (Eds.): CHES 2003, LNCS 2779, pp. 426–440, 2003.

incorporate the second type of countermeasures. The story of the attack ended here by incorporating appropriate warnings in appropriate standards [9], [10], [12], and [15]. Security architects were especially instructed not to allow an attacker to know whether the plaintext P being decoded has the prescribed mandatory structure marks or not.

Besides being warned to carry out the above-mentioned countermeasure, architects were also instructed to carefully verify all possible marks of P that are specific for the SSL/TLS protocols. In particular, they were told to check the correctness of a version number (c.f. §5.2 and [12]), which is stored in the two left-most bytes of the *premaster-secret*. Unfortunately, it has not been properly specified how such a test may be combined with the countermeasure mentioned above and what to do if the "version number test" fails. Designers may be very tempted to simply issue an error message. In reality, however, such a message opened up a Pandora's box bringing a new variant of side channel attack. In this paper we present this attack. It turns out that the version number, which was initially believed to rule out the original attack [1], even allows a relatively optimized variant of the attack if the version number check is badly implemented. Our practical tests showed that among hundreds of SSL/TLS servers randomly chosen from the Internet, two thirds of them were vulnerable to our attack (for details see §4.3).

We note that the TLS protocol may be historically viewed as an SSL bearing the version number 3.1 [12], while the SSL with the version number 3.0 is often referred to as a "plain" SSL. There are some minor changes between SSL and TLS, but these changes are unimportant for the purpose of this paper, since we rely on the general properties, which are common to both SSL v. 3.0 and TLS. Therefore, we will talk about them as about the SSL/TLS protocols. We note that SSL protocols with version numbers less than 3.0 will not be considered here, since they have already been proven to have several serious weaknesses [10], [16].

The rest of the paper is organized as follows: in §2 we introduce a *bad-version oracle* (BVO), which is a construction that mathematically encapsulates side information leaking from the decoding process. The BVO is then used for mounting our attack in §3. The attack is based on an extended variant of Bleichanbacher's algorithm from [1]. The complexity of the attack together with the statistics of the vulnerable servers found on the Internet are given in §4. We note that due to page constraints, paragraphs 5 (Technical details) and 6 (Countermeasures) are fully elaborated in the extended version of the paper [19]. The conclusions are made in §7. In the appendix we recall a slightly generalized version of the original Bleichenbacher's algorithm [1].

Proposition 1 (Connection and session). *Unless stated otherwise, the term* connection *means the communication carried out between a client and a server. It lasts from when the client opened up a networked pipe with the server, until the pipe is closed. The term* session *is used to refer to a particular part of this connection which is protected under the same value of symmetrical encryption keys.*

Proposition 2 (RSA-based session). *We say that the session is RSA-based if it uses the RSA scheme to establish its symmetrical keys.*

2 Bad-Version Oracle

We start by recalling the definition of PKCS-conforming plaintext [1]. Unless stated otherwise, the term plaintext means an *RSA* plaintext. Furthermore, we denote RSA instance parameters as (N, e, d), where N is a public modulus, e is a public exponent, and d is a private exponent, such that for all $x, x \in <0, N - 1>$ it holds that $x = (x^e \bmod N)^d \bmod N$. We denote as k the length of the modulus N in bytes, i.e. $k = \lceil (\log_2 N)/8 \rceil$, and the boundary B as $B = 256^{k-2}$.

Definition 1 (PKCS-conforming plaintext). *Let us denote the plaintext as P, P =* $\sum_{i=1}^{k}(P_i * 256^{k-i})$, *$0 \le P_i \le 255$, where P_1 is the most significant byte of the plaintext. We say that P is PKCS-conforming if the following conditions hold:*

    i) $P_1 = 0$
    ii) $P_2 = 2$
    iii) $P_j \neq 0$ *for all $j \in <3, 10>$*
    iv) $\exists j, j \in <11, k>, P_j = 0$; *the string $P_{j+1}||...||P_k$ is then called as a message M or a data payload*

The definition describes the set of all valid plaintexts for the given modulus of the length k bytes. In the case of SSL/TLS protocols, however, only the subset of this set is allowed, since these protocols introduce several extensions to the basic PKCS#1 (v. 1.5) format. Therefore, we define the term S-PKCS-conforming plaintext as follows.

Definition 2 (S-PKCS-conforming plaintext). *We say that P is S-PKCS-conforming if it is PKCS-conforming and the following conditions hold:*

    i) $P_j \neq 0$ *for all $j \in <3, k - 49>$*
    ii) $P_{k-48} = 0$

The main restriction introduced here is the constant number of data bytes (which is equal to 48). The number of padding bytes equals $k - 51$. Furthermore, SSL/TLS protocols introduce a special interpretation for the first two data bytes P_{k-47} and P_{k-46}, which are respectively regarded as major and minor version numbers. This extension was introduced to thwart so-called version rollback attacks. The *data payload*, which is the concatenation of $P_{k-47} || P_{k-46} || P_{k-45} || ... || P_k$, is called a *premaster-secret* here. It is the only secret used in the key derivation process that produces the session keys used by the client and the server in the given session. An attacker, who is able to discover the *premaster-secret*, can decrypt the whole communication between the client and server which has been carried out in the session. The value of $P_{k-45} || ... || P_k$ is generated randomly by the client who then adds the version number P_{k-47} and P_{k-46}, encrypts the whole value of the *premaster-secret* by the server's public RSA key, and sends the resulting ciphertext C to the server. The server decrypts it and creates its own copy of the *premaster-secret*.

It is widely known that the server shall not report whether the plaintext $P, P = C^d \bmod N$, is PKCS-conforming or not. In practice, a server is recommended to continue with a randomly chosen value of the *premaster-secret* if the value of P is not S-PKCS-conforming. Obviously, the communication breaks down soon after sending a Finished message, since the client and the server will both use different values for the session keys. However, the client (attacker) does not know whether the communication has broken down due to an invalid format of P or due to incorrect value of the *premaster-secret*. So, the attack is effectively defeated in this way. Of course, the attacker still gains some information from such an interaction with the server. She may at least try to confirm her guesses of the correct value of the *premaster-secret*. However, it has been shown by Jonsson and Kaliski [4] that it is infeasible to exploit this information for an attack.

Let us suppose that the server incorporates the above-mentioned countermeasure, the primary aim of which is to thwart Bleichenbacher's attack [1]. Furthermore, let all S-PKCS-conforming plaintexts be processed by the server to check the validity of proprietary SSL/TLS extensions according to the following proposition.

Proposition 3 (Conjectured server's behavior).

i) *The server checks if the deciphered plaintext P is S-PKCS-conforming. If the plaintext is not S-PKCS conforming, the server generates a new premaster-secret randomly, thereby breaking down the communication soon, after receiving the client's* Finished *message.*

ii) *The server checks each S-PKCS-conforming plaintext P to see whether P_{k-47}* = major *and P_{k-46}* = minor, *where* major.minor *is the expected version number which is known to the attacker. For instance, the most usual version numbers at the time of writing this paper were 3.0 and 3.1. If the test fails, the server issues a distinguishable error message. The test is never done for plaintexts that are not S-PKCS-conforming.*

Practical tests showed that it is reasonable to assume Proposition 3 is fulfilled in many practical realizations of SSL/TLS servers.

Definition 3 (Bad-Version Oracle - BVO). *BVO is a mapping BVO: $Z_N \rightarrow \{0, 1\}$. BVO(C) = 1 iff $C = P^e$ mod N, where e is the server's public exponent, N is the server's modulus, and P is an S-PKCS conforming plaintext, such that either $P_{k-47} \neq$* major *or $P_{k-46} \neq$* minor, *where* major.minor *is the expected version number. BVO(C) = 0 otherwise.*

BVO can be easily constructed for any SSL/TLS server that acts according to Proposition 3. We send the ciphertext C to the server and if we receive the distinguished message from (ii), we set BVO(C) = 1. Otherwise, we set BVO(C) = 0.

Theorem 1 (Usage of BVO). *Let us have a BVO for given (e, N) and* major.minor *and let C be an RSA ciphertext. Then BVO(C) = 1 implies that $C = P^e$ mod N, where P is an S-PKCS-conforming plaintext.*

Proof. Follows directly from Definition 3.

■

Because S-PKCS-conforming plaintext is also PKCS-conforming, it follows from Theorem 1 that we can use BVO to mount Bleichenbacher's attack. We discuss the details in §3. Now we introduce several definitions that will be useful in the rest of this paper. We use a similar notation to the one used in [1].

Definition 4 (Probabilities concerning BVO). *Let Pr(A) = B/N be the probability of the event A that the conditions (i-ii) of Definition 1 hold for randomly chosen plaintext P. Let Pr(S-PKCS|A) be the conditional probability that the plaintext P is S-PKCS-conforming assuming that A occurred for P. Let Pr(BVO|S-PKCS) be the* *conditional probability that $BVO(P^e \mod N) = 1$ assuming that P is S-PKCS-conforming.*

For $Pr(A)$ we have $256^{-2} < Pr(A) < 256^{-1}$ as stated in [1]. The probability $Pr(S\text{-}PKCS|A)$ can be expressed as $Pr(S\text{-}PKCS|A) = (255/256)^{(k-51)}*256^{-1}$, since the length of the non-zero padding bytes must be equal to k-51. There is usually one value of the version number that is expected by BVO. Therefore, $Pr(BVO|S\text{-}PKCS) = 1\text{-}256^{-2}$. Note that the value of $Pr(BVO|S\text{-}PKCS)*Pr(S\text{-}PKCS|A)*Pr(A)$ is the probability that for a randomly chosen ciphertext C we get $BVO(C) = 1$.

3 Attacking *Premaster-Secret*

3.1 Mounting and Extending Bleichenbacher's Attack

This attack allows us to compute the value $x = y^d \mod N$ for any given integer y, where d is an unknown RSA private exponent and N is an RSA modulus. This attack works under the condition that an attacker has an oracle that for any ciphertext C tells her whether the corresponding RSA plaintext $P = C^d \mod N$ is PKCS-conforming or not. Theorem 1 shows that BVO introduced in the previous part can be used as such an oracle. In the case of the SSL/TLS protocols this means that we can mount this attack to either disclose a *premaster-secret* for an arbitrary captured session or to forge a server's signature. In the following text, we mainly focus on the *premaster-secret* disclosure. Forging of signatures is discussed briefly in §3.4.

The main idea here is to employ Bleichenbacher's attack with several changes related to the specific properties of S-PKCS and BVO (§3.2). Furthermore, we employed particular optimizations, which we have tested in our sample programs, and which generally help an attacker (§3.3).

3.2 S-PKCS and BVO Properties

We show how to modify Bleichenbacher's original RSA inversion algorithm for use with the BVO and to increase its efficiency. For the sake of completeness we repeat the necessary facts from [1] in the appendix together with a brief generalization of it.

Recall that PKCS-conforming plaintext P satisfies the following system of inequalities

$$E \leq P \leq F,$$

where $E = 2B$, $F = 3B-1$, and $B = 256^{k-2}$. The boundaries E, F are extensively used through the whole RSA inversion algorithm. Since BVO as well as the SSL/TLS protocols deal only with S-PKCS-conforming plaintexts, we may refine the boundaries as

$$E' \leq P \leq F',$$

where the value of E' is obtained by incorporating the minimum value of the padding and the value of F' is computed with respect to the fixed position of the zero delimiter in the plaintext P:

$E' = 2B + 1*256^{k-3} + 1*256^{k-4} + \ldots + 1*256^{49} = 2B + 256^{49}(256^{k-51} - 1)/255$ and

$F' = 2B + 255*(256^{k-3} + 256^{k-4} + \ldots + 256^{49}) + 0 + 255*(256^{47} + 256^{46} + \ldots + 256^{0}) = 3B - 255*256^{48} - 1$.

Substituting E', F' in place of E, F in the original algorithm (see the appendix) increases its effectiveness.

It follows from the definition of the protocol SSL/TLS that the attacker knows the expected value of the version number, which is checked by BVO. Therefore, when attacking the ciphertext C_0, such that $BVO(C_0) = 0$, carrying the *premaster-secret*, the attacker knows exactly the two bytes $P_{0,k-47}$ and $P_{0,k-46}$ of the S-PKCS-conforming plaintext $P_0 = C_0^d \bmod N$. She also knows that $P_{0,k-48} = 0$. We used this knowledge in our program to further trim the interval boundaries $<a, b>$ computed in step 3 of the algorithm (see the appendix).

3.3 Basic General Optimizations

Besides the optimizations that follow directly from §3.2, we also used the generally applicable methods described in the following subparagraphs.

Definition 5 (Suitable multiplier). *Let us have an integer C. The integer s is said to be a suitable multiplier for C if it holds that $C' = s^e C \bmod N = (P')^e \bmod N$, where P' is a S-PKCS-conforming plaintext.*

3.3.1 Beta Method

The following method (β-method) follows from a generalization of the remark mentioned in [1], pp.7 - 8.

Lemma 1 (On linear combination). *Let us have two ciphertexts C_i and C_j, such that $C_i = (s_i)^e C_0 \bmod N$, $C_j = (s_j)^e C_0 \bmod N$, where s_i and s_j are suitable multipliers for C_0. I.e. $P_i = C_i^d \bmod N = 2B + 256^{49} PS_i + D_i$ and $P_j = C_j^d \bmod N = 2B + 256^{49} PS_j + D_j$, where $0 < PS_{i,j}$ and $0 \le D_{i,j} < 256^{48}$. Then for C, $C = s^e C_0 \bmod N$ and $\beta \in \mathbf{Z}$, where $s = [(1-\beta)s_i + \beta s_j] \bmod N$, it holds that $C^d \bmod N = P$, such that $P = [2B + 256^{49}((1-\beta)PS_i + \beta PS_j) + (1-\beta)D_i + \beta D_j] \bmod N$.*

Proof. It suffices to observe that $P = [(1-\beta)s_i + \beta s_j]P_0 \bmod N = [(1-\beta)P_i + \beta P_j] \bmod N$, where $P_0 = C_0^d \bmod N$. ∎

It follows from the lemma written above that once we have suitable multipliers $s_{i,j}$ for a ciphertext C, we can try to search for the next suitable multiplier s as for a linear combination of s_i and s_j. In practice, we can try small positive and negative values of β and test whether the particular linear combination s gives S-PKCS-conforming plaintext or not. Working in this way, we may hope to accelerate the algorithm in step 2b (c.f. the appendix). Since we can reasonably assume that $\gcd(s_j - s_i, N) = 1$, there is a particular value of β for every triplet of suitable multipliers (s_i, s_j, s). However, experiments have shown that there are also differences in how much information can be obtained from such s depending on the size of β. For small values of β, it has been observed that the obtained values of s do not reduce the size of M_i as fast as the values of s obtained for β close to $N/2$. The reason is perhaps a linear dependency on Z, which is stronger for small β. On the other hand, β close to $N/2$ clearly cannot be directly found by "brute force" searching. More precisely, we may find such β directly, but we cannot assure that obtained s will be of moderate size for further processing by the RSA inversion algorithm. Therefore, it remains to extract as much information as possible from reasonably small values of β and then to either continue with incremental searching used in the original version of the algorithm [1] or to use the Parallel-Threads (PT) method described in §3.3.2. In advance of the following discussion, we note that the source for the next incremental searching or for the PT-method is the maximum suitable multiplier s_j found, such that $s_j < N/2$.

When using the above-mentioned method with negative values of β, we may get a multiplier s that is close to N (it can be regarded as a small negative value modulo N). Such an s cannot be directly processed, since it induces a very large interval for r in the original algorithm (see step 3 in the appendix). We will show how the algorithm can be adjusted to process small positive values of s as well as small negative values of s modulo N.

Theorem 2 (On symmetry). *Let us have integers s, P, and N satisfying*
$$E_1 \leq sP \bmod N \leq F_1, \text{ where } E_1, F_1 \in \mathbf{Z}.$$
Then there is the integer v, $v = N - s$, satisfying
$$E_2 \leq vP \bmod N \leq F_2, \text{ where } E_2 = N - F_1, F_2 = N - E_1.$$

Proof. We have that $vP \bmod N = (N - s)P \bmod N = (-sP) \bmod N = N - (sP \bmod N)$. The upper boundary of $(sP \bmod N)$ is F_1, therefore, the lower boundary E_2 of $(vP \bmod N)$ is $E_2 = N - F_1$. Analogically, the upper boundary F_2 of $(vP \bmod N)$ is given by the lower boundary E_1 as $F_2 = N - E_1$. ∎

We use the theorem as follows: if we get a high value of s using the β-method described above, then we convert it to the corresponding symmetric value $v = N - s$ which is then processed in a modified version of step 3 of the algorithm (see the appendix). The core of the modification is using boundaries E_2, F_2 instead of the original boundaries $E_1 = E'$, $F_1 = F'$ (c.f. §3.2).

3.3.2 Parallel-Threads (PT) Method

Recall that the complexity of step 2 of the algorithm (see the appendix) for $i > 1$ depends on the size of M_{i-1}. Generally, the step is expected to be much faster if $|M_{i-1}| = 1$ than if $|M_{i-1}| > 1$. The reason is that $|M_{i-1}| = 1$ means there is only one interval approximating the value of P_0 left and therefore certain rules can be used when searching for the next suitable multiplier s_i. Experimenting with our test program, we observed that even if $|M_{i-1}| > 1$, the number of intervals was usually small enough that it was better to start a parallel thread T for each $I \in M_{i-1}$ as if it was the only interval left, i.e. it starts its own thread in step 2c of the algorithm. These threads $T_1, ..., T_w$, where $w = |M_{i-1}|$, were precisely multitasked on a per BVO call basis. They were arranged in the cycle $T_1 \rightarrow T_2 ... \rightarrow T_w \rightarrow T_1$ and stepping was done in the cycle after each one BVO call. The results obtained when thread T_j found a suitable multiplier were projected on the whole current set of intervals for all threads. After that, the threads belonging to the intervals that disappeared were discarded. We observed that the PT-method increased the effectiveness of the original algorithm.

Using a certain amount of heuristics we set the condition that directs whether we should use the PT-method or not. The PT-method is started in step i if the following inequality holds

$$| M_{i-1}| < (2\varepsilon \Pr(A))^{-1} + 1.$$

The value of ε estimates the number of passes it takes from the start of the PT-method until there is only one interval left, i.e. $|M_{i+\varepsilon-1}| = 1$, where the PT-method started in pass i. In our programs, we used $\varepsilon = 2$ which was the ceiling of the mean value observed for ε.

3.4 Note on Forging a Server's Signature

The BVO construction allows us to mount Bleichenbacher's attack without any restrictions on its functionality. As noted above, we can compute the RSA inverse for any integer y, thereby obtaining the value $x = y^d$ mod N for the particular server's private exponent d and the modulus N. Discussing the so-called semantics of the attack, there are only two cases in which it would be reasonable to compute this inversion.

In the first case we compute the RSA inverse for a captured ciphertext carrying an encrypted value of the *premaster-secret*. This approach allows us to decrypt the whole communication that was carried out in a given session between a client and the server. This is the main approach of this paper, which we have practically tested and optimized.

In the second case we compute an RSA signature of a message m on behalf of the server. The whole attack runs in a similar way, which means that the main activity between an attacker and the server is still concerned on the phase of passing the *premaster-secret* value during the handshake procedure of the SSL/TLS protocols. However, this is only because we need to build up a BVO (c.f. §2) for computing the RSA inversion. The source of this inversion (the ciphertext C) will no longer be an encrypted *premaster-secret* itself, but the formatted value of $h(m)$, where h is an appropriate hash function. Currently, the SSL/TLS protocols sets $h(m) =^{\text{def}}$ MD5(m) || SHA-1(m) and the value of $h(m)$ is further formatted according to the EMSA-PKCS1-v1_5 method from PKCS#1 ([9], [10], [12], [13], [17]). At the end of the attack we obtain C^d mod N which is the signature of our input C. It further depends on the keyUsage property [18] of the certificate of the server's RSA key, whether such a signature can be used for further attacks or not. At first the server's RSA key must be attributed for signing purposes. Secondly, it depends on the specific system as to how far the faked signature is important, directly implying how dangerous the attack is. From the basic properties of SSL/TLS ([10], [12]) it follows that such a signature may be abused to certify an ephemeral RSA or D-H [11] public key of a faked server. The faked server can then be palmed on an ordinary user to elicit some secret information from her. Generally speaking, this would be an attack on the authentication of a server. The necessary condition here is that the user is willing to use either the so-called export RSA key or the ephemeral Diffie-Hellman key agreement [11]. The practical situation is that some clients will - some clients will not. It strongly depends on the attention paid to the configuration of such a client. Unfortunately, these "minor" details are very often neglected in a huge amount of applications. Moreover, we emphasize that the attack described here may not be the only one possible, since the particular importance of a server's signature depends on the role that the server plays in a particular information system. The best way to avoid all these attacks is to not attribute the server's RSA key for signing purposes, unless it is absolutely necessary.

From the effectiveness viewpoint, we can estimate that using the RSA inversion based on BVO for signature forging will require more BVO calls, since we need to insert an extra masking zero-step (see appendix, step 1 of the algorithm). The number of additional BVO calls may be calculated as $[\Pr(BVO|S\text{-}PKCS)* \Pr(S\text{-}PKCS|A)* \Pr(A)]^{-1}$, which is given by the probability that for a randomly chosen ciphertext C we get $BVO(C) = 1$. Adding this value to the number of BVO calls in the former attack on *premaster-secret* (c.f. §4) gives an estimate of the overall complexity of signature forging.

4   Complexity Measurements

Basing on the elaboration from [1], we can estimate the number of BVO calls for decrypting a plaintext C_0 belonging to a S-PKCS-conforming plaintext P_0 as

$$2*\Pr(P)^{-1} + (16k - 32)*\Pr(P|A)^{-1}, \text{ where } \Pr(P|A) = \Pr(BVO|S\text{-}PKCS)*\Pr(S\text{-}PKCS|A),$$

$$\Pr(P) = \Pr(P, A) = \Pr(P|A)*\Pr(A),$$

where $\Pr(P)$ is the probability that for a randomly chosen ciphertext C we get $BVO(C) = 1$.

This estimation does not cover the optimization described in §3.2 and §3.3. Therefore we treat it as the worst-case estimation for a situation when these optimizations are not notably helping an attacker. Experiments show that the optimized algorithm is practically almost two times faster than this estimation (c.f. §4.1) for the most widely used RSA key lengths. Let us comment on the expression of the estimation now.

The first additive factor corresponds with our assumption that the attacker wants to decipher C_0 belonging to a properly formatted plaintext carrying a value of the *premaster-secret*. In such a situation, she does not have to carry out initial blinding (c.f. the appendix, step 1). According to [1], we can estimate that she needs to find two suitable multipliers $s_{1,2}$ for C_0, until she can proceed with the generally faster step 2c. This gives the first factor as $2*\Pr(P)^{-1}$. Note that, heuristically speaking, the optimizations (§3) mainly reduce the necessity of finding s_2 in the "hard" way, thereby decreasing the first factor closely to the value $\Pr(P)^{-1}$. This hypothesis corresponds well with the results of our measurements.

The second factor is a slightly modified expression presented in [1]. It corresponds to the number of expected BVO calls for the whole number of passes through step 2c. Recall that $C_0 = (P_0)^e \mod N$, where $2B \le P_0 \le 3B - 1$, so P_0 lays in the interval of the length B, $B = 256^{k-2}$. Conjecturing that each pass through step 3 roughly halves the length of the interval for P_0, we may estimate that we need $8(k - 2)$ passes.

Furthermore, it is conjectured [1] that each pass through step 2c takes $2*\Pr(P|A)^{-1}$ BVO calls. From here follows the estimation of BVO calls as $(16k - 32)*\Pr(P|A)^{-1}$.

Finally, we note that the complexity of the attack is mainly determined by the amount of necessary BVO calls. This amount actually limits the attack in the three ways. The first one is that an attacked server must bear such a number of corrupted Handshakes [12] (i.e. not collapse due to a log overflow, etc.). The second limitation comes from a total network delay that increases linearly with the number of BVO calls. The third limit is determined by the computational power of the server itself, which mainly means how fast it can carry out the RSA operation with a private key. Other computations during the attack are essentially faster and therefore we do not discuss them here.

4.1 Simulated Local BVO

In this paragraph, we present the measured complexity of the attack with respect to the total amount of BVO calls. The data of our experiment was obtained for the four particular randomly generated RSA moduli of 1024, 1025, 2048 and 2049 bits in length. For every such modulus we implemented a local simulation of BVO that we linked together with the optimized algorithm discussed in this paper. We then measured the number of BVO calls for 1200 ciphertexts of the randomly generated and encrypted values of the *premaster-secret*.

Due to the strong dependence of the number of BVO calls on $\Pr(A)$ we see that the complexity of the attack is not strictly increasing with respect to the length of the modulus N. This discrepancy was already mentioned in [1]. It follows that one should use moduli with a bit length in the form $8r$, where r is an integer, mainly avoiding the moduli with the length $8r + 1$.

Table 1. Basic statistics of a measured attack complexity in BVO calls

Modulus length (bits)	BVO calls				
	Originally estimated (without optimizations)	Practically measured (with optimizations from §3)			
		Min	Max	Median	Mean
1024	36 591 001	815 835	278 903 416	13 331 256	20 835 297
1025	979 488	630 589	105 122 011	1 197 380	1 422 176
2048	48 054 328	2 824 986	354 420 492	19 908 079	28 728 801
2049	2 794 937	1 413 005	475 298 397	3 462 557	3 896 432

Analyzing the measured data, we observed that the distribution of the amount of BVO calls can be approximated by a log-normal Gaussian distribution, i.e. the logarithm of the amount of BVO calls roughly follows a normal Gaussian distribution. Heuristically speaking, this means that the most basic random events governing the complexity of the attack primarily combine together in a multiplicative manner. The values of median, and mean are presented in Table 1. These values were obtained using the log-normal approximation of the data samples measured. These approximations are plotted in Fig. 1. We can see that all the distributions skew to the right. Therefore, the most interesting values are perhaps given by the medians. For example, in the case of a 1024 bits long modulus, we can expect that the one half of all attacks succeed in less then 13.34 million BVO calls. Furthermore, the data in Table 1 supports our conjecture that the optimizations proposed in §3 mainly speed up the first "hard" part of the algorithm. Therefore, this speeding up is clearly notable for moduli of 1024 and 2048 bits, while there is no observable effect for the moduli of 1025 and 2049 bits.

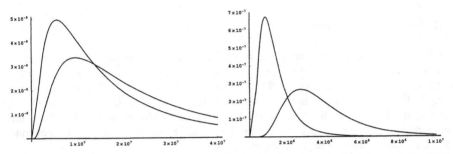

Fig. 1. Log-normal approximation of BVO calls density functions: in the left graph for 1024 (the higher peak) and 2048, and in the right graph for 1025 (the higher peak) and 2049 bits long moduli

4.2   Real Attack

We successfully tested the attack on a real SSL server (AMD Athlon/1 GHz, 256MB RAM) using 1024 bits long RSA key. The total number of BVO calls for decryption of a randomly selected *premaster-secret* was 2 727 042 and the whole attack took 14 hours 22 minutes and 45 seconds. It gives an estimated speed of 52.68 BVO invocations per second. The server and the attacking client were locally connected via a 100 Mb/s Ethernet network without any other notable traffic. With respect to the whole conditions of this experiment, we can conclude that this is probably one of the best practically achievable results. Therefore, we can expect that there would be few practical attacks succeeding in less then 14 hours of sustained high effort (for a 1024 bits long RSA key). Using the value of the median for 1024 bits modulus from Table 1, we can roughly expect one half of all attacks in our setup to succeed in less than 70 hours and 18 minutes. For 2048 bits long RSA key in the same setup we get an estimated speed of 11.47 BVO calls per second. Therefore, one half of all attacks should then succeed in less than 21 days.

The experiment setup described above could be slightly improved by using a more powerful server. Plugging in such a server (2x Pentium III/1.4 GHz, 1 GB RAM, 100 Mb/s Ethernet, OS RedHat 7.2, Apache 1.3.27), it was possible to achieve a speed of 67.7 BVO calls per second for a 1024 bits RSA key. The median time for a whole attack on the *premaster-secret* could be then estimated as 54 hours and 42 minutes.

Note that all these estimates assume achieving and sustaining high communication and computation throughput on the server's side.

4.3 Real Vulnerability

To assess the practical impacts of the attack presented here, we had randomly generated a list of 611 public Internet SSL/TLS servers (we accepted servers providing SSL v. 3.0 or TLS v. 1.0) and then tested these servers to see whether it was possible to construct a BVO for them or not. We found that two thirds of these servers were vulnerable to our attack. We emphasize that it does not necessarily mean that the attack would always succeed on every such server. Despite the fact that all these servers can be regarded as broken from a pure cryptanalytic viewpoint, the complexity of the attack may still render it impractical in a large amount of cases. We expect that a properly administrated server (e.g. log messages are often inspected, suspicious clients are added to black-lists, etc.) should withstand the attack. Under such an administration, the attack should be recognized and the attacking client would soon be blocked. Of course, the cryptographic strength of all these SSL/TLS implementations should definitely be improved. We strongly recommend applying appropriate patches as soon as possible.

We observed an interesting anomaly for 110 out of 611 tested servers. All of them provided both SSL v. 3.0 and TLS. 26 of them were *primarily* vulnerable only through the SLL v3.0 protocol, while the remaining 84 servers were *primarily* vulnerable only through the TLS protocol. We advisedly used the word "primarily", since if these servers share the same RSA key for both protocols, which is a very common practice, then an attacker can easily assault one protocol through an interaction with the other one. Moreover, the format of the ciphertext carrying the *premaster-secret* is the same for both protocols, so this cross-attacking actually does not increase the complexity of the whole attack.

5 Technical Details

Please refer to the extended version of this paper [19].

6 Countermeasures

Due to the compatibility demands, it does not seem possible to simply leave the EME-PKCS1-v1_5 method and use its successor EME-OAEP. Note that even the EME-OAEP method must be implemented carefully (c.f. [5], [6]). On the other hand, it has been recently shown by Jonsson and Kaliski in [4] that the EME-PKCS1-v1_5 can offer reasonable security (the proof was carried out for the TLS protocol) assuming that it is implemented properly – i.e. mainly that side channels are avoided. What remains is to show what a proper implementation should look like. The current guidelines in [12] together with [15] are obviously insufficient and should be updated to avoid weaknesses like the one discussed in this paper. Moreover, it seems that the edge between secure and insecure implementation of EME-PKCS1-v1_5 is very sharp. This implies that the standards regarding its implementation must really be very precise.

We propose to keep generating P_{k-45}, ..., P_0 randomly, if P is not S-PKCS-conforming. Furthermore, we propose to replace P_{k-47} and P_{k-46} with the expected version number in either case (i.e. if P is or is not S-PKCS-conforming). A more detailed description and discussion of this subject is provided in the extended version of the paper [19].

7 Conclusions

We have presented a new practically feasible side channel attack against the SSL/TLS protocols. When Bleichenbacher presented his attack on PKCS#1 (v. 1.5) in 1998 [1], it was generally assumed that the attack was impractical for the SSL/TLS protocols, since these protocols add several proprietary restrictions on the plaintext format, which increase the complexity of the attack. Of course, the protocols could not be called secure from a pure cryptographical viewpoint. Therefore, a special countermeasure was introduced and generally adopted [10], [12]. However in this paper, we have shown that problems with Bleichenbacher's-like attacks on the SSL/TLS protocols are still not properly solved. We have identified a new possibility of a substantial side channel occurring during an SSL/TLS Handshake. The side channel originates when a receiver checks a version number value stored in the two left-most bytes of the *premaster-secret*. Based on the receiver's behavior during this check, we have defined its mathematical encapsulation as a *bad-version oracle* (BVO, c.f. §2). Such a check is widely recommended for SSL/TLS servers, but unfortunately it is not properly specified how it should be performed. Practical tests showed that two thirds of randomly chosen Internet servers carried out the test wrongly, thereby allowing the construction of BVO resulting in a new attack on RSA-based sessions. The attack itself may be viewed as an optimized and generalized variant of the original Bleichenbacher's attack [1]. The most obvious target of our attack would probably be discovering the *premaster-secret*, thereby decrypting a captured RSA-based session. It is also possible (with an additionally increased complexity, c.f. §3.4) to compute the signature of any arbitrary message on behalf of the server.

The attack was carried out in practice and its efficiency was measured (§4). The amount of time the attack takes in practice is mainly determined by the amount of BVO calls. Each BVO call corresponds to one attempt to establish a SSL/TLS connection with an attacked server. If the server uses a typical 1024 bits long RSA key, then we can expect that roughly 50% of attacks succeed in less than 13.34 million BVO calls. This load may be further spread in time and even distributed to many computers. The main aim would not be speeding up the attack, but making its localization and blocking harder. Although the complexity presented here is definitely very low from a pure cryptographic viewpoint, there may still be technical measures that can thwart the attack in a practice. For instance, each BVO call should produce at least one log record on the server's side. If these logs are well maintained and appropriately inspected, then the attack should be recognized in time. Unfortunately, there also seem to be poorly administrated servers where SSL/TLS audit messages are almost ignored. These servers remain protected solely by their network and computational throughput, which is obviously alarming.

Acknowledgements. We are grateful to Jiří Hejl for technical support and consultations. We also appreciate technical help of Roman Kalač and Libor Kratochvíl. The third author is grateful to his postgraduate supervisor Dr. Petr Zemánek for continuous support in research projects.

References

1. Bleichenbacher, D.: Chosen Ciphertexts Attacks Against Protocols Based on the RSA Encryption Standard PKCS#1, *in Proc. of CRYPTO '98*, pp. 1–12, 1998
2. Canvel, B.: Password Interception in a SSL/TLS Channel, http://lasecwww.epfl.ch/memo_ssl.shtml, February, 2003
3. Håstad, J., Näslund M.: The Security of Individual RSA Bits, *in Proc. of FOCS '98*, pp. 510–521, 1998
4. Jonsson, J., Kaliski, B., S., Jr.: On the Security of RSA Encryption in TLS, *in Proc. of CRYPTO '02*, pp. 127–142, 2002
5. Klíma, V., Rosa, T.: Further Results and Considerations on Side Channel Attacks on RSA, *in Proc. of CHES '02*, August 13–15, 2002
6. Manger, J.: A Chosen Ciphertext Attack on RSA Optimal Asymmetric Encryption Padding (OAEP) as Standardized in PKCS #1, *in Proc. of CRYPTO'01*, pp. 230–238, 2001
7. OpenSSL: OpenSSL ver. 0.9.7, http://www.openssl.org/, December 31, 2002
8. PKCS#5 ver. 2.0: Password-Based Cryptography Standard, RSA Laboratories, March 25, 1999
9. PKCS #1: RSA Encryption Standard, An RSA Laboratories Technical Note, Version 1.5, Revised November 1, 1993
10. Rescorla, E.: SSL and TLS: Designing and Building Secure Systems, Addison-Wesley, New York, 2000
11. RFC 2631: Rescorla, E.: Diffie-Hellman Key Agreement Method, June 1999
12. RFC 2246: Allen, C., Dierks, T.: The TLS Protocol, Version 1.0, January 1999
13. RFC 1321: Rivest, R.: The MD5 Message-Digest Algorithm, April 1992
14. Rivest, R., L., Shamir, A., Adleman L.: A Method for Obtaining Digital Signatures and Public-Key Cryptosystems, *Communications of the ACM*, 21, pp. 120–126, 1978
15. RSA Labs: Prescriptions for Applications that are Vulnerable to the Adaptive Chosen Ciphertext Attack on PKCS #1 v1.5, RSA Laboratories, http://www.rsasecurity.com/rsalabs/pkcs1/prescriptions.html
16. Schneier, B., Wagner, D.: Analysis of the SSL 3.0 Protocol, *The Second USENIX Workshop on Electronic Commerce Proceedings*, USENIX Press, November 1996, pp. 29–40
17. Secure Hash Standard, FIPS Pub 180-1, 1995 April 17
18. X509: ITU-T Recommendation X.509 (06/97) - Information Technology - Open System Interconnection – The Directory: Authenticantion Framework, ITU, 1997
19. Klima V., Pokorny O., Rosa T.: Attacking RSA-based Sessions in SSL/TLS, *Cryptology ePrint Archive: Report 2003/052*, http://eprint.iacr.org/2003/052/

Appendix

For the sake of completeness we enclose here the algorithm from [1]. For our purposes we define directly a slight generalization and modification of it. Recall that in the original text $E = 2B$, $F = 3B{-}1$, where $B = 256^{k-2}$. In our variant, we will use the refined values E' and F' (c.f. §3). According to Definition 1 and the original notation used bellow, we note that a ciphertext C is said to be PKCS conforming iff $C = P^e$ mod N, where P is PKCS-conforming plaintext. The modified algorithm is as follows.

Step 1: Blinding. Given an integer c, choose different random integers s_0; then check, by accessing the oracle, whether $c(s_0)^e$ mod N is PKCS conforming.
For the first successful s_0, set

$$c_0 \leftarrow c(s_0)^e \bmod N$$
$$M_0 \leftarrow \{[E, F]\}$$
$$i \leftarrow 1.$$

Step 2: Searching for PKCS conforming messages.

Step 2.a: Starting the search. If $i = 1$, then search for the smallest positive integer $s_1 \geq N/(F{+}1)$, such that the ciphertext $c(s_1)^e$ mod N is PKCS conforming.

Step 2.b: Searching with more than one interval left. Otherwise, if $i > 1$ and the number of intervals in M_{i-1} is at least 2, then search for the smallest integer $s_i > s_{i-1}$, such that the ciphertext $c(s_i)^e$ mod N is PKCS conforming.

Step 2.c: Searching with one interval left. Otherwise, if M_{i-1} contains exactly one interval (i.e. $M_{i-1} = \{[a, b]\}$), then choose small integer values r_i, s_i such that

$$r_i \geq \lceil 2(bs_{i-1} - E)/N \rceil$$

and

$$(E + r_iN)/b \leq s_i < (F + r_iN)/a$$

until the ciphertext $c(s_i)^e$ mod N is PKCS conforming.

Step 3: Narrowing the set of solution. After s_i has been found, the set M_i is computed as

$$M_i \leftarrow \cap_{(a,b,r)} \{[\max (a, \lceil (E + rN)/s_i \rceil), \min (b, \lfloor (F + rN)/s_i \rfloor)]\}$$

for all $[a, b] \in M_{i-1}$ and $(as_i - F)/N \leq r \leq (bs_i - E)/N$.

Step 4: Computing the solution. If M_i contains only one interval of length 1 (i.e., $M_i = \{[a, a]\}$), then set $m \leftarrow a(s_0)^{-1}$ mod N, and return m as solution of $m \equiv c^d \pmod{N}$. Otherwise, set $i \leftarrow i + 1$ and go to step 2.

Author Index

Lecture Notes in Computer Science

For information about Vols. 1–2698
please contact your bookseller or Springer-Verlag